D1179544

Forensic Toxicology

Drug Use and Misuse

Forensic Toxicology

Drug Use and Misuse

Edited by

Susannah Davies

LTG, London, UK
Email: ltg.committee@gmail.com

Atholl Johnston

Queen Mary University of London, UK
Email: atholl.johnston@bioanalytics.co.uk

David Holt

St George's, University of London, UK
Email: david.holt@bioanalytics.co.uk

ROYAL SOCIETY
OF **CHEMISTRY**

THE QUEEN'S AWARDS
FOR ENTERPRISE:
INTERNATIONAL TRADE
2013

Print ISBN: 978-1-78262-156-0

A catalogue record for this book is available from the British Library

© The Royal Society of Chemistry 2016

Published by The Royal Society of Chemistry,
Thomas Graham House, Science Park, Milton Road,
Cambridge CB4 0WF, UK

Registered Charity Number 207890

Visit our website at www.rsc.org/books

Printed in the United Kingdom by CPI Group (UK) Ltd, Croydon, CR0 4YY, UK

Preface

This is a book about Forensic Toxicology, focusing on the issues surrounding drug use and misuse. Everyone seems to think they know about forensic toxicology, but the depth of knowledge possessed is often very superficial. In my experience, students wishing to enter the field have garnered most of their opinions on the topic from popular TV programmes. So, it's good to have the latest authoritative insights from a set of experts on a range of problems that consume significant parts of our medical and social budgets.

Alas, the terms drug use and misuse do tend to go together. Almost every drug for pharmaceutical use that I've ever worked on has ended up being misused in some way and, of course, a high proportion of the drugs we work on in forensic toxicology have been manufactured for or diverted into the illicit drug market.

In forensic toxicology we set out to detect and measure these drugs and then to use the measurements as a guide to medical intervention, to interpret findings after death, to support legal investigations or to give substance to epidemiological surveys on drug use and changing patterns of drug use.

There have indeed been some significant changes in the pattern of illicit drug use of late. Just which drugs are being sold on the illicit market has become an increasingly difficult question to answer, with the burgeoning of a large number of compounds with psychoactive properties. From just a few new compounds emerging onto the market each year about a decade ago, by 2014 the European Monitoring Centre for Drugs and Drug Addiction (EMCDDA) Early Warning System reported 101 new psychoactive compounds in that year alone. A high proportion of these compounds were the synthetic cannabinoids, more accurately described as synthetic cannabinoid receptor

Forensic Toxicology: Drug Use and Misuse
Edited by Susannah Davies, Atholl Johnston and David Holt
© The Royal Society of Chemistry 2016
Published by the Royal Society of Chemistry, www.rsc.org

agonists; by the end of 2014, 134 had been notified to the Early Warning System.

For the laboratory these rapid changes pose substantial methodological problems in terms of sourcing reliable calibrators and in developing methods for the detection and quantification of these new compounds, in drug seizures and in biological samples. Happily, improvements in the availability of mass spectrometric equipment suitable for clinical laboratories has greatly facilitated the development of methodology to address these issues, but these techniques require high levels of investment in personnel with analytical skills for their optimal use.

For law enforcement a major problem is the ease with which these compounds can be made available to users *via* the internet and poorly regulated courier systems. This, in turn, has been a driving force for some focused entrepreneurs to make small changes to the molecular structures of some starting materials with known psychoactive properties. These compounds have appeared at a faster rate than they can be controlled by legislation, and they have been sold on the so-called "legal highs" market. Of course, by the nature of their production and marketing, these compounds have not been subject to any pharmaceutical regulatory procedure, so that their clinical toxicology remains largely unknown until patients arrive in hospital, or individual users post their own, unreliable, reports on social media sites.

Forensic toxicology plays an important role in defining which drugs are being consumed and matching these data with clinical harm. For some time, we have known that finding out what people are taking is not as simple as asking them, or relying on drug seizures. Too often, packaging is designed to obfuscate the contents and purchasers on the illicit market may not receive what they were under the impression they had bought. Nor should it be overlooked that there is a similar problem in the purchase of pharmaceutical preparations. Patients, both within and outside socialised medical settings, do resort to purchasing drugs online, or in open market places in poorly regulated countries. Many of these drugs, for diseases such as HIV, malaria and tuberculosis, are falsified or substandard, posing a global health risk. In this book some of the approaches that help to define the changing patterns of drug use, in areas such as driving, drug use in the workplace and in sport, are explored.

If we are going to bring any order to the problems outlined in this book, either by defining them or introducing regulatory controls, then the careful and thoughtful approaches encapsulated in these chapters, backed by apposite case examples, are a prerequisite. Those working in laboratories must keep abreast of developments in a constantly changing environment by maintaining an active dialogue with those using their services. In the clinical field, drug measurements must be integrated with both pharmacokinetic and pharmacodynamic measurements if they are to be of value in supporting treatment regimens or to interpret toxicological findings. For forensic legal applications, innovative approaches to sampling may be needed, backed by analyses and interpretation that pass the ever increasing demands of international regulatory strictures.

These chapters will be of interest and value to a wide range of readers, not only those with a day-to-day involvement in the topics covered, but also those who have a peripheral involvement in the issues dealt with and wish to extend their own knowledge of the highly topical issues covered.

David W. Holt

Contents

Forensic Toxicology: Drug Use and Misuse
Edited by Susannah Davies, Atholl Johnston and David Holt
© The Royal Society of Chemistry 2016
Published by the Royal Society of Chemistry, www.rsc.org

3 The Role of Amnesty Bins in Understanding the Pattern of Recreational Drugs and Novel Psychoactive Substances Being Used Within a Locality   44
David M. Wood and Paul I. Dargan

4 Contamination of Water with Drugs and Metabolites   54
Victoria Hilborne

13   Drugs and Driving   262
Kim Wolff

14   Alcohol Technical Defences in Road Traffic Casework   276
Mike Scott-Ham

15   New Psychoactive Substances and the Criminal Law   297
Rudi Fortson

16   Scheduling of Drugs in the United States   343

Jeffery Hackett

17   Drug Legislation in New Zealand   356

Keith Bedford

20 Hair Testing in Forensic Toxicology 411

Donna M. Cave and Robert Kingston

21 Drugs in Oral Fluid 426

Peter Akrill

1 Introduction to Forensic Toxicology and the Value of a Nationwide Database

Alan Wayne Jones

Linköping University, Linköping, Sweden
Email: wayne.jones@liu.se

1.1 Introduction

Forensic toxicology, formerly known as forensic chemistry, is a multidisciplinary subject concerned with various aspects of chemistry, physiology, pharmacology and toxicology, as well as other branches of science and technology.[1,2] Forensic toxicology practitioners are first and foremost trained in analytical chemistry because their principal task is extraction, detection, identification and quantitative analysis of a plethora of drugs and poisons in biological specimens.[3,4] Another important duty of the forensic toxicologist or forensic pharmacologist is to provide expert testimony in criminal and civil cases involving the use and abuse of drugs in society and also in drug-related crimes, such as when drunk and drugged drivers are prosecuted.[5]

Knowledge about the disposition and fate of drugs in the body and how psychoactive substances alter normal functioning of the brain are important considerations when forensic toxicologists interpret their analytical findings in a legal context.[6] The relationships between the various clinical signs and symptoms of impairment and the

Forensic Toxicology: Drug Use and Misuse
Edited by Susannah Davies, Atholl Johnston and David Holt
© The Royal Society of Chemistry 2016
Published by the Royal Society of Chemistry, www.rsc.org

concentrations of psychoactive substances in blood is other relevant information available from TOXBASE, a forensic toxicology database.

The various sub-disciplines of forensic toxicology and the types of information stored in TOXBASE are listed below:

- Unnatural death investigations involving poisoning and over-dosing with drugs.
- Driving under the influence of alcohol and/or other drugs.
- Drug facilitated crimes (incapacitation), such as date-rape.
- Control of illicit drugs in society, especially by people detained in prisons and other sectors of the criminal justice system.
- Child welfare and custody cases where suspicion arises that the parents or care-givers abuse drugs or administer drugs to the infants or elderly for which they are responsible.
- Monitoring the use of banned drugs by people enrolled in treatment and rehabilitation programs for substance abuse disorder.
- Drug testing in the workplace.
- Use of doping agents in sports.

Forensic pharmacovigilance is a subject of increasing interest and importance when it comes to the safety of medicines and the dangers of prescribing certain combinations of drugs to patients.[7,8] The information in TOXBASE can help to flag for the emergence of new recreational drugs of abuse (designer drugs), and help to decide whether these should be banned or classified as controlled substances. Other information stored in TOXBASE deals with the drugs commonly encountered in poisoning deaths.[9,10] By cross-linking information in TOXBASE with other databases (*e.g.* prescription registers) or safety of medicines records the potential dangers of certain drugs and drug combinations become easier to document.[11] The post-mortem section of TOXBASE can be complemented with information about the cause and manner of death (*e.g.* suicide, homicide, accident) according to findings at autopsy and the official death certificate.[12]

1.2 Forensic Toxicology in Sweden

Geographically, Sweden is roughly the same size as California or twice the size of the UK. However, the population of Sweden is only 9.5 million (2014), which means that one central forensic toxicology laboratory provides analytical services for the whole country. This centralization has several advantages when it comes to the choice of analytical methods and the availability of modern state-of-the-art equipment, quality assurance procedures and laboratory accreditation.

The annual workload in terms of numbers and types of forensic cases submitted for analysis to the Swedish National Laboratory of Forensic Toxicology are summarized in Table 1.1. Official records show that the number of forensic autopsies has remained remarkably constant, averaging 5000 per year over the past 20 years. The laboratory workload is strongly influenced by new government legislation dealing with abuse of alcohol and drugs in society, as exemplified by enactment of a zero-tolerance law for driving under the influence of drugs (DUID) in 1999.[13] The forensic toxicology database available in Sweden contains information about hundreds of drugs and their metabolites and the concentrations in blood and other biological specimens from living and deceased persons.[14] Senior scientists at the Swedish National Laboratory of Forensic Toxicology are closely associated with the faculty of medicine at the University of Linköping, particularly clinical pharmacology. A good collaboration between these two organizations has resulted in considerable research activity and many joint publications.[15]

Forensic autopsies in Sweden are carried out at the six university teaching hospitals, which are located in the cities of Umeå, Uppsala, Stockholm, Gothenburg, Linköping and Lund. During an autopsy, the forensic pathologists take blood and other biological specimens for toxicological analysis. The preferred source of blood is from a femoral

Table 1.1   Number and type of cases submitted to the Swedish Board of Forensic Medicine for toxicological analysis during the years 2012, 2013 and 2014.

Type of investigation	Year 2012	Year 2013	Year 2014
Post-mortem toxicology	5051	5084	5310
Traffic cases (drunk and drugged drivers)[a]	14 867	14 769	16 037
Abuse of illicit drugs in the community	38 119	37 860	37 426
Abuse of illicit drugs by prison inmates	29 105	24 819	25 709
Sexual assault cases that might involve drugs	1531	1377	1565
Patients and others in drug rehabilitation programs	7100	7656	8991

[a]Arrested drivers submitting to an evidential breath-alcohol test are not included in this table.

vein and, in addition, specimens of urine and eye fluid (vitreous humor) are taken and sent for analysis of drugs and poisons.

The sampling of biological specimens is done in a highly standardized way and chemical preservatives such as sodium or potassium fluoride (1–2%) are added as enzyme inhibitor to stabilize drug concentrations after sampling. If blood from a femoral vein is not available, owing to massive trauma or decomposition of the body, then cardiac blood is usually sent for toxicological analysis. The biological specimens are shipped to the central forensic toxicology laboratory by special delivery and are refrigerated during transport and storage. The results of the toxicological analysis are of prime importance when unnatural deaths are investigated and when the anatomical and histological evidence fails to reveal the cause and manner of death.[16]

Figure 1.1 is a flow chart showing the forensic toxicology routines operational when nothing remarkable was discovered after performing the autopsy. A negative or inconclusive autopsy shifts attention towards poisoning as a possible cause of death, which requires a close collaboration and discussion between the pathologist and the toxicologist.[17] Witness statements and discoveries made by the police authorities at the death scene, especially any drugs or drug paraphernalia found near the body, is other important information. Likewise, interviews with relatives and friends about the deceased's alcohol and/or drug habits and general state of health is other relevant information to consider when poisoning deaths are investigated.

1.3   Analytical Methods

Forensic toxicology is essentially analytical chemistry applied to the determination of drugs and toxins in biological specimens.[4,18,19] The accuracy, precision and specificity of the methods used, including limit of detection (LOD) and limit of quantitation (LOQ), are important to document. In this connection, one needs to differentiate between a drug screening analysis and a more sophisticated verification and quantitative analysis of a particular drug.[20] Traditionally, the preliminary screening analysis involves some type of enzyme immunoassay procedure, which is generally targeted towards groups of drugs rather than individual substances.[21] Typical drug categories encountered in forensic toxicology include opiates, cannabinoids, amphetamines, cocaine metabolites and benzodiazepines. The objective of the screening test is to eliminate "drug negative" cases from

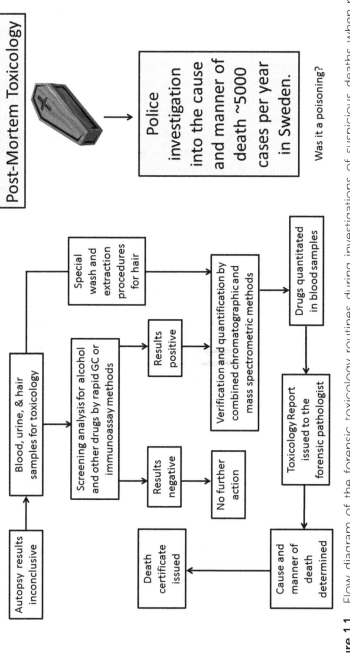

Figure 1.1   Flow diagram of the forensic toxicology routines during investigations of suspicious deaths when nothing remarkable was found at autopsy.

further consideration and to identify presumptive positive cases. More modern approaches to drug screening involve some type of hyphenated technique such as liquid chromatography (LC) or gas chromatography (GC) combined with time-of-flight (TOF) mass spectrometry (MS) or LC-TOF-MS, which is now used routinely in many laboratories.[22–24]

Urine is the preferred biological specimen for screening analysis of drugs and their metabolites, because concentrations excreted in the urine are higher than in the blood.[25] Furthermore, drugs and metabolites are excreted for a longer time in the urine, which lengthens the window of detection considerably.[26] If urine is unavailable, then screening analysis of drugs is done on a blood sample after precipitation of the proteins. When immunoassay methods are used, the choice of analytical cut-off concentrations become important because these influence the proportion of false positives and false negatives after a verification analysis is done.[27]

Another useful biological matrix gaining ground in forensic toxicology and widely used for analysis of drugs and toxins is hair strands (also see Chapter 20).[28,29] Depending on the rate of growth of the hair, this biological matrix can detect the intake of drugs spanning back weeks or months.[30] By means of segmental analysis of the hair strands this furnishes an approximate timeline reflecting exposure to drugs, which is helpful to verify or challenge compliance with prescribed medication.[31] Hair is viable as a specimen for drug analysis even if the body is decomposed and when other specimens might not be available or unsuitable for toxicological analysis.[32] Another unusual matrix used in toxicology is finger or toe nail clippings (also see Chapter 19), which can be taken from exhumed bodies and the analytical results furnish information about drug intake during life.[33]

The verification methods used in forensic toxicology laboratories generally involve a four-step procedure:

(i) Use of liquid–liquid or solid-phase extraction to "clean up" the sample and extract drugs and metabolites for further analysis.
(ii) After extraction into an organic solvent, the parent drugs and metabolites are separated by use of some type of gas or liquid chromatography technique.
(iii) Qualitative identification is achieved based on GC retention times or by running a mass spectrum and then comparing the data with known authentic substances or libraries of known compounds.

(iv) Quantitative analysis of the amount of substance present is obtained from the response of one or more detectors, either flame ionization, nitrogen–phosphorus, electron-capture or a mass selective detector.[34]

If verification analysis shows concentrations of drugs above the method LOQ, then descriptive statistics (mean, median and upper percentiles) can be justifiably calculated and entered into the database for future reference.[35] During statistical analysis of data, because drug concentrations in blood tend to be skewed to the right, the median of the frequency distribution is a better indication of the central tendency than the mean value.

1.4   Post-mortem Toxicology

In post-mortem (PM) toxicology the condition of the specimen received for analysis is highly variable, depending on the circumstances surrounding the death (*e.g.* extensive trauma and blood loss) and whether the body was decomposed, which might impact on the concentrations of certain drugs present.[36] Pre-analytical sources of variation tend to dominate over analytical sources of variation, as outlined in a recent review article[37] and by analysis of methadone in blood from different sampling sites.[38] Some practitioners advocate that toxicology results should be reported as the concentration of drug or toxin in the specimen as received, whether blood, urine or vitreous humor.

The word autopsy means to see for oneself and entails making an external and internal examining of a corpse to discover the cause of death.[39] During an autopsy the pathologist has to document any anatomical, pathological or medical conditions that might account for the person's sudden and unexpected death. Most practitioners make use of the well-proven triad of autopsy findings (including histopathology), discoveries at the scene, and toxicological analysis of alcohol and/or other drugs in blood samples.[40] The known drinking and/or drug habits of the deceased is obviously relevant when intoxication deaths are considered.

Forensic autopsies are usually requested by the police authorities when a suspicious out-of-hospital death is investigated. The coroner or medical examiner is charged with assembling all available evidence before issuing a death certificate in which the cause and manner of death is reported or simply given as unascertained.[41] Deaths caused by overdosing with drugs, either accidentally or intentionally in a suicide attempt, rarely leave tell-tale signs for the pathologist, although froth in the mouth and airways is a characteristic of heroin or opioid-related deaths.[42]

A poisoning death might be the result of suicide, homicide or an accidental (unintentional) overdose with drugs and is not always easy to determine, hence "inconclusive", "uncertain" or "undetermined" often appear on death certificates.[43] The concentrations of drugs in blood are of prime importance when intoxication deaths are investigated, because these give the best indication of the effects on vital body functions, such as respiratory and circulatory centers in the brain.

Poly-drug use (at least five different medications) is increasingly common in today's society, especially in the elderly, and use of multiple drugs increases the propensity for fatal drug–drug interactions.[44] The physiological changes that occur during ageing, including renal and hepatic functioning, body composition and gastric emptying, also impact on uptake, distribution, metabolism and excretion of drugs that deserve consideration when toxicology results are interpreted.[45] Many practitioners find it useful to compare the post-mortem concentrations of drugs with concentrations encountered during therapeutic drug monitoring (TDM). However, such comparisons are complicated for various reasons, such as the composition of post-mortem blood (sometimes clotted) and the fact that TDM analysis is usually done on specimens of plasma or serum, which tend to contain higher drug concentrations compared with whole blood.[46]

The concentrations of drugs in post-mortem blood should not be converted into the dose of drug a person might have ingested during life based on body weight and the drug's volume of distribution.[47] The phenomenon of post-mortem redistribution of drugs refers to time-dependent changes in concentration in blood after death and before samples are time-dependent taken at autopsy. The magnitude of this redistribution phenomenon seems to differ for different drugs, depending on lipid solubility, tissue uptake, distribution volume and other properties.[41,48]

Table 1.2 shows descriptive statistics for the concentrations of 20 drugs most often identified in femoral blood in connection with forensic autopsies in Sweden and Finland.[35,49] Both these nations use highly standardized routines for sampling femoral blood for toxicological analysis and modern and comparable analytical methods are used.

Table 1.2 Descriptive statistics of the concentrations of the 20 drugs most often identified in post-mortem femoral blood in all causes of death. A comparison is made for TOXBASES available in Finland and Sweden.

Drugs identified in femoral blood	Country	Number of cases (N)	LOQ (mg L^{-1})	Mean (mg L^{-1})	Median (mg L^{-1})	Upper percentiles (mg L^{-1}) 90th	95th	97.5th
Alprazolam	Sweden	716	0.02	0.09	0.06	0.18	0.30	0.40
	Finland	940	0.02	0.09	0.05	0.20	0.30	0.40
Amitriptyline	Sweden	233	0.05	0.77	0.40	2.0	2.94	3.96
	Finland	1589	0.10	1.5	0.40	2.9	5.5	8.8
Amphetamine	Sweden	558	0.03	1.54	0.5	2.9	4.5	6.3
	Finland	565	0.04	0.91	0.28	2.1	3.7	6.2
Carbamazepine	Sweden	566	0.5	6.4	4.8	12	15	20
	Finland	1482	0.3	7.9	6.2	13	19	27
Citalopram	Sweden	1302	0.05	0.72	0.40	1.1	1.7	3.75
	Finland	3542	0.10	0.97	0.40	1.4	2.4	5.3
Codeine	Sweden	843	0.005	0.32	0.05	0.7	1.3	2.4
	Finland	1903	0.02	0.72	0.16	1.8	3.2	5.1
Dextropropoxyphene	Sweden	694	0.10	2.0	0.8	4.4	6.9	10.2
	Finland	249	0.10	6.5	2.6	12.0	17.0	38.0
Diazepam	Sweden	1223	0.05	0.23	0.10	0.5	0.7	1.0
	Finland	7404	0.02	0.17	0.09	0.4	0.6	0.8
Fluoxetine	Sweden	193	0.10	0.63	0.30	1.58	1.94	2.64
	Finland	649	0.20	0.80	0.50	1.60	2.50	3.60
Hydroxyzine	Sweden	266	0.05	0.45	0.20	0.86	1.9	3.4
	Finland	159	0.20	0.80	0.30	1.6	2.8	5.3
Levomepromazine	Sweden	242	0.05	1.02	0.2	1.2	2.9	5.5
	Finland	1602	0.10	0.99	0.4	1.9	3.2	5.0

Table 1.2 (Continued)

Drugs identified in femoral blood	Country	Number of cases (N)	LOQ (mg L^{-1})	Mean (mg L^{-1})	Median (mg L^{-1})	Upper percentiles (mg L^{-1})		
						90th	95th	97.5th
Methadone	Sweden	503	0.10	0.51	0.30	1.1	1.5	1.9
	Finland	207	0.05	0.59	0.40	1.3	1.8	2.2
Mirtazapine	Sweden	659	0.05	0.34	0.1	0.6	1.1	1.8
	Finland	2179	0.05	0.49	0.20	0.80	1.7	2.9
Morphine	Sweden	906	0.005	0.30	0.11	0.5	0.86	1.54
	Finland	1094	0.02	0.20	0.07	0.37	0.67	1.1
Sertraline	Sweden	508	0.05	0.34	0.2	0.7	1.0	1.6
	Finland	445	0.10	0.55	0.30	1.0	1.7	2.3
THC (cannabis or marijuana use)	Sweden	467	0.0003	0.003	0.001	0.007	0.011	0.016
	Finland	347	0.001	0.005	0.002	0.008	0.013	0.024
Tramadol	Sweden	716	0.05	2.64	0.60	5.0	10.2	19.8
	Finland	1581	0.10	3.1	0.90	6.5	13.0	21.0
Venlefaxine	Sweden	451	0.05	4.0	0.4	4.1	18.5	48.5
	Finland	824	0.10	3.7	0.6	5.7	17.0	33.0
Zolpidem	Sweden	417	0.05	0.60	0.20	1.6	2.8	3.6
	Finland	287	0.10	0.59	0.30	1.4	2.0	2.9
Zopiclone	Sweden	994	0.02	0.30	0.09	0.7	1.3	1.9
	Finland	2577	0.02	0.34	0.10	0.8	1.5	2.4

The large sample sizes (*N*) in Table 1.2 make this useful reference data to compare with the drug concentrations reported in individual cases when the same drugs are identified. Note that all-cause mortality was included in the cases reported in Table 1.2 and not just poisoning deaths. A particular problem with interpreting drug concentrations in post-mortem cases is that the deceased seldom used a single drug or medication. The high prevalence of poly-drug use in today's society heightens the risk of encountering a drug-related intoxication death.[50] Nevertheless, it is often possible to identify the major drug culprit among several other drugs identified in blood. In Finland, pathologists found that three drugs in particular were over-represented in poisoning deaths and these were dextropropoxyphene (propoxyphene), buprenorphine and methadone, all widely used opioid prescription pain medications, but also drugs subject to abuse.[49] This confirms the trends seen in many US states, showing an upsurge of overdose deaths involving prescription pain medication, such as fentanyl, methadone, oxycodone and hydrocodone.[51,52] The drug most often identified in mono-intoxication deaths in Sweden was the legal drug ethanol followed by the illicit drug heroin.[53]

As mentioned earlier, when overdose deaths are interpreted a major problem is site-to-site variation in drug concentrations, which might vary by a factor of 10 for some drugs.[54,55] A lot depends on the condition of the body, the degree of trauma, the post-mortem interval, the ambient temperature (refrigeration) and whether decomposition and putrefaction processes had commenced, all of which needs to be noted and made clear when interpretations are made.[45,56] The movement of drugs from high concentrations in the stomach contents or sequestered to body organs and tissues into the vascular system after death is appreciated by pathologists and toxicologists and is considered when the analytical results are interpreted.[48,57]

1.5 Alcohol and Drug Impaired Driving

Drunk and drugged drivers are responsible for many road-traffic crashes and impaired drivers are over-represented among drivers killed in such collisions (also see Chapter 13).[58,59] The police authorities throughout Sweden send specimens of blood and urine from apprehended drivers to the National Laboratory of Forensic Toxicology for analysis. TOXBASE therefore contains analytical results from all impaired driving cases, including the demographics of offenders, the types of drugs used and the concentrations in blood. Also available in some cases are signs and/or symptoms of impairment observed by arresting police officers or when a driver was examined by a physician and clinical tests of intoxication were administered.[60]

Punishable (statutory) limits of blood–alcohol concentration exist in most nations, although these vary four-fold from 20 to 80 mg per 100 mL (0.02–0.08 g%), probably reflecting national politics rather than traffic-safety research.[61] Moreover, many countries now enforce zero-tolerance laws for driving under the influence of drugs other than alcohol.[58,62] Besides a ban on driving after use of illicit recreational drugs, many psychoactive prescription drugs are included, such as benzodiazepines, because they impair driving ability when taken in overdose and this medication is subject to abuse.[63] Information gleaned from the Swedish version of TOXBASE has been used to document the demographics of traffic offenders and the concentrations of drugs (both licit and illicit) determined in blood in arrested drivers and in drivers killed in traffic crashes.[60,64]

Table 1.3 presents the age and gender of people arrested in Sweden for drunk or drugged driving offences. The results show a clear predominance of male offenders (85–90%) and a relationship between choice of drug and mean age of the individuals. Alcohol impaired drivers were eldest (mean 40 years) and those taking γ-hydroxybutyrate (GHB) were youngest (mean 24 years). The cases selected for display in Table 1.3 were those when only a single drug was present in the blood sample analyzed, even though multiple drugs are a

Table 1.3 Demographics of people arrested for drug-impaired driving in Sweden and the drugs identified in blood arranged according to mean age of traffic offenders.

Drug present in venous blood samples	Number of cases	Men (%)	Women (%)	Age of suspects (mean ± SD) (years)
Ethanol	32 814	90%	10%	40.0 ± 8.7
Amphetamine	9162	85%	15%	36.5 ± 9.0
Morphine/6-MAM[a]	52	89%	11%	33.2 ± 8.7
Cannabis/marijuana (THC in blood)[b]	7750	94%	6%	32.6 ± 8.4
Cocaine and/or BZE[c]	160	96%	4%	28.6 ± 7.0
Ecstasy	483	92%	8%	26.4 ± 7.2
GHB	548	96%	4%	26.0 ± 6.8

[a]6-MAM = 6-acetylmorphine, a metabolite of heroin.
[b]THC = tetrahydrocannabinol, the active constituent of marijuana/cannabis.
[c]BZE = benzoylecgonine, an inactive metabolite of cocaine.

common finding in people arrested for impaired driving in Sweden and other countries.[13,65,66]

The relationship between the concentration of a certain drug in blood and the degree of impairment of cognitive and psychomotor functioning varies widely, owing to the development of tolerance and other individual characteristics.[67–70] Accordingly, impairment-based drugged driving laws were not very effective, because tangible evidence of impairment was not easy to obtain, even after clinical tests were made by physicians or police officers trained to recognize drug impairment. The legal framework behind concentration *per se* DUID laws is different, because the main evidence for a prosecution is essentially the toxicology result and concentration of a banned substance in the driver's blood and not the effects of the drug or impairment caused by a particular substance.[71,72]

Table 1.4 shows descriptive statistics for the concentrations of drugs determined in venous blood from people suspected of DUID in Sweden.

Although the statutory blood–alcohol limit for driving (since 1989) is 20 mg per 100 mL (0.02 g%), the median blood–alcohol concentration (BAC) in apprehended drivers is 170 mg per 100 mL, which is more than eight times above the legal limit.[73] Reaching such a high BAC requires binge drinking and most of the traffic delinquents suffer

Table 1.4 Mean, median and highest concentrations of the drugs most commonly identified in blood from people arrested for driving under the influence of drugs in Sweden.

Drug present in blood	Number of cases	Mean conc. (mg L^{-1})	Median conc. (mg L^{-1})	Highest conc. (mg L^{-1})
Ethanol	32 814	1740	1780	5180
Amphetamine	9162	0.77	0.60	22.3
Methamphetamine	644	0.34	0.20	3.7
Ecstasy (MDMA)	493	0.23	0.10	3.5
Tetrahydrocannabinol (THC)	7750	1.9	1.0	36
γ-Hydroxybutyrate (GHB)	548	89	82	340
Cocaine	160	0.069	0.05	0.31
Benzoylecgonine[a]	160	0.80	0.60	3.0
Morphine[b,c]	2029	0.046	0.03	1.13
6-MAM[b]	52	0.016	0.01	0.10
Codeine[b,c]	1391	0.047	0.01	2.4
Diazepam[c]	1950	0.36	0.20	6.2

[a]6-MAM = 6-acetylmorphine, a metabolite of heroin.
[b]THC = tetrahydrocannabinol, the active constituent of marijuana/cannabis.
[c]BZE = benzoylecgonine, an inactive metabolite of cocaine.

from an alcohol abuse problem and clinically they might be diagnosed as alcoholics (also see Chapter 14).[74] This makes it tempting to suggest that mandatory treatment and rehabilitation for substance abuse might be more worthwhile than conventional punishments for drunken driving, such as monetary fines, revocation of the driving permit or short terms of imprisonment.[75,76]

Likewise, the median concentration of amphetamine in the blood of drivers apprehended by the police in Sweden was 0.6 mg L^{-1}, which verifies that large doses of this central stimulant were ingested some time before driving.[77] Amphetamine has a few legitimate therapeutic uses, such as attention deficit disorder, but when used for this reason the peak concentrations in blood are rarely higher than 0.1–0.2 mg L^{-1}.[77] Among drivers killed in traffic crashes with amphetamine determined in blood at autopsy, the median concentrations was 1.0 mg L^{-1} and 75% of the deceased had been arrested previously for use of illicit drugs or DUID.[78]

1.6   Drug-facilitated Sexual Assault

Drug-related violent crimes, such as drug-facilitated sexual assault (DFSA), are common in most communities and the news media sometimes refer to this behavior as date-rape.[79,80] When complaints of sexual assault are brought to the attention of the police authorities, the sampling of body fluids for toxicology should be done without delay. A typical case scenario might involve a perpetrator adding (spiking) a victim's food or drink with an incapacitating drug to cause drowsiness and inability to resist sexual advances, and in this drug-induced state a crime was committed.[80,81]

The retrieval of forensic evidence in DFSA cases is often hampered because some victims do not report a crime to the police or do not seek treatment for injuries until many hours, sometimes days, afterwards. This delay in reporting means that alcohol and/or other drugs in the body are eliminated by metabolism and excretion processes, so forensic toxicology evidence of drug exposure is often lacking or negative. In this connection, obtaining hair strands or nail clippings several weeks or months after an offence can furnish useful information about drug use when the alleged crime was committed.[82]

Table 1.5 shows the drugs identified and the concentrations in blood in two studies of alleged sexual assault of females in Sweden spanning a period of 8 years.[83,84] Many of the drugs identified represent ordinary prescription medications that might have been

Table 1.5 The top 10 drugs identified in the blood of female victims of sexual assault based on two epidemiological studies done in 2003–2007 and 2008–2010, representing 3260 cases.[83,84]

Drug	Period	N	Age (mean ± SD) (years)	Mean conc. (mg L^{-1})	Median conc. (mg L^{-1})	Highest conc. (mg L^{-1})
Ethanol	2003–2007	882	25 ± 9.9	1240	1190	3700
	2008–2010	791	25 ± 10.3	1230	1220	4300
Cannabis (THC)	2003–2007	100	25 ± 9.3	0.0012	0.0006	0.006
	2008–2010	85	25 ± 8.3	0.0010	0.0007	0.006
Diazepam	2003–2007	88	32 ± 12.7	0.35	0.20	2.7
	2008–2010	86	33 ± 12.1	0.24	0.10	1.0
Paracetamol	2003–2007	68	20 ± 10.4	7.7	5.0	48
	2008–2010	88	29 ± 12.6	6.5	5.0	25
Amphetamine	2003–2007	86	29 ± 11.1	0.22	0.10	1.8
	2008–2010	55	30 ± 11.2	0.35	0.24	1.7
Alprazolam	2003–2007	55	27 ± 11.4	0.05	0.04	0.17
	2008–2010	33	32 ± 12.3	0.06	0.04	0.19
Zopiclone	2003–2007	35	31 ± 10.9	0.05	0.03	0.13
	2008–2010	33	34 ± 14.3	0.10	0.05	0.40
Citalopram	2003–2007	30	32 ± 10.0	0.09	0.08	0.30
	2008–2010	26	28 ± 10.7	0.11	0.10	0.30
Fluoxetine	2003–2007	22	29 ± 11.8	0.25	0.20	0.50
	2008–2010	18	23 ± 6.7	0.28	0.20	0.70
Codeine	2003–2007	29	36 ± 11.8	0.06	0.05	0.35
	2008–2010	17	39 ± 8.2	0.07	0.08	0.16

used by the victim at the time when the offence occurred. The police investigation needs to document the victim's recreational drug habits and what medication, if any, was being taken and in what dosage at the time the crime was committed. The commonest drug detected in blood at elevated concentrations was ethanol (mean 124 mg per 100 mL) with a range from <20 mg per 100 mL (negligible) to >300 mg per 100 mL (incapacitating).

The forensic pharmacokinetics of ethanol has been extensively studied and is well documented in the literature (see Chapter 14).[85] This sometimes justifies making a back-calculation of the victim's BAC from the time of sampling blood to the time the alleged crime was committed.[86] Many studies have found that a good average elimination rate of ethanol from blood is 15 mg per 100 mL per h, with a range from 10 to 25 mg per 100 mL per h.[87] However, making a back-calculation is not advisable if only a urine sample is available for analysis from the victim, because converting urine alcohol concentration (UAC) to BAC is a dubious practice.[88] The UAC/BAC ratio is highly variable, depending on many factors including time of last emptying the bladder, before voiding urine for toxicological analysis.[89]

The most prominent illicit drugs identified in blood of sexual assault victims in Sweden were amphetamine and cannabis (tetrahydrocannabinol, THC), although neither of these psychoactive substances are known to cause drowsiness and incapacitation after acute recreational doses. Most of the drugs listed in Table 1.5 were prescription medication, especially benzodiazepines (diazepam and alprazolam), antidepressants (citalopram and fluoxetine) or pain killers (codeine and paracetamol), which confirms the results from many previous DFSA studies.[90]

Several potentially incapacitating drugs were identified in blood samples, however, including the fast acting hypnotic zopiclone,[91] which therefore might have been added to the victim's food or drink with criminal intent. However, any police investigation needs to ascertain whether the victim had a prescription for this medication and might have been medicating with the drug at the time the crime was committed. Accordingly, obtaining a positive toxicology report does not prove a drug was used to incapacitate a victim.

1.7 Concluding Remarks

The creation of a nationwide forensic toxicology database, as described here for Sweden, has permitted a large number of pharmacoepidemiological projects and, over the years, scores of original publications have been published.[15] TOXBASE contains reference concentrations for hundreds of drugs and metabolites, both licit and illicit, and these can be compared with future cases involving the same drugs with which the laboratory might be involved.[92]

Several compilations of therapeutic, toxic and lethal concentrations of drugs are available for scrutiny in the literature, although these should be used and interpreted cautiously. The methods of drug analysis differ between laboratories and analytical cut-off concentrations for reporting positive results are not necessarily the same. Whether a single intoxicating substance or multiple drugs were involved is not always clarified in these compilations. This makes a big difference when toxicity is assessed, owing to the potential for adverse drug–drug or drug–alcohol interactions.[53] Moreover, a critical review shows that the concentrations listed for many drugs in these compilations are simply copied from one list to another without any explanations given.[93] Older analytical methods from the 1960–1970s sometimes failed to distinguish a parent drug from its major metabolite, which would not be acceptable in today's forensic toxicology laboratory.

The validity and acceptance of analytical toxicology results are highly dependent on the accuracy, precision, specificity of the method and the condition of the sample received for analysis. The quality of post-mortem blood can vary widely in terms of water content and the proportion of red cells to serum, and the specimen sometimes contains clots. The participation of laboratories in external proficiency tests of toxicological drug analysis is a pre-requisite for accreditation and the results need to be well documented and made available for inspection if and when required.

Sudden and unexpected death in an otherwise healthy individual with nothing remarkable found at autopsy draws attention to a possible poisoning or drug intoxication death. Under these circumstances the toxicology report is highly relevant when cause and manner of death are determined, whether self-administration (suicide), administration by others (homicide), self-administration for special purposes (drug abuse, abortion) or accidental overdosing with medication. Such information can also be gleaned from a search of TOXBASE.[94,95]

As discussed earlier, pre-analytical (sampling) variations dominate over analytical variations, which underscores the need for using reliable and reproducible sampling procedures to enhance integrity of the specimen analyzed.[96,97] A well-known post-mortem artifact is redistribution of drugs into the vascular system after death, thereby increasing concentrations in blood. Many drugs tend to concentrate in lipid compartments or tissue depots and diffuse into the blood-stream as the time after death increases. For some drugs, especially those with large distribution volumes, concentrations in autopsy blood might be appreciably higher than the concentration at the time of death.[98] The best choice of blood for toxicology is from a femoral vein in the legs and not central or cardiac blood for which drug concentrations are usually higher.[99] Indeed, the cardiac/femoral concentration ratio is often used as an indicator of post-mortem redistribution.[100]

Some drugs are notoriously unstable in autopsy blood, as exemplified by the nitrobenzodiazepines, such as clonazepam, flunitrazepam and nitrazepam.[101] Evidence of taking these sedative hypnotics during life is best obtained by analysis and quantification of their 7-amino metabolites.[102] Furthermore, the illicit drug cocaine is degraded *in vitro* to its main metabolites benzoylecgonine and ecgonine methyl ester. The hypnotic drug zopiclone is also unstable and is transformed into 2-amino-5-chloropyridine (ACP). The analytical methods used by forensic toxicologists should be able to identify the parent drugs and the main metabolites as evidence of ante-mortem ingestion or exposure.[103,104]

The concentrations of some abused drugs, such as ethanol and GHB, can increase in the body after death, depending on conditions of storage (*e.g.* temperature and use of fluoride preservative) and whether any decomposition of the corpse had commenced.[105,106] This requires using a higher analytical cut-off concentrations to report positive results, such as 30 mg L^{-1} for GHB in blood and 20 mg per 100 mL for ethanol, otherwise people might be falsely accused of ante-mortem ingestion of these substances. Some countries (*e.g.* Norway, Sweden and Poland) enforce low statutory BAC limits for driving (20 mg per 100 mL). Reporting an autopsy BAC above this limit is often interpreted to mean that the driver had consumed alcohol before the crash and was above the legal limit. Such a conclusion would have serious consequences for next of kin when insurance claims are made. The routine analytical methods used in toxicology laboratories to determine BAC cannot distinguish ethanol derived from drinking alcoholic beverages during life from ethanol generated by bacteria and micro-organisms after death.[106,107]

Interpretation of ethanol concentrations in autopsy blood is made more reliable if alternative specimens are analyzed, such as urine, vitreous humor and blood from different sampling sites, such as heart, iliac, subclavian and femoral veins.[106,108] Another approach to distinguish ante-mortem ingestion from post-mortem synthesis of ethanol is to identify the non-oxidative metabolites ethyl glucuronide or ethyl sulfate, along with ethanol.[109,110] Because ethanol produced by fermentation is not conjugated with glucuronic acid, finding the presence of ethyl glucuronide in post-mortem blood supports ante-mortem ingestion of ethanol.[111]

The practice of comparing the concentrations of drugs determined in post-mortem blood with therapeutic concentrations in plasma or serum from living subjects is not recommended.[57] Many drugs bind to plasma proteins and the concentrations in plasma or serum are higher than in erythrocytes and consequently also higher than in whole blood. Examples of serum/blood concentration ratios range from 1.15 : 1 for ethanol[112] and 1.8 : 1 for diazepam[46] to 2.0 : 1 for THC.[113] Furthermore, post-mortem blood is highly variable in composition, which also impacts on the concentrations of drugs present.[114]

When results of the forensic autopsy and histopathology are inconclusive, attention shifts towards the possibility of drug intoxication as a cause of death. The information in TOXBASE becomes very important in such cases and functions as a reference material for drug concentrations.[49] If the reference database of drug concentrations is sufficiently large, the information might be arranged according to age and gender of the deceased or the number of co-ingested drugs identified in blood samples.[115,116] The TOXBASE available in Sweden classifies poisoning deaths as those involving a single intoxicating substance, a single drug together with an elevated blood–ethanol level or multiple drug deaths.[9] As a reference material in the living, the post-mortem concentrations are compared with impaired drivers and, for many medicinal drugs, TDM cases can be used.[115,116]

TOXBASE was used to compile a list of drugs and drug concentrations determined in femoral blood representing all causes of death.[35] These data provide a quick reference source to compare and contrast with future toxicology cases dealt with by the laboratory and analyzed by the same methods. The median and upper percentile points of the frequency distribution can be compared with the concentrations of the same drugs in new autopsy cases. For example, if a new case contained a concentration above the 97.5th percentile, then only 2.5% of all previous cases reached such a high concentration. This should raise a warning flag for the pathologist to consider, along with other information such as age and gender of the deceased and the types and concentrations of other drugs in the same blood sample.[117]

Whether evidence-based post-mortem toxicology will ever become a reality remains to be seen, although whatever happens TOXBASE will prove very useful for developing this project. Recommendations and general guidelines for good laboratory practice in forensic toxicology already exist in the UK and Ireland[118] as well the USA, as published by the Society of Forensic Toxicologists (SOFT) organization (www.SOFT. org). More recently in the USA a Scientific Working Group for Forensic Toxicology (www.SWGTOX.org) was organized with the mission to develop and disseminate consensus standards for analytical methods suitable for toxicological analysis, including training and certification of analysts.

Both clinical and forensic toxicology laboratories have much to gain by creating a nationwide toxicology database.[119,120] The treatment of poisoned patients, including the best available antidotes and their effectiveness, is other information that can be retrieved from clinical toxicology databases.[121] Ethical questions often arise when large databases of information are available and searched on-line, and it is important to ensure that individual patients or police suspects are not identifiable by name.[120]

The Swedish TOXBASE was created in 1992 and contains enormous amounts of information about drugs and metabolites identified in blood samples from living and deceased persons. The material is homogeneous and the information is used for research and routine purposes when questions arise about the harmfulness of drugs and various drug combinations. Interpreting the meaning of drug concentrations in blood and the risk for drug-induced intoxication or toxicity are often contentious issues when expert testimony is presented to the court.[122] The probability of encountering critical concentration of drugs in blood can be gleaned by a search of TOXBASE. The results can sometimes function as an early warning system to spot trends in drug abuse in society, such as the so-called "legal highs", so that measures can be taken to classify them as controlled substances.

References

1. N. G. Coley, Forensic chemistry in 19th century Britain, *Endeavour*, 1998, **22**(4), 143–147.
2. O. H. Drummer, Forensic toxicology, *EXS*, 2010, **100**, 579–603.
3. A. Poklis, Analytical/forensic toxicology, in *Casarett and Doull's Toxicology the Basic Science of Poisons*, ed. C. D. Klaassen,

McGraw Hill Co., Inc., New York, 7th edn, 2008, pp. 1237–1255.

4. H. H. Maurer, Analytical toxicology, *EXS*, 2010, **100**, 317–337.
5. R. E. Ferner, Toxicological evidence in forensic pharmacology, *Int. J. Risk Saf. Med.*, 2012, **24**(1), 13–21.
6. M. R. Meyer and H. H. Maurer, Absorption, distribution, metabolism and excretion pharmacogenomics of drugs of abuse, *Pharmacogenomics.*, 2011, **12**(2), 215–233.
7. I. R. Edwards and D. Body, Forensic pharmacovigilance, *Int. J. Risk Saf. Med.*, 2012, **24**(1), 1–2.
8. R. K. Sewal, V. K. Saini and B. Medhi, Forensic pharmacovigilance: newer dimensions of pharmacovigilance, *J. Forensic Leg. Med.*, 2015, **34**, 113–118.
9. H. Druid and P. Holmgren, A compilation of fatal and control concentrations of drugs in postmortem femoral blood, *J. Forensic Sci.*, 1997, **42**(1), 79–87.
10. R. J. Flanagan and G. Connally, Interpretation of analytical toxicology results in life and at postmortem, *Toxicol. Rev.*, 2005, **24**(1), 51–62.
11. D. N. Bateman and A. M. Good, Five years of poisons information on the internet: the UK experience of TOXBASE, *Emerg. Med. J.*, 2006, **23**(8), 614–617.
12. R. A. Lahti and E. Vuori, Fatal drug poisonings: medico-legal reports and mortality statistics, *Forensic Sci. Int.*, 2003, **136**(1–3), 35–46.
13. A. Holmgren, P. Holmgren, F. C. Kugelberg, A. W. Jones and J. Ahlner, Predominance of illicit drugs and poly-drug use among drug-impaired drivers in Sweden, *Traffic Inj. Prev.*, 2007, **8**(4), 361–367.
14. H. Druid, P. Holmgren and P. Lowenhielm, Computer-assisted systems for forensic pathology and forensic toxicology, *J. Forensic Sci.*, 1996, **41**(5), 830–836.
15. A. W. Jones, Research work published by the Department of Forensic Toxicology, National Board of Forensic Medicine 1956-2013, Rättsmedicinalverket, Stockholm, 2014, pp. 1–140.
16. M. C. Kennedy, Post-mortem drug concentrations, *Intern. Med. J.*, 2010, **40**(3), 183–187.
17. A. F. Manini, L. S. Nelson, D. Olsen, D. Vlahov and R. S. Hoffman, Medical examiner and medical toxicologist agreement on cause of death, *Forensic Sci. Int.*, 2011, **206**(1–3), 71–76.
18. R. J. Flanagan, Developing an analytical toxicology service: principles and guidance, *Toxicol. Rev.*, 2004, **23**(4), 251–263.

19. *Forensic Drug Analysis*, ed. O. H. Drummer and D. Gerostamoulos, Future Science, London, 2013.

20. M. H. Huang, R. H. Liu, Y. L. Chen and S. L. Rhodes, Correlation of drug-testing results – Immunoassay versus gas chromatography-mass spectrometry, *Forensic Sci. Rev.*, 2006, **18**, 9–41.

21. M. N. Cox and C. R. Baum, Toxicology reviews: immunoassay in detecting drugs of abuse, *Pediatr. Emerg. Care*, 1998, **14**(5), 372–375.

22. P. Marquet, Is LC-MS suitable for a comprehensive screening of drugs and poisons in clinical toxicology?, *Ther. Drug Monit.*, 2002, **24**(1), 125–133.

23. H. H. Maurer, Position of chromatographic techniques in screening for detection of drugs or poisons in clinical and forensic toxicology and/or doping control, *Clin. Chem. Lab. Med.*, 2004, **42**(11), 1310–1324.

24. H. H. Maurer, Perspectives of liquid chromatography coupled to low- and high-resolution mass spectrometry for screening, identification, and quantification of drugs in clinical and forensic toxicology, *Ther. Drug Monit.*, 2010, **32**(3), 324–327.

25. M. Ceelen, T. Dorn, M. Buster, J. Stomp, P. Zweipfenning and K. Das, Post-mortem toxicological urine screening in cause of death determination, *Hum. Exp. Toxicol.*, 2011, **30**(9), 1165–1173.

26. A. G. Verstraete, Detection times of drugs of abuse in blood, urine, and oral fluid, *Ther. Drug Monit.*, 2004, **26**(2), 200–205.

27. G. M. Reisfield, B. A. Goldberger and R. L. Bertholf, 'False-positive' and 'false-negative' test results in clinical urine drug testing, *Bioanalysis*, 2009, **1**(5), 937–952.

28. G. A. Cooper, Hair testing is taking root, *Ann. Clin. Biochem.*, 2011, **48**(Pt 6), 516–530.

29. F. Pragst and M. A. Balikova, State of the art in hair analysis for detection of drug and alcohol abuse, *Clin. Chim. Acta*, 2006, **370**(1–2), 17–49.

30. S. Vogliardi, M. Tucci, G. Stocchero, S. D. Ferrara and D. Favretto, Sample preparation methods for determination of drugs of abuse in hair samples: A review, *Anal. Chim. Acta*, 2015, **857**, 1–27.

31. J. Barbosa, J. Faria, F. Carvalho, M. Pedro, O. Queiros, R. Moreira, *et al.*, Hair as an alternative matrix in bioanalysis, *Bioanalysis*, 2013, **5**(8), 895–914.

32. R. Agius, P. Kintz and European Workplace Drug Testing Society, Guidelines for European workplace drug and alcohol testing in hair, *Drug Test. Anal.*, 2010, **2**(8), 367–376.

33. A. Palmeri, S. Pichini, R. Pacifici, P. Zuccaro and A. Lopez, Drugs in nails: physiology, pharmacokinetics and forensic toxicology, *Clin. Pharmacokinet.*, 2000, **38**(2), 95–110.
34. O. H. Drummer, Post-mortem toxicology, *Forensic Sci. Int.*, 2007, **165**(2–3), 199–203.
35. A. W. Jones and A. Holmgren, Concentration distributions of the drugs most frequently identified in post-mortem femoral blood representing all causes of death, *Med. Sci. Law*, 2009, **49**(4), 257–273.
36. R. W. Byard and M. Tsokos, The challenges presented by decomposition, *Forensic Sci., Med., Pathol.*, 2013, **9**(2), 135–137.
37. G. Skopp, Preanalytic aspects in postmortem toxicology, *Forensic Sci. Int.*, 2004, **142**(2–3), 75–100.
38. K. Linnet, S. S. Johansen, A. Buchard, J. Munkholm and N. Morling, Dominance of pre-analytical over analytical variation for measurement of methadone and its main metabolite in postmortem femoral blood, *Forensic Sci. Int.*, 2008, **179**(1), 78–82.
39. M. J. Clark, Historical Keyword Autopsy, *Lancet*, 2005, **366**, 1767.
40. C. M. Milroy and J. L. Parai, The histopathology of drugs of abuse, *Histopathology*, 2011, **59**(4), 579–593.
41. O. H. Drummer, B. Kennedy, L. Bugeja, J. E. Ibrahim and J. Ozanne-Smith, Interpretation of postmortem forensic toxicology results for injury prevention research, *Inj. Prev.*, 2013, **19**(4), 284–289.
42. G. G. Davis, Complete republication: National Association of Medical Examiners position paper: Recommendations for the investigation, diagnosis, and certification of deaths related to opioid drugs, *J. Med. Toxicol.*, 2014, **10**(1), 100–106.
43. M. Warner, L. J. Paulozzi, K. B. Nolte, G. G. Davis and L. S. Nelson, State variation in certifying manner of death and drugs involved in drug intoxication deaths, *Acad. Forensic Pathol.*, 2013, **3**, 231–237.
44. I. A. Scott, S. N. Hilmer, E. Reeve, K. Potter, D. Le Couteur, D. Rigby, *et al.*, Reducing inappropriate polypharmacy: the process of deprescribing, *JAMA Intern. Med.*, 2015, **175**(5), 827–834.
45. R. W. Byard, Post mortem toxicology in the elderly, *Forensic Sci., Med., Pathol.*, 2013, **9**(2), 254–255.
46. A. W. Jones and H. Larsson, Distribution of diazepam and nordiazepam between plasma and whole blood and the influence of hematocrit, *Ther. Drug Monit.*, 2004, **26**(4), 380–385.
47. R. E. Ferner, Post-mortem clinical pharmacology, *Br. J. Clin. Pharmacol.*, 2008, **66**(4), 430–443.

48. A. L. Pelissier-Alicot, J. M. Gaulier, P. Champsaur and P. Marquet, Mechanisms underlying postmortem redistribution of drugs: a review, *J. Anal. Toxicol.*, 2003, **27**(8), 533–544.

49. T. Launiainen and I. Ojanpera, Drug concentrations in postmortem femoral blood compared with therapeutic concentrations in plasma, *Drug Test. Anal.*, 2014, **6**(4), 308–316.

50. K. Wester, A. K. Jonsson, O. Spigset, H. Druid and S. Hagg, Incidence of fatal adverse drug reactions: a population based study, *Br. J. Clin. Pharmacol.*, 2008, **65**(4), 573–579.

51. W. A. Ray, C. P. Chung, K. T. Murray, W. O. Cooper, K. Hall and C. M. Stein, Out-of-hospital mortality among patients receiving methadone for noncancer pain, *JAMA Intern. Med.*, 2015, **175**(3), 420–427.

52. M. R. Larochelle, F. Zhang, D. Ross-Degnan and J. F. Wharam, Rates of opioid dispensing and overdose after introduction of abuse-deterrent extended-release oxycodone and withdrawal of propoxyphene, *JAMA Intern. Med.*, 2015, **175**(6), 978–987.

53. A. W. Jones, F. C. Kugelberg, A. Holmgren and J. Ahlner, Drug poisoning deaths in Sweden show a predominance of ethanol in mono-intoxications, adverse drug-alcohol interactions and polydrug use, *Forensic Sci. Int.*, 2011, **206**(1–3), 43–51.

54. A. M. Langford and D. J. Pounder, Possible markers for postmortem drug redistribution, *J. Forensic Sci.*, 1997, **42**(1), 88–92.

55. D. J. Pounder, The nightmare of postmortem drug changes, *Leg. Med.*, 1993, 163–191.

56. D. M. Butzbach, The influence of putrefaction and sample storage on post-mortem toxicology results, *Forensic Sci., Med., Pathol.*, 2010, **6**(1), 35–45.

57. J. B. Leikin and W. A. Watson, Post-mortem toxicology: what the dead can and cannot tell us, *J. Toxicol. Clin. Toxicol.*, 2003, **41**(1), 47–56.

58. J. Ahlner, A. Holmgren and A. W. Jones, Prevalence of alcohol and other drugs and the concentrations in blood of drivers killed in road traffic crashes in Sweden, *Scand J Public Health*, 2014, **42**(2), 177–183.

59. J. E. Brady and G. Li, Trends in alcohol and other drugs detected in fatally injured drivers in the United States, 1999-2010, *Am. J. Epidemiol.*, 2014, **179**(6), 692–699.

60. A. S. Christophersen and J. Morland, Frequent detection of benzodiazepines in drugged drivers in Norway, *Traffic Inj. Prev.*, 2008, **9**(2), 98–104.

61. A. W. Jones, Effects of alcohol on fitness to drive, in *Handbook of Forensic Medicine*, ed. B. Madea, John Wiley & Sons, London, 2014, pp. 1057–1073.
62. A. W. Jones, F. C. Kugelberg, A. Holmgren and J. Ahlner, Five-year update on the occurrence of alcohol and other drugs in blood samples from drivers killed in road-traffic crashes in Sweden, *Forensic Sci. Int.*, 2009, **186**(1–3), 56–62.
63. K. K. Ojaniemi, T. P. Lintonen, A. O. Impinen, P. M. Lillsunde and A. I. Ostamo, Trends in driving under the influence of drugs: a register-based study of DUID suspects during 1977-2007, *Accid. Anal. Prev.*, 2009, **41**(1), 191–196.
64. H. Gjerde, A. S. Christophersen, P. T. Normann and J. Morland, Toxicological investigations of drivers killed in road traffic accidents in Norway during 2006-2008, *Forensic Sci. Int.*, 2011, **212**(1–3), 102–109.
65. V. Vindenes, D. Jordbru, A. B. Knapskog, E. Kvan, G. Mathisrud, L. Slordal, *et al.*, Impairment based legislative limits for driving under the influence of non-alcohol drugs in Norway, *Forensic Sci. Int.*, 2012, **219**(1–3), 1–11.
66. A. W. Jones and A. Holmgren, What non-alcohol drugs are used by drinking drivers in Sweden? Toxicological results from ten years of forensic blood samples, *J. Safety Res.*, 2012, **43**(3), 151–156.
67. A. W. Jones, A. Holmgren and F. C. Kugelberg, Driving under the influence of central stimulant amines: age and gender differences in concentrations of amphetamine, methamphetamine, and ecstasy in blood, *J. Stud. Alcohol Drugs*, 2008, **69**(2), 202–208.
68. A. W. Jones, A. Holmgren and F. C. Kugelberg, Driving under the influence of cannabis: a 10-year study of age and gender differences in the concentrations of tetrahydrocannabinol in blood, *Addiction*, 2008, **103**(3), 452–461.
69. A. W. Jones, A. Holmgren and F. C. Kugelberg, Concentrations of cocaine and its major metabolite benzoylecgonine in blood samples from apprehended drivers in Sweden, *Forensic Sci. Int.*, 2008, **177**(2-3), 133–139.
70. A. W. Jones, A. Holmgren and F. C. Kugelberg, Driving under the influence of opiates: concentration relationships between morphine, codeine, 6-acetyl morphine, and ethyl morphine in blood, *J. Anal. Toxicol.*, 2008, **32**(4), 265–272.
71. V. Vindenes, F. Boix, P. Koksaeter, M. C. Strand, L. Bachs, J. Morland, *et al.*, Drugged driving arrests in Norway before and after the implementation of per se law, *Forensic Sci. Int.*, 2014, **245C**, 171–177.

72. A. W. Jones, Driving under the influence of drugs in Sweden with zero concentration limits in blood for controlled substances, *Traffic Inj. Prev.*, 2005, **6**(4), 317–322.

73. A. W. Jones and A. Holmgren, Age and gender differences in blood-alcohol concentration in apprehended drivers in relation to the amounts of alcohol consumed, *Forensic Sci. Int.*, 2009, **188**(1–3), 40–45.

74. M. A. Schuckit, Alcohol-use disorders, *Lancet*, 2009, 373(9662), 492–501.

75. E. Day, A. Copello and M. Hull, Assessment and management of alcohol use disorders, *Br. Med. J.*, 2015, **350**, h715.

76. R. B. Voas and J. C. Fell, Preventing alcohol-related problems through health policy research, *Alcohol Res. Health*, 2010, 33(1–2), 18–28.

77. A. W. Jones and A. Holmgren, Amphetamine abuse in Sweden: subject demographics, changes in blood concentrations over time, and the types of coingested substances, *J. Clin. Psycho-pharmacol.*, 2013, **33**(2), 248–252.

78. A. W. Jones, A. Holmgren and J. Ahlner, High prevalence of previous arrests for illicit drug use and/or impaired driving among drivers killed in motor vehicle crashes in Sweden with amphetamine in blood at autopsy, *Int. J. Drug Policy*, 2015, **26**(8), 790–793.

79. S. Kerrigan, The use of alcohol to facilitate sexual assault, *Forensic Sci. Rev.*, 2010, **22**, 15–32.

80. J. Du Mont, S. Macdonald, N. Rotbard, D. Bainbridge, E. Asllani, N. Smith, *et al.*, Drug-facilitated sexual assault in Ontario, Canada: toxicological and DNA findings, *J. Forensic Leg. Med.*, 2010, **17**(6), 333–338.

81. S. A. Papadodima, S. A. Athanaselis and C. Spiliopoulou, Toxicological investigation of drug-facilitated sexual assaults, *Int. J. Clin. Pract.*, 2007, **61**(2), 259–264.

82. G. A. Cooper, R. Kronstrand, P. Kintz and Society of Hair Testing, Society of Hair Testing guidelines for drug testing in hair, *Forensic Sci. Int.*, 2012, **218**(1–3), 20–24.

83. A. W. Jones, F. C. Kugelberg, A. Holmgren and J. Ahlner, Occurrence of ethanol and other drugs in blood and urine specimens from female victims of alleged sexual assault, *Forensic Sci. Int.*, 2008, **181**(1-3), 40–46.

84. A. W. Jones, A. Holmgren and J. Ahlner, Toxicological analysis of blood and urine samples from female victims of alleged sexual assault, *Clin. Toxicol.*, 2012, **50**(7), 555–561.

85. A. W. Jones, Pharmacokinetics of ethanol - issues of forensic importance, *Forensic Sci. Rev.*, 2011, **23**, 91–136.
86. K. O. Lewis, Back calculation of blood alcohol concentration, *Br. Med. J. (Clin. Res. Ed.)*, 1987, **295**(6602), 800–801.
87. A. W. Jones, Evidence-based survey of the elimination rates of ethanol from blood with applications in forensic casework, *Forensic Sci. Int.*, 2010, **200**, 1–20.
88. A. W. Jones and P. Holmgren, Urine/blood ratios of ethanol in deaths attributed to acute alcohol poisoning and chronic alcoholism, *Forensic Sci. Int.*, 2003, **135**(3), 206–212.
89. A. W. Jones, Urine as a biological specimen for forensic analysis of alcohol and variability in the urine-to-blood relationship, *Toxicol. Rev.*, 2006, **25**(1), 15–35.
90. J. Hall, E. A. Goodall and T. Moore, Alleged drug facilitated sexual assault (DFSA) in Northern Ireland from 1999 to 2005. A study of blood alcohol levels, *J. Forensic Leg. Med.*, 2008, **15**(8), 497–504.
91. M. Villain, M. Cheze, A. Tracqui, B. Ludes and P. Kintz, Testing for zopiclone in hair application to drug-facilitated crimes, *Forensic Sci. Int.*, 2004, **145**(2-3), 117–121.
92. C. L. Winek, Tabulation of therapeutic, toxic, and lethal concentrations of drugs and chemicals in blood, *Clin. Chem.*, 1976, **22**(6), 832–836.
93. M. Schulz and A. Schmoldt, Therapeutic and toxic blood concentrations of more than 800 drugs and other xenobiotics, *Pharmazie*, 2003, **58**(7), 447–474.
94. J. Hedlund, J. Ahlner, M. Kristiansson and J. Sturup, A population-based study on toxicological findings in Swedish homicide victims and offenders from 2007 to 2009, *Forensic Sci. Int.*, 2014, **244**, 25–29.
95. A. W. Jones, A. Holmgren and J. Ahlner, Toxicology findings in suicides: concentrations of ethanol and other drugs in femoral blood in victims of hanging and poisoning in relation to age and gender of the deceased, *J. Forensic Leg. Med.*, 2013, **20**(7), 842–847.
96. G. Hoiseth, B. Fjeld, M. L. Burns, D. H. Strand and V. Vindenes, Long-term stability of morphine, codeine, and 6-acetylmorphine in real-life whole blood samples, stored at -20 degrees C, *Forensic Sci. Int.*, 2014, **239**, 6–10.
97. F. T. Peters, Stability of analytes in biosamples - an important issue in clinical and forensic toxicology?, *Anal. Bioanal. Chem.*, 2007, **388**(7), 1505–1519.

98. G. Skopp, Postmortem toxicology, *Forensic Sci. Med. Pathol.*, 2010, **6**(4), 314–325.
99. *Postmortem Toxicological Redistribution*, ed. O. H. Drummer, Springer-Verlag, London, 2008.
100. E. Han, E. Kim, H. Hong, S. Jeong, J. Kim, S. In, *et al.*, Evaluation of postmortem redistribution phenomena for commonly encountered drugs, *Forensic Sci. Int.*, 2012, **219**(1–3), 265–271.
101. O. H. Drummer, Methods for the measurement of benzodiazepines in biological samples, *J. Chromatogr. B: Biomed. Sci. Appl.*, 1998, **713**(1), 201–225.
102. M. D. Robertson and O. H. Drummer, Postmortem drug metabolism by bacteria, *J. Forensic Sci.*, 1995, **40**(3), 382–386.
103. D. S. Isenschmid, B. S. Levine and Y. H. Caplan, A comprehensive study of the stability of cocaine and its metabolites, *J. Anal. Toxicol.*, 1989, **13**(5), 250–256.
104. R. Jantos, A. Vermeeren, D. Sabljic, J. G. Ramaekers and G. Skopp, Degradation of zopiclone during storage of spiked and authentic whole blood and matching dried blood spots, *Int. J. Leg. Med.*, 2013, **127**(1), 69–76.
105. A. S. Korb and G. Cooper, Endogenous concentrations of GHB in postmortem blood from deaths unrelated to GHB use, *J. Anal. Toxicol.*, 2014, **38**(8), 582–588.
106. F. C. Kugelberg and A. W. Jones, Interpreting results of ethanol analysis in postmortem specimens: a review of the literature, *Forensic Sci. Int.*, 2007, **165**(1), 10–29.
107. V. A. Boumba, V. Economou, N. Kourkoumelis, P. Gousia, C. Papadopoulou and T. Vougiouklakis, Microbial ethanol production: experimental study and multivariate evaluation, *Forensic Sci. Int.*, 2012, **215**(1–3), 189–198.
108. A. L. Pelissier-Alicot, N. Coste, C. Bartoli, M. D. Piercecchi-Marti, A. Sanvoisin, J. Gouvernet, *et al.*, Comparison of ethanol concentrations in right cardiac blood, left cardiac blood and peripheral blood in a series of 30 cases, *Forensic Sci. Int.*, 2006, **156**(1), 35–39.
109. G. Hoiseth, R. Karinen, A. Christophersen and J. Morland, Practical use of ethyl glucuronide and ethyl sulfate in postmortem cases as markers of antemortem alcohol ingestion, *Int. J. Leg. Med.*, 2010, **124**(2), 143–148.
110. H. Krabseth, J. Morland and G. Hoiseth, Assistance of ethyl glucuronide and ethyl sulfate in the interpretation of postmortem ethanol findings, *Int. J. Leg. Med.*, 2014, **128**(5), 765–770.

111. M. Sundstrom, A. W. Jones and I. Ojanpera, Utility of urinary ethyl glucuronide analysis in post-mortem toxicology when investigating alcohol-related deaths, *Forensic Sci. Int.*, 2014, **241**, 178–182.

112. R. C. Charlebois, M. R. Corbett and J. G. Wigmore, Comparison of ethanol concentrations in blood, serum, and blood cells for forensic application, *J. Anal. Toxicol.*, 1996, **20**(3), 171–178.

113. E. W. Schwilke, E. L. Karschner, R. H. Lowe, A. M. Gordon, J. L. Cadet, R. I. Herning, *et al.*, Intra- and intersubject whole blood/plasma cannabinoid ratios determined by 2-dimensional, electron impact GC-MS with cryofocusing, *Clin. Chem.*, 2009, **55**(6), 1188–1195.

114. R. W. Byard and D. M. Butzbach, Issues in the interpretation of postmortem toxicology, *Forensic Sci., Med., Pathol.*, 2012, **8**(3), 205–207.

115. M. Reis, T. Aamo, J. Ahlner and H. Druid, Reference concentrations of antidepressants. A compilation of postmortem and therapeutic levels, *J. Anal. Toxicol.*, 2007, **31**(5), 254–264.

116. A. K. Jonsson, C. Soderberg, K. A. Espnes, J. Ahlner, A. Eriksson, M. Reis, *et al.*, Sedative and hypnotic drugs–fatal and non-fatal reference blood concentrations, *Forensic Sci. Int.*, 2014, **236**, 138–145.

117. T. Launiainen, E. Vuori and I. Ojanpera, Prevalence of adverse drug combinations in a large post-mortem toxicology database, *Int. J. Leg. Med.*, 2009, **123**(2), 109–115.

118. G. A. Cooper, S. Paterson and M. D. Osselton, The United Kingdom and Ireland Association of Forensic Toxicologists Forensic toxicology laboratory guidelines (2010), *Sci. Justice*, 2010, **50**(4), 166–176.

119. D. N. Bateman, A. M. Good, C. A. Kelly and W. J. Laing, Web based information on clinical toxicology for the United Kingdom: uptake and utilization of TOXBASE in 2000, *Br. J. Clin. Pharmacol.*, 2002, **54**(1), 3–9.

120. A. M. Good and D. N. Bateman, TOXBASE on the Internet, *J. Accid. Emerg. Med.*, 1999, **16**(6), 399.

121. R. E. Ferner, J. W. Dear and D. N. Bateman, Management of paracetamol poisoning, *Br. Med. J.*, 2011, **342**, d2218.

122. E. O'Brien, N. Nic Daeid and S. Black, Science in the court: pitfalls, challenges and solutions, *Philos. Trans. R. Soc., London, Ser. B*, 2015, **370**, 1674.

2 Forensic Pharmacology

Nigel J. Langford

University of Leicester and Leicester Royal Infirmary, Leicester, UK
Email: nigellangford@nhs.net

2.1 Introduction

Substances that come into contact with the body have the ability to affect its normal physiological workings. Some agents may be beneficial, whilst others are deleterious. The manner in which these effects arise is varied. Certain agents have physiochemical properties that have a direct chemical reaction, affecting the areas with which they come into contact. For environmental agents this will most commonly affect the skin and mucus membranes; for medicinal agents given parenterally this may affect the injection site or surrounding tissue, particularly if extravasation has occurred. Alternatively, such reactions can arise following oral ingestion of a substance as it passes to the stomach, reacting with the gastric acid present. Indeed, such an effect is used therapeutically with the administration of antacids for the treatment of indigestion. However, for many substances to exert a significant physiological effect the drug is required to enter the systemic circulation. Substances may then exert a widespread non-specific effect owing to their physicochemical properties or have a targeted effect at specific receptors. For those substances acting at receptors, the substance is required to link to the receptor, causing the molecular structure of the receptor to undergo a conformational change in either a temporary or a permanent manner to allow further

Forensic Toxicology: Drug Use and Misuse
Edited by Susannah Davies, Atholl Johnston and David Holt
© The Royal Society of Chemistry 2016
Published by the Royal Society of Chemistry, www.rsc.org

reactions to occur. Alternatively for some compounds, by linking to the receptor they prevent other compounds from interacting with the receptor, thereby stopping the normal physiological processes from happening. Such compounds are known as antagonists. For other compounds, only a partial response is achieved. With such substances the full response is often prevented from occurring. Such compounds are known as partial agonists.

Although there is an almost infinite number of different chemical compounds that can be manufactured within the body, there are six major types of drug–receptor interactions. These consist of transmembrance ion channels, transmembrance receptors coupled to intracellular G proteins, transmembrance receptors with enzymatic cytosolic domains, intracellular receptors including enzymes, transcriptional regulators and structural proteins, and extracellular enzymes and cell surface adhesion receptors.[1] The amount of interaction a substance has with these receptors may be highly variable, varying from full stimulation of the receptor producing a maximal response to having no discernible stimulatory response to the receptor itself but binding to the receptor and blocking its ability to interact with other substances. The effect of such substances on the body is known as pharmacodynamics.

In contrast, pharmacokinetics is the study of what the body does to a drug, describing the drug's movement into, through and out of the body, namely the processes of absorption, distribution, metabolism and excretion (collectively known as the ADME processes).[2]

2.2 Drug Dose

"What is there that is not poison? All things are poison and nothing [is] without poison. Solely the dose determines that a thing is not a poison." This is a quote commonly ascribed to Philip Theophrastus Bombast von Hohenheim, probably better known as Paracelsus (1493–1541). As stated, the effects of a drug on the body depend on how much has been taken. Generally a small dose of the drug often has a lesser effect compared to a larger dose of the drug. However, the effects seen are often not linear, such that doubling the dose does not double the effect. Obviously there are a wide range of doses that may be prescribed and taken. This lack of linearity can be a significant issue for individuals in trying to calculate the amount of drug they require to achieve the required drug action without precipitating significant adverse effects.

2.3 Drug Absorption

For a drug to be effective it has to come into contact with the body. Some routes of administration are obvious and ensure complete administration of the drug into the body (such as with intravenous administration); however, other routes are more convoluted (such as rectal administration) and may achieve a very different concentration of the drug reaching the systemic circulation. For example, with oxycodone a 5 mg oral dose is thought to have similar efficacy to a 30 mg suppository. Drugs that are absorbed through the gastro-intestinal system will usually involve transport through the portal system and the liver. This exposes the drug to a variety of metabolic enzyme systems, which may result in a significant reduction in the amount of drug reaching the systemic circulation (first-pass metabolism). The amount of first-pass metabolism can be influenced by a variety of factors, including the amount of metabolic work that the liver is performing.

Consequently, alternative methods may be employed in order to avoid this, allowing an adequate drug concentration to reach the systemic circulation. The most effective route in ensuring the drug reaches the systemic circulation is *via* intravenous injection, delivering the drug directly into the systemic circulation. However, delivering the drug intravenously can be difficult and is not always desirable. Other parenteral routes may be used such as intramuscular and sub-cutaneous injections. Both avoid first-pass metabolism but are limited by the volume of substance to be delivered as well as the physio-chemical properties of the drug in some circumstances. Very strong alkali or acidic substances may cause local tissue damage. Alternative routes, again avoiding the vagaries of first-pass metabolism, include the buccal route, such as with buprehorphine, and certain nitrates and benzodiazepines, such as midazolam and alprazolam. Alternatively, the transdermal route is becoming more commonly used pharmaceutically and includes transdermal patches being designed for a variety of drugs, including opiates, nitrates and hormones, amongst others. Vaginal and ocular routes may also be available, although they are not commonly used with respect to drugs of abuse.

For most drugs the oral route is the most commonly used. However, in designing a drug it can be problematic to ensure that an appropriate amount of the drug is consistently absorbed. The stomach is a hostile environment for many chemical agents. It is a strongly acidic and may chemically interact with a drug, leading to its breakdown or the formation of insoluble non-absorbable compounds. Furthermore, most medicines are weak acids or bases. In a soluble form they commonly exist in a partially ionized form. The proportion of the drug that is ionized is dependent on the pH of the solution it is in. For acid-based compounds the amount ionized is less at a low pH and more at a high pH, and *vice versa* for alkali substances. Drugs are more likely to be absorbed across a membrane if they are in a non-ionized state.

Although some drugs are absorbed within the stomach, for all practical purposes most drugs are absorbed in the small intestine. The rate a drug is absorbed is partly dependent on how quickly it can pass through the stomach. For example, we are aware that drinking alcohol on an empty stomach causes rapid inebriation and the well-known phrase advocating the "lining of your stomach" if you want to stay sober prior to going out drinking. The science behind such advice is quite sound, namely that drinking alcohol on an empty stomach allows rapid transport through the stomach to the small intestine, allowing quicker absorption of the alcohol and higher peak alcohol concentrations. This compares to drinking alcohol with food ("lining your stomach"), causing a slower transit time and a more gentle increase in the alcohol concentration and a subsequently lower peak alcohol level. However, too rapid transit through the bowel can also result in problems, leading to reduced drug absorption as the drug fails to be in contact with the wall of the small intestine for a longer enough period of time to allow absorption to occur.

2.4 Drug Bioavailability

The amount of drug entering the body and reaching the systemic circulation is known as bioavailability. If a drug is given intravenously, then all of that drug will reach the systemic circulation. However, if a drug is given by an alternative route, such as orally, then the amount reaching the systemic circulation may be significantly reduced owing to the potential problems of absorption across the gut wall, and then metabolism along its route either in the gut wall mucosa, the portal circulation or *via* the liver. Therefore the amount of drug reaching the systemic circulation may be far less then that which had been taken, and the overall effect correspondingly reduced. Bioavailability may be defined quantitatively by eqn (2.1):

$$\text{Bioavailability} = \frac{\text{Quantity of drug reaching the systemic circulation}}{\text{Quantity of drug administered}} \quad (2.1)$$

Table 2.1   Different dose equivalences for morphine, depending on route used.[15,16]

Route	Dose equivalence
Epidural	3
Intrathecal	30
Intramuscular	1
Intravenous	1
Subcutaneous	1
Rectal	0.4
Oral	0.4

Such effects can be appreciated clinically, as the dose of the drug given is varied according to the route used. Opiates are widely used, both medicinally and recreationally. Morphine undergoes significant first-pass metabolism; consequently the bioavailability of oral morphine is only approximately 20–30% of the initial dose administered. Thus a 10 mg intravenous dose is equivalent to 20–40 mg of an oral dose. When low doses of a drug are administered, all the drug may be metabolized by first-pass metabolism, leading an individual to experience no appreciable pharmacological effect. This first-pass metabolism is the explanation for why, when eating foodstuffs such as poppy seeds which contain natural opiates including morphine, no pharmacological effect is seen despite detectable amounts of morphine products being found in the urine for up to 12 hours afterwards.[3] Approximate dose equivalences for morphine can be seen in Table 2.1.

2.4.1   Volume of Distribution

On reaching the systemic circulation the medication is distributed to its site of action. Once a sufficient concentration is reached and the required number of receptors stimulated, a resultant pharmacological action may occur. The concentration of the drug in the systemic circulation is often measured to allow an approximation to be made as to the amount of drug present within the body. However, such measurements are often only a rough calculation as drugs are rarely distributed uniformly throughout the body. The volume of distribution, V_d, describes the extent to which a drug is distributed between the plasma and the tissue compartments. Mathematically it is defined by eqn (2.2):

$$\text{Volume of distribution} = \frac{\text{Dose absorbed}}{\text{Concentration of drug in plasma}} \quad (2.2)$$

The volume of distribution for a drug varies, depending on the physiochemical properties of the drug. Water-soluble drugs that remain in the plasma component of the body have a low volume of distribution, whereas fat-soluble drugs have a high volume of distribution and are distributed throughout the tissue and lipid compartments of the body.

Calculating a drug's volume of distribution can have useful clinical applications. One such application is in deciding whether a drug's rate of elimination can be enhanced by dialysis. Aspirin, a water-soluble drug, which has a low V_d (0.2 l kg^{-1}) remains predominately in the plasma and is therefore potentially dialyzable. Alternatively, for highly fat-soluble drugs that have a high V_d and are found largely within the tissues of the body, dialysis is unlikely to be a useful measure to achieve enhanced elimination; thus cannabis, a highly lipid-soluble drug with a V_d of <10 L kg^{-1}, cannot be dialyzed. Clinically, drugs with large V_d are steadily taken from the plasma and stored in tissue. For such drugs the tissue stores often have to be saturated first before appreciable increases in plasma concentration can occur. This further adds to the difficulty of trying to relate plasma concentrations of the drug to clinical effects. Any such readings should be correlated with the clinical effect and not looked at in isolation.

2.5 Clearance

Drug clearance is a term which defines the rate at which the active drug is removed from the body. The pattern over which the drug disappears from the body usually follows one of two distinct forms. For some drugs a set amount of drug is cleared per unit time, whilst in the other a set proportion of drug is removed per unit time. These processes are known as zero-order and first-order kinetics, respectively. The majority of drugs demonstrate first-order kinetics when used at standardized therapeutic doses. However, at toxic concentrations, standard pharmacokinetic pathways may no longer apply. With drugs that are metabolized and eliminated following first-order elimination patterns the enzyme pools work within normal parameters and have spare capacity (*i.e.* they are non-saturated). Consequently, increases in drug doses are matched with increases in the rate of drug metabolism.

However, some enzyme systems have only a limited or no reserve and work at maximum capacity. Therefore no matter how much drug enters the system, the amount of drug that is removed remains at a set value. Although such systems are rare they do include a number of both therapeutic drugs, such as phenytoin, aspirin and theophylline,[4] and recreational drugs, such as alcohol and γ-hydroxybutyrate (GHB), when these drugs are used at therapeutic or social doses.

The extent to which an organ contributes to the elimination or metabolism of a drug can be quantified using its extraction ratio. This simply looks at the concentration of the drug entering an organ compared to the concentration leaving the organ. The closer the ratio is to one, the more effective that organ has been in clearing that compound. Consequently, for drugs undergoing extensive first-pass metabolism, the liver extraction ratio will be close to unity.

2.6　Half-life

According to Drummer, the most commonly used pharmacokinetic term in court is "half-life".[5] The elimination half-life defines the time required for the body to remove half of the drug absorbed, with the plasma concentration decreasing to one half of its original concentration. Knowledge of a drug's half-life allows one to estimate not only the length of time for which a pharmacological effect can be seen, but also how long the drug will be in the body and for how long after a drug has been taken it will be detectable. For most drugs which are eliminated following first-order kinetics, then the half-life can be estimated as given in eqn (2.3):

$$t_{1/2} = \frac{0.693 \times V_d}{\text{Clearance}} \qquad (2.3)$$

where $t_{1/2}$ is the calculated half-life and V_d is the volume of distribution. Consequently, any factors which affect the volume of distribution of the drug and/or the clearance of the drug will also affect the half-life of the drug. Relevant factors may include the age and sex of an individual as well as whether they have other medical conditions such as congestive heart failure, liver disease or acute or chronic renal failure. The half-life of a drug may also be altered depending on whether other substances have been taken. For example, the apparent half-life for cocaine is short, at approximately 0.8 ± 0.2 hours; however, when cocaine is taken with ethanol (alcohol), cocaethylene (ethyl benzoylecgonine) may be formed having a reported half-life in man of 2.5 hours.

Although the half-life of a drug measures the amount of drug within the circulation, it does not always correlate directly with the clinical effect of the drug. The effect of the drug represents what has been happening at the drug–receptor interface. Such areas are usually impossible to measure and hence the blood concentration is used as a surrogate. However, because of this mismatch, and particularly where the action of the drug is intracellular or the drug is predominantly found in the lipid areas of the body, it is difficult to be able to interpret the clinical effect of the drug despite having a known plasma/ blood concentration.

2.7 Drug Biotransformation

Once a drug is absorbed into the body it will usually undergo some form of transformation or metabolism. Although there are many potential sites within the body that can alter a drug's composition, the liver is the organ predominantly responsible for this. There are two main processes involved: one whereby the drug undergoes chemical manipulation and the other whereby the drug commonly forms conjugates. These processes can occur independently of each other.

The process of chemical manipulation (also known as Phase 1 processes) primarily involves the processes of hydroxylation, oxidation and dealkylation and occurs predominantly but not exclusively within the liver. The majority of these reactions are catalyzed by enzymes of the heme protein monooxygenases of the cytochrome P450 class. The P450-mediated reactions are responsible for approximately 95% of oxidative biotransformations.[6]

There are over 50 different types of cytochrome isoform that have been discovered. The principle enzymes involved in the cytochrome P450 system with respect to drug metabolism include CYP1A2, CYP2C19, CYP2C9, CYP2D6, CYP3E1 and CYP3A4. CYP3A4 is the most abundant enzyme. Although the liver is the main site of drug metabolism, these enzymes are found in many different tissues throughout the body. For example, CYP3A4 is found in the gut mucosa and is partially responsible for metabolism of drugs during absorption and contributes to some of the first-pass metabolism seen. Drugs which undergo extensive first-pass metabolism by CYP3A4 include nimodipine and simvastatin. Others that are less highly metabolized but are still affected include amiodarone, carbamazepine, nifedipine and sildenafil.[7]

Although not classically thought of as drugs, it has become apparent that foodstuffs can also significantly interfere with drug metabolism. Grapefruit juice was serendipitously found to increase felodipine concentrations when used as a masking agent in a clinical trial.[8] Since then it has been shown that it interacts with gut mucosal CYP3A4 and can produce significant effects on drug bioavailability. The effects of a single ingestion of grapefruit can last up to 24 hours. Drugs that can be significantly affected by grapefruit other than the calcium channel blockers include cyclosporin, saquinavir and midazolam. Similarly, herbal drugs can also interfere significantly with drug metabolism, altering both the amount of drug that can be absorbed as well as the metabolic activity of the cytochrome enzyme system.

The relative proportions of the cytochrome P450 enzymes present is variable and may be influenced by a variety of factors, including other medication as well as environmental pollutants. This can result in alterations in the mount of enzymes present, leading to changes in the rate of drug metabolism. Therefore enzyme activity can either be enhanced, a process known as enzyme induction and leading to an increase in the metabolic processing power, or alternatively be reduced and slowing the metabolic process down, known as enzyme inhibition. However, a similar metabolic effect may also be seen when two drugs have the same metabolic route with the enzyme working at maximal capacity. Consequently, the drug with the more potent affinity for the receptor will be metabolized preferentially compared to the other. This may result in an increased concentration of one drug or a reduction in its ability to be eliminated, leading to increased concentrations and possible enhanced pharmacological action as well as the clinical effect being prolonged. Failure to appreciate such effects has led to patient deaths, an issue previously seen with alcohol and co-ingestion with γ-butyrolactone.

Within mankind there is significant variability in the enzymes that are responsible for metabolizing drugs. Genetic polymorphisms are common and can significantly alter the rate and hence biological effects of a substance. The classical example is seen with codeine, which undergoes metabolism involving the CYP2D6 enzyme during the process of chemical manipulation. Several molecular genetic mechanisms are involved, resulting in a wide variation in the different rates of drug metabolism. The ability to metabolize the drug in an ultrarapid way, resulting in increased levels of morphine production, has significant differences amongst different ethnic groups such that large numbers of ultra-metabolizers can be found in Ethiopians and those from Saudi Arabia but is rare in northern Europeans (see Table 2.2). However, up to 10% of Caucasians may be classified as poor metabolizers, whereby very little or no morphine is produced and subsequently the drug fails to provide the appropriate level of analgesic action. From a forensic

Table 2.2 Different rates of metabolism for CYP2D6.[17-20]

Ultrarapid	2–6%
Extensive/normal	50–70%
Intermediate	10–15%
Diminished intermediate/poor	1–2%
Poor	5–10%

perspective this difference in the ability to metabolize drugs has led to a number of deaths. In one case a breast-fed neonate who was found at post-mortem to have high concentrations of morphine in their body allegedly died owing to their mother, who was taking codeine, being an ultra-metabolizer. Tramadol, a synthetic analogue of codeine, is metabolized in a similar manner.[9,10]

Other important relevant non-mediated P450 reactions include alcohol metabolism. This occurs predominantly by the alcohol dehydrogenase pathway converting alcohol to acetaldehyde. Other important pathways include the metabolism of many amine containing endogenous compounds such as the catecholamines, which occur *via* the monoamine oxidase enzyme pathways.

Conjugation reactions include formation of glucuronides and sulfate esters as well as conjugates with glycine, sulfate and glutathione and processes of O, N and S methylation. Glucuronidation involves the enzyme UDP-glucuronyltransferease. This enzyme is predominantly found in the liver but is also present elsewhere in the body. The activity of the enzyme varies with age. In early life the activity is low, and can be one of the reasons why neonates develop jaundice. It also places them at risk of increased drug sensitivity. However, for drugs that are metabolized *via* this route, once conjugated, their pharmacological activity is usually severely curtailed. The resultant conjugate is commonly filtered by the kidney and the drug metabolite is eliminated. A notable exception of this is morphine 6-glucuronide (M6G). This conjugate retains significant pharmacological activity, and hence why in patients with renal impairment can still cause significant morphine toxicity.

2.8 Excretion

Drugs or their metabolites are commonly excreted in the urine. However, a significant number of drugs or their metabolites may be excreted *via* the gastrointestinal tract. Faecal excretion includes passage of unabsorbed oral drugs. However, drug excretion in the gastrointestinal tract also includes biliary excretion. Large molecular weight compounds and conjugates commonly utilize this route. For some substances which are excreted in the bile and pass into the intestine, as they pass through the gut they may be reabsorbed back into the systemic circulation, leading to a prolonged pharmacological effect, a process known as enterohepatic recirculation. Both morphine and its metabolite M6G undergo enterohepatic recycling. With morphine the amount re-absorbed back into the body may be further increased owing to bacterial deconjugation. Measurable drug concentrations may also be present in other bodily secretions, including saliva, sweat and breath.

2.9   Tolerance

Repeated administration of a drug may lead to an alteration in its effect on the body, often reducing its impact. This diminution of the effect of a drug despite the same dose being used is known as tolerance. For example, with morphine, tolerance occurs quickly to the effects on an individual's mood, such as itching, urinary retention and respiratory depression. However, tolerance to one effect of a drug may occur before tolerance to another effect, such that with morphine the tolerance to its analgesic effects occurs more slowly. With some effects such as constipation or miosis (the constriction of the pupil of the eye), tolerance fails to occur. As a result of tolerance, individuals taking morphine often require increased doses of the drug to achieve the same amount of pain-relief. Such dose increases are not associated with any relationship to misuse or addiction. Once the need for morphine is finished and an individual stops taking the medication, then their ability to tolerate the drug is lost. Loss of tolerance occurs rapidly. This is a common problem seen amongst opiate abusers who have their usual habit disturbed for some reason, such as incarceration or hospital admission.[11] These individuals are at risk of overdose if they re-start using opiates at a similar amount as previously.

2.10   Adverse Drug Reactions

All drugs have the ability to produce adverse effects in certain individuals. Various definitions and classifications of adverse drug reactions exist. Rawlins and Thompson[12] originally classified adverse drug reactions into two main groups: Type A and Type B. Type A reactions (Augmented) are drug reactions that can be extrapolated from the known pharmacological actions of a drug arising in a dose-related manner. For example, warfarin, a drug used as an anticoagulant, can cause bleeding. Small changes in either dose of the drug or other external factors, including an individual's diet, can cause large changes in the degree of anticoagulation an individual has and hence their propensity to have a haemorrhage. Type B reactions (Bizarre) were for the group of reactions that were thought to be unpredictable, often appearing to occur in a sporadic manner and not obviously related to the dose of the drug taken. Classically this has been described as being the anaphylaxis associated with penicillin administration. Another example had been the rhabdomyolysis seen with statin therapy. However, it has now been shown that there is a genetic disposition for certain individuals to develop muscle breakdown following statin therapy. As it has become apparent that there are multiple different factors involved in an adverse drug event occurring, so additional classes of adverse drug reactions had been added to the original two. More recently the classification has been overhauled as it has been realized that for an adverse event to happen, various factors are required.

Aronson and Ferner[13] comment that the three crucial factors for an adverse drug reaction to arise are: that the drug at a certain dose has to be taken, that the reaction occurs within a certain timeframe and for the individual in contact with the substance to have susceptibility for that reaction to occur. For some drugs the adverse effect is highly predictable. For example, if an excess of paracetamol is taken, at a certain point liver damage will arise in a set timeframe though the dose and the susceptibility of an individual to the toxicity seen may be highly variable and depend on a wide variety of addition variables such as body mass index, gender, nutritional status, concomitant medication and co-morbid disease states, as well as their genetic make-up. In the latter case we are increasingly able to reduce an individual's risk with genotyping. For example, with the retroviral drug Abacavir the incidence of adverse effects is approximately 2.7%. With human leukocyte antigen (HLA) screening this incidence can be reduced to almost zero.[14] Similar genetic-type testing has also been used/considered for other drugs such as carbamazepine, allopurinol and mercaptopurine.

2.11 Conclusion

Knowledge of the dose ingested and the way a drug is metabolized can help a clinician in anticipating the potential problems that may arise following an ingestion of a substance. It can also allow an estimate as to the time a patient remains at risk, as well as the potential problems that may subsequently arise. However, the impact of taking the same amount of drug may vary between individuals owing to differences in genetic make-up as well as the impact of having taken concomitant medication.

References

1. *Principles of Pharmacology: The Pathophysiologic Basis of Drug Therapy*, ed. D. E. Golan, Lippincott Williams & Wilkins, Baltimore, 3rd edn, 2012.
2. R. E. Ferner, *Forensic Pharmacology. Medicines, Mayhem and Malpractice*, Oxford University Press, Oxford, 1996.
3. S. Karch, *Karch's Pathology of Drug Abuse*, CRC Press, Boca Raton, 4th edn, 2009.
4. M. Dukes and J. Aronson, *Meyler's Side Effects of Drugs*, Elsevier Science, Amsterdam, 15th edn, 2006.
5. O. H. Drummer, *The Forensic Pharmacology of Drugs of Abuse*, Arnold, London, 2001.
6. J. Martin and M. Fay, Cytochrome P450 drug interactions: are they clinically relevant?, *AustPrescr*, 2001, **24**, 10–12.
7. C. C. Ogu and J. L. Maxa, Drug Interactions due to cytochrome P450, *BUMC Proc.*, 2000, **13**, 422–423.
8. D. G. Bailey, M. O. Arnold and D. J. Spence, Grapefruit juice-drug interaction, *Br. J. Pharmacol.*, 1998, **46**, 101–110.
9. S. Bernard, K. A. Neville, A. t. Nguyen *et al.*, Interethnic Differences in Genetic Polymorphisms of CYP2D6 in the U.S. Population: Clinical Implications, *Oncologist*, 2006, **11**, 126–135.
10. Y. Gasche, Y. Daali, M. Fathi *et al.*, Codeine Intoxication Associated with Ultrarapid CYP2D6 Metabolism, *N. Engl. J. Med.*, 2004, **351**, 2827–2831.
11. P. Oliver and J. Keen, Concomitant drugs of misuse and drug using behaviours associated with fatal opiate-related poisonings in Sheffield, UK, 1997–2000, *Addiction*, 2003, **98**, 191–197.
12. M. D. Rawlins and J. W. Thompson, Pathogenesis of adverse drug reactions, in *Textbook of Adverse Drug Reactions*, ed. D. M. Davies, Oxford University Press, Oxford, 1977, p. 10.
13. J. K. Aronson and R. E. Ferner, Joining the DoTs: new approach to classifying adverse drug reactions, *Br. Med. J.*, 2003, **327**, 1222–1225.

14. S. Mallal, E. Phillips, G. Carosi *et al.*, PREDICT-1 Study Team. HLA-B*5701 screening for hypersensitivity to abacavir, *N. Engl. J. Med.*, 2008 Feb 7, **358**(6), 568–579.
15. British National Formulary Accessed at https://www.medicinescomplete.com/mc/bnf/current/PHP2740-morphine.htm?q=morphine&t=search&ss=text&p=2#_hit on 22 November 2015.
16. Martindale the extra pharmacopoeia accessed at https://www.medicinescomplete.com/mc/bnf/current/PHP2740-morphine.htm?q=morphine&t=search&ss=text&p=2#_hit on 22 November 2015.
17. M. Ingelman-Sundberg, Genetic polymorphisms of cytochrome P450 2D6 (CYP2D6): clinical consequences, evolutionary aspects and functional diversity, *Pharmacogenom. J.*, 2005, **5**, 6–13.
18. Stephen Bernard, Kathleen A. Neville, Anne T. Nguyen and David A. Flockhart, Evaluation of CYP2D6 Polymorphic Types and Their Effect on Tamoxifen Efficacy Among Turkish Tamoxifen Users with Breast Cancer, *Int. J. Hematol. Oncol.*, 2014, **24**, 157–162.
19. S. Bernard, K. A. Neville, A. T. Nguyen and D. A. Flockhart, Interethnic Differences in Genetic Polymorphisms of CYP2D6 in the U.S. Population: Clinical Implications, *Oncologist*, 2006, **11**, 126–135.
20. L. D. Bradford, CYP2D6 allele frequency in European Caucasians, Asians, Africans and their descendants, *Pharmacogenomics*, 2002, **3**, 229–243.

3 The Role of Amnesty Bins in Understanding the Pattern of Recreational Drugs and Novel Psychoactive Substances Being Used Within a Locality

David M. Wood and Paul I. Dargan*

Guy's and St Thomas' NHS Foundation Trust and King's College London, London, UK
*Email: paul.dargan@gstt.nhs.uk

3.1 Introduction

There is currently no single information source that is able to provide a robust picture of the pattern of use of recreational drugs and novel psychoactive substances (NPS). Understanding the patterns of use, trends over time and the emergence of new NPS is essential not only for legislative authorities, but also to ensure that appropriate targeted information is provided to healthcare professionals and the wider community. Typically, population (*e.g.* Crime Survey England and Wales) and sub-population (*e.g.* Global Drugs Survey and *in situ* surveys within night-time economy venues) level surveys are used to determine national and/or local patterns of use of recreational drugs and/or NPS.[1-4] These surveys are based on self-reported recreational drugs/NPS use, rather than using analytical confirmation of the drugs used. Several studies have shown that there is variability in the

Forensic Toxicology: Drug Use and Misuse
Edited by Susannah Davies, Atholl Johnston and David Holt
© The Royal Society of Chemistry 2016
Published by the Royal Society of Chemistry, www.rsc.org

contents and the consistency of contents of both classical established recreational drugs and NPS.[5-12] There is increasing interest in attempting to confirm the actual drugs/NPS that individuals are using to provide more robust and valid data on the patterns of use of recreational drugs and NPS. Data reported to international bodies such as the European Monitoring Centre for Drugs and Drug Addiction (EMCDDA) includes analysis of seizures at border control level to determine drugs entering an individual country.[1] Whilst seizures from an individual at a local law enforcement level are often analysed as part of the criminal investigation, this information is often not collated and reported within the public domain; furthermore, such analyses are often targeted to a range of compounds which carry higher penalties. In addition, where an individual admits to possession, then the seized substances may not actually be analysed.

3.2 What are Amnesty Bins?

It is the responsibility of night-time economy venues to prevent drugs entering their premises, which in turn will limit their use and supply within the venue. Typically as a condition of entry into larger night-time economy venues, such as nightclubs and dance venues, individuals have to agree to a "search" as a condition of entry into the venue.[13] Whilst door security staff do not have a statutory right to search individuals, those who refuse to undergo a search are usually refused entry into the venue. Prior to this search some venues encourage individuals to deposit recreational drugs and/or NPS into "amnesty boxes". In addition, anything found during the search which is deemed to have been for "personal" consumption can be deposited by the venue security staff into these amnesty bins. Identification of individuals with larger amounts of recreational drugs/NPS which would appear to be possessed with the intent to supply will be reported to the police, who will then deal with the substances. Amnesty bins/boxes are typically brightly coloured, well-placed secure boxes that enable drugs to be deposited into them, but not easily removed.[14] Best practice guidance would recommend that substances deposited by staff should be contained within numbered sealed police evidence bags, and deposited in view of the individual they were seized from and/or one additional member of staff.[14] Removal of substances from amnesty bins should be by the relevant law enforcement agencies; samples can then either be taken away and destroyed or alternatively can be sent to an approved laboratory for subsequent analysis. This latter option enables a better understanding of local patterns of drugs being used, which can enable targeted educational and public health messages to the local communities. In the subsequent sections we will discuss some of the data that have been published from amnesty bin sample analysis and the implications of these data.

3.3 Data from Amnesty Bins Studies

The first reported study analysing samples collected in amnesty bins at a large London dance venue was from samples collected over a one-year period up to February 1999.[15] Over this time, there were 299 items discarded into the amnesty bin, consisting of 156 tablets, 90 powders, 26 cannabis items, 11 capsules, 10 knives, 5 "snorters" and 1 CS gas canister. Tablets and powders were compared to existing items on the TICTAC® database and then this was confirmed or identified using the Marquis test and gas chromatography with mass spectrometry (GC-MS) detection. An active ingredient was detected in 105 (67%) of the tablets and 79 (88%) of the powders; the breakdown of detected active ingredients is shown in Table 3.1.

In a follow-up study, samples were collected from one large central London dance club between January and March 2004 and from seven smaller Manchester dance clubs between July 2003 and September 2004.[16] After collection, the items were sorted and tablets that appeared identical (*e.g.* same colour, shape and logo/markings) were assumed to all be identical. Identification of the potential "active ingredients" was undertaken by comparison of solid items (tablets or capsules) with the TICTAC® database. To confirm this identification, one tablet from each batch was analysed by the Marquis test and GC-MS detection. Tablets identified as "pharmaceutical" based on comparison with the TICTAC® database were not further analysed unless there was concern about the validity of this identification. Overall, there were nine sealed police evidence bags with a total of 1016 items. The London dance venue bin contained 753 tablets, 48 bags of powders and 4 capsules; the Manchester dance venue bins contained 166 tablets, 35 bags of powders and 2 capsules. In addition, there was one police evidence bag from London which contained 42 individual cannabis joints. Of the 753 tablets in London, the majority (714, 94.8%) contained MDMA (lone MDMA, 712; MDMA with MDEA, 2); similarly, in Manchester the majority (140, 84%) of the 166 tablets were MDMA (lone MDMA, 139; MDMA with caffeine, 1).

Table 3.1 Identity of compounds/drugs detected in 156 tablets and 90 powders retrieved from an amnesty bin in a London dance venue for the year up to February 1999.[15]

Compound	Tablets percentage (number)	Powders percentage (number)
Unidentified	33 (51)	12 (11)
MDMA[a]	29 (46)	6 (5)
MDEA[b]	2 (3)	0
4-MTA[c]	1 (2)	0
Amphetamine	3 (5)	47 (42)
Cocaine	0	8 (7)
Ketamine	4 (6)	4 (4)
Ephedrine	2 (3)	0
Caffeine	3 (5)	31 (28)
Paracetamol	6 (9)	9 (8)
Other over-the-counter medications	8 (12)	4 (4)
Other prescription-only medications	6 (9)	1 (1)
Confectionary	4 (6)	0

[a]3,4-Methylenedioxymethamphetamine.
[b]3,4-Methylenedioxyethylamphetamine.
[c]4-Methylthioamphetamine.

The authors did not provide exact numbers for the other active compounds detected, but these included benzodiazepines, sildenafil, analgesics, anti-diarrhoeals, plaster of Paris, confectionary, "zinc containing tablets" and ephedrine. In terms of the 48 powders in London, drugs detected included cocaine (14 powders), amphetamine (12), MDMA (9), methamphetamine (1), heroin (1), caffeine (11), lidocaine (3) and phenacetin (4). The 35 powders from Manchester contained cocaine (13 powders), ketamine (7), amphetamine (9), MDMA (4), caffeine (3), lidocaine (1) and phenacetin (1). Caffeine was found as the only active ingredient in one powder in both London and Manchester; it was co-detected with amphetamine (3 powders), cocaine (5) and MDMA (4). The lidocaine and phenacetin in both London and Manchester were detected only in powders that contained cocaine. There were three liquid samples in the dance club bins (London, 2; Manchester, 1); these were found to contain legitimate eye drops (1 sample) and volatile nitrites (poppers) (2 samples).

The use of analysis of samples from amnesty bins can be useful for informing both local and national training, public information and law enforcement strategies and policy around recreational drugs and NPS. Our group has used analysis of these types of samples to inform in these areas on both methamphetamine, γ-hydroxybutyrate (GHB) and γ-butyrolactone (GBL), and the role of the amnesty bin analysis is summarised below.[17,18]

Given the previous increasing interest in the potential threat from methamphetamine in the UK, our group used a multi-layered approach to show that there was no evidence to support these concerns related to increasing use or acute harm related to its use when compared to the classical recreational drug MDMA.[17] As part of this multi-layered approach, we reviewed the frequency of detection of methamphetamine in crystalline substances deposited in amnesty bins at men who have sex with men (MSM) nightclubs in South East London. There were 418 samples deposited into these amnesty bins between August 2006 and January 2007, of which 12 were "crystalline" in nature. Of the crystalline substances, only one was found to contain methamphetamine; the remaining 11 contained MDMA (6 samples), benzocaine/cocaine (3), ketamine (1) and benzocaine/ketamine/cocaine (1).

In 2006, when there was a difference in the legal status of GHB (Class C under the UK Misuse of Drugs Act, 1971) compared to GBL (uncontrolled), there was increasing concern that there was a shift from the use of GHB to GBL whilst it remained legal. We demonstrated that in 2006 the majority of individuals presenting to our hospital emergency department with acute toxicity self-reported the use of GHB (150 patients, 94.9% of "GHB/GBL attendances") compared to GBL (8 patients, 5.1% of "GHB/GBL attendances"). However, analysis of 225 liquid samples from amnesty bins in the nightclubs where the majority of those presenting to the emergency department had been, showed that in fact there was a higher frequency of detection of GBL (140 samples, 62.2%) compared to GHB (85 samples, 37.8%). This clearly demonstrates the role that analysing samples from amnesty bins in the night-time economy can have in showing that users may not be aware of exactly what they are using. This type of information was then provided to the relevant legislative authorities (the UK Advisory Council on the Misuse of Drugs) as part of their subsequent consideration for the control of GBL under the UK Misuse of Drugs Act, 1971.

In addition, data from analysis of samples deposited in nightclub amnesty bins can be compared to *in situ* surveys undertaken within the same nightclubs. In September 2011 we undertook a collection and analysis of samples deposited into amnesty bins in two large MSM nightclubs in South East London.[19] There were a total of 544 samples, consisting of 240 liquids, 220 powders, 42 herbal products and 41 tablets/capsules. All of the herbal products were cannabis. Of the

Table 3.2 Comparison of compounds detected in amnesty bin sample analysis to self-reported drugs used in *in situ* surveys within the same nightclub.

Drug	Amnesty bin frequency[19]	*In situ* survey frequency[3,20]
GBL	29.4%	24%
Mephedrone	19.3%	41.0%
Volatile nitrites	13.2%	N/R[d]
Cannabis	7.9%	14%
MDMA	6.3%	5.8%
Ketamine	5.1%	13%
Cocaine	3.9%	16.7%
Methamphetamine	1.1%	N/R
Piperazines	1.1%	0.6%
4-MEC[a]	0.6%	N/R
PMMA[b]	0.6%	N/R
Amphetamine	0.4%	N/R
Benzodiazepine	0.2%	N/R
6-APB[c]	0.2%	N/R
Aminoindanes	0.2%	0%

[a]4-Methylethcathinone.
[b]*para*-Methoxymethamphetamine.
[c]6-(2-Aminopropyl)benzofuran.
[d]N/R = not reported.

240 liquids, 160 were GBL; interestingly, when compared to the amnesty bin analysis in 2006, where almost 40% of the "GHB/GBL samples" contained GHB, there was no detected GHB in any of the samples.[18] The most common compound detected in the non-liquid samples was mephedrone, which was detected in 105 of the non-liquid samples. The data from the amnesty bin analysis was compared to an *in situ* survey undertaken in the same nightclub in early summer 2011.[3,20] This showed that the pattern of self-reported drugs being used on the night of the *in situ* survey was similar to the pattern of drugs detected in the amnesty bin sample analysis (Table 3.2)

3.4 Expansion to Settings Outside of the Night-time Economy

In addition to collecting drug samples in amnesty bins within night-time economy venues, it is possible to collect samples from individuals when they present to the emergency department with acute

recreational drug toxicity. Whilst collection and analysis of a sample
in relation to an individual is not able to influence their management,
with informed consent it can be useful in confirming the drug(s) used
so that new drugs detected/novel patterns of harm can be published.[21] In addition, it is possible to collect samples in a similar
manner to the amnesty bin studies previously described. Typically,
these samples are collected from patients and then destroyed without
analysis. We have undertaken a pilot study in our emergency department where these samples are collected and sent for analysis,
rather than being destroyed.[22] This pilot study collected 33 samples
(12 liquid and 21 non-liquid). Only seven of the liquid samples were
suitable for analysis, and these were either GBL (5 samples) or iso-
propyl nitrite (2 samples); the remaining five were labelled as "popper
bottles" and it is likely that they contained a volatile nitrite at the time
of collection but that the contents had evaporated between the time of
collection and subsequent analysis. The 21 non-liquid samples were
seven powders, six crystalline substances and eight tablets; the
contents of these non-liquid samples are shown in Table 3.3.

Table 3.3 Compounds detected in the analysis of non-liquid samples
obtained from individuals presenting to a central London
emergency department with acute recreational drug toxicity.[22]

Type of sample	Compound detected	Number of samples
Powder	MDMA	1
	Ketamine	1
	Cocaine/phenacetin	1
	Cocaine/phenacetin/lidocaine	1
	Paracetamol/codeine	1
	No active ingredient	2
Tablets	Codeine	1
	Caffeine	1
	Ephedrine	1
	Kamagra[a]	1
	Chlorophenylpiperazine	1
	1-Benzylpiperazine	1
	Heroin	1
	No active ingredient	1
Crystalline	Frankincense	1
	Ketamine	1
	MDMA	1
	Methamphetamine	1
	No active ingredient	1

[a]Herbal erectile dysfunction medication.

This pilot study highlighted that whilst non-liquid samples are relatively stable, there is a need for a short timeframe between collection and analysis to reduce the loss of more volatile liquid substances prior to analysis.

3.5 Conclusion

The information on the pattern of drug use collected on a local level is usually based on subpopulation surveys of self-reported recreational drugs and/or novel psychoactive substances used. Analysis of suspected drugs deposited into night-time economy venue amnesty bins either by individuals or those seized during door entry searches is complementary to these subpopulation surveys. Analysis of these drugs from amnesty bins enables a better understanding of patterns of the actual recreational drugs and novel psychoactive substances being used on a local basis. Previous studies have been able to detect trends over time in the pattern of use, with changes reflecting data from other information sources such as subpopulation and population level surveys. In addition, there is the potential to compare the pattern of use of drugs between discrete localities within a region, country or in a larger international area. There is the potential for further work to develop systematic protocols for utilisation of drug amnesty bins and analysis of their contents across not only an individual country but also across multiple countries (*e.g.* across Europe). This could then provide robust data on local use patterns of recreational drugs and novel psychoactive substances for national and international authorities such as the UK Advisory Council for the Misuse of Drugs (ACMD) and the European Monitoring Centre for Drugs and Drug Addiction (EMCDDA).

References

1. EMCDDA EDR: EMCDDA. European Drug Report: Trends and Developments. Available at: http://www.emcdda.europa.eu/ attachements.cfm/att_228272_EN_TDAT14001.pdf.
2. CSEW 2013-14: Drug misuse: findings from the 2013 to 2014 CSEW 2nd edn. Available at: https://www.gov.uk/government/ publications/drug-misuse-findings-from-the-2013-to-2014-csew/ drug-misuse-findings-from-the-201314-crime-survey-for-england-and-wales.

3. D. M. Wood, L. Hunter, F. Measham and P. I. Dargan, Limited use of novel psychoactive substances in South London nightclubs, *QJM*, 2012, **105**, 959–964.

4. GDS 2014: Global Drugs Survey. The Global Drug Survey 2014 findings. Available at: http://www.globaldrugsurvey.com/facts-figures/the-global-drug-survey-2014-findings/.

5. T. M. Brunt and R. J. Niesink, The Drug Information and Monitoring System (DIMS) in the Netherlands: implementation, results, and international comparison, *Drug Test. Anal.*, 2011, **3**, 621–634.

6. N. Vogels, T. M. Brunt, S. Rigter, P. van Dijk, H. Vervaeke and R. J. Niesink, Content of ecstasy in the Netherlands: 1993-2008, *Addiction*, 2009, **104**, 2057–2066.

7. J. Ramsey, P. I. Dargan, M. Smyllie, S. Davies, J. Button, D. W. Holt and D. M. Wood, Buying 'legal' recreational drugs does not mean that you are not breaking the law, *QJM*, 2010, **103**, 777–783.

8. S. Davies, T. Lee, J. Ramsey, P. I. Dargan and D. M. Wood, Risk of caffeine toxicity associated with the use of 'legal highs' (novel psychoactive substances), *Eur. J. Clin. Pharmacol.*, 2012, **68**, 435–439.

9. S. D. Brandt, H. R. Sumnall, F. Measham and J. Cole, Analyses of second-generation 'legal highs' in the UK: initial findings, *Drug Test. Anal.*, 2010, **2**, 377–382.

10. S. D. Brandt, S. Freeman, H. R. Sumnall, F. Measham and J. Cole, Analysis of NRG 'legal highs' in the UK: identification and formation of novel cathinones, *Drug Test. Anal.*, 2011, **3**, 569–575.

11. S. Davies, D. M. Wood, G. Smith, J. Button, J. Ramsey, R. Archer, D. W. Holt and P. I. Dargan, Purchasing "Legal Highs" on the Internet – Is there consistency in what you get?, *QJM*, 2010, **103**, 489–493.

12. H. A. Spiller, M. L. Ryan, R. G. Weston and J. Jansen, Clinical experience with and analytical confirmation of "bath salts" and "legal highs" (synthetic cathinones) in the United States, *Clin. Toxicol.*, 2011, **49**, 499–505.

13. Safer Nightlife: London Drug Policy Forum. Safer Nightlife: Best practice for those concerned about drug use and the night-time economy, 2008. Available from: https://www.cityoflondon.gov.uk/services/health-and-wellbeing/drugs-and-alcohol/london-drug-and-alcohol-policy-forum/Documents/SS_LDPF_safer_nightlife.pdf.

14. LDAPF: London Drug and Alcohol Policy Forum. Drugs at the door (an annex to Safer nightlife): Guidance for venues and staff on handling drugs. 2011. Available from: https://www.cityoflondon.

gov.uk/services/health-and-wellbeing/drugs-and-alcohol/london-drug-and-alcohol-policy-forum/Documents/drugs-at-the-door.pdf.

15. J. D. Ramsey, M. A. Butcher, M. F. Murphy, T. Lee, A. Johnston and D. W. Holt, A new method to monitor drugs at dance venues, *Br. Med. J.*, 2001, **323**, 603.

16. S. L. Kenyon, J. D. Ramsey, T. Lee, A. Johnston and D. W. Holt, Analysis for identification in amnesty bin samples from dance venues, *Ther. Drug Monit.*, 2005, **27**, 793–798.

17. D. M. Wood, J. Button, T. Ashraf, S. Walker, S. L. Greene, N. Drake, J. Ramsey, D. W. Holt and P. I. Dargan, What evidence is there that the UK should tackle the potential emerging threat of methamphetamine toxicity rather than established recreational drugs such as MDMA ('ecstasy')?, *QJM*, 2008, **101**, 207–213.

18. D. M. Wood, C. Warren-Gash, T. Ashraf, S. L. Greene, Z. Shather, C. Trivedy, S. Clarke, J. Ramsey, D. W. Holt and P. I. Dargan, Medical and legal confusion surrounding gamma-hydroxy-butyrate (GHB) and its precursors gamma-butyrolactone (GBL) and 1,4-butanediol (1,4BD), *QJM*, 2008, **101**, 23–29.

19. T. Yamamoto, A. Kawsar, J. Ramsey, P. I. Dargan and D. M. Wood, Monitoring trends in recreational drug use from the analysis of the contents of amnesty bins in gay dance clubs, *QJM*, 2013, **106**, 1111–1117.

20. D. M. Wood, F. Measham and P. I. Dargan, 'Our Favourite Drug': Prevalence of use and preference for mephedrone in the London night time economy one year after control, *J Subst. Use*, 2012, **68**, 853–856.

21. J. Ward, S. Rhyee, J. Plansky and E. Boyer, Methoxetamine: a novel ketamine analog and growing health-care concern, *Clin. Toxicol.*, 2011, **49**, 874–875.

22. D. M. Wood, P. Panayi, S. Davies, D. Huggett, U. Collignon, J. Ramsey, J. Button, D. W. Holt and P. I. Dargan, Analysis of recreational drug samples obtained from patients presenting to a busy inner-city emergency department: a pilot study adding to knowledge on local recreational drug use, *Emerg. Med. J.*, 2011, **28**, 11–13.

4 Contamination of Water with Drugs and Metabolites

Victoria Hilborne

London South Bank University, London, UK
Email: hilborv@lsbu.ac.uk

4.1 Introduction

Across the world, vast quantities of drugs and lifestyle products such as caffeine and nicotine are consumed every day. The majority of these drugs are legal; however, substantial amounts of illegal substances are ingested orally, by nasal insufflation or injected, usually for recreational purposes. The European Monitoring Centre for Drugs and Drug Addiction (EMCDDA) reports over 3 million users of illegal amphetamines, including methylenedioxymethamphetamine (MDMA), and over 18 million users of cannabis in Europe each year.[1] Pharmaceutical consumption is substantially higher, particularly for the synthetic analgesic and non-steroid anti-inflammatory drugs (NSAIDs). Euromonitor International estimated the global retail selling price value of acetaminophen (paracetamol) to be £3095 million for the year 2013.[2] It is not a simple task to convert monetary value into quantities consumed. The illegal trade in pharmaceuticals, including internet sales, is an increasing issue. These are sold as legitimate products and are usually counterfeit or expired drug stock. After consumption, drug metabolites together with smaller amounts of the unchanged drug are excreted in sweat, urine and faeces. These then pass into sewage systems and on into sewage treatment plants or

Forensic Toxicology: Drug Use and Misuse
Edited by Susannah Davies, Atholl Johnston and David Holt
© The Royal Society of Chemistry 2016
Published by the Royal Society of Chemistry, www.rsc.org

directly into the aquatic or soil environments. A not insignificant contribution to the aquatic environment is from veterinary medicines. Both the large size and numbers of farm animals, in relatively small areas, can result in increased concentrations of drugs and metabolites. Wastewater treatment effectively removes much of the drugs and metabolites, but not all. The treated wastewater, effluent, is reintroduced into local rivers, leading to traces of drugs being detected in surface, ground and tap waters. The persistence and ecotoxicity of drugs and drug metabolites in the aquatic environment is of concern and needs further investigation. Knowledge of the impact of legal and illegal drugs on the environment, including their potential acute or chronic toxicity, is often limited and varies, particularly for the new psychoactive substances. Concentrations of illicit drugs measured in sewage wastewaters are used to estimate the degree of illegal drug use in a city, town or local area such as a prison. This provides evidence of an illegal drugs trade and supports social and economic studies on drug use habits. High levels of uncertainty are inevitable when estimating drug use from sewage. Methods used for sample collection and *in situ* drug detection and quantification need further development to reduce uncertainty. Predictions of regional drug use patterns from the mass balance of drugs in sewage influent are compared and validated against drug consumption data.

4.2 Drug Consumption

Levels of pharmaceutical consumption are determined from sales figures, the number of prescriptions dispensed and the therapeutic defined daily dose (DDD). These indicators have a degree of uncertainty: not all bought or prescribed medicines are consumed and the DDD may not be used in a given country. The number of illegal drug seizures by customs and other law enforcement agencies gives some indication of use levels. Other measures include drug rehabilitation studies and self-reporting in behaviour surveys. National pharmaceutical consumption grows with increasing access or with an aging population, whereas illegal drug consumption follows trends in attitudes and availability.

4.2.1 Trends in Pharmaceutical Consumption

Annual surveys conducted by the Organisation for Economic Cooperation and Development (OECD)[3] compare national health

Comparison of antidepressant consumption and national populations 2013.

(data taken from from OECD (2013) 'Pharmaceutical Consumption')

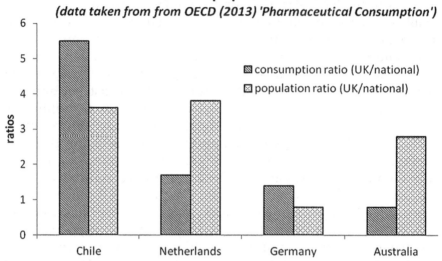

Figure 4.1 Ratios of UK and national population consumption of antidepressants.[3]

statistics. Examples include national consumption of hypertension and antidepressant pharmaceuticals, where the DDD per 1000 people per day are compared. Figure 4.1 illustrates ratios of antidepressant consumption for the year 2012 in the UK against that in Chile, the Netherlands, Germany and Australia. These are compared against population ratios. In the UK, 42% more antidepressants were consumed than in Germany and 81% more than in Chile.

The UK National Statistics, Health and Social Care Information Centre publishes the levels of pharmaceutical prescriptions dispensed annually.[4] Reports include dispensing trends over a decade, for example from 2003 to 2013. The top three UK drugs dispensed in 2013 were reported as the angiotensin converting enzyme (ACE) inhibitor ramipril at 24.9 million prescriptions, the gastric medicine omeprazole at 27.3 million prescriptions and the treatment for blood pressure and heart disease drug amlodipine at 23.1 million prescriptions. Over the period 2003 to 2013, prescription supply had increased by over 10%, corresponding with a population increase of 7.8%. The types of drugs prescribed and the associated prescription patterns were thought to be a consequence of the population age profile, improvements in diagnosis, increased prevalence of long-term conditions and decreased mortality. Additional contributions are the availability of new medicines, and increased prescribing for prevention or risk reduction of serious illness. Non-opioid analgesic painkillers and NSAIDs were and are currently consumed more than any other pharmaceutical type. Paracetamol, particularly the 500 mg tablets and capsules, was the most dispensed non-opioid analgesic, and codeine was the most dispensed opioid analgesic when combining data on all forms such as codeine phosphate in Co-Codamol, dihydrocodeine in Co-Dydramol and dihydrocodeine tartrate.[4] From 2003 to 2013, prescription of all analgesics increased, with the exception of aspirin. It is difficult and often costly to obtain data on consumption of over-the-counter medicines. Data are gathered through independent or government led surveys or from national and global market research commissioned by the pharmaceutical companies. The Spanish Ministry for Health and Social Policy publishes the number of packets of pharmaceuticals sold annually, including both with and without a prescription.[5] These data, together with the average number of pills per packet and the mass of active ingredient per dose, is used to estimate the quantities of pharmaceutically active compounds consumed.[6] Using these data it was calculated that over 1000 tonnes of paracetamol and 200 tonnes of ibuprofen were consumed by the Spanish population in the year 2012.[6] Estimated paracetamol consumption with examples of other popular pharmaceuticals is illustrated in Figure 4.2.

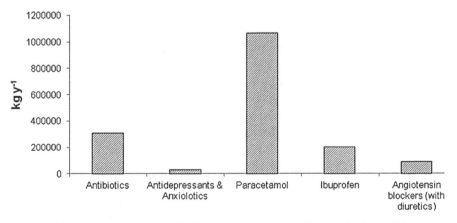

Figure 4.2 Consumption of pharmaceuticals in Spain during 2012, estimated using data from the Spanish Ministry of Health and Social Policy[5] on the number of packets sold, the number of pills per packet and the mass of active ingredient per typical dose.[6]

In 2012 the population of Spain was 47.27 million and in the UK it was 63.23 million. It is therefore reasonable to assume that significantly greater amounts of active pharmaceuticals were consumed in the UK. A survey of the amounts of analgesics stored in UK homes found that of the 1013 respondents questioned, 957 said they had paracetamol and 726 said they had ibuprofen.[7] Using the anecdotal data given on the frequency of paracetamol use and assuming that this referred to a single tablet per use, a conservative estimate of monthly paracetamol consumption by 521 respondents aged 16 to 65+ is 3725 tablets. Uncertainty includes 33% of the respondents claiming to take paracetamol less often than once a month and 17% claiming to not know how many were taken. Data on the number of tablets taken, the tablet dose or total dose amounts were not published. Survey data on illegal drug consumption have even greater uncertainty. Respondents are less likely to accurately estimate the amounts they have taken, particularly if they are poly-drug users. In addition, drug purity is not likely to be known.

4.2.2 Trends in Illegal Drug Consumption

A survey of trends, extent and frequency in illicit drug use in England and Wales 2013/14 was reported by the UK government Home Office[8] under the National Statistics code of practice. Using these data, the percentage individual drug use claimed by 16–59 year olds is illustrated in Figure 4.3 and the frequency of use in terms of more or less than once a month is illustrated in Figure 4.4. Frequency of heroin and crack cocaine use was notably omitted, indicating either low frequency of use or low levels of reporting of repeat use. The annual proportion of 16–59 year olds claiming to have used illicit drugs was around 12% from 1996 to 2003/4. This then decreased and remained fairly stable at around 8–9% up to and including the year 2014. Salvia and nitrous oxide were described in the UK Home Office survey as "legal emerging drugs"; their use was reported as 2.3% and 0.5%, respectively. Over 2011/12, levels of khat use by 16–59 year olds was recorded as 0.2%. Cocaine, MDMA, lysergic acid diethylamide (LSD) and ketamine use increased from 2012/13 to 2014/15. Data specific to new psychoactive substances were not given except for mephedrone. Self-reported consumption of new psychoactive substances may have been captured under the "any amphetamine" heading.

Other surveys include those conducted by individuals or groups of researchers for the purposes of social work, drug use rehabilitation, drug control and environmental monitoring. Willingness to report

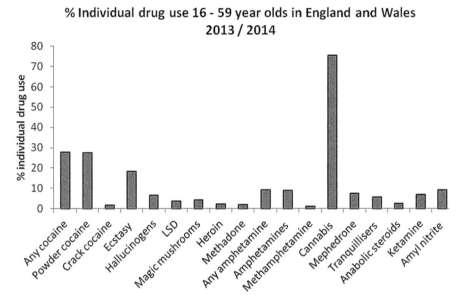

Figure 4.3 Trends in individual drug use of 16–59 year olds in England and Wales, UK.[8]

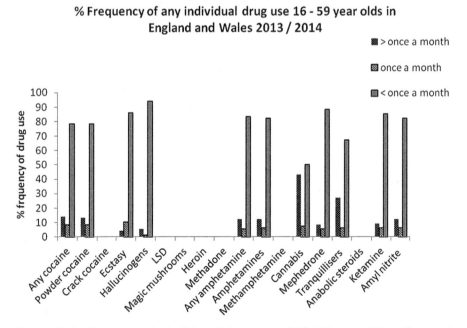

Figure 4.4 Frequency in individual drug use of 16–59 year olds in England and Wales, UK.[8]

illegal drug use can depend on the levels of stigma attributed by a society. Government commissioned surveys may be reluctant to critique records of falling drug consumption as this gives an opportunity to claim positive progress on the "war on drugs". The validity of self-reporting of illegal drug use to evaluate the effectiveness of US national drug control policy was tested by comparing self-reporting against any drugs present in urine samples.[9] It was shown that self-reporting alone substantially underestimates use compared to levels of drugs found from urine tests. When the subjects knew their urine was being tested, self-reporting increased by around 35% for marijuana users and by 82% for cocaine users. Subsequent agreement between reported cocaine use over 30 days and urine analysis was 98.5% and agreement for marijuana use was 90%. Uncertainties were not quoted; however, the possibility of drug use despite negative test results was acknowledged. Self-reporting of amphetamine, cocaine and heroin drug consumption in three Norwegian cities was found to be highly variable; confidence intervals of the usual dose of cocaine ranged from 80 to 488 mg.[10] Although the median dose for all three illicit substances was 250 mg, doses were reported at significantly higher levels than the usual 200–250 mg. Quantities of amphetamines could be 2.7 times higher, 2.1 times higher for cocaine and 2.3 times higher for heroin.[10] In addition to estimating the quantity and frequency of illicit drug consumption, the drug purity must also be considered if sewage epidemiology is to be an indicator of prevalence of drug use within a population. Analysis of drugs seized by law enforcement agents can provide information on the purity around the likely time of consumption. The rates and variability of drug metabolism and excretion must be considered when using concentrations of drug contaminants in wastewaters to assess population wide consumption.

4.3   Sources of Drugs in Water

Human urine, faeces and other body fluids carry the greatest net amount of drugs and drug metabolites into wastewaters. Inappropriate disposal of unwanted pharmaceuticals, such as flushing down toilets and sinks in domestic homes, contributes to their presence in their original unmetabolised forms. Once in the sewage influent there is the potential for reaction with other chemicals present and with bacteria or sunlight. Natural bacterial and photodegradation processes remove some of the pharmaceutical contaminants, particularly those of natural origin such as mor-phine.[11] Sewage waters flowing from hospitals are likely to have greater concentrations of drugs and their metabolites, as many of the occupants will be taking a variety of medicines. Antibiotics in farm animal feed are another potential source of increased concentration in sewage influent and surface water run-off. Large animals require larger doses of medication. If these animals then excrete in a relatively small space, such as a barn or a milking shed, it follows that the drugs and metabolites will be concentrated in wastewaters from that area. Human waste is still the greatest source of drugs in the aquatic environment.

4.3.1   Human Waste

Each day in the UK, over 11 billion litres of wastewater runs from domestic homes, municipal, commercial and industrial premises and rainwater run-off.[12] This is treated at 9000 sewage treatment works under the Urban Waste Water Treatment Directive 91/271/EEC.[13] The treated water effluent is then discharged to inland waters, estuaries and out to sea. The directive mainly focuses on the removal of organic material, including harmful pathogens. It does not generally include drugs and their metabolites. Depending on the water solubility of the drugs in sewage influent, they will partition between the aqueous and sludge solid phases. In the UK in 2010, over 1 118 000 tonnes of dry solids were reused as soils for public recreational areas and for agriculture, 8787 tonnes went to landfill and 259 642 tonnes were incinerated.[12] Secondary treatment of wastewater influent consists of bacterial breakdown where the majority of drug contaminants are removed, but not all. Tertiary treatments are employed where the treated water effluent is to be discharged to ecologically sensitive aquatic environments. Types of tertiary treatment are typically phosphorus or nitrate reduction and disinfection by UV or filter membranes. Lack of regulation can result in poor methods of disposal of bulk pharmaceutical waste. If disposed in landfill, these chemicals can leach over time into local water systems.

4.3.2   Pharmaceutical Waste Disposal

Disposal of bulk stock of waste pharmaceutical products is controlled in Europe under the Waste Framework Directive 2008/98/EC.[14] In England and Wales, their destruction and disposal are subject to the

Waste Management Licensing Regulations 1994 and the Hazardous Waste Regulations 2005. Controls may be less well regulated or controls may not even exist in other parts of the world. There may not be the facilities at local pharmacies for public return of unused pharmaceuticals for collection and appropriate disposal. Pharmaceutical companies are increasingly reporting that out-of-date pharmaceuticals are being repackaged and resold over the internet or through local distributors. This occurs more often in poorer parts of the world where medicines are less accessible.[15,16] Counterfeit medicine can contain too much active ingredient, too little or none at all. Dangerous contaminants such as dried lead paint, phenacetin and boric acid have been found mixed in counterfeit products.[15] These contaminant issues are also associated with illicit, recreational drugs and are additional aquatic pollutants, including at the sites of manufacture.[17]

4.3.3   Illegal Drug Manufacture

Clandestine laboratories that manufacture illegal drugs on an industrial scale pose a localised contamination hazard. Contaminants include raw materials used in the manufacture of synthetic or semi-synthetic drug products. Significant production of new psychoactive substances occurred in Poland, the Netherlands, Belgium and the Baltic countries in 2013. India and China were also major suppliers of substances packaged and labelled as "research chemicals" purchased for recreational use.[18] Clandestine laboratories are generally messy, dangerous places where principles of good manufacturing processes are far from observed. Equipment washings and waste products are poured down local drains or onto fields, where precipitation run-off and soil percolation will wash the contaminants into water systems. Legal protection of water supplies is generally focused on limits of pesticide, polychlorinated biphenyl, polyaromatic hydrocarbon and heavy metal contaminants.

4.3.4   Legislation

Contamination of aquatic environments is controlled *via* the European water framework directive, 2000/60/EC.[19] The chemical status of surface and ground waters must not be ecologically harmful and limit values of contaminants depend on persistence and bioaccumulation. Improvements in water treatment processes must be coordinated, with close monitoring of drinking water abstraction sites such as groundwater sources. The accompanying priority substances decision 2455/2001/EC[20] considers the degree of acute and chronic harm of so-called "emerging" contaminants over trophic levels. Of the 33 emerging contaminants considered, only four are pharmaceuticals and their metabolites; these are ibuprofen, diclofenac, 17α-ethynylestradiol and 17β-estradiol. Lifestyle products and illegal drugs are not considered.

4.4 Ecotoxicology

Although many drugs and metabolites degrade quickly, their input is continuous. Many drug compounds are unknown toxicants in unknown amounts under a variety of environmental conditions. Toxicological effects on ecosystems, including human health, may be subtle at parts per billion (ppb, $\mu g\ L^{-1}$) levels. There is potential for cumulative effects and for the exposure toxicity of multiple drugs or in combination with other contaminants to be additive or synergistic. Reaction with chlorine in wastewater treatment increases the toxicity of some drug contaminants.[21] There are several studies on the ecological toxicity of pharmaceuticals in water systems; however, information on illegal drugs is limited.[22] The human and ecotoxicology of new psychoactive substances is not yet fully understood. In contrast, the toxicologies of antibiotics and birth control hormones on aquatic ecologies are well studied. Data from many ecotoxicological studies on pharmaceuticals have been compiled in the freely available database "WikiPharma".[22] Over half of the data is from studies of short-term exposure and acute responses.

4.4.1 Ecotoxicology of Pharmaceuticals

The pharmaceuticals most studied for their potential harm to the aquatic environment are antibiotics and hormones in contraceptive pills. The concern with antibiotics is the encouragement of antibiotic resistant bacteria and the destruction of symbiotic, helpful microorganisms. The hormones 17α-ethynylestradiol and 17β-estradiol are endocrine disruptors that reduce reproductive success in fish, inhibit testicular growth and bioaccumulate.[23] Under the water framework directive 2000/60/EC and priority substances decision 2455/2001/EC the annual average environmental quality standard (AA-EQS) for 17α-ethynylestradiol is set at $3.5\times10^{-5}\ \mu g\ L^{-1}$ and for 17β-estradiol is $4\times10^{-4}\ \mu g\ L^{-1}$. These are among the low and no observed effect levels

(LOEL and NOEL).[23,24] There are an increasing number of studies on ecotoxicology of synergistic, additive or even antagonistic combinations of contaminants.[23,24] A study on the effect of mixtures of NSAIDs and metabolites of diclofenac and ibuprofen with lipid lowering clofibric acid showed a two-fold increased toxic effect to *Daphnia magna* after acute and chronic exposure.[24] An additive effect was found with caffeine and sulfamethoxazole on *Carassius auratus* (goldfish) on the induction of acetylcholinesterase and activation of glutathione *S*-transferase.[25] There are potentially vast numbers of both qualitative and quantitative combinations of drugs and other contaminants such as pesticides. Concurrent chronic exposure adds another layer of complexity. Exposure studies should be representative of real or most likely scenarios and therefore should use aqueous matrices such as urban wastewater effluent. The effects of physical environmental stressors such as heat and drought should be included as worst case scenarios. Such conditions can result in the concentration of contaminants in water and encourage their accumulation through water recycling processes. Studies on the ecotoxicology of illicit drugs and their metabolites are few compared to that of pharmaceuticals.

4.4.2   Illicit Drugs and the Environment

The negative effect of illicit drugs and metabolites on aquatic biota and organisms is likely to be comparable with legal pharmaceutical products. Many illicit drugs are chiral compounds; hence the toxicity of the optical isomers needs consideration. Of the few ecotoxicology studies published, most are of illicit stimulants such amphetamines and cocaine. *Danio renio* (zebrafish) showed mutation in genes that affect dopamergic signalling in the retina and brain when exposed to cocaine.[26] MDMA in wastewater influent was found to react with chlorine in wastewater treatment to produce 3-chlorocatechol.[27] Chlofibric acid pesticide is converted into 4-chlorocatechol, which is toxic to the fish *Pimephales promelas* (fathead minnows) and with acute toxicity LC_{50} of 1.58 mg L^{-1} at 96 hours. As with assessment of the ecotoxicology of pharmaceuticals, the mode of action of the drug and the degree of comparable metabolic pathways between human and aquatic organisms will guide understanding of the toxicity of illicit drugs and their metabolites.[28] Identifying the type and degree of contamination in a water system depends on the sampling and analysis methods used.

4.5 Sampling and Analysis

Automated samplers use peristaltic pumps and valves programmed to collect sequential samples of water in glass bottles. They require weekly maintenance visits and sample collections. Grab sampling methods such as these provide limited information on contaminant loadings over time. The ability to collect samples over extended periods and to concentrate traces of contaminants is improving through the development of passive sampling devices.[29–31] Examples of these include lipid filled semipermeable membrane devices (SPMDs) and silicone rubber (SR)–polydimethylsiloxane (PDMS) or low density polyethylene (LDPE) for non-polar contaminants. For polar substances such as pesticides and pharmaceuticals the passive samplers contain a hydrophilic reverse-phase wettable polymer such as Oasis® HLB or styrene–divinylbenzene (SDB) disks sandwiched between poly(ether sulfone) (PES) membranes. The samplers are transported to laboratory benchtop instruments for identification and quantification of drugs and metabolites. Interpretation of quantification data in terms of time-weighted average concentrations requires robust determination of sampling rates (R_s), which can typically range from 2×10^{-4} to 75 litres per day.[29–31] To determine R_s the passive samplers are pre-loaded with performance reference compounds (PRCs) that do not interfere analytically and do not occur in the environment. The uptake rate of analytes is assumed to be proportional to the dissipation rate of the PRCs and hence provides an *in situ* estimation of R_s. These correct for variable water flow rates and *in situ* bio-fouling; however, uptake and release was not found to be isotropic for SDB disks with PES membranes.[30,32] Correlation between water concentration and sampled mass on an SBD disk was poor, with highly fluctuating water flow rates.[30] Uptake rates are chemical species dependent and therefore R_s values need to be determined for each substance.[30] Knowledge of a diffusion coefficient for each analyte in water and through the sampler semipermeable membrane is crucial for accuracy in determining water concentrations.[29] Complexity in the chemical composition of the water sample matrix, the amount of biofouling and environmental variables, particularly flow velocities, all contribute to uncertainty in R_s estimation.

4.5.1 Environment Variables

Wastewater influent is by far the most complex matrix as it consists of mixtures of many substances in liquid or solid forms that are chemically ionic, polar or organic. The $\log K_{\mathrm{ow}}$ partition and $\log D_{\mathrm{ow}}$ distribution coefficient values of legal and illegal drugs give some indication of their degree of separation into aqueous and organic phases. This includes uptake in biota, bioaccumulation and bio-magnification. Many of the pharmaceuticals, illegal drugs and their metabolites are polar compounds and hence will partition to a greater extent into the aqueous phase. Partition of water-based chemicals is affected by pH, salinity, temperature and even chemical reaction through exposure to sunlight at water surfaces. Their pharmacological activity indicates potential for uptake by organisms, leading to re-excretion by aquatic species. These variables effect the chemical composition of drugs and metabolites in water matrices. Physical variables such as water volume flow rates, degree of turbulent mixing, temperature, geology and patterns of precipitation vary contaminant concentrations over space and time. Non-reactive contaminant transport models often employ a numerical or an analytical interpretation of the generalised dispersion equation. The main horizontal flow direction is defined as longitudinal flow and the degree of longitudinal dispersion of a contaminant is defined by its mass transfer coefficient (longitudinal dispersivity). Vertical transverse flow is often assumed to be negligible; however, downward turbulent mixing causes contaminants to "smear" in the horizontal direction perpendicular to the main flow. Vertical mixing can be a critical physical factor that controls the spread length of a contaminant plume. Each physiochemical variable contributes to uncertainty in estimation of contaminant exposure and drug use from sewage.

4.5.2 Uncertainty

Uncertainty in monitoring environmental contaminant concentrations, including drugs and their metabolites, is driven by variability in wastewater treatment efficiency, the weather, associated hydrology, drug consumption profiles and environmental persistence or accumulation. Measurement and sampling methods introduce their own quantifiable uncertainty. Whether samples are collected *via* Cartesian grab or Lagrangian in-stream systems will influence how uncertainty is determined. Owing to the high number of physiochemical variables, stochastic modelling approaches have been developed to quantify the propagation of uncertainty, for example below the land surface, through groundwater systems and downstream surface waters.[33,34] Assuming continuous release of pharmaceuticals and their metabolites, variability in travel time will not affect dilution of a contaminant plume at steady state. Uncertainty in predictive models of pollutant dispersion is dependent on the model assumptions as these may introduce bias and therefore they should be validated against field measurement data. Total daily maximum load (TDML) models for water quality management have been criticised for inadequate estimation of uncertainty due to lack of substance sampling rates and missing performance reference compounds in passive samplers.[34] Monte Carlo simulation of probability density functions of input variables and model output has been used to assess uncertainty and to analyse model sensitivity.[34,35] In Bayesian Monte Carlo methods the model output probability density functions such as grams of drug consumed per 1000 head of population per day are conditional on excretion rates, influent concentrations and population size.[35] Using Bayesian methods to quantify model prediction and its uncertainty are limited by the lack of observational data of drugs and their metabolites in water systems.

4.6 Sewage Epidemiology

Studies on the occurrence of pharmaceutical and lifestyle products, including caffeine, nicotine and personal care products, in aquatic environments across Europe, China and the USA have increased over recent years. The pharmaceuticals found in greatest amounts in UK and Spanish wastewater influent and effluent were acetaminophen glucuronide, 2-hydroxyibuprofen metabolites and amoxicillin.[7,36,37] This is likely to be replicated across the developed world. Examples of influent and effluent concentrations of the commonly used lifestyle products caffeine and nicotine and the pharmaceuticals acetaminophen and ibuprofen are listed in Table 4.1. Several of the water samples were collected from the fast flowing centre of water streams, avoiding days of heavy rainfall.[38]

Relatively high median concentrations of 1.26 µg L^{-1} of codeine were also measured in the UK wastewater influent as well as many prescribed pharmaceuticals such as other opiate analgesics, benzodiazepines, antidepressants and sildenafil. Measured quantities ranged from the hundreds of pg to hundreds of ng L^{-1} in both influent and effluent.[38] Concentrations of pharmaceuticals in hospital wastewaters are significantly higher. Acetaminophen in wastewaters from several Norwegian hospitals was around 12% above the general domestic contribution to sewage influent.[39] By far the greatest

Table 4.1 Concentrations of caffeine, nicotine and the NSAID pharmaceuticals acetaminophen and ibuprofen measured in wastewater effluent and influent over recent years in Spain, Sweden or the UK.

Lifestyle product or pharmaceutical	European country	Influent ($\mu g\ L^{-1}$)	Effluent ($\mu g\ L^{-1}$)	Ref.
Caffeine	Spain	59 (maximum)	<8	36
	UK	23.8 (median)	1.7 (median)	39
Nicotine	Spain	>1	<0.5	36
	UK	0.4 (median)	0.085 (median)	39
Acetaminophen	Spain	185 (maximum)	13.8 (maximum)	36
	Sweden	59 (maximum)	0.36 (maximum)	37
Total ibuprofen	Spain	4.4 (median)	0.8 (median)	38
	Sweden	10.2 (maximum)	0.48 (maximum)	37

pharmaceutical contribution from hospitals is antibiotics.[40] Levels of the antineoplastic chemotherapy drug vincristine and the antibiotics tazobactam and piperacillin in Australian hospital wastewaters were estimated to be present at ecotoxicology effect thresholds.[40] Illicit drugs in the aquatic environment are of increasing concern as they are highly bioactive and have psychoactive properties. Studies on population-wide trends in illicit drug use are increasingly using data on drug concentrations in wastewaters.

4.6.1 Illicit Drug Use and Sewage Epidemiology

Sewage epidemiology studies show trends in national and international illegal drug consumption and provide evidence of their supply and demand.[41–53] Discharge of these drugs and their metabolites is likely to be somewhere between pulsed and continuous, depending on the time frame considered, as drug availability and use is known to vary over time. An increase in the cocaine metabolite benzoylecgonine was measured in wastewater influent from the city of Florence, Italy, over the Easter and Christmas holidays. Samples were collected at wastewater treatment plants serving the city and the drug concentrations were measured monthly from July 2006 to June 2007.[42] Peaks in cocaine and benzoylecgonine concentrations might be due to an increase in use by the resident population; however, it is more likely that the greater numbers of visitors to the city during the holidays included cocaine users.

Comparison of drug types and concentrations in wastewaters across Europe over the years 2007–2010 shows distinct geographical variation in drug use trends.[43] Mean cocaine use in Milan, Italy, was estimated as 0.91 g day^{-1} per 1000 inhabitants,[45] in Lugano, Switzerland, as 0.62,[45] in Dublin, Ireland, as 0.2,[46] in London, UK, as 0.6–1.2[45,47] and in north east Spain as 0–5.4 g day^{-1} per 1000 inhabitants.[48] Repeated sampling over a one-week period from multiple wastewater treatment plants along a river basin in England demonstrated significant variability in wastewater cocaine and benzoylecgonine loads, which were 396–670 and 997–1721 g day^{-1}, respectively.[49] Variation in estimated cocaine use from influent concentrations of cocaine and benzoylecgonine measured over 19 consecutive days in a wastewater treatment plant in Brussels-Noord, Belgium, ranged 0.3 to 0.8 g day^{-1} per 1000 inhabitants.[50] Distinct peaks in consumption could be seen over the two weekends. This indicates that transit of the drug and metabolite from the domestic waste sources to the sewage treatment plant was rapid, within a few hours. Mass loading of cocaine and metabolites in sewage influent is back-calculated using generalised data on the proportions of parent drug and metabolites in excreted human waste. Only 1–9% of a cocaine dose is excreted unchanged, and 35–54% and 32–49% is excreted as benzoylecgonine and ecgonine methyl ester, respectively.[49,51] Methods for back-calculation of consumption in g day^{-1} per 1000 people use a ratio of estimated load and population size multiplied by excretion.[44,49] Other workers later adjusted for sorption of drugs and metabolites to suspended particulate matter and for stability in wastewaters.[49] Using percentage loads of both unchanged cocaine and metabolites with the wastewater concentrations multiplied by water volume flow rate gave higher consumption results than if benzoylecgonine data alone were used.[46,47] The higher estimation was attributed to disposal of unused cocaine down sinks and toilets. Benzoylecgonine is also more stable in water and hence uncertainty in estimation due to variability in degradation rate is reduced. Consumption of the amphetamine group of illegal drugs is estimated using the following proportions of unchanged drug in excreted urine and faeces: amphetamine 30%, methamphetamine 43%, MDMA 65% and ephedrine 75%.[45,51,52]

Trends in the use of these illegal drugs across France prior to the year 2012 showed clear geographical variation.[53] Cocaine consumption, based on benzoylecgonine data, ranged from 3 to 2434 mg day^{-1} per 1000 inhabitants. The highest levels of cocaine and cannabis consumption were in the north of France. Cannabis consumption was determined from concentrations of 11-nor-Δ^9-tetrahydrocannabinol-9-carboxylic acid (THC-COOH). Cannabis use ranged from 28 to 1000 mg day^{-1} per 1000 inhabitants and its use was more widespread than other types of illegal drug. Conversely, MDMA consumption was predominant in the south of France and ranged from 5 to 167 mg day^{-1} per 1000 inhabitants. Peaks in consumption were evident at weekends for all of the illegal drug types except for methadone and opiates.[53] Assuming cocaine purity of 30–60% and a typical dose of 100 mg for MDMA,[54] the estimated levels of cocaine consumption in France are comparable with other European cities. The low levels of MDMA consumption may be due to low drug purities and should be compared to common MDMA adulterants such as amphetamine and ketamine.[55]

International trends in illegal drug consumption are evident from sewage epidemiology studies. Heroin and cannabis consumption was relatively high in the city of Lugano compared to London and Milan.[43] Cannabis doses/day per 1000 inhabitants were estimated at 53 for a population of 0.12 million in Lugano compared to 24 doses/day per 1000 inhabitants for a population of 1.25 million in Milan.[45] Cannabis consumption of 43 mg day^{-1} per 1000 people in Lugano and 20 mg day^{-1} per 1000 people in Milan[45] was used to estimate doses, plus the assumption that that 6 μg of THC-COOH is excreted in urine per dose.[43] Consumption levels of amphetamine in European cities were estimated from concentrations in wastewater and compared.[43,49] Converse to the trends in heroin and cannabis use, the doses/day per 1000 inhabitants of amphetamine from a population of 0.14 million in South Wales, UK, was estimated at 83, whereas none was detected in wastewaters in Lugano with a population of similar size.[43] The number of amphetamine doses in South Wales was determined from the estimate of the average daily mass consumption of 2.5 g day^{-1} per 1000 people, which implies 30 mg of amphetamine per dose. This, however, was suspected to be an overestimate as there was no distinction made between legal and illegal use of amphetamine.[49]

Wastewater analysis can provide real time estimates of drug use within a population, profile drug use patterns over time and indicate emerging drug use trends. Uncertainties associated with periods and frequency of wastewater sampling, stability of drug residues, population size estimation and back-calculation parameters need better definition and reduction. Some of these uncertainties were defined and included in a "best practice protocol" to produce comparable data.[41]

4.6.2 Estimation Assumptions

Consumption prediction models use eqn (4.1) for back-calculations to estimate drug use in mass/day per 1000 inhabitants of a population:

$$\text{Drug use} = [(C \times F) \times \text{Cf}]/P \tag{4.1}$$

where C is the measured concentration of the target drug residue (ng L^{-1}), F is the corresponding sewage flow rate (L day^{-1}), Cf is a drug-specific correction factor associated with drug excretion rates and P is the population served by the sewage treatment plant.[41,44] Others expanded upon this method by including estimates of drug residue stability and loss through sorption to solid phase material to the correction factor.[49] Loads of parent drug per day (g day^{-1}) are first estimated from mass ratios of parent drug and excreted drug target residue (eqn (4.2)):[49]

$$\text{Load} = C \times F \times [100/(100 + \text{Stability})] \times [100/(100 - \text{Sorption})] \tag{4.2}$$

where C and F are as above, "Stability" is the stability change of each compound % and "Sorption" is the sorption % of each compound to solid phase material. Consumption $(\text{mg day}^{-1}$ per 1000 people) was then calculated (eqn (4.3)):[49]

$$\text{Consumption} = 1000 \times \text{Load} \times (1/\text{Excretion}) \times (\text{MW}_{\text{par}}/\text{MW}_{\text{DTR}})$$
$$\times (1000/\text{population}) - \text{OS} \tag{4.3}$$

where "Excretion" is the % excreted of the drug target residue (DTR) after associated forms of administration, MW_{par} is the molecular weight of the parent drug and MW_{DTR} is the molecular weight of the drug target residue; OS is the amount of DTR present in wastewater due to other sources. The proportions of drugs excreted unchanged and metabolised are highly variable. In addition to the examples given in Section 4.6.1, 37–62% of methamphetamine can be excreted unchanged in urine.[56,57]

Measured ranges in DTR concentrations are influenced by the water sample matrix, the sampling type, water volume flow rate, sampling frequency and choice of sampling point. Loss through leakage from pipes and at sewage treatment works was assumed to be negligible. Uncertainty in other variable parameters and estimation of their propagation in sewage epidemiology models was discussed in detail in Section 4.5.2. Removal of pharmaceuticals and illicit drugs at wastewater treatment plants is not wholly effective.[58] Tertiary treatments such as oxidation, ozonation and osmosis can efficiently remove organic contaminants, but not all treatment plants use these technologies. As a consequence, traces of drugs and drug metabolites find their way into surface, ground and even drinking waters.

4.7 Drugs in Surface Waters

Wastewater effluent concentrations of the more persistent drugs, such as those that are resilient to bio- and photo-degradation, will increase over time. Pulsed increases in drug and metabolite concentrations in river water were evident after wastewater treatment plant effluent discharge.[38] Drug concentrations accumulate as a river flows downstream and passes additional wastewater treatment plants.[38] Over recent decades in the UK, traces of the pharmaceuticals ibuprofen, mefenamic acid and paracetamol were identified in the river Thames at Hampton Court,[59] clotrimazole, ibuprofen and erythromycin in the river Tyne at Newcastle,[60] erythromycin, ibuprofen and codeine in the rivers Taff and Ely in South Wales[61] and 17β-estradiol, carbamazepine and diclofenac in the river Ouse in Sussex.[62] The pharmaceuticals are not exclusive to these rivers. Wastewater effluent samples from seven wastewater treatment plants that fed into the same river were collected over a 12 month period and showed high variability in concentrations of legal and illegal drugs.[38] Caffeine and nicotine levels ranged from 148 to 34 198 ng L^{-1} and from 52 to 56 945 ng L^{-1}, respectively. NSAIDs were not measured, so of the pharmaceuticals identified the greatest concentration were of the opioids tramadol and codeine, ranging from 86 to 1603 ng L^{-1} and from 9.7 to 1502 ng L^{-1}, respectively. Of the illegal drugs, benzylpiperazine (BZP) and MDMA were found in the greatest concentrations, ranging from 8.6 to 318 ng L^{-1} and from 0.6 to 178 ng L^{-1}.[38] Table 4.2 lists the concentrations of the cocaine metabolite benzoylecgonine in various European rivers during the years 2004–2010.

Through natural hydrological processes, surface waters are recycled into drinking water treatment plants including *via* groundwater abstraction. Benzoylecgonine, methadone, its metabolite EDDP and many other illegal drugs, pharmaceuticals and their metabolites have survived the treatment plants.[63–65]

4.8 Drugs in Groundwater and Drinking Water

Lifestyle products and pharmaceuticals such as caffeine, nicotine, antibiotics, NSAIDs and even barbiturates in landfill have

Table 4.2 Concentrations of benzoylecgonine measured in European rivers between the years 2004 and 2010.

Country	River	Concentration (ng L^{-1})	Year	Ref.
UK	Thames	16	2005	64
	Unknown[a]	61.8	ns[b]	39
Belgium	Aa	10	2007	65
	Demer	47.6	2007	65
	Grote Molenbeek	190.5	2007	65
Italy	Olona	183	2006	64
	Arno	21.8	2006	64
Spain	Ebro	11.4	2007/8	66
	Henares	142	2010	67
Switzerland	Unknown	1.6	2009	68

[a]Somewhere in Yorkshire, UK.
[b]ns = not stated.

contaminated groundwater that is below or down gradient.[66,67] Other pathways into groundwater are leaking or overflow from flooded sewers, septic tanks, surface run-off from industrial discharge and surface water infiltration. An addition source is leaching from sewage sludge used for agricultural fertiliser and soils for recreational areas, although recent screening found that contribution from this source was minimal.[26] Landfill and septic tanks are point-localised sources, but the diffuse source of surface water infiltration is the greatest gross contributor of legal and illegal drugs to groundwater. Samples of groundwater collected from monitoring stations at drinking water abstraction sites across Europe during 2008 found average levels of caffeine of 13 ng L^{-1} and average and maximum ibuprofen concentrations of 3 and 395 ng L^{-1}, respectively.[68] Groundwater samples collected in south-east France over the years 2007–2008 found no ibuprofen in groundwater but mean groundwater concentrations of 10 ng L^{-1} of both carbamazepine and acetaminophen. Of the emerging contaminants controlled under European water framework directive 2000/60/EC, mean concentrations of 9.7 ng L^{-1} of diclofenac, 0.4 ng L^{-1} of 17β-estradiol and 1.2 ng L^{-1} of 17α-ethynylestradiol were measured.[69] Studies of levels of illegal drugs in groundwater are minimal; their presence, however, has been identified in tap waters. Concentrations of caffeine, nicotine and examples of illegal drugs in tap waters supplying 12 million people in Spanish cities, including Barcelona, are listed in Table 4.3.

Table 4.3   Concentrations of some illicit drugs and their metabolites measured in Spanish tap waters in the years 2010 and 2012.[64,70,71]

Drug (metabolite)	Concentration maximum^a (ng L^{-1})		
	Across Spain 2010[64]	Barcelona 2010[70]	Toledo 2012[71]
Caffeine	342	—	—
Nicotine (conicotine)	141 (9.8)	—	—
Cocaine	2.3	3.8	2.1
(Benzoylecgonine)	(3.1)	(1.5)	(2.5)
(Cocaethylene)	(0.9)	(0.1)	(n/d)
Methadone	2.7	7.4	0.99
(EDDP)	(3.5)	(0.7)	(0.3)
Morphine	1.4	—	n/d
Amphetamine	1.7	—	n/d
Methamphetamine	1.4	—	n/d
MDMA	n/d	3.9	1.5
(MDA)	(0.9)	—	—
(MDEA)	(0.6)	—	—

^{a}n/d = none detected.

Much of Spain has a semi-arid climate where summer temperatures are commonly over 30 °C and rainfall below 20–30 mm, therefore potentially concentrating any traces of drugs and lifestyle products in the drinking water supply.

4.9   Conclusion

As population densities and age profiles increase in towns and cities, so does the supply of drugs and their metabolites into sewage, wastewater treatment plants and surface waters. The potential for traces of drug residues and lifestyle products such as caffeine and nicotine to find their way into drinking water is being realised. This is of particular concern in arid, hot, dry climates where drug and metabolite residues can be concentrated though water evaporation and wastewater recycling. Ecotoxicology studies in aquatic environments include synergistic, additive and even antagonistic consequences of contaminant combinations. Available data on ecotoxicology of drugs is limited and greater focus has been on acute rather than chronic effects. Concurrent exposure and toxicity of products from reaction with chlorine in wastewater treatment processes adds complexity. Studies on illicit drugs and metabolites are few compared to pharmaceuticals. Sewage epidemiology clearly shows geographical variation in levels of drugs consumed.[43–48] Uncertainty in estimation of daily consumption of illicit drugs can be reduced by comparing concentrations measured in wastewater with purities of drugs seized by law enforcement agencies. Studies on ecotoxicology and sewage epidemiology need reliable estimation of uncertainties in population size, drug excretion rates, stability of drug residues in aquatic matrices and sorption to suspended particulate matter. Efficiency of drug and metabolite removal by wastewater treatment will ultimately determine the concentrations detected in surface waters. Drug residues can accumulate downstream of multiple urban wastewater treatment plants. Water samples collected from fast flowing centres of rivers, avoiding days of heavy rainfall, will show different drug supply and removal rates than those collected from slower moving waters near river banks. Passive samplers preloaded with performance reference compounds have been used to assess time-weighted average sampling rates at varying flow velocities.[29–31] Corrections to reference compound dissipation and drug residue uptake rates depend on flow velocities and degree of biofouling *in situ*. Uptake and release rates are not isotropic for all types of passive sampling devices and correlation between concentration and sampled mass is poor for highly fluctuating flows.[30,32] The sampling methods used such as in-stream or grab sampling influence how drug concentrations over time and space are interpreted. Physiochemical variables such as pH, temperature, photochemistry, precipitation, geology, variable flow velocities and turbulent mixing all contribute to uncertainty in estimation of drug residue concentrations over time and space. Stochastic models have been developed to quantify propagation of uncertainty in drug residue mass transport in surface and ground waters.[33,34] The lack of observational data of drugs and metabolites in water systems limits further development and validation of drug transport and epidemiology models.

References

1. European Monitoring Centre for Drugs and Drug Addiction (EMCDDA), *European Drug Report: Trends and Developments*, EMCDDA, Luxemburg, 2014, http://www.emcdda.europa.eu/publications/edr/trends-developments/2014 [accessed 26th February 2015].

2. Euromonitor International, *Adult systemic analgesics options around the world*, published Online, Euromonitor International, 2014, http://www.euromonitor.com/analgesics [accessed 26th February 2015].

3. Organisation for Economic Cooperation and Development (OECD), *Pharmaceutical Consumption in Health at a Glance: Europe 2014*, OECD Publishing, 2014, DOI: 10.1787/health_glance_eur-2014-35-en [accessed 5th Sept 2014].

4. Prescribing and Primary Care Health and Social Care Information Centre (hscic), *Prescriptions dispensed in the community: England 2003–2013*, online, National Statistics V.1.0, 2014, http://www.hscic.gov.uk/catalogue/PUB14414/pres-disp-com-eng-2003-2013-rep.pdf [accessed 3rd November 2014].

5. Spanish Agency of Medicines and Health Products (SAMHP), *Use of nonsteroidal anti-inflammatory drugs in Spain, 2000–2012*, online, SAMHP, 2014, http://www.aemps.gob.es/medicamentosUsoHumano/observatorio/docs/AINE.pdf [accessed 26th February 2015].

6. S. Ortiz de Garcia, G. Pinto Pinto, P. Garcia Encina and R. I. Mata, *Sci. Total Environ.*, 2013, **444**, 451.

7. A. D. Shah, D. M. Wood and P. I. Dargan, *QJM*, 2013, **106**(3), 253.

8. Home Office, National Statistics, Drug misuse: findings from the 2013/2014 crime survey For England and Wales, online, Home Office, 2014, http://www.gov.uk/government/publications/drug-misuse-findings-from-the-2013-to-2014-csew [accessed 5th November 2014].

9. S. Magura, *Eval. Program Plann.*, 2010, **33**, 234.

10. E. J. Amundsen and M. J. Reid, *Sci. Total Environ.*, 2014, **487**, 740.

11. I. Senta, I. Krizman, M. Ahel and S. Terzic, *Sci. Total Environ.*, 2014, **487**, 659.

12. Department for Environment, Food and Rural Affairs (DEFRA), *UK Implementation of the Urban Wastewater Directive*, DEFRA, London, 2012, http://www.defra.gov.uk/environment/quality/water/sewage/sewage-treatment [accessed 2nd December 2014].

13. Directive 91/271/EEC of 21 May 1991 of the Council of the European Communities concerning urban waste water treatment [1991] OJ L135/40.

14. Directive 2008/98/EC of 19 November 2008 of the European Parliament and of the Council of the European Union on waste and repealing certain directives [2008] OJ L312/3.

15. Pfizer Inc, *Counterfeit Pharmaceuticals, A Serious Threat to Patient Safety*, 2007, http://www.pfizer.com/files/products/CounterfeitBrochure.pdf [accessed 19 March 2015].

16. P. E. Chaudhry and S. A. Stumf, *Business Horizons*, 2013, **56**, 189.

17. A. Janusz, K. P. Kirkbride, T. L. Scott, R. Naidu, M. V. Perkins and M. Megharaj, *Forensic Sci. Int.*, 2003, **134**, 62.
18. United Nations Office on Drugs and Crime (UNDOC), *The Challenge of New Psychoactive Substances*, UNDOC, Vienna, 2013, [accessed 20th March 2015] http://www.unodc.org/documents/scientific/NPS_2013_SMART.pdf.
19. Directive 2000/60/EC of 23 October 2000 of the European Parliament and of the Council establishing a framework for community action in the field of water policy [2000] OJ L327/1.
20. Directive 2455/200/EC of 20 November 2001 establishing the list of priority substances in the field of water policy and amending directive 2000/60/EC. OJ L331/1.
21. M. Huerta-Fontela, O. Pineda, F. Ventura and M. T. Galceran, *Water Res.*, 2012, **46**, 3304.
22. L. Molander, M. Ågerstrand and C. Rudén, *Regul. Toxicol. Pharmacol.*, 2009, **55**, 367.
23. R. Länge, T. H. Hutchinson, C. P. Croudace, F. Seigmund, H. Schweinfurth, P. Hampe, G. H. Panteranf and J. P. Sumpter, *Environ. Toxicol. Chem.*, 2001, **20**(6), 1216.
24. P. K. Jjemba, *Ecotoxicol. Environ. Saf.*, 2006, **63**, 113.
25. G. H. Han, H. G. Hur and S. D. Kim, *Environ. Toxicol. Chem.*, 2006, **25**, 265.
26. M. Stuart, D. Lapworth, E. Crane and A. Hart, *Sci. Total Environ.*, 2012, **416**, 1.
27. Z. Li, G. Lu, X. Yang and C. Wang, *Ecotoxicology*, 2011, **21**, 1.
28. V. Christen, S. Hickmann, B. Rechenberg and K. Fent, *Aquat. Toxicol.*, 2010, **96**, 167.
29. E. S. Emelogu, P. Pollard, C. D. Robinson, L. Webster, C. McKenzie, F. Napier, L. Steven and C. F. Moffat, *Sci. Total Environ.*, 2013, **445–446**, 261.
30. C. Moschet, E. L. M. Vermeirssen, H. Singer, C. Stamm and J. Hollender, *Water Res.*, 2015, **71**, 306.
31. C. Moschet, E. L. M. Vermeirssen, R. Seiz, H. Pfefferli and J. Hollender, *Water Res.*, 2014, **66**, 411.
32. M. Shaw and J. F. Mueller, *Environ. Sci. Technol.*, 2009, **43**(5), 1443.
33. K. Persson and G. Destouni, *J. Hydrol.*, 2009, **377**, 434.
34. Y. Zheng, W. Wang, F. Han and J. Ping, *Adv. Water Resour.*, 2011, **34**, 887.
35. H. E. Jones, M. Hickman, B. Kaspryzk-Hordern, N. J. Welton, D. R. Baker and A. E. Ades, *Sci. Total Environ.*, 2014, **487**, 642.
36. Department for Environment, Food and Rural Affairs (DEFRA), *Desk based Review of, Current Knowledge of Pharmaceuticals in*

Drinking Water and Estimation of Potential Levels, Cranfield University, Watts and Crane Associates, Cranfield, 2007, http://dwi.defra.gov.uk/research/completed-research/reports/dwi70-2-213.pdf [accessed 2nd December 2014].

37. M. J. Martinez Bueno, M. J. Gomez, S. Herrera, M. D. Hernando and A. Agüera, *Environ. Pollut.*, 2012, **164**, 267.
38. D. R. Baker and B. Kaspryzk-Hordern, *Sci. Total Environ.*, 2013, **454–455**, 442.
39. K. V. Thomas, C. Dye, M. Schlabach and K. H. Langford, *J. Environ. Monit.*, 2007, **9**(12), 1410.
40. K. S. Le Corre, C. Ort, D. Kateley, B. Allen, B. I. Escher and J. Keller, *Environ. Int.*, 2012, **45**, 911.
41. S. Castiglioni, K. V. Thomas, B. Kasprzyk-Hordern, L. Vandam and P. Griffiths, *Sci. Total Environ.*, 2014, **487**, 613.
42. F. Mari, L. Politi, A. Biggeri, G. Accetta, C. Trignano, M. Di Padua and E. Bertol, *Forensic Sci. Int.*, 2009, **189**, 88.
43. A. L. N. van Nuijs, S. Castiglioni, I. Tarcomnicu, C. Postigo, M. Lopez de Alda, H. Neels, E. Zuccato, D. Barcelo and A. Covaci, *Sci. Total Environ.*, 2011, **409**, 3564.
44. E. Zuccato, C. Chiabrando, C. Castiglioni, S. Calamari, D. Bagnati and R. Schiarea, *Environ. Health*, 2005, **4**, 14.
45. E. Zuccato, C. Chiabrando, C. Castiglioni, R. Bagnati and R. Fanelli, *Environ. Health Perspect.*, 2008, **116**, 1027.
46. J. Bones, K. V. Tomas and B. Paul, *J. Environ. Monit.*, 2007, **9**, 701.
47. B. Kasprzyk-Hordern, R. M. Dinsdale and A. J. Guwy, *Environ. Pollut.*, 2009, **157**, 1773.
48. M. Huert-Fontela, M. T. Galceren, J. Martin-Alonso and F. Ventura, *Sci. Total Environ.*, 2008, **397**, 31.
49. D. R. Baker, L. Barron and B. Kasprzyk-Hordern, *Sci. Total Environ.*, 2014, **487**, 629.
50. A. L. N. van Nuijs, B. Pecceu, L. Theunis, N. Dubois, C. Charlier, P. G. Jorens, L. Bervoets, R. Blusy, H. Neels and A. Covaci, *Water Res.*, 2009, **43**, 1341.
51. R. C. Baselt, *Disposition of Toxic Drugs in Man*, Biomedical Publications, Foster City, 8th edn, 2008.
52. C. Postigo, M. J. Lopez de Alda and D. Barceló, *Environ. Int.*, 2009, **36**, 75.
53. T. Nefau, S. Karolak, L. Castillo, V. Boireau and Y. Levi, *Sci. Total Environ.*, 2013, **461–462**, 712.
54. United Nations Office on Drugs and Crime (UNDOC), *World Report 2012*, UNDOC, Vienna, 2012, https://www.undoc.org/documents/data-and-analysis/WDR2012_web_small.pdf [accessed 6th January 2015].

55. B. Zhu, L. Meng and K. Zheng, *Forensic Sci. Int.*, 2014, **242**, e44.
56. K. Shimosato, M. Tomita and I. Ijiri, *Arch. Toxicol.*, 1986, **59**, 135.
57. C. E. Cook, A. R. Jeffcoat, B. M. Sadler, J. M. Hill, R. D. Voyksner, D. E. Pugh, W. R. White and M. Perez-Reyes, *Drug Metab. Dispos.*, 1992, **20**(6), 856.
58. R. Pal, M. Megharaj, K. P. Kirkbride and R. Naidu, *Sci. Total Environ.*, 2013, **463–464**, 1079.
59. J. P. Bound and N. Voulvoulis, *Water Res.*, 2006, **40**(15), 2885.
60. P. H. Roberts and K. V. Thomas, *Sci. Total Environ.*, 2006, **356**(1–3), 143.
61. B. Kasprzyk-Hordern, R. M. Dinsdale and A. J. Guwy, *Water Res.*, 2008, **42**(13), 3498.
62. Z. Zhang, A. Hibbard and J. L. Zhou, *Anal. Chim. Acta*, 2008, **607**, 37.
63. M. Huerta-Fontela, M. T. Galceren and F. Ventura, *Environ. Sci. Technol.*, 2008, **42**(18), 6809.
64. M. Rosa Boleda, M. Huerta-Fontela, F. Ventura and M. T. Galceran, *Chemosphere*, 2011, **84**, 1601.
65. K. K. Barnes, E. T. Kolpin, Furlong, S. D. Zaugg, M. T. Meyer and L. B. Barber, *Sci. Total Environ.*, 2008, **402**, 192.
66. K. K. Barnes, S. C. Christenson, E. T. Kolpin, M. J. Focazio, E. T. Furlong and S. D. Zaugg, *Groundwater Monit. Rem.*, 2004, **24**, 119.
67. P. D. Buszka, Yeskis, D. W. Kolpin, E. Furlong, S. Zaugg and M. Meyer, *Bull. Environ. Contam. Toxicol.*, 2009, **82**, 653.
68. R. Loos, G. Locoro, S. Comero, S. Contini, D. Schwesig, F. Werres, P. Balsaa, O. Gans, S. Weiss, L. Blaha, M. Bolchi and B. M. Gawlik, *Water Res.*, 2010, **44**, 4115.
69. E. Vulliet and C. Cren-Olivé, *Environ. Pollut.*, 2011, **159**, 2929.
70. A. Jurado, N. Mastroianni, E. Vazquez-Sune, J. Carrera, I. Tubau and E. Pujades, *Sci. Total Environ.*, 2012, **424**, 280.
71. Y. Valcarcel, F. Martinez, S. Gonzales-Alonso, Y. Segura, M. Catala, R. Molina, J. C. Montero-Rubio, N. Mastrioanni, M. Lopez de Alda, C. Postigio and D. Barcelo, *Environ. Int.*, 2012, **41**, 35.

5 Understanding the Utility of Analysis of Anonymous Pooled Urine from Standalone Urinals in Detecting and Monitoring Recreational Drug Use

David M. Wood and Paul I. Dargan*

Guy's and St Thomas' NHS Foundation Trust and King's College London, London, UK
*Email: paul.dargan@gstt.nhs.uk

5.1 Introduction

Over the last decade there has been a change in the pattern of use of drugs, with an increase in the use of novel/new psychoactive substances (NPS, also known incorrectly colloquially as "legal highs").[1] There has been a year-on-year increase in the number of NPS reported through the Early Warning System (EWS) at the European Monitoring Centre for Drugs and Drug Addiction (EMCDDA) over the last five years, with 83 substances reports in 2013 and 1–2 new substances reported every week across Europe in 2014/15.[2] The detection of these substances is often in seizures from law enforcement and/or border agencies. Whilst this indicates potential availability of these new substances to users across Europe, this does not provide information on the actual pattern(s) of use of these substances or indeed of classical recreational drugs such as cocaine, MDMA and amphetamine.

Forensic Toxicology: Drug Use and Misuse
Edited by Susannah Davies, Atholl Johnston and David Holt
© The Royal Society of Chemistry 2016
Published by the Royal Society of Chemistry, www.rsc.org

Current information on the actual pattern(s) of use often is obtained from population and/or subpopulation level surveys of potential users.[2-5] Whilst population level surveys, such as the Crime Survey England and Wales (formerly known as the British Crime Survey), can provide robust data over time on classical recreational drugs (*e.g.* cocaine), they are often not positioned to be able to monitor NPS in the same manner. Typically this is because individual NPS may only emerge onto the recreational drug scene for a short period of time and therefore a population-level survey may not be able to adapt sufficiently rapidly enough. As a result, the Crime Survey England and Wales currently only collects data from respondents on the use of a limited range of NPS: 1-benzylpiperazine, "spice" (synthetic cannabinoid receptor agonists, SCRAs) and mephedrone (4-methylmethcathinone).[3] Smaller localised population or discrete sub-population surveys, such as the Global Drugs Survey (formerly known as the MixMag survey) or surveys of the men who have sex with men (MSM) clubbing communities, have the ability to adapt more rapidly to collate information on newly emerging NPS.[5,6] One limitation is that they are not true population level surveys, and therefore it is not possible to extrapolate and generalise the data from these surveys to the wider total population. In addition, data from both population and sub-population surveys are based on self-reported use rather than the actual drugs that were used. Data from the Drug Information and Monitoring System (DIMS) drugs screening programme in the Netherlands for classical recreational drugs,[7,8] and UK studies analysing substances purchased from street-level drug dealers and internet suppliers for NPS, have shown that there is significant variability in the content and consistency of these substances.[9-13] Therefore the data obtained in these user surveys are based on intended, rather than actual, drug(s) used, which may be unreliable.

In light of these limitations, the technique of wastewater analysis has been developed to enable analytical confirmation of the drugs used in a population.

5.2 Wastewater Analysis

Wastewater analysis has been developed over the last decade (also see Chapter 4 for further details). This technique involves the analysis of samples collected at wastewater (sewage) treatment plants and it can be used to detect a range of different drugs, including prescription medicines and recreational drugs.[14] Statistical techniques have been developed which enable back-calculation of an estimation of the population level consumption of recreational drugs based on the concentration of the drugs in the wastewater analysis.[14,15] Wastewater analysis has been used to confirm patterns of drug use, for example confirming higher use of cocaine in Western/Central Europe than Northern/Eastern Europe,[16] and greater use of cocaine in large cities and at weekends.[17] Recently, wastewater analysis has been employed to detect a small number of NPS.[18] However, the detection of NPS using wastewater analysis is generally limited by the poor knowledge of the stability of these substances in wastewater, uncertainty regarding the impact of bacterial metabolism and limited knowledge of human metabolism to enable back-calculations to be performed.[14] Furthermore, prevalence of the use of new emerging NPS is often low when they first start to be used on the recreational drug scene and therefore concentrations of these NPS within wastewater from large populations may be too low to enable reliable detection.[14]

5.3   Anonymous Pooled Urine Sample Analysis

To mitigate some of the limitations described with wastewater analysis, there has been interest in collecting urine samples from closer to the point at which they are passed. The ideal scenario would be individually collected samples with detailed information on self-reported drug(s) used. This, however, requires informed consent for collection [even without the information on reported drug(s) used], which often is not practical in the setting of the night-time economy where an individual is likely to be "intoxicated" and is also labour intensive. In addition, there may be associated dehydration related to the drug(s), which may mean collection at the time of informed consent may not be possible and it requires storage of the individually collected samples prior to analysis.

Standalone street urinals are used in many cities across Europe to prevent street urination. These urinals can either be directly draining through street drains/sewers into existing wastewater systems or can be self-contained units in which the urine is collected in a central reservoir (often capable of holding several hundreds of litres of "pooled" urine). It is possible to easily collect samples from the latter types of street urinals using simple hand-held vacuum pumps to remove aliquots of pooled urine.[19] The use of these standalone urinals within city centres, festivals or other locations is anonymous and voluntary. Individuals can choose to use the urinal or other toilet facilities; often the urinals are complementary to other toilet facilities available to individuals, particularly when they are used at festivals and within nightclubs or other night-time economy settings. It is not possible to trace the results back to an individual as pooled urine is collected anonymously over a period of many hours from numerous individuals. Currently, owing to the design of these standalone urinals, it is only possible for males to use them; therefore the results are only representative of males. This technique has the advantage that only urine is collected and therefore it mitigates the concerns regarding bacterial metabolism of drugs by faecal bacteria as is the case with wastewater analysis.[14]

5.4 Ethical Considerations

Similar to analysis of samples from wastewater treatment plants, analysis of samples from pooled urinals is anonymous because individuals using the urinals are not identified (nor are the number of individuals using the urinals identified). Hall *et al.* reviewed in detail potential ethical concerns related to the analysis of samples from these types of studies and concluded that, particularly when samples are collected from large cohorts/populations, there are no significant ethical concerns as the data are aggregated and anonymous.[20] When our group have undertaken pooled urinal collection from single nightclub or festival venues we have always ensured that information is available to individuals within the venue to inform them that samples are being collected from the urinal in order that they can decide whether to use the study urinal or an alternative toilet facility.[21]

There are several considerations when interpreting the results of the analysis undertaken on pooled urine samples collected from standalone urinals. The detection of one or more metabolite is important, since this indicates that the parent drug has been used by one or more individual, and that the detection is not simply due to the parent drug being discarded into the urinal. In addition, the detection of commonly used substances such as caffeine and/or nicotine in samples is important to validate the sample collection and analysis. Given the frequency of use of both of these, the lack of detection of both would suggest that there were issues with the sample and caution would be required when interpreting the results. Finally, the non-detection of an individual drug/NPS in the urinal dose not mean that it is not being use. For drugs, particularly NPS, where prevalence of use is low, it is possible that non-detection reflects dilution by urine from other non-users to a concentration below the lower limit of detection for that particular compound. Furthermore, drugs that are of higher potency will be present at much lower concentrations and so there is the potential that once again dilution will further decrease the concentration of these drugs below the lower limit of detection.

5.5 Results from Analysis of Anonymous Pooled Urine Samples

To date, analysis of samples collected from standalone portable urinals has been undertaken in a variety of different settings and shown to be useful in a number of circumstances.[21–26]

5.5.1 Nightclub Study

The first portable urinal study was undertaken by our group and utilised a single standalone urinal placed in the outside "chill out"/smoking area of a south east London nightclub catering predominately for the MSM community.[21] Use of the study urinal was voluntary and anonymous, with individuals able to use other toilet facilities within the venue if they preferred; information on the study and the urinal was provided in advance through the nightclub internet site and on the night throughout the nightclub to all potential users. In addition, during these promotions a general survey of attendees was undertaken to understand the patterns of drug use within the nightclub on that night.[4] Over the course of two promotions at the venue, four samples each of 500 mL were collected and stored at 4 °C until analysis. Liquid chromatography coupled to accurate high-resolution mass spectrometry (Orbitrap and time of flight) was undertaken independently in two laboratories. The data were processed automatically against databases containing over 900 drugs and metabolites (prescription, over-the-counter, anabolic steroids, classical and novel recreational drugs). A total of 72 drugs and their metabolites were detected in the four pooled urine samples collected. These could be divided into: (i) classical recreational drugs; (ii) novel psychoactive substances; (iii) potential adulterants; and (iv) prescription/over-the-counter (OTC) medications. Since we are unable to know the intent of use in relation to detected substances, it is not possible to determine whether the latter group of substances detected were being used for legitimate purposes or being misused. The classical recreational drugs detected were amphetamine*, cocaine*, ketamine*, methamphetamine, morphine, MDMA* and GHB (γ-hydroxybutyrate) (* denotes detection of metabolites in addition to the parent compound). Novel psychoactive substances detected were

2-aminoindane (2-AI), (3-trifluoromethylphenyl)piperazine (TFMPP)*, mephedrone* and methylhexanamine.

In this initial study, the absolute concentration of drugs detected was reported. This was not standardised to creatinine or any other biological marker. Since it is not possible to determine the number of individuals who have used a urinal during the course of the collection or the volume of urine passed per individual, it is not possible to understand the true significance of these measured concentrations. Therefore in subsequent studies utilising this technique, we have focused on qualitative rather than quantitative analysis for the detection (or absence) of individual compounds.

5.5.2 City Centre Projects

As previously noted, standalone urinals are often utilised by local councils at weekends to prevent male street urination. In conjunction with the City of Westminster council we collected urine from 12 standalone urinals that are each deposited in the same location within the night-time economy area on Friday and Saturday nights.[24,25] In an initial study on a single Saturday night in March 2012, a total of 109 parent drugs and/or their metabolites were detected in the samples collected from the 12 urinals (the same analytical techniques were used as described for the earlier study). The frequency of detection of classical recreational drugs and NPS is shown in Table 5.1. Paired parent drug and relevant metabolite(s) were detected for amphetamine, cocaine, ketamine, MDMA, methiopropamine and methoxetamine, demonstrating that detection is due to use and subsequent metabolism and excretion rather than detecting drugs/NPS discarded into the urinals. In this study, hordenine was detected in all of the urinals; this compound is a phenethylamine alkaloid which is naturally found in some psychedelic cacti and other plants, although typically these also contain mescaline. The lack of detection of mescaline in the urinal samples suggests that the origin of the hordenine detected is from an alternative source rather than the use of psychedelic plant material. It is more likely that it has occurred as a breakdown product of beer brewed from barley.[27] This highlights the need to have a wide understanding of potential alternative sources for individual compounds detected, rather than simply assuming that the detection simply reflects recreational use of the substance.

Collection of urine from these central London street urinals has continued on a monthly basis (first Saturday of each month) to enable some understanding of potential trends in classical recreational drug and NPS use with time. Data from the first six months (July 2012 to

Table 5.1 Frequency of detection of classical recreational drugs and NPS in pooled urine samples collected from 12 individual standalone portable urinals on a single night in central London in March 2012.[24]

	Drug/NPS	Number of urinals
Classical drugs	Cathine	11
	Cocaine[a]	11
	Cannabis	11
	MDMA[a]	11
	Amphetamine[a]	10
	3,4-Methylenedioxyamphetamine (MDA)	9
	Ketamine[a]	6
	Methamphetamine	5
NPS	Methylhexanamine	9
	4-Methylmethcathinone (mephedrone)	6
	Methiopropamine[a]	2
	Methoxetamine[a]	1

[a]Denotes detection of metabolite(s).

December 2012) of time-trend analysis has been published.[25] The overall trends in detection of individual classical recreational drugs and NPS are shown in Table 5.2. This study showed that there was consistency over time in the detection of certain classical recreational drugs (cocaine, morphine and MDMA) and the NPS mephedrone. There was greater variability on a month-to-month basis in the detection of other NPS, with some such as 1-(4-methoxyphenyl)piperazine and 4-fluoroephedrine being only detected in one urinal in one month during the six-month study. In addition, there were no synthetic cannabinoid receptor agonists detected, and whilst methoxetamine was detected in March 2012,[24] this was not detected in any samples collected in this subsequent study.[25]

Finally, in addition to detecting time-trends within an individual location, it is possible to detect geographical trends across a region or a country. We have undertaken a study collecting urine from street urinals across nine different UK cities on a single night in April 2014.[26] There was variation in the number of classical recreational drugs/NPS detected from only one in Leeds to 14 in London (all cities apart from Leeds had at least five compounds detected). The frequency of detection of classical recreational drugs and NPS is shown in Table 5.3. Overall there was a three-fold greater detection of

Table 5.2 Frequency of detection of classical recreational drugs and NPS in pooled urine samples collected from 12 individual standalone portable urinals in central London between July and December 2012.[25]

Group	Drug	July	Aug.	Sept.	Oct.	Nov.	Dec.
NPS	4-Methylmethcathinone (mephedrone)	5	8	8	5	4	2
	Methylhexanamine	3	7	8	12	4	3
	Methcathinone[a]	5	0	0	0	0	0
	4-Ethylmethcathinone	3	0	0	0	0	0
	Methiopropamine	1	2	1	1	0	2
	Pipradrol	1	4	1	2	0	0
	Cathinone	3	2	2	3	0	0
	5-(2-Aminopropyl)benzofuran (5-APB)	1	0	0	0	1	1
	1-(3-Trifluoromethylphenyl)piperazine (TFMPP)	0	0	0	1	2	0
	4-Methylbuphedrone	0	1	4	0	0	0
	4-Methylethcathinone	0	1	1	0	1	0
	1-(4-Methoxyphenyl)piperazine	0	0	0	0	1	0
	4-Fluoroephedrine	0	0	1	0	0	0
Classical	Cathine	12	12	7	7	7	5
	Cocaine[a]	12	12	12	12	8	12
	Morphine	12	12	12	12	12	12
	MDMA[a]	9	11	12	9	11	12
	Methadone	5	11	11	7	7	7
	Amphetamine[a]	7	9	9	5	2	9
	Ketamine[a]	5	7	5	8	4	5
	3,4-Methylenedioxyamphetamine	3	2	5	6	5	2
	Cannabis[a]	8	12	12	12	2	1
	Methamphetamine	4	2	1	1	3	6

[a]Detection of metabolite(s).

Table 5.3 Frequency of detection of classical recreational drugs and NPS in pooled urine samples collected across nine UK cities in April 2014.[26]

Group	Drug	Number of cities detected in	Edinburgh	Newcastle	Leeds
NPS	4-Methylmethcathinone (mephedrone)	5		Y	
	Methiopropamine	3		Y	
	Methylhexanamine	2			
	Methoxetamine[a]	2			
	1-(3-Trifluoromethylphenyl)piperazine (TFMPP)	1			
	1-Benzylpiperazine (BZP)	1			
	2-(Methylamino)-1-phenylpentan-1-one (pentedrone)	1			
	3,4-Methylenedioxy-N-methylcathinone (methylone)	1			
	Glaucine	1			
Classical	Cocaine[a]	9	Y	Y	Y
	MDMA (3,4-methylenedioxymethamphetamine)	8	Y	Y	
	3,4-MDA (3,4-methylenedioxyamphetamine)	7	Y	Y	
	Morphine	7	Y	Y	
	Methadone[a]	6	Y	Y	
	Amphetamine	5	Y		
	Cannabis (carboxy-THC)	5	Y	Y	
	Ketamine[a]	3			
	Cathine	2			
	Methamphetamine	1			

[a]Denotes detection of metabolite(s).

Manchester	Liverpool	Birmingham	Bristol	Brighton	London
Y	Y	Y			Y
		Y			Y
			Y		Y
Y	Y	Y			
		Y			
		Y			
		Y			
					Y
					Y
Y	Y	Y	Y	Y	Y
Y	Y	Y	Y	Y	Y
Y	Y	Y	Y		Y
Y	Y		Y	Y	Y
	Y		Y	Y	Y
		Y	Y	Y	Y
Y	Y	Y			
Y	Y				Y
		Y			Y
					Y

classical recreational drugs compared to NPS. Similar to previous studies, although they were screened for, there were no synthetic cannabinoid receptor agonists detected.

5.5.3 Festivals

The final setting in which our group have undertaken pooled urinal studies is music festivals.[22,23] In the first reported study, samples were collected from two standalone urinals at a music festival in the north west of England in July 2012.[23] In total, nine NPS were detected in these two urinals. For both urinals these were 1-(4-methoxyphenyl)piperazine, 1-(3-trifluoromethylphenyl)piperazine, 4-ethylmethcathinone, 4-methylmethcathinone, 4-fluoroephedrine, 5-(2-aminopropyl)benzofuran (5-APB) and methiopropamine. For one urinal only they were 4-fluoronorephedrine and methylhexaneamine. In a subsequent study in December 2012, four urinals were placed at a music festival in Oslo, Norway.[22] Owing to the outside temperature, the contents of two urinals froze, meaning that samples could only be collected from two. Although a range of classical recreational drugs (amphetamine, cannabis, cocaine, MDMA and methamphetamine) was detected in the two urinals that remained unfrozen, the only NPS detected was methylhexanamine. These small studies suggest that the pattern of drugs detected from standalone urinals at festivals differs from that seen in standalone urinals placed within city centre night-time economy locations.[24–26] However, further work is required to confirm that this is the case, as data have only been reported on pooled urinal collections from two festivals.[22,23] As opposed to the collection from city centre urinals which is only from males, there is also the potential that at festivals samples could be collected from females, as there is increasing use of "femurinals" at these types of events. This would then allow reporting of a more complete picture representing both males and females.

5.6 Conclusions

The analysis of pooled urine collected from standalone urinals in nightclubs, city centres and music festivals can be used to detect the use of both classical recreational drugs and NPS. The detection of metabolites means that there is detection following individual use and subsequent metabolism and excretion, rather than drugs/NPS being discarded into the urinals. Using this technique, it is possible to detect trends in the patterns of use of both classical recreational drugs and NPS, not only over time but also between different geographical regions. Further work is needed to develop this technique to determine whether it is possible to move from qualitative detection reporting currently undertaken to standardisation with one or more biological markers to attempt to quantify the use of different classical recreational drugs and/or NPS. Currently the technique can only be used to look at drug use in males, but the use of "femurinals" may allow expansion of the technique to monitor drug use in females. This novel technique compliments information gained from other sources (in particular, self-report drug use surveys) and will help increase the robustness of data on the epidemiology of drug use to ensure that reliable information is available to national and international authorities such as the UK Advisory Council for the Misuse of Drugs (ACMD) and the European Monitoring Centre for Drugs and Drug Addiction (EMCDDA).

References

1. UNODC NPS: The United Nations Office on Drugs and Crime (UNODC). The Challenge of the New Psychoactive Substances. Available at: http://www.unodc.org/documents/scientific/NPS_2013_SMART.pdf.
2. EMCDDA EDR: EMCDDA. European Drug Report: Trends and Developments. Available at: http://www.emcdda.europa.eu/attachements.cfm/att_228272_EN_TDAT14001.pdf.
3. CSEW 2013-14: Drug misuse: findings from the 2013 to 2014 CSEW second edition. Available at: https://www.gov.uk/government/publications/drug-misuse-findings-from-the-2013-to-2014-csew/drug-misuse-findings-from-the-201314-crime-survey-for-england-and-wales.
4. D. M. Wood, F. Measham and P. I. Dargan, 'Our favourite drug': prevalence of use and preference for mephedrone in the London night-time economy one year after control, *J. Subst. Use*, 2012, **17**, 91.
5. GDS 2014: Global Drugs Survey. The Global Drug Survey 2014 findings. Available at: http://www.globaldrugsurvey.com/facts-figures/the-global-drug-survey-2014-findings/.
6. D. M. Wood, L. Hunter, F. Measham and P. I. Dargan, Limited use of novel psychoactive substances in South London nightclubs, *QJM*, 2012, **105**(10), 959–964.

7. T. M. Brunt and R. J. Niesink, The Drug Information and Monitoring System (DIMS) in the Netherlands: implementation, results, and international comparison, *Drug Test. Anal.*, 2011, **3**(9), 621–634.

8. N. Vogels, T. M. Brunt, S. Rigter, P. van Dijk, H. Vervaeke and R. J. Niesink, Content of ecstasy in the Netherlands: 1993-2008, *Addiction*, 2009, **104**(12), 2057–2066.

9. J. Ramsey, P. I. Dargan, M. Smyllie, S. Davies, J. Button, D. W. Holt and D. M. Wood, Buying 'legal' recreational drugs does not mean that you are not breaking the law, *QJM*, 2010, **103**(10), 777–783.

10. S. Davies, T. Lee, J. Ramsey, P. I. Dargan and D. M. Wood, Risk of caffeine toxicity associated with the use of 'legal highs' (novel psychoactive substances), *Eur. J. Clin. Pharmacol.*, 2012, **68**(4), 435–439.

11. S. D. Brandt, H. R. Sumnall, F. Measham and J. Cole, Analyses of second-generation 'legal highs' in the UK: initial findings, *Drug Test. Anal.*, 2010, **2**(8), 377–382.

12. S. D. Brandt, S. Freeman, H. R. Sumnall, F. Measham and J. Cole, Analysis of NRG 'legal highs' in the UK: identification and formation of novel cathinones, *Drug Test. Anal.*, 2011, **3**(9), 569–575.

13. S. Davies, D. M. Wood, G. Smith, J. Button, J. Ramsey, R. Archer, D. W. Holt and P. I. Dargan, Purchasing 'legal highs' on the Internet–is there consistency in what you get?, *QJM*, 2010, **103**(7), 489–493.

14. EMCDDA 2008: EMCDDA Insights – Assessing illicit drugs in wastewater. Potential and limitations of a new monitoring approach. Available at: http://www.emcdda.europa.eu/attachements.cfm/att_139185_EN_emcdda-insights-wastewater.pdf.

15. A. L. van Nuijs, S. Castiglioni, I. Tarcomnicu, C. Postigo, M. Lopez de Alda, H. Neels, E. Zuccato, D. Barcelo and A. Covaci, Illicit drug consumption estimations derived from wastewater analysis: a critical review, *Sci. Total Environ.*, 2011, **409**(19), 3564–3577.

16. K. V. Thomas, L. Bijlsma, S. Castiglioni, A. Covaci, E. Emke, R. Grabic, F. Hernández, S. Karolak, B. Kasprzyk-Hordern, R. H. Lindberg, M. Lopez de Alda, A. Meierjohann, C. Ort, Y. Pico, J. B. Quintana, M. Reid, J. Rieckermann, S. Terzic, A. L. van Nuijs and P. de Voogt, Comparing illicit drug use in 19 European cities through sewage analysis, *Sci. Total Environ.*, 2012, **432**, 432–439.

17. A. L. van Nuijs, B. Pecceu, L. Theunis, N. Dubois, C. Charlier, P. G. Jorens, L. Bervoets, R. Blust, H. Neels and A. Covaci, Cocaine and metabolites in waste and surface water across Belgium, *Environ. Pollut.*, 2009, **157**(1), 123–129.

18. M. J. Reid, L. Derry and K. V. Thomas, Analysis of new classes of recreational drugs in sewage: synthetic cannabinoids and amphetamine-like substances, *Drug Test. Anal.*, 2014, **6**(1–2), 72–79.
19. J. R. Archer, S. Hudson, D. M. Wood and P. I. Dargan, Analysis of urine from pooled urinals – a novel method for the detection of novel psychoactive substances, *Curr. Drug Abuse Rev.*, 2013, **6**(2), 86–90.
20. W. Hall, J. Prichard, P. Kirkbride, R. Bruno, P. K. Thai, C. Gartner, F. Y. Lai, C. Ort and J. F. Mueller, An analysis of ethical issues in using wastewater analysis to monitor illicit drug use, *Addiction*, 2012, **107**, 1767–1773.
21. J. R. H. Archer, P. I. Dargan, S. Hudson, S. Davies, M. Puchnarewicz, A. T. Kicman, F. Measham, M. Wood, A. Johnstonn and D. M. Wood, Taking the pissoir – a novel and reliable way of knowing what drugs are being used in nightclubs, *J. Subst. Use*, 2014, **19**, 103–107.
22. F. Heyerdahl, P. I. Dargan, J. Kleven, K. E. Hovda, J. R. Archer, S. Hudson and D. M. Wood, Anonymous pooled urine samples from festival urinals to detect classical recreational drug and novel psychoactive substance use: Is it possible in when it's freezing outside?, *Clin. Toxicol.*, 2013, **51**, 664–665.
23. D. M. Wood, J. R. H. Archer, F. Measham, S. Hudson and P. I. Dargan, Detection of use of novel psychoactive substances by attendees at a music festival in the North West of England, *Clin. Toxicol.*, 2013, **51**, 340–341.
24. J. R. Archer, P. I. Dargan, S. Hudson and D. M. Wood, Analysis of anonymous pooled urine from portable urinals in central London confirms the significant use of novel psychoactive substances, *QJM*, 2013, **106**(2), 147–152.
25. J. R. Archer, P. I. Dargan, H. M. Lee, S. Hudson and D. M. Wood, Trend analysis of anonymised pooled urine from portable street urinals in central London identifies variation in the use of novel psychoactive substances, *Clin. Toxicol.*, 2014, **52**(3), 160–165.
26. J. R. Archer, S. Hudson, O. Jackson, T. Yamamoto, C. Lovett, H. M. Lee, S. Rao, L. Hunter, P. I. Dargan and D. M. Wood, Analysis of anonymised pooled urine in nine UK cities: variation in classical recreational drug, novel psychoactive substance and anabolic steroid use, *QJM*, 2015, **108**(12), 928–933.
27. A. K. Singh, K. Granley, U. Misrha, K. Naeem, T. White and Y. Jiang, Screening and confirmation of drugs in urine: interference of hordenine with the immunoassays and thin layer chromatography methods, *Forensic Sci. Int.*, 1992, **54**(1), 9–22.

6 Mephedrone and New Psychoactive Substances

Simon Elliott

ROAR Forensics, Malvern, UK
Email: simon.elliott@roarforensics.com

6.1 Introduction

The reader of this chapter may rightly ask: why highlight mephedrone and why new psychoactive substances? In the context of use and misuse of drugs and forensic toxicology, substance changes that have taken place in the last decade have arguably had more impact on toxicology than anything in the last five decades.[1]

Following many years of stimulant drug chemistry largely based on amphetamine and related derivatives (including 3,4-methylenedioxymethamphetamine, MDMA or "Ecstasy"), in the 21st century a relatively minor alteration of the phenylethylamine (phenethylamine) framework gave rise to the β-ketone versions, collectively referred to as cathinones.[2,3] Of these, mephedrone (4-methylmethcathinone) became a popular drug of abuse for its purported MDMA-like effects in users and, at the time, legal status.[4] *In lieu* of MDMA's availability, controlled status and changing tablet strength, mephedrone's popularity was such that, at its peak, it was challenging that of cocaine, which itself had seen a rise in popularity in preceding decades.[5] Primarily sold through websites, the concept of a legally available stimulant was attractive to many users, despite some reports of significant adverse "come-down" effects, both psychological and

Forensic Toxicology: Drug Use and Misuse
Edited by Susannah Davies, Atholl Johnston and David Holt
© The Royal Society of Chemistry 2016
Published by the Royal Society of Chemistry, www.rsc.org

physical.[4] Mephedrone became widely known within the media and popular culture in a very short space of time, which had not really been seen since γ-hydroxybutyrate (GHB) around 10 years earlier. Its fall was as rapid as its rise and in April 2010, only a couple of years after initial appearance, mephedrone became a Class B substance in the UK under the Misuse of Drugs Act 1971.[6] Whilst other cathinone derivatives were controlled at the same time, this was the beginning of a proliferation in drug types that is continuing to this day. This chapter will discuss this proliferation and subsequent effect on forensic toxicology.

The topic of terminology and definition of such drugs could be a chapter in its own right. Terms including "designer drugs", "new designer drugs", "new synthetic drugs", "novel psychoactive substances" or "new psychoactive substances" (NPS) have been used over the years, with each term having multiple "definitions" that vary between the organisation or researcher that refers to it.[7] Nevertheless, the over-arching concept is that a drug or substance (see the problem!) has been purposefully developed to avoid existing drug legislation whilst maintaining (or attempting to maintain) the effects of the drug/substance it is mimicking. For the purpose of this chapter, compounds that are encompassed by this concept will be discussed. Outside of formal definitions, vendors and users have a similar variable terminology for the products sold and used, often with geographical variation as well. For instance, initial terms such as "party pills" or "legal ecstasy" evolved into "legal highs", especially within the predominant websites as well as so-called bricks and mortar "head shops". Furthermore, there have been attempts to make reference to supposed uses as the compounds were (and still are) often marketed as "research chemicals – not for human consumption", "plant food", "pond cleaner" or "bath salts". Within these terms it is important to note that whilst products may have identical branding and packaging, the constituent drugs are changeable.[8–10] Although this is not a new occurrence (for example, not all "Ecstasy" tablets contain MDMA and, as referred to above, not at the same amount) it is an important consideration to both forensic toxicologists and prospective users alike. As such, case investigations should not rely and assume the drug involved solely on the basis of product information, although data from local drug officers and drug scientists may help inform the toxicologist of possible constituents. In the case of synthetic cannabinoids, the "herbal product" contained in the often brightly coloured or elaborate packaging may indeed contain a synthetic cannabinoid, but due to the many, many potential compounds (*e.g.* 5F-PB-22, AKB-48, AB-FUBINACA, *etc.*) it is not pos-sible to know from the product name/brand which are involved.[11,12] This is even more problematic with the occurrence of white powder products, packaged and labelled or not, as associated with the aforementioned "research chemicals" or "plant food", *etc.* It is no longer the case that a white powder may be cocaine or ketamine. Produced in clandestine laboratories or research chemical com-panies, the source and synthetic route usually derives from pre-existing publications, patents or research data/notes that have been resurrected, or exploited, by a new generation of chemists. This may result in very pure single-compound products. However, powders (as well as tablets) are also often supplied as a mixture of more than one drug as well as possible adulterants (*e.g.* caffeine, ephedrine, lignocaine, *etc.*).

6.2 Analysis

Although this is not a focus for the chapter, owing to the wide vari-ation of drugs/compound chemistries, new psychoactive substances are definitely an analytical challenge for toxicologists and drug sci-entists alike.[13] This is a particular problem when analysing biological fluid in casework; however, they should be detectable as part of general screening if modern methods are used, such as high-resolution accurate mass spectrometry, gas chromatography–mass spectrometry and liquid mass spectrometry, with additional analytical evidence through the use of liquid chromatography with ultraviolet diode-array detection (which may also provide positional isomeric determination).[13–16] Traditional immunoassay (*e.g.* general drugs of abuse) screening is not designed to detect many of these substances, although there may be indications of their presence (*e.g.* "amphet-amines = positive").[17] Even if using modern techniques, the growing trend for "targetted" (restricted) analysis may not detect such drugs if they are not included in the analytical targeting (*i.e.* if you don't look, you won't find).[13] However, this approach may be a requirement to overcome potential sensitivity problems for the potent drugs (*e.g.* synthetic cannabinoids and NBOMes) due to the likely lower dose taken/involved.[18–21] Nevertheless, it should be borne in mind that identification is based on library comparison but initial detection is not, allowing for detection of "unidentified" compounds based on chemical characteristics if a general ("non-targeted") screening methodology is used.

Absolute identification requires comparison with pure reference material/standards.[22] Consequently, lack of reference material may hinder identification/quantitation (or result in significant delays due to the time required to obtain such material – even if available). It should also be noted that some of the drugs (especially cathinones like mephedrone and 4-methylethcathinone) may be unstable in blood (in particular) and urine and may have degraded prior to toxicology analysis or been affected by storage conditions and container type, thereby resulting in an artificially low or undetectable concentration.[23–25] Stability is an important consideration when interpreting any measured concentration but is especially important in new psychoactive substance investigation where the stability profile of a newly developed compound may not be known. It would therefore be beneficial that the investigating toxicologist undertake such work to ensure appropriate interpretation of the findings.

6.3 Drug Types

It has been apparent that since the early years of "designer drugs" (*e.g.* fentanyl derivatives in the late 1970s), the evolution of substances has derived from discovery or amendment of a chemical framework with subsequent substitutions and additions to the core structure.[1,26] This creates a myriad of alternative compounds with variations in effects, sometimes subtle differences, sometimes with vast differences in, for example, potency. It is also this development and chemical tweaking that has the biggest impact on drug legislation and scheduling of substances, as even a minor change may result in the drug being excluded from international or national drug control (which is of course the intention of the producer).[27] Aspects of drug control are discussed in more detail in Chapters 15–17, but below is an indication of how rapidly changing new psychoactive substance evolution has been, with a proliferation of drug types over the last few decades, most being within the last decade.[28,29] Whilst their effects can involve variable neuropharmacology, they can be categorised as being stimulants, hallucinogens and sedatives (see Table 6.1; although it should be borne in mind that this is a simplification of complex activity).

In addition to the general drug classes, Table 6.2 shows some selected substances of significance that have been encountered in casework in the last five years.[30–56]

The following sections describe the common classes of new psychoactive substances based around certain chemical frameworks,

Table 6.1 NPS categorised as being stimulants, hallucinogens or sedatives.

Decade	Drug type/class	General description (predominant effect)
1970s	Amphetamines	Stimulants
1980s	Phenethylamine derivatives	Stimulants (some hallucinogens)
	Tryptamines	Hallucinogens
1990s	Further phenethylamine derivatives	Stimulants (some hallucinogens)
	Further tryptamines	Hallucinogens
2000s	Piperazines	Stimulants
2010s	Pyrrolidinopropiophenones	Stimulants
	Cathinones	Stimulants
	Aminoindanes	Stimulants
	Arylcyclohexylamines	Hallucinogens
	Synthetic cannabinoids	Hallucinogens
	Arylalkylamines (inc. benzofurans)	Stimulants
	NBOMe phenethylamines	Hallucinogens
	Piperidines (inc. phenidates)	Stimulants
	"Designer" benzodiazepines	Sedatives
	Pyrrolidines	Stimulants
	Opioids (inc. AH-7921, MT-45)	Sedatives

with particular emphasis on (a) dose, formulation and route of administration, (b) effects and (c) toxicity/safety.

6.4 Amphetamines/Phenylethylmines

2-Phenylethylamine (or phenethylamine) is a naturally occurring chemical which is used as the basis for a large number of compounds. Many so-called "designer drugs" come from deliberate alteration of the chemical structure of phenylethylamines.[26,27] Chemical methodology and the outcome of experimentation with hundreds of phenyl-ethylamines is described in "PiHKAL" (Phenethylamines I Have Known And Loved) by Shulgin and Shulgin.[57]

Amphetamine (α-methylphenylethylamine) is available as Dexe-drine® (dexamphetamine) by prescription only for narcolepsy and attention deficit and hyperactivity disorder (ADHD) in children. It is a stimulant which has been abused for many years. It has various associated street names including "speed", "base" and "whizz". Amphetamine powder can be administered by rubbing it into the

Table 6.2 Some selected substances of significance that have been encountered in recent casework.

Drug	Drug type/class	General description (predominant effect)
Methiopropamine (MPA)	Arylalkylamine	Stimulant
p-Methyl-4-methylaminorex (4,4'-DMAR)	Aminorex derivative	Stimulant
2-Methoxydiphenidine (MXP)	Piperidine derivative	Hallucinogen
Camfetamine	Atypical	Stimulant
Ethylphenidate	Piperidine	Stimulant
Etaqualone	Quinazolinone	Sedative
Desoxypipradrol (2-DPMP)	Pyrrolidine	Stimulant
Diphenylprolinol (D2PM)	Pyrrolidine	Stimulant
Methoxetamine	Arylcyclohexylamine	Hallucinogen
5- & 6-(2-Aminopropyl)-benzofuran (5- & 6-APB)	Arylalkylamine	Stimulant
Diclazepam	Designer benzodiazepines	Sedative
Phenazepam		
Etizolam		
Flubromazepam		
Pyrazolam		
5-(2-Aminopropyl)indole (5-IT)	Arylalkylamine	Stimulant

gums, eating or snorting. Intravenous injection has also been described, sometimes along with heroin "speedball". Prolonged use of amphetamine can lead to paranoia and psychosis.[58]

A related drug, methamphetamine (*N*-methylamphetamine, *N*-methyl-α-methylphenylethylamine) is more prevalent in some areas of the world (Asia, USA) but less so in the UK[59] and wider Europe, although this appears to be changing.[60] It is a potent and addictive compound which can be ingested, "snorted" (nasally insufflated), injected or smoked. It has various associated street names including "crystal meth", "ice", "crank" and "yaba". It is not a prescription drug in the UK but is in some countries and L-methamphetamine (levomethamphetamine) may be present in some healthcare products (*e.g.* in the USA).

3,4-Methylenedioxymethamphetamine (MDMA) has been used as a stimulant by people typically at dance events or parties since the mid to late 1980s.[58,61] It has the street name "Ecstasy" or "E". There are numerous types of tablets, manufactured illegally, which contain

Table 6.3   Other phenylethylamine-based NPS.

Psychoactive substance	Also known as
4-Fluoroamphetamine	4-FA
4-Methylamphetamine	4-MA
4-Methylthioamphetamine	4-MTA
2-(4-Bromo-2,5-dimethoxyphenyl)ethylamine	2C-B
2-(4-Iodo-2,5-dimethoxyphenyl)ethylamine	2C-I
2,5-Dimethoxy-4-ethylthiophenethylamine	2C-T-2
2,5-Dimethoxy-4-(*n*)-propylthiophenethylamine	2C-T-7
p-Methoxyamphetamine	PMA
p-Methoxymethylamphetamine	PMMA
p-Methoxyethylamphetamine	PMEA
2,5-Dimethoxy-4-iodoamphetamine	DOI
2,5-Dimethoxy-4-methylamphetamine	DOM
N-Methyl-1-(1,3-benzodioxol-5-yl)-2-butanamine	MBDB
2,4,5-Trimethoxyamphetamine	TMA-2

MDMA at various amounts, in addition to "adulterants" such as amphetamine, ketamine, ephedrine and caffeine. The different types of tablets usually have different names based on the style or marking (logo) on the tablet, *e.g.* "dove", "mitsubishi", "rolex", "ferrari", "molly" or "dolphin".[62] Other phenylethylamine-based new psychoactive substances of casework note are shown in Table 6.3.

An issue associated with many new psychoactive substances, but particularly phenylethylamines, is the invariable presence of other substances within the consumed product. These may be bulking agents (*e.g.* lactose, cellulose), pharmaceutical (*e.g.* caffeine, ephedrine, paracetamol) or other new psychoactive drug adulterants (*e.g.* other phenylethylamines). New psychoactive substances may also be adulterants themselves in "traditional" products such as MDMA, amphetamine, cocaine, ketamine, *etc.*[63] All this results in an unknown safety profile and requires consideration by the investigating toxicologist to ensure the toxicological significance of each drug and the combination is appropriately assessed. As such, the following sections provide a general overview of this class of drugs but polydrug use is a likely circumstance.

6.4.1   Dose, Formulation and Route of Administration

Phenylethylamines typically exist in powder and tablet form and rarely as a liquid. Because of this, such drugs are often ingested orally or "snorted" by the user, as well as being injected. They are rarely smoked although users have described (especially in internet drug fora) numerous routes of administration, but certainly the former routes are the most common.

In terms of dose, "Ecstasy" tablets may typically contain 80–160 mg MDMA, but this is very changeable with lower and higher doses in some batches and geographies.[64] This also applies to other substituted phenylethylamines that may be present or sold as "Ecstasy" (*e.g.* PMA, PMMA) or other formulations available as a phenylethylamine-based stimulant product. Conversely, products (including tablets, powders, pellets, *etc.*) containing phenylethylamine-based hallucinogens are associated with lower doses, generally 1–50 mg. Of course, as many effects are dose dependent, this is an important consideration for users and toxicologists, and is a problem due to the variable nature of the products both from a constituent and dose perspective. As referred to elsewhere, it cannot be assumed that a particular product has such consistency, whether it is branded, packaged, labelled or associated with a certain logo.

6.4.2 Effects

Phenylethylamines are generally stimulant in nature (*e.g.* amphetamine) with euphoric effects (*e.g.* MDMA). However, some have hallucinogenic properties (*e.g.* 2C-B). The neuropharmacology of this can be complicated, but overall they typically affect, in one way or another, monoaminergic systems in the brain and especially serotonin and dopamine through alteration in the release, reuptake or receptor transport of these neurotransmitters.

6.4.3 Toxicity/Safety

The toxicity of phenylethylamines can be idiosyncratic (*e.g.* MDMA), whereby one user may be affected more than another user or equally a user cannot predict they will react the same each time the drug is used.[61] Some phenylethylamines may have a delayed effect (*e.g.* 4-MTA, PMA and PMMA), whereby a user may consume multiple doses of the drug following the perception it was not having an effect. This is especially dangerous, since when the drug eventually does take effect the cumulative impact on the body can be serious and often fatal.[65–69]

Initial adverse effects of phenylethylamines (when taken by whatever route) may be nausea/vomiting, headache, confusion and increased heart rate (tachycardia), and it is the potential cardiac effects of phenylethylamines (and stimulants in general) that present the biggest risk to the user.[61] They may also precipitate a pre-existing heart condition. Following excessive use, phenylethylamines may result in a "serotonin syndrome" through the overstimulation of the serotonergic system, which can be characterised by increased body temperature (hyperpyrexia), clenched jaw or muscle rigidity and convulsions.[70] An increased body temperature coupled with potential risks if taken during physical exertion (i.e. dancing) in a hot or enclosed environment (e.g. dance club/venue) can lead to increased water intake by the user which may cause hyponatraemia (reduction in electrolyte sodium levels). Separate from body stimulation effects, any hallucinations (auditory or visual) experienced by the user may result in accidental injury.[71]

6.5 Piperazines

Piperazine itself is an anti-worming agent that is not abused. However, derivatives of piperazine have been abused in the last 15 years due to their stimulant effects (e.g. benzylpiperazine, BZP; and phenylpiperazines, e.g. trifluoromethylphenylpiperazine, TFMPP and m-chlorophenylpiperazine, mCPP).[72,73] Some have hallucinogenic properties as well. Although many are now largely controlled, piperazines were referred to as "legal ecstasy", "legal highs" or "herbal ecstasy". As such, they are typically supplied in tablet form, mimicking "Ecstasy" tablets. The vast majority of reported fatal cases involving these compounds typically involve an alternative cause of death or additional drugs,[72,74] but some cases have involved very high concentrations that have been associated with fatal toxicity.[30] Piperazines are also discussed in Chapter 7.

6.5.1 Dose, Formulation and Route of Administration

As for phenylethylamines, piperazines typically exist in powder and tablet form, with the primary route of administration being oral consumption. Other routes of administration have been explored by users (e.g. "snorting" and injecting) although these are uncommon. The typical dose ingested by users appears to be between 50–200 mg but this can vary considerably, depending on the product used.

6.5.2 Effects

In the brain, piperazines stimulate the release and inhibit the reuptake of noradrenaline but mainly dopamine and serotonin neurotransmitters. Studies have found piperazines to be less potent than MDMA, methamphetamine or amphetamine. This results in piperazines generally producing mild stimulant effects[75] with moderate euphoria (*e.g.* BZP), although some have hallucinogenic properties (*e.g.* TFMPP). TFMPP is commonly used in conjunction with BZP in order to seek the entactogenic effects of MDMA, but this can also be unintentional due to the presence of both compounds in a particular product, something that the user may not be aware of.

6.5.3 Toxicity/Safety

Adverse effects may occur when piperazines are co-ingested with other drugs (in particular MDMA and other serotonergic/dopaminergic compounds), but toxic effects with piperazines alone (*e.g.* BZP and mCPP) have also been reported. Agitation, tachycardia and seizures have been recorded, with additional adverse effects such as nausea, headache, dizziness, sweating and cardiovascular symptoms (*e.g.* increased heart rate and blood pressure).[72,76]

6.6 Tryptamines

Compounds used include dimethyltryptamine (DMT), α-methyltryptamine (AMT), diethyltryptamine (DET), 5-methoxy-*N*,*N*-diisopropyltryptamine (5-MeO-DiPT, "foxy methoxy"), dipropyltryptamine (DPT) and *N*,*N*-diallyltryptamine derivatives (DALTs) such as 5-MeO-DALT.[49,77–79] They can be derived from plants or can be synthesised directly. Plant extracts are usually treated as snuff and inhaled or can be smoked, as they are generally not active if eaten directly. Alternatively, a liquid concoction can be produced. As for phenylethylamines, the chemical methodology and the outcomes of experimentation with hundreds of tryptamines are described in "TiHKAL" (Tryptamines I Have Known And Loved) by Shulgin and Shulgin.[80] A related compound, psilocybin, is an active component of widely known "magic mushrooms" and produces hallucinogenic effects similar to other tryptamines.

6.6.1 Dose, Formulation and Route of Administration

Tryptamine derivatives are largely available as powders or pellets and sometimes as liquid concoctions. They are usually smoked or injected but are sometimes snorted as most are not orally active, being degraded by the enzyme monoamine oxidase (MAO). However, some users may purposefully mix tryptamines with monoamine oxidase inhibitors (MAOIs) to facilitate their activity. In fact, AMT has monoamine oxidase inhibitory properties itself,[81] so can be taken orally and this is a problem if the drug is taken with other drugs that affect the monoamine neurotransmitters noradrenaline, dopamine and serotonin, which relates to many drugs. Overall, owing to their varying potency and route of administration, there is a wide dose range (1–400 mg) associated with tryptamines.

6.6.2 Effects

Owing to their chemical structure and similarity to serotonin (5-hydroxytryptamine, 5-HT), tryptamines largely affect the serotonergic systems in the brain but many also have effects on dopaminergic and noradrenergic systems. They are generally hallucinogenic or psychedelic in nature rather than being stimulants, producing profound auditory and visual effects. However, some (*e.g.* AMT) do result in stimulatory effects such as euphoria and amphetamine-like mood elevation. There may be a slow onset of these actions (about 3–4 hours) but with an extended duration of effect (12–24 hours or more). Physical effects may include mild increases in blood pressure or respiration rate, tachycardia, mydriasis, diaphoresis, salivation, severe nausea, severe vomiting, deep tendon reflexes, impaired coordination, visual and auditory disturbances and distortions.

6.6.3 Toxicity/Safety

Smoking tryptamines can be irritating to the throat and lungs and use may cause stomach cramps, anxiety, nausea/vomiting, headache and confusion, as mentioned above. From a toxicity point of view, not all tryptamines will produce adverse cardiovascular effects although some do (*e.g.* AMT),[82] but it is the hallucinatory effects that present a significant risk to users, especially from accidental injury or alterations in behaviour. The practice of mixing tryptamines with MAOIs can further increase the psychedelic effect but also produces an additional toxicity risk. As with all hallucinogens, the effects will depend on the environment in which they are taken and may be worse in people with psychological or emotional problems.

6.7 Cathinones

Cathinone itself is a β-ketone (bk) derivative of amphetamine (additional oxygen atom). Along with cathine (norpseudoephedrine, a phenylethylamine-like compound), they are found in khat leaves (from the plant *Catha edulis*) in East Africa and Arabia (also see Chapter 9). Chewing the leaves produces a mild euphoria.[2] As for phenylethylamine, cathinone can be used as a basis to synthesise altered compounds such as methcathinone (cathinone version of methamphetamine) and methylone (3,4-methylenedioxymethcathinone), which is the β-ketone or "cathinone version" of MDMA. A relatively wide number of the cathinone derivatives are now controlled in the UK by the Misuse of Drugs Act 1971, but not all. Legislative descriptions also have to consider the added complication of positional isomers such as 2-methylmethcathinone (2-MMC), 3-methylmethcathinone (3-MMC) and 4-methylmethcathinone (4-MMC, mephedrone).[83]

Initially sold as research chemicals on the internet, cathinones (including propiophenones) were quickly commercialised/branded by being referred to as "plant food" or "bath salts" and "not for human consumption".[2,8,84–104] Examples of casework note are shown in Table 6.4.[30,44,85] Example street/product names include maiow maiow, meow meow, bubbles, NRG-1, NRG-2, NRG-3, top cat, ivory wave and MCAT.

Table 6.4 Cathinones.

Cathinone	Also known as
4-Methylmethcathinone	Mephedrone, 4-MMC
4-Methylethcathinone	4-MEC
4-Methoxymethcathinone	Methedrone, bk-PMMA
Butylone	bk-MBDB
Methylone	bk-MDMA
Ethylone	bk-MDEA
β-Ethylmethcathinone	Pentedrone
β-Ketomethylbenzodioxolylpentanamine	bk-MBDP, pentylone
Methcathinone	bk-Methamphetamine
4-Fluoromethcathinone	Flephedrone, 4-FMC
3,4-Methylenedioxypyrovalerone	MDPV
Methylenedioxy-α-pyrrolidinobutyrophenone	MDPBP
Napthylpyrovalerone	Naphyrone
4-Chloromethcathinone	Clephedrone, 4-CMC
α-(Methylamino)butyrophenone	Buphedrone, MABP
α-Pyrrolidinopentiophenone	α-PVP
4-Fluoro-α-pyrrolidinopentiophenone	4F-α-PVP

6.7.1 Dose, Formulation and Route of Administration

Cathinones and related propiophenones are typically available as powders and, less commonly, as tablets or liquids. As such they are predominantly ingested orally or snorted, but have been reported to be injected. Smoking is rarely reported. The "average" (*e.g.* mephedrone) dose is around 1 g but some (*e.g.* naphyrone) are advised on internet websites and user forums to be less (for example, around 20 mg). Again, as for phenylethylamines, adverse impacts of inappropriate dosage are significant and in the case of cathinones and propiophenones, where they may be taken as a nondescript white powder, the chances of this occurring are high. Equally the same problems of lack of product consistency and multi-component contents for phenylethylamines are also an issue.

6.7.2 Effects

Cathinones (including propiophenones) generally produce stimulant effects but some have hallucinogenic properties. They largely affect (release/inhibition) dopamine, noradrenaline and/or serotonin neurotransmitters in the brain. Perceived effects by users are dose related and include euphoria, increased energy, feelings of empathy, increased libido, sweating, tachycardia, headache and teeth grinding. Users may exhibit features of excitement, hyperactivity, talkativeness and dilated pupils. Also, the reported stimulant highs can be associated with subsequent lows as well as aggressive behaviour, psychosis, paranoia, insomnia, hallucinations, chaotic behaviour, nausea, lightheadedness and loss of appetite.[105]

6.7.3 Toxicity/Safety

Hospital admissions have involved individuals suffering increased heart rate, agitation, shortness of breath and increased blood pressure.[106–109] Deaths have been reported from acute toxicity, but also often with other drugs being involved.[30,86–91] There are also reports of possible suicidal ideation and this has resulted in a number of suicides associated with cathinone use.[30]

6.8 Aminoindanes

Also referred to as aminoindans, based on 2-aminoindane, these compounds were developed in the 1990s by a team of researchers for

Table 6.5 Aminoindanes.

Aminoindane	Also known as
2-Aminoindane	2-AI
5,6-Methylenedioxy-2-aminoindane	MDAI
5-Iodo-2-aminoindane	5-IAI
5-Methoxy-6-methyl-2-aminoindane	MMAI
N-Methyl-2-aminoindane	N-methyl-2-AI

studying MDMA-like serotonergic agonists.[110,111] They are totally synthetic, derived from a cyclized ring form of the phenylethylamine amine chemical grouping. Therefore similar substituents for phenylethylamines can be produced but only a few have become available.[112] Examples are shown in Table 6.5.

Aminoindane derivatives have been sold as or along with cathinones on the internet, but are typically encountered as "research chemicals" rather than "products" (*e.g.* plant food).[113]

6.8.1 Dose, Formulation and Route of Administration

Aminoindanes are most often available as powders and rarely tablets or liquids. The route of administration is typically oral or snorting. Whilst they can be injected, other routes of administration such as smoking have not been widely reported. The "average" dose reported by users is around 100–200 mg and is therefore comparable to many stimulant drugs. They are predominantly white powder formulations and were initially associated with so-called "fake cocaine" products and packaged as such with the additional presence of other drugs that are commonly found in cocaine (*e.g.* lignocaine, benzocaine). Aminoindanes are currently only controlled in a limited number of countries and are not included in the UK Misuse of Drugs Act 1971.[112]

6.8.2 Effects

Aminoindanes are generally stimulant in nature with some euphoric effects but not necessarily hallucinogenic effects. Although there are limited user reports, perceived effects are dose related and include euphoria, increased energy, feelings of empathy/affection, sweating and tachycardia.

6.8.3 Toxicity/Safety

As indicated from the original research intention, aminoindanes are selective serotonin releasing agents in the brain. When taken to excess or along with other drugs possessing monoaminergic activity, they have the propensity to induce a "serotonin syndrome". This is characterised by high body temperature, clenched jaw and seizures, potentially resulting in death. There are no published hospital admission reports but some fatalities have been reported, including direct acute toxicity.[30,114,115]

6.9 Synthetic Cannabinoids

Δ^9-Tetrahydrocannabinol (THC) has been known to be the primary active component in cannabis for decades, which in the 1970s led to the development and investigation of synthetic cannabinoids as pharmaceutical products, *e.g.* for nausea and vomiting following chemotherapy. Synthetic cannabinoids were later found to bind much more strongly and selectively to cannabinoid receptor type 1 (CB_1) and 2 (CB_2), compared to THC.[116,117] Synthetic cannabinoids are full agonists of the CB_1 and CB_2 receptors, whereas THC is a partial agonist. CB_1 receptors are believed to be responsible for the euphoric and psychoactive effects by mediating the inhibition of transmitter release. CB_2 receptors are thought to play a minor role in pain control.

Aside from some radiolabelling applications, synthetic cannabinoids have not become pharmaceutical products and were first encountered as drugs of abuse in 2008 when associated with "herbal incense" products that had been sold since around 2006.[11,118] The packaging was initially branded as "Spice" and this continues to be used as a term associated with synthetic cannabinoids; however, an increasing number of branded products have been produced, including "Annihilation", "Black Mamba", "Pandora's Box", "Clockwork Orange", "Exodus", "Damnation", *etc.*[11,12] Users include those who would normally use cannabis as well as those that are seeking different effects or to avoid random drug tests (*e.g.* in prisons) that may not incorporate synthetic cannabinoids. The first synthetic cannabinoid to be characterised in a drug of abuse product was JWH-018, an aminoalkylindole cannabinoid named after John W. Huffman who synthesised these THC analogues and metabolites.[118] Since 2008, there has been a proliferation of compounds often referred to as "generations" involving an addition of a halogen atom (*e.g.* fluorine) or chemical grouping as well as changing the chemical framework on which to base each generation of synthetic cannabinoid.[119–123] There are now hundreds of synthetic cannabinoids and a selection of notable ones are shown in Table 6.6.

Table 6.6 A selection of notable cannabinoids in circulation.

Cannabinoid	Generation
JWH-018	I
JWH-073	I
AM-2201	II
AM-694	II
RCS-8	II
UR-144	III
XLR-11	III
STS-135	III
MAM-2201	III
5F-PB-22 (5-fluoro-QUPIC)	IV
PB-22 (QUPIC)	IV
5F-AKB48 (5-fluoro-APINACA)	IV
AKB48 (APINACA)	IV
AB-FUBINACA	IV
AB-CHMINACA	IV
BB-22	IV
ADB-FUBINACA	IV
MDMB-CHMICA	V

Cannabinoids and synthetic cannabinoids are also discussed in more detail in Chapter 8.

6.9.1 Dose, Formulation and Route of Administration

As they are entirely synthetic, some cannabinoids are available as a powder in a similar formulation to internet sold "research chemicals", but they are more commonly available as herbal or herbal incense products. Along with a perfumed fragrance, the synthetic cannabinoid compound itself is applied or sprayed on herbal/plant material (*e.g. Turnera diffusa*, also known as damiana) to resemble a cannabis-based product with the intention for it to be smoked. However, with the availability of white powder products this is another example of dose risk, especially if the powder is ingested/snorted mistakenly for cocaine, a phenylethylamine or a cathinone product. Products are usually available in 1–3 g sachets with a typical amount of herbal material smoked being around 300 mg. However, studies have shown a significant variance in the composition and amount of actual synthetic cannabinoid(s) being present in the plant material products, ranging from less than 1 mg per g to greater than

40 mg per g. With the typical intended dose being 2–5 mg for a desired effect, this further presents a dose risk as the user would not be aware of the actual dose used, irrespective of measuring the actual herbal material ingested (by whatever route).

6.9.2   Effects

The intention of synthetic cannabinoid use is to mimic the effects of cannabis, namely producing relaxation, disinhibition and mind-altering effects. Some users may become drowsy and have reduced co-ordination. Panic and feelings of paranoia may also occur. Although controversial, as many cases are not analytically confirmed, synthetic cannabinoid use has been associated with acute psychosis, bizarre and aggressive behaviour.[124–129]

6.9.3   Toxicity/Safety

Reports arising out of hospital admission suggest that, compared to cannabis use, there is more tachycardia, agitation, hallucinations and hypertension seen with the synthetic cannabinoids.[130,131] It has also been suggested that there is a potential for rhabdomyolysis (the breakdown of damaged muscle) and subsequent acute kidney injury.[132,133] There are also reports of patients presenting with chest pain and ischemic changes on ECG, although this is rare. Instances of cerebral ischemia have also been reported.[134,135] Currently, there have been a few reported deaths associated with synthetic cannabinoid use, but the exact circumstances and ante-mortem symptoms are unclear with the majority of the decedents being found dead at home after use.[136–140] Some cases where synthetic cannabinoids have been found have also involved suicides and injuries resulting from dangerous activities.

6.10   New Psychoactive Substances in Investigations

Owing to the wide range of possible new psychoactive substances and various factors involved, the investigation of these cases is a challenging aspect of forensic toxicology. The first indication of the potential involvement of new psychoactive substances can be the mention of their use either by the user, witnesses or evidence at the scene. However, investigators should not assume that a particular

Table 6.7 Possible symptoms associated with drug types.

Symptoms	Drug type
Euphoria	Stimulants
Confusion	Hallucinogens, sedatives
Disinhibition	Stimulants, sedatives
Hallucinations	Stimulants, hallucinogens
Unsteadiness	Hallucinogens, sedatives
Hyperthermia	Stimulants
Vomiting	Stimulants, hallucinogens, sedatives
Cardiac effects	Stimulants, hallucinogens
Drowsiness	Sedatives
Odd behaviour	Stimulants, hallucinogens
Coma	Sedatives
Respiratory depression	Sedatives

brand or product is associated with a particular substance. Any actual seized products should be analysed to determine the true contents and prospective substances to aid analytical strategies. These issues are also risks to users and the ingestion (by whatever route) of an unexpected substance may have adverse dose outcomes. For example, snorting of a white powder thought to be mephedrone but actually being 25I-NBOMe would result in a significant dose of 25I-NBOMe. Equally, the possibility of multi-component products presents a similar risk to the user and should be taken into account by the toxicologist in terms of expected effects or toxicity.

If there is no scene evidence, no specific drug history or no obvious indication of new psychoactive drug use, it is important to be aware of the possible signs and symptoms, particularly in relation to the case circumstances available. Within the case history or any documented clinical presentation there may be the symptoms shown in Table 6.7 associated with drug types.

With the hallucinogens the individual may have received injuries due to an hallucinatory-induced accident and behaved in an erratic and violent way to themselves or to others. With sedatives such as benzodiazepines (*e.g.* diclazepam, pyrazolam, phenazepam, etizolam, flubromazepam) and opioids (*e.g.* AH-7921, MT-45) they may be used in the same context as traditional benzodiazepines (*e.g.* diazepam, temazepam, nitrazepam) and opiates/opioids (*e.g.* heroin, morphine, oxycodone).[30,141-144] In particular, owing to potential dose and potency differences, the risks of toxicity increase, for instance if an individual has used phenazepam instead of diazepam, especially if concomitantly used with heroin.

Whether the possible use of new psychoactive substances is suspected or not, it is important that they remain a consideration in any laboratory analytical strategy due to their prolific use in many different circumstances (*e.g.* fatalities, hospital admissions, driving under the influence and criminal casework: homicide, sexual offences). Typical analytical toxicology approaches of targeted and non-targeted screening should be used appropriately to ensure a suitable coverage of possible substances. Targeted screening can be used for commonly encountered new psychoactive substances and those that are expected to be present at low concentrations and require high sensitivity. However, owing to the continuing evolution of these drugs, a non-targeted approach should be considered to at least provide an indication of their presence and ideally allow identification. Aforementioned techniques such as high-resolution accurate mass spectrometry can be a useful tool in determining a potential molecular formula and chemical structure through analysis of precursor and product ions. Furthermore, chromatographic separation, differences in ion abundance ratios and UV spectral differences can be used to determine particular isomers (*e.g.* 2-, 3-, 4-positional isomers), especially in biological material (where IR and NMR cannot be used) which can be relevant for legislative purposes.

Investigators should also consider that the evolution of new psychoactive substances is not merely a chemical evolution but an evolution of their use, the latter resulting in the potential for previously encountered (common or otherwise) substances to return to the drug markets. Primary examples of this are *p*-methoxyamphetamine (PMA) and *p*-methoxymethylamphetamine (PMMA), which having originally been encountered in the 1970s and then in the late 1990s in "Ecstasy" tablets, their use reduced significantly before once again being encountered in "Ecstasy" tablets and casework in recent years. Whilst this may be sporadic due to specific batches, the possibility of "older" new psychoactive substances being involved in cases should not be excluded.

Once detected, confirmation and quantitative analysis requires the use of reference materials. This is another challenge for analytical investigation, as although there may be various commercial sources of certified reference material for known and relatively common compounds, there can be a lack of such material for very new drugs, as mentioned previously. Toxicologists should therefore be aware of any published or available data from researchers or investigators who have characterised any synthesised material or acquired product material.[145–148]

Even where the concentration of a new psychoactive substance has been determined (especially in blood), it can be difficult if not impossible to attribute such a concentration to "recreational" or "excessive" use that would normally be performed when assessing prescription medications (*i.e.* therapeutic use *versus* excessive use or overdosage). This would apply to all types of casework but is of particular relevance when determining cause of death. Similarly, if little information regarding the pharmacology of a new psychoactive substance is known or available (particularly as administration studies are not often performed), this further hinders evaluating toxicological significance. Consequently, an assessment of the circumstances and significance of other drugs involved where information is known can assist. For example, if there is no other cause of death and other drugs are of little or no significance (if involved at all), then the role of the new psychoactive substance(s) may be of importance. As described elsewhere, it is necessary to be mindful of factors that affect the concentration and type of compounds found, including possible drug instability (and subsequent impact on the measured concentration or lack of parent drug) and possible metabolism (*e.g.* synthetic cannabinoids may be present solely as metabolites in urine, with parent compounds primarily only being detected in blood).

Finally, it is also important to bear in mind that new psychoactive substances may be present in an individual's biological specimen from an alternative source rather than ingestion of the substance directly. For example:

- Methamphetamine – known metabolite of selegiline (anti-Parkinson's medication).
- *p*-Methoxyethylamphetamine (PMEA) and PMA – known metabolites of mebeverine (anti-spasmodic medication).
- Tryptamine and 2-phenylethylamine – known putrefactants. Both are metabolites of the amino acids tryptophan and phenylalanine, respectively.
- *m*-Chlorophenylpiperazine (mCPP) – known metabolite of the anti-depressant drug trazodone.

Overall, the key considerations when investigating new psychoactive substances in forensic toxicology are:

- The case circumstances and drug intelligence available.
- Analyse the suspected drug product if available.
- The branding or the product name will not consistently pertain to a particular NPS.
- The route of administration.

- The user may have taken the wrong product.
- The user may have got the dose wrong.
- The user may have taken other drugs (poly-drug use).
- Ensure analysis is appropriate (both in coverage and in analytical sensitivity).
- Older NPS may return to the drug market.
- Some NPS (in particular cathinones) are prone to instability.
- NPS and drug metabolism (*i.e.* only metabolites may be present).
- Delayed effect of the NPS (especially 4-MTA, PMA, PMMA, PMEA).
- NPS have varying pharmacology but may be categorised as stimulant, hallucinogenic or sedative in nature.

References

1. D. Favretto, J. P. Pascali and F. Tagliaro, New challenges and innovation in forensic toxicology: focus on the "New Psychoactive Substances", *J. Chromatogr. A*, 2013, **1287**, 84–95.
2. M. J. Valente, P. Guedes de Pinho, M. de Lourdes Bastos, F. Carvalho and M. Carvalho, Khat and synthetic cathinones: a review, *Arch. Toxicol.*, 2014, **88**(1), 15–45.
3. J. P. Kelly, Cathinone derivatives: a review of their chemistry, pharmacology and toxicology, *Drug Test. Anal.*, 2011, **3**(7–8), 439–453.
4. F. Schifano, A. Albanese, S. Fergus, J. L. Stair, P. Deluca, O. Corazza, Z. Davey, J. Corkery, H. Siemann, N. Scherbaum, M. Farre', M. Torrens, Z. Demetrovics and A. H. Ghodse, Mephedrone (4-methylmethcathinone; 'meow meow'): chemical, pharmacological and clinical issues. Psychonaut Web Mapping; ReDNet Research Groups, *Psychopharmacology*, 2011, **214**(3), 593–602.
5. K. Moore, P. I. Dargan, D. M. Wood and F. Measham, Do novel psychoactive substances displace established club drugs, supplement them or act as drugs of initiation? The relationship between mephedrone, ecstasy and cocaine, *Eur. Addict. Res.*, 2013, **19**(5), 276–282.
6. K. McElrath and C. O'Neill, Experiences with mephedrone pre- and post-legislative controls: perceptions of safety and sources of supply, *Int. J. Drug Policy*, 2011, **22**(2), 120–127.
7. S. D. Brandt, L. A. King and M. Evans-Brown, The new drug phenomenon (editorial perspective), *Drug Test. Anal.*, 2014, **6**, 587–597.

8. M. M. Schmidt, A. Sharma, F. Schifano and C. Feinmann, "Legal highs" on the net-Evaluation of UK-based Websites, products and product information, *Forensic Sci. Int.*, 2011, **206**(1–3), 92–97.

9. S. Davies, D. M. Wood, G. Smith, J. Button, J. Ramsey, R. Archer, D. W. Holt and P. I. Dargan, Purchasing 'legal highs' on the Internet–is there consistency in what you get?, *QJM*, 2010 Jul, **103**(7), 489–493.

10. M. Baron, M. Elie and L. Elie, An analysis of legal highs: do they contain what it says on the tin?, *Drug Test. Anal.*, 2011, 3(9), 576–581.

11. L. K. Brents and P. L. Prather, The K2/Spice phenomenon: emergence, identification, legislation and metabolic character-ization of synthetic cannabinoids in herbal incense products, *Drug Metab. Rev.*, 2014, **46**(1), 72–85.

12. S. Hudson, J. Ramsey, L. King, S. Timbers, S. Maynard, P. I. Dargan and D. M. Wood, Use of high-resolution accurate mass spectrometry to detect reported and previously unreported cannabinomimetics in "herbal high:" products, *J. Anal. Toxicol.*, 2010, **34**(5), 252–260.

13. S. Elliott, Cat and mouse: the analytical toxicology of designer drugs, *Bioanalysis*, 2011, 3(3), 249–251.

14. F. T. Peters, Recent developments in urinalysis of metabolites of new psychoactive substances using LC-MS, *Bioanalysis*, 2014, **6**(15), 2083–2107.

15. A. H. Wu, R. Gerona, P. Armenian, D. French, M. Petrie and K. L. Lynch, Role of liquid chromatography-high-resolution mass spectrometry (LC-HR/MS) in clinical toxicology, *Clin. Toxicol.*, 2012, **50**(8), 733–742.

16. M. Paul, J. Ippisch, C. Herrmann, S. Guber and W. Schultis, Analysis of new designer drugs and common drugs of abuse in urine by a combined targeted and untargeted LC-HR-QTOFMS approach, *Anal. Bioanal. Chem.*, 2014, **406**(18), 4425–4441.

17. M. J. Swortwood, W. L. Hearn and A. P. DeCaprio, Cross-reactivity of designer drugs, including cathinone derivatives, in commercial enzyme-linked immunosorbent assays, *Drug Test. Anal.*, 2014, **6**(7–8), 716–727.

18. J. L. Poklis, D. J. Clay and A. Poklis, High-performance liquid chromatography with tandem mass spectrometry for the de-termination of nine hallucinogenic 25-NBOMe designer drugs in urine specimens, *J. Anal. Toxicol.*, 2014, **38**(3), 113–121.

19. R. D. Johnson, S. R. Botch-Jones, T. Flowers and C. A. Lewis, An evaluation of 25B-, 25C-, 25D-, 25H-, 25I- and 25T2-NBOMe via

LC-MS-MS: method validation and analyte stability, *J. Anal. Toxicol.*, 2014, **38**(8), 479–484.

20. R. Karinen, S. S. Tuv, E. L. Øiestad and V. Vindenes, Concentrations of APINACA, 5F-APINACA, UR-144 and its degradant product in blood samples from six impaired drivers compared to previous reported concentrations of other synthetic cannabinoids, *Forensic Sci. Int.*, 2015, **246**, 98–103.

21. K. B. Scheidweiler and M. A. Huestis, Simultaneous quantification of 20 synthetic cannabinoids and 21 metabolites, and semi-quantification of 12 alkyl hydroxy metabolites in human urine by liquid chromatography-tandem mass spectrometry, *J. Chromatogr. A*, 2014, **1327**, 105–117.

22. R. P. Archer, R. Treble and K. Williams, Reference materials for new psychoactive substances, *Drug Test. Anal.*, 2011, 3(7–8), 505–514.

23. Y. N. Soh and S. Elliott, An investigation of the stability of emerging new psychoactive substances, *Drug Test. Anal.*, 2014, **6**(7–8), 696–704.

24. R. D. Johnson and S. R. Botch-Jones, The stability of four designer drugs: MDPV, mephedrone, BZP and TFMPP in three biological matrices under various storage conditions, *J Anal. Toxicol.*, 2013, **37**(2), 51–55.

25. S. Kneisel, M. Speck, B. Moosmann and V. Auwärter, Stability of 11 prevalent synthetic cannabinoids in authentic neat oral fluid samples: glass versus polypropylene containers at different temperatures, *Drug Test. Anal.*, 2013, **5**(7), 602–606.

26. L. A. King and A. T. Kicman, A brief history of 'new psychoactive substances', *Drug Test. Anal.*, 2011, 3(7–8), 401–403.

27. L. A. King, New phenethylamines in Europe, *Drug Test. Anal.*, 2014, **6**(7–8), 808–818.

28. http://www.emcdda.europa.eu/publications/2015/new-psychoactive-substances.

29. http://www.unodc.org/wdr2014.

30. S. Elliott and J. Evans, A 3-year review of new psychoactive substances in casework, *Forensic Sci. Int.*, 2014, **243**, 55–60.

31. S. Cosbey, S. Kirk, M. McNaul, L. Peters, B. Prentice, A. Quinn, S. P. Elliott, S. D. Brandt and R. P. Archer, Multiple fatalities involving a new designer drug: para-methyl-4-methylaminorex, *J Anal. Toxicol.*, 2014, **38**(6), 383–384.

32. H. M. Lee, D. M. Wood, S. Hudson, J. R. Archer and P. I. Dargan, Acute toxicity associated with analytically confirmed recreational use of methiopropamine (1-(thiophen-2-yl)-2-methylaminopropane), *J. Med. Toxicol.*, 2014, **10**(3), 299–302.

33. S. P. Elliott, S. D. Brandt, J. Wallach, H. Morris and P. V. Kavanagh, First Reported Fatalities Associated with the 'Research Chemical' 2-Methoxydiphenidine, *J. Anal. Toxicol.*, 2015, **39**(4), 287–293.
34. J. Krueger, H. Sachs, F. Musshoff, T. Dame, J. Schaeper, M. Schwerer, M. Graw and G. Roider, First detection of ethylphenidate in human fatalities after ethylphenidate intake, *Forensic Sci. Int.*, 2014, **243**, 126–129.
35. J. M. Corkery, S. Elliott, F. Schifano, O. Corazza and A. H. Ghodse, 2-DPMP (desoxypipradrol, 2-benzhydrylpiperidine, 2-phenylmethylpiperidine) and D2PM (diphenyl-2-pyrrolidin-2-yl-methanol, diphenylprolinol): A preliminary review, *Prog. Neuro-Psychopharmacol. Biol. Psychiatry*, 2012, **39**(2), 253–258.
36. P. Kriikku, L. Wilhelm, J. Rintatalo, J. Hurme, J. Kramer and I. Ojanperä, Prevalence and blood concentrations of desoxypipradrol (2-DPMP) in drivers suspected of driving under the influence of drugs and in post-mortem cases, *Forensic Sci. Int.*, 2013, **226**(1–3), 146–151.
37. D. M. Wood, M. Puchnarewicz, A. Johnston and P. I. Dargan, A case series of individuals with analytically confirmed acute diphenyl-2-pyrrolidinemethanol (D2PM) toxicity, *Eur. J. Clin. Pharmacol.*, 2012, **68**(4), 349–353.
38. P. Adamowicz and D. Zuba, Fatal intoxication with methoxetamine, *J. Forensic Sci.*, 2015, **60**(Suppl 1), S264–S268.
39. M. Wiergowski, J. S. Anand, M. Krzyżanowski and Z. Jankowski, Acute methoxetamine and amphetamine poisoning with fatal outcome: a case report, *Int. J. Occup. Med. Environ. Health*, 2014, **27**(4), 683–690.
40. L. Imbert, A. Boucher, G. Delhome, T. Cueto, M. Boudinaud, J. Maublanc, S. Dulaurent, J. Descotes, G. Lachâtre and J. M. Gaulier, Analytical findings of an acute intoxication after inhalation of methoxetamine, *J. Anal. Toxicol.*, 2014, **38**(7), 410–415.
41. A. A. Elian and J. Hackett, A polydrug intoxication involving methoxetamine in a drugs and driving case, *J. Forensic Sci.*, 2014, **59**(3), 854–858.
42. M. Wikström, G. Thelander, M. Dahlgren and R. Kronstrand, An accidental fatal intoxication with methoxetamine, *J. Anal. Toxicol.*, 2013, **37**(1), 43–46.
43. I. M. McIntyre, R. D. Gary, A. Trochta, S. Stolberg and R. Stabley, Acute 5-(2-aminopropyl)benzofuran (5-APB) intoxication and

fatality: a case report with postmortem concentrations, *J. Anal. Toxicol.*, 2015, **39**(2), 156–159.

44. P. Adamowicz, D. Zuba and B. Byrska, Fatal intoxication with 3-methyl-N-methylcathinone (3-MMC) and 5-(2-aminopropyl)benzofuran (5-APB), *Forensic Sci. Int.*, 2014, **245C**, 126–132.

45. W. L. Chan, D. M. Wood, S. Hudson and P. I. Dargan, Acute psychosis associated with recreational use of benzofuran 6-(2-aminopropyl)benzofuran (6-APB) and cannabis, *J. Med. Toxicol.*, 2013, **9**(3), 278–281.

46. R. Kronstrand, M. Roman, M. Dahlgren, G. Thelander, M. Wikström and H. Druid, A cluster of deaths involving 5-(2-aminopropyl)indole (5-IT), *J. Anal. Toxicol.*, 2013, **37**(8), 542–546.

47. L. N. Seetohul and D. J. Pounder, Four fatalities involving 5-IT, *J. Anal. Toxicol.*, 2013, **37**(7), 447–451.

48. M. Bäckberg, O. Beck, P. Hultén, J. Rosengren-Holmberg and A. Helander, Intoxications of the new psychoactive substance 5-(2-aminopropyl)indole (5-IT): a case series from the Swedish STRIDA project, *Clin. Toxicol.*, 2014, **52**(6), 618–624.

49. S. P. Elliott, S. D. Brandt, S. Freeman and R. P. A. M. T. Archer, (3-(2-aminopropyl)indole) and 5-IT (5-(2-aminopropyl)indole): an analytical challenge and implications for forensic analysis, *Drug Test. Anal.*, 2013, **5**(3), 196–202.

50. B. Moosmann, P. Bisel and V. Auwärter, Characterization of the designer benzodiazepine diclazepam and preliminary data on its metabolism and pharmacokinetics, *Drug Test. Anal.*, 2014 Jul–Aug, **6**(7–8), 757–763.

51. T. Nakamae, T. Shinozuka, C. Sasaki, A. Ogamo, C. Murakami-Hashimoto, W. Irie, M. Terada, S. Nakamura, M. Furukawa and K. Kurihara, Case report: Etizolam and its major metabolites in two unnatural death cases, *Forensic Sci. Int.*, 2008, **182**(1–3), e1–e6.

52. C. W. O'Connell, C. A. Sadler, V. M. Tolia, B. T. Ly, A. M. Saitman and R. L. Fitzgerald, Overdose of etizolam: the abuse and rise of a benzodiazepine analog, *Ann. Emerg. Med.*, 2015, **65**(4), 465–466.

53. P. Kriikku, L. Wilhelm, J. Rintatalo, J. Hurme, J. Kramer and I. Ojanperä, Phenazepam abuse in Finland: findings from apprehended drivers, post-mortem cases and police confiscations, *Forensic Sci. Int.*, 2012, **220**(1–3), 111–117.

54. J. M. Corkery, F. Schifano and A. H. Ghodse, Phenazepam abuse in the UK: an emerging problem causing serious adverse health problems, including death, *Hum. Psychopharmacol.*, 2012, **27**(3), 254–261.

55. J. B. Stephenson, D. E. Golz and M. J. Brasher, Phenazepam and its effects on driving, *J. Anal. Toxicol.*, 2013, **37**(1), 25–29.
56. B. Moosmann, L. M. Huppertz, M. Hutter, A. Buchwald, S. Ferlaino and V. Auwärter, Detection and identification of the designer benzodiazepine flubromazepam and preliminary data on its metabolism and pharmacokinetics, *J. Mass Spectrom.*, 2013, **48**(11), 1150–1159.
57. A. Shulgin, A. Shulgin. *PIHKAL: A Chemical Love Story*. Transform Press, Berkeley, CA, 1991.
58. M. Carvalho, H. Carmo, V. M. Costa, J. P. Capela, H. Pontes, F. Remião, F. Carvalho and L. Bastos Mde, Toxicity of amphetamines: an update, *Arch. Toxicol.*, 2012, **86**(8), 1167–1231.
59. D. M. Wood, J. Button, T. Ashraf, S. Walker, S. L. Greene, N. Drake, J. Ramsey, D. W. Holt and P. I. Dargan, What evidence is there that the UK should tackle the potential emerging threat of methamphetamine toxicity rather than established recreational drugs such as MDMA ('ecstasy')?, *QJM*, 2008, **101**(3), 207–213.
60. http://www.emcdda.europa.eu/publications/emcdda-papers/exploring-methamphetamine-trends-in-Europe.
61. A. P. Hall and J. A. Henry, Acute toxic effects of 'Ecstasy' (MDMA) and related compounds: overview of pathophysiology and clinical management, *Br. J. Anaesth.*, 2006, **96**(6), 678–685.
62. M. Duterte, C. Jacinto, P. Sales and S. Murphy, What's in a label? Ecstasy sellers' perceptions of pill brands, *J. Psychoact. Drugs*, 2009, **41**(1), 27–37.
63. C. V. Giné, I. F. Espinosa and M. V. Vilamala, New psychoactive substances as adulterants of controlled drugs. A worrying phenomenon?, *Drug Test. Anal.*, 2014, **6**(7–8), 819–824.
64. N. Vogels, T. M. Brunt, S. Rigter, P. van Dijk, H. Vervaeke and R. J. Niesink, Content of ecstasy in the Netherlands: 1993-2008, *Addiction*, 2009, **104**(12), 2057–2066.
65. S. P. Elliott, Fatal poisoning with a new phenylethylamine: 4-methylthioamphetamine (4-MTA), *J. Anal. Toxicol.*, 2000, **24**(2), 85–89.
66. R. W. Byard, N. G. Rodgers, R. A. James, C. Kostakis and A. M. Camilleri, Death and paramethoxyamphetamine - an evolving problem, *Med. J. Aust.*, 2002, **176**, 496.
67. S. S. Johansen, A. C. Hansen, I. B. Müller, J. B. Lundemose and M.-B. Franzmann, Three fatal cases of PMA and PMMA poisoning in Denmark, *J. Anal. Toxicol.*, 2003, **27**, 253.
68. S. Refstad, Paramethoxyamphetamine (PMA) poisoning; a 'party drug' with lethal effects, *Acta Anaesthesiol. Scand.*, 2003, **47**, 1298.

69. Y. Lurie, A. Gopher, O. Lavon, S. Almog, L. Sulimani and Y. Bentur, Severe paramethoxymethamphetamine (PMMA) and paramethoxyamphetamine (PMA) outbreak in Israel, *Clin. Toxicol.*, 2012, **50**(1), 39–43.

70. A. Bosak, F. LoVecchio and M. Levine, Recurrent seizures and serotonin syndrome following "2C-I" ingestion, *J. Med. Toxicol.*, 2013, **9**(2), 196–198.

71. B. V. Dean, S. J. Stellpflug, A. M. Burnett and K. M. Engebretsen, 2C or not 2C: phenethylamine designer drug review, *J. Med. Toxicol.*, 2013, **9**(2), 172–178.

72. S. Elliott, Current awareness of piperazines: pharmacology and toxicology, *Drug Test. Anal.*, 2011, **3**(7–8), 430–438.

73. M. D. Arbo, M. L. Bastos and H. F. Carmo, Piperazine compounds as drugs of abuse, *Drug Alcohol Depend.*, 2012, **122**(3), 174–185.

74. S. Elliott and C. Smith, Investigation of the first deaths in the United Kingdom involving the detection and quantitation of the piperazines BZP and 3-TFMPP, *J. Anal. Toxicol.*, 2008, **32**(2), 172–177.

75. H. Campbell, W. Cline, M. Evans, J. Lloyd and A. W. Peck, Comparison of the effects of dexamphetamine and 1-benzylpiperazine in former addicts, *Eur. J. Clin. Pharmacol.*, 1973, **6**(3), 170–176.

76. L. J. Schep, R. J. Slaughter, J. A. Vale, D. M. Beasley and P. Gee, The clinical toxicology of the designer "party pills" benzylpiperazine and trifluoromethylphenylpiperazine, *Clin. Toxicol.*, 2011, **49**(3), 131–141.

77. J. M. Corkery, E. Durkin, S. Elliott, F. Schifano and A. H. Ghodse, The recreational tryptamine 5-MeO-DALT (N,N-diallyl-5-methoxytryptamine): a brief review, *Prog. Neuro-Psychopharmacol. Biol. Psychiatry*, 2012, **39**(2), 259–262.

78. E. Tanaka, T. Kamata, M. Katagi, H. Tsuchihashi and K. Honda, A fatal poisoning with 5-methoxy-N,N-diisopropyltryptamine, Foxy, *Forensic Sci. Int.*, 2006, **163**(1–2), 152–154.

79. J. M. Wilson, F. McGeorge, S. Smolinske and R. Meatherall, A foxy intoxication, *Forensic Sci. Int.*, 2005, **148**(1), 31–36.

80. A. Shulgin, A. Shulgin. *TIHKAL: The Continuation*. Transform Press, Berkeley, CA, 1997.

81. A. W. Lessin, R. F. Long and M. W. Parkes, Central Stimulant Actions of Alpha-Alkyl Substituted Tryptamines in Mice, *Br. J. Pharmacol. Chemother.*, 1965, **24**, 49–67.

82. D. M. Boland, W. Andollo, G. W. Hime and W. L. Hearn, Fatality due to acute alpha-methyltryptamine intoxication, *J. Anal. Toxicol.*, 2005, **29**(5), 394–397.

83. R. Christie, E. Horan, J. Fox, C. O'Donnell, H. J. Byrne, S. McDermott, J. Power and P. Kavanagh, Discrimination of cathinone regioisomers, sold as 'legal highs', by Raman spectroscopy, *Drug Test. Anal.*, 2014, **6**(7–8), 651–657.
84. M. E. Musselman and J. P. Hampton, "Not for human consumption": a review of emerging designer drugs, *Pharmacotherapy*, 2014, **34**(7), 745–757.
85. M. Bäckberg, E. Lindeman, O. Beck and A. Helander, Characteristics of analytically confirmed 3-MMC-related intoxications from the Swedish STRIDA project, *Clin. Toxicol.*, 2015, **53**(1), 46–53.
86. S. H. Cosbey, K. L. Peters, A. Quinn and A. Bentley, Mephedrone (methylmethcathinone) in toxicology casework: a Northern Ireland perspective, *J. Anal. Toxicol.*, 2013, **37**(2), 74–82.
87. H. Torrance and G. Cooper, The detection of mephedrone (4-methylmethcathinone) in 4 fatalities in Scotland, *Forensic Sci. Int.*, 2010, **202**(1–3), e62–e63.
88. P. D. Maskell, G. De Paoli, C. Seneviratne and D. J. Pounder, Mephedrone (4-methylmethcathinone)-related deaths, *J. Anal. Toxicol.*, 2011, **35**(3), 188–191.
89. E. Gerace, M. Petrarulo, F. Bison, A. Salomone and M. Vincenti, Toxicological findings in a fatal multidrug intoxication involving mephedrone, *Forensic Sci. Int.*, 2014, **243**, 68–73.
90. A. J. Dickson, S. P. Vorce, B. Levine and M. R. Past, Multiple-drug toxicity caused by the coadministration of 4-methylmethcathinone (mephedrone) and heroin, *J. Anal. Toxicol.*, 2010, **34**(3), 162–168.
91. F. Schifano, J. Corkery and A. H. Ghodse, Suspected and confirmed fatalities associated with mephedrone (4-methylmethcathinone, "meow meow") in the United Kingdom, *J. Clin. Psychopharmacol.*, 2012, **32**(5), 710–714.
92. B. M. Cawrse, B. Levine, R. A. Jufer, D. R. Fowler, S. P. Vorce, A. J. Dickson and J. M. Holler, Distribution of methylone in four postmortem cases, *J. Anal. Toxicol.*, 2012, **36**(6), 434–439.
93. J. M. Pearson, T. L. Hargraves, L. S. Hair, C. J. Massucci, C. C. Frazee 3rd, U. Garg and B. R. Pietak, Three fatal intoxications due to methylone, *J. Anal. Toxicol.*, 2012, **36**(6), 444–451.
94. P. N. Carbone, D. L. Carbone, S. D. Carstairs and S. A. Luzi, Sudden cardiac death associated with methylone use, *Am. J. Forensic Med. Pathol.*, 2013, **34**(1), 26–28.
95. B. J. Warrick, J. Wilson, M. Hedge, S. Freeman, K. Leonard and C. Aaron, Lethal serotonin syndrome after methylone and butylone ingestion, *J. Med. Toxicol.*, 2012, **8**(1), 65–68.

96. S. Rojek, M. Kłys, M. Maciów-Głąb, K. Kula and M. Strona, Cathinones derivatives-related deaths as exemplified by two fatal cases involving methcathinone with 4-methylmethcathinone and 4-methylethcathinone, *Drug Test. Anal.*, 2014, **6**(7–8), 770–777.

97. D. Zuba, P. Adamowicz and B. Byrska, Detection of buphedrone in biological and non-biological material–two case reports, *Forensic Sci. Int.*, 2013, **227**(1–3), 15–20.

98. L. J. Marinetti and H. M. Antonides, Analysis of synthetic cathinones commonly found in bath salts in human performance and postmortem toxicology: method development, drug distribution and interpretation of results, *J Anal. Toxicol.*, 2013, **37**(3), 135–146.

99. J. F. Wyman, E. S. Lavins, D. Engelhart, E. J. Armstrong, K. D. Snell, P. D. Boggs, S. M. Taylor, R. N. Norris and F. P. Miller, Postmortem tissue distribution of MDPV following lethal intoxication by "bath salts", *J. Anal. Toxicol.*, 2013, **37**(3), 182–185.

100. B. L. Murray, C. M. Murphy and M. C. Beuhler, Death following recreational use of designer drug "bath salts" containing 3,4-Methylenedioxypyrovalerone (MDPV), *J. Med. Toxicol.*, 2012 Mar, **8**(1), 69–75.

101. K. Kesha, C. L. Boggs, M. G. Ripple, C. H. Allan, B. Levine, R. Jufer-Phipps, S. Doyon, P. Chi and D. R. Fowler, Methylene-dioxypyrovalerone ("bath salts"), related death: case report and review of the literature, *J. Forensic Sci.*, 2013, **58**(6), 1654–1659.

102. M. Sykutera, M. Cychowska and E. Bloch-Boguslawska, A Fatal Case of Pentedrone and α-Pyrrolidinovalerophenone Poisoning, *J. Anal. Toxicol.*, 2015, **39**(4), 324–329.

103. J. L. Knoy, B. L. Peterson and F. J. Couper, Suspected impaired driving case involving α-pyrrolidinovalerophenone, methylone and ethylone, *J. Anal. Toxicol.*, 2014, **38**(8), 615–617.

104. A. Wurita, K. Hasegawa, K. Minakata, K. Gonmori, H. Nozawa, I. Yamagishi, O. Suzuki and K. Watanabe, Postmortem distribution of α-pyrrolidinobutiophenone in body fluids and solid tissues of a human cadaver, *Leg. Med.*, 2014, **16**(5), 241–246.

105. M. V. Stoica and A. R. Felthous, Acute psychosis induced by bath salts: a case report with clinical and forensic implications, *J. Forensic Sci.*, 2013, **58**(2), 530–533.

106. P. Mas-Morey, M. H. Visser, L. Winkelmolen and D. J. Touw, Clinical toxicology and management of intoxications with synthetic cathinones ("bath salts"), *J. Pharm. Pract.*, 2013, **26**(4), 353–357.

107. D. M. Wood, S. L. Greene and P. I. Dargan, Clinical pattern of toxicity associated with the novel synthetic cathinone mephedrone, *Emerg. Med. J.*, 2011, **28**(4), 280–282.
108. D. M. Wood, S. Davies, S. L. Greene, J. Button, D. W. Holt, J. Ramsey and P. I. Dargan, Case series of individuals with analytically confirmed acute mephedrone toxicity, *Clin. Toxicol.*, 2010, **48**(9), 924–927.
109. K. J. Lusthof, R. Oosting, A. Maes, M. Verschraagen, A. Dijkhuizen and A. G. Sprong, A case of extreme agitation and death after the use of mephedrone in The Netherlands, *Forensic Sci. Int.*, 2011, **206**(1–3), e93–e95.
110. M. P. Johnson, X. M. Huang and D. E. Nichols, Serotonin neurotoxicity in rats after combined treatment with a dopaminergic agent followed by a nonneurotoxic 3,4-methylenedioxy-methamphetamine (MDMA) analogue, *Pharmacol. Biochem. Behav.*, 1991, **40**(4), 915–922.
111. R. Oberlender and D. E. Nichols, (+)-N-methyl-1-(1,3-benzo-dioxol-5-yl)-2-butanamine as a discriminative stimulus in studies of 3,4-methylenedioxy-methamphetamine-like behavioral activity, *J. Pharmacol. Exp. Ther.*, 1990, **255**(3), 1098–1106.
112. P. D. Sainsbury, A. T. Kicman, R. P. Archer, L. A. King and R. A. Braithwaite, Aminoindanes–the next wave of 'legal highs'?, *Drug Test. Anal.*, 2011, **3**(7–8), 479–482.
113. C. T. Gallagher, S. Assi, J. L. Stair, S. Fergus, O. Corazza, J. M. Corkery and F. Schifano, 5,6-Methylenedioxy-2-aminoindane: from laboratory curiosity to 'legal high', *Hum. Psychopharmacol.*, 2012, **27**(2), 106–112.
114. J. M. Corkery, S. Elliott, F. Schifano, O. Corazza and A. H. Ghodse, MDAI (5,6-methylenedioxy-2-aminoindane; 6,7-dihydro-5H-cyclopenta[f][1,3]benzodioxol-6-amine; 'sparkle'; 'mindy') toxicity: a brief overview and update, *Hum. Psychopharmacol.*, 2013, **28**(4), 345–355.
115. M. Coppola and R. Mondola, 5-Iodo-2-aminoindan (5-IAI): chemistry, pharmacology, and toxicology of a research chemical producing MDMA-like effects, *Toxicol. Lett.*, 2013, **218**(1), 24–29.
116. B. K. Atwood, J. Huffman, A. Straiker and K. Mackie, JWH018, a common constituent of 'Spice' herbal blends, is a potent and efficacious cannabinoid CB receptor agonist, *Br. J. Pharmacol.*, 2010, **160**(3), 585–593.
117. B. K. Atwood, D. Lee, A. Straiker, T. S. Widlanski and K. Mackie, CP47,497-C8 and JWH073, commonly found in 'Spice' herbal blends, are potent and efficacious CB(1) cannabinoid receptor agonists, *Eur. J. Pharmacol.*, 2011, **659**, 139–145.

118. V. Auwärter, S. Dresen, W. Weinmann, M. Müller, M. Pütz and N. Ferreirós, 'Spice' and other herbal blends: harmless incense or cannabinoid designer drugs?, *J. Mass Spectrom.*, 2009, **44**(5), 832–837.

119. K. G. Shanks, T. Dahn, G. Behonick and A. Terrell, Analysis of first and second generation legal highs for synthetic cannabinoids and synthetic stimulants by ultra-performance liquid chromatography and time of flight mass spectrometry, *J. Anal. Toxicol.*, 2012, **36**(6), 360–371.

120. K. G. Shanks, G. S. Behonick, T. Dahn and A. Terrell, Identification of novel third-generation synthetic cannabinoids in products by ultra-performance liquid chromatography and time-of-flight mass spectrometry, *J. Anal. Toxicol.*, 2013, **37**(8), 517–525.

121. R. Kikura-Hanajiri, N. U. Kawamura and Y. Goda, Changes in the prevalence of new psychoactive substances before and after the introduction of the generic scheduling of synthetic cannabinoids in Japan, *Drug Test. Anal.*, 2014, **6**(7–8), 832–839.

122. N. Uchiyama, M. Kawamura, R. Kikura-Hanajiri and Y. Goda, URB-754: a new class of designer drug and 12 synthetic cannabinoids detected in illegal products, *Forensic Sci. Int.*, 2013, **227**(1-3), 21–32.

123. T. Takayama, M. Suzuki, K. Todoroki, K. Inoue, J. Z. Min, R. Kikura-Hanajiri, Y. Goda and T. Toyo'oka, UPLC/ESI-MS/MS-based determination of metabolism of several new illicit drugs, ADB-FUBINACA, AB-FUBINACA, AB-PINACA, QUPIC, 5F-QUPIC and α-PVT, by human liver microsome, *Biomed. Chromatogr.*, 2014, **28**(6), 831–838.

124. J. van Amsterdam, T. Brunt and W. van den Brink, The adverse health effects of synthetic cannabinoids with emphasis on psychosis-like effects, *J. Psychopharmacol.*, 2015, **29**(3), 254–263.

125. D. Papanti, F. Schifano, G. Botteon, F. Bertossi, J. Mannix, D. Vidoni, M. Impagnatiello, E. Pascolo-Fabrici and T. Bonavigo, "Spiceophrenia": a systematic overview of "spice"-related psychopathological issues and a case report, *Hum. Psychopharmacol.*, 2013, **28**(4), 379–389.

126. S. Every-Palmer, Synthetic cannabinoid JWH-018 and psychosis: an explorative study, *Drug Alcohol Depend.*, 2011, **117**(2–3), 152–157.

127. S. Every-Palmer, Warning: legal synthetic cannabinoid-receptor agonists such as JWH-018 may precipitate psychosis in vulnerable individuals, *Addiction*, 2010, **105**(10), 1859–1860.

128. S. Peglow, J. Buchner and G. Briscoe, Synthetic cannabinoid induced psychosis in a previously nonpsychotic patient, *Am. J. Addict.*, 2012, **21**(3), 287–288.

129. A. L. Patton, K. C. Chimalakonda, C. L. Moran, K. R. McCain, A. Radominska-Pandya, L. P. James, C. Kokes and J. H. Moran, K2 toxicity: fatal case of psychiatric complications following AM2201 exposure, *J. Forensic Sci.*, 2013, **58**(6), 1676–1680.

130. E. W. Gunderson, H. M. Haughey, N. Ait-Daoud, A. S. Joshi and C. L. Hart, "Spice" and "K2" herbal highs: a case series and systematic review of the clinical effects and biopsychosocial implications of synthetic cannabinoid use in humans, *Am. J. Addict.*, 2012, **21**(4), 320–326.

131. M. Hermanns-Clausen, S. Kneisel, B. Szabo and V. Auwärter, Acute toxicity due to the confirmed consumption of synthetic cannabinoids: clinical and laboratory findings, *Addiction*, 2013, **108**(3), 534–544.

132. D. Durand, L. L. Delgado, D. M. de la Parra-Pellot and D. Nichols-Vinueza, Psychosis and severe rhabdomyolysis associated with synthetic cannabinoid use: A case report, *Clin. Schizophr. Relat. Psychoses*, 2015, **8**(4), 205–208.

133. G. L. Buser, R. R. Gerona, B. Z. Horowitz, K. P. Vian, M. L. Troxell, R. G. Hendrickson, D. C. Houghton, D. Rozansky, S. W. Su and R. F. Leman, Acute kidney injury associated with smoking synthetic cannabinoid, *Clin. Toxicol.*, 2014, **52**(7), 664–673.

134. M. Takematsu, R. S. Hoffman, L. S. Nelson, J. M. Schechter, J. H. Moran and S. W. Wiener, A case of acute cerebral ischemia following inhalation of a synthetic cannabinoid, *Clin. Toxicol.*, 2014, **52**(9), 973–975.

135. M. J. Freeman, D. Z. Rose, M. A. Myers, C. L. Gooch, A. C. Bozeman and W. S. Burgin, Ischemic stroke after use of the synthetic marijuana "spice", *Neurology*, 2013, **81**(24), 2090–2093.

136. G. Behonick, K. G. Shanks, D. J. Firchau, G. Mathur, C. F. Lynch, M. Nashelsky, D. J. Jaskierny and C. Meroueh, Four postmortem case reports with quantitative detection of the synthetic cannabinoid, 5F-PB-22, *J. Anal. Toxicol.*, 2014, **38**(8), 559–562.

137. D. Gerostamoulos, O. H. Drummer and N. W. Woodford, Deaths linked to synthetic cannabinoids, *Forensic Sci. Med. Pathol.*, 2015, **11**(3), 478.

138. R. Kronstrand, M. Roman, M. Andersson and A. Eklund, Toxicological findings of synthetic cannabinoids in recreational users, *J. Anal. Toxicol.*, 2013, **37**(8), 534–541.

139. K. G. Shanks, T. Dahn and A. R. Terrell, Detection of JWH-018 and JWH-073 by UPLC-MS-MS in postmortem whole blood casework, *J. Anal. Toxicol.*, 2012, **36**(3), 145–152.
140. T. Saito, A. Namera, N. Miura, S. Ohta, S. Miyazaki, M. Osawa and S. Inokuchi, A fatal case of MAM-2201 poisoning, *Forensic Toxicol.*, 2013, **31**, 333–357.
141. R. Kronstrand, G. Thelander, D. Lindstedt, M. Roman and F. C. Kugelberg, Fatal intoxications associated with the designer opioid AH-7921, *J. Anal. Toxicol.*, 2014, **38**(8), 599–604.
142. R. Karinen, S. S. Tuv, S. Rogde, M. D. Peres, U. Johansen, J. Frost, V. Vindenes and Å. M. Øiestad, Lethal poisonings with AH-7921 in combination with other substances, *Forensic Sci. Int.*, 2014, **244**, e21–e24.
143. S. P. Vorce, J. L. Knittel, J. M. Holler, J. Magluilo Jr, B. Levine, P. Berran and T. Z. Bosy, A fatality involving AH-7921, *J. Anal. Toxicol.*, 2014, **38**(4), 226–230.
144. A. Helander, M. Bäckberg and O. Beck, MT-45, a new psychoactive substance associated with hearing loss and unconsciousness, *Clin. Toxicol.*, 2014, **52**(8), 901–904.
145. G. McLaughlin, N. Morris, P. V. Kavanagh, J. D. Power, B. Twamley, J. O'Brien, B. Talbot, G. Dowling, O. Mahony, S. D. Brandt, J. Patrick, R. P. Archer, J. S. Partilla and M. H. Baumann, Synthesis, characterization, and monoamine transporter activity of the new psychoactive substance 3′,4′-methylenedioxy-4-methylaminorex (MDMAR), *Drug Test. Anal.*, 2014, DOI: 10.1002/dta.1732. [Epub ahead of print]
146. S. D. Brandt, S. Freeman, H. R. Sumnall, F. Measham and Cole, Analysis of NRG 'legal highs' in the UK: identification and formation of novel cathinones, *J. Drug Test. Anal.*, 2011, **3**(9), 569–575.
147. J. Wallach, P. V. Kavanagh, G. McLaughlin, N. Morris, J. D. Power, S. P. Elliott, M. S. Mercier, D. Lodge, H. Morris, N. M. Dempster and S. D. Brandt, Preparation and characterization of the 'research chemical' diphenidine, its pyrrolidine analogue, and their 2,2-diphenylethyl isomers, *Drug Test. Anal.*, 2014, DOI: 10.1002/dta.1689. [Epub ahead of print]
148. S. D. Brandt, M. H. Baumann, J. S. Partilla, P. V. Kavanagh, J. D. Power, B. Talbot, B. Twamley, O. Mahony, J. O'Brien, S. P. Elliott, R. P. Archer, J. Patrick, K. Singh, N. M. Dempster and S. H. Cosbey, Characterization of a novel and potentially lethal designer drug (±)-cis-para-methyl-4-methylaminorex (4,4′-DMAR, or 'Serotoni'), *Drug Test. Anal.*, 2014, **6**(7–8), 684–695.

7 Novel Psychoactives in New Zealand

Samantha J. Coward* and Hilary J. Hamnett

Institute of Environmental Science and Research Limited, Porirua,
New Zealand
*Email: samantha.coward@esr.cri.nz

7.1 Introduction: The New Zealand Drug Scene

Nearly one in two adults (49%) in New Zealand (NZ) have used drugs
(excluding alcohol, tobacco and party pills) for recreational purposes
in their lifetime. The most recent data, from a survey carried out in NZ
between 2007 and 2008, reports that cannabis is the most prevalent
drug used, followed by legal or illegal party pills and hallucinogens.[1]
These modern findings are mainly due to events in the history of the
drug trade in NZ (Figure 7.1). Following the collapse of the "Mr Asia"
drug syndicate in 1979, who were responsible for the importation of
cannabis products and heroin into NZ, the availability of heroin
declined rapidly. Cannabis plants and products were locally produced.
An increase in the use of psychotropic medicines (by people known to
be heroin users) was noted by police. Users of heroin also had to make
their own drug. As the 1990s progressed, the demand for stimulating
drugs increased, resulting in locally produced methamphetamine.[2]

Drug abuse in NZ is heavily influenced by the fact that it is geographically isolated.[3] As a result, drugs of abuse that are common in
other Western countries, such as cocaine and amphetamine, are rarely
encountered in NZ. According to the most recent Illicit Drug Monitoring

Forensic Toxicology: Drug Use and Misuse
Edited by Susannah Davies, Atholl Johnston and David Holt
© The Royal Society of Chemistry 2016
Published by the Royal Society of Chemistry, www.rsc.org

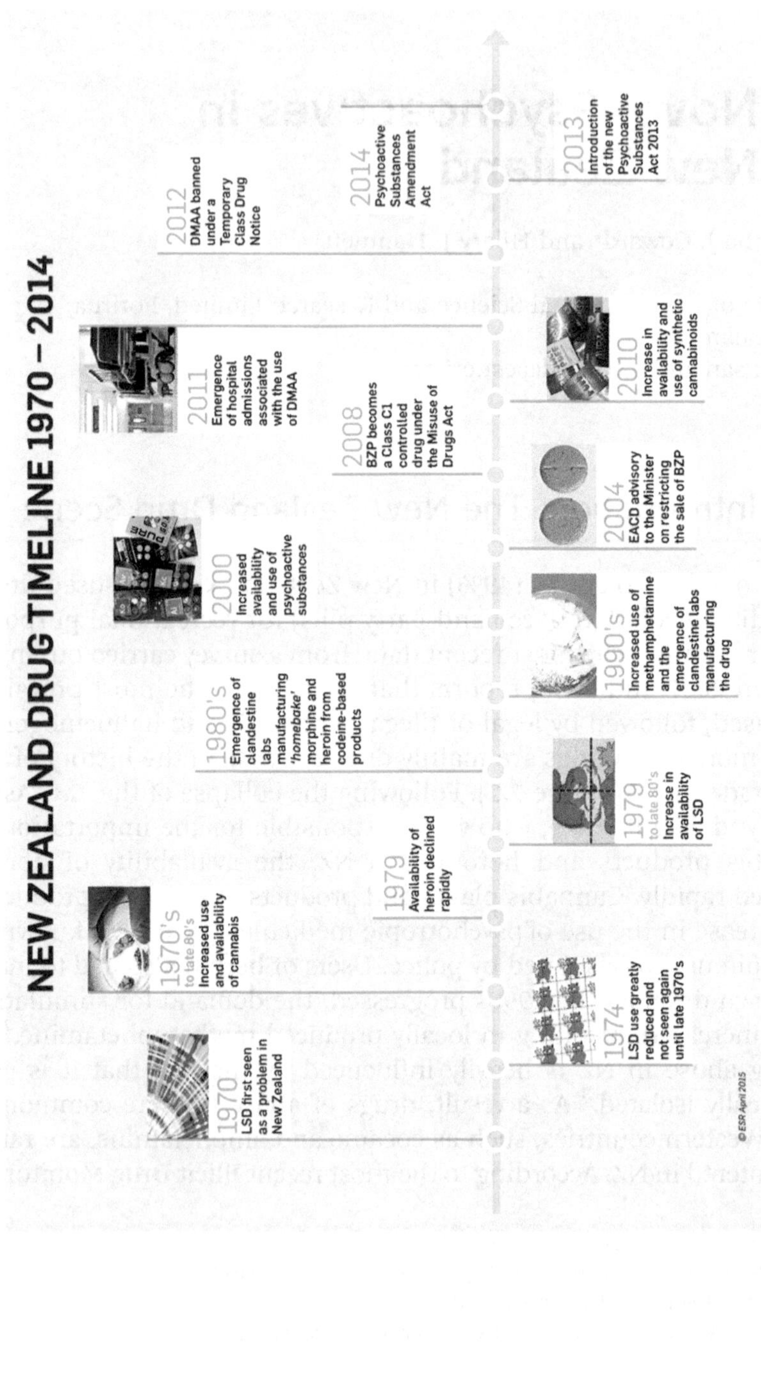

Figure 7.1 Timeline of events in New Zealand's drug scene history.

Survey (IDMS), cocaine and heroin markets continue to be small in NZ.[4] The heroin that is encountered in NZ is usually called "homebake" heroin, produced on a small scale from pharmaceutical morphine and codeine tablets.[3] Methamphetamine is largely produced in NZ from ephedrine and pseudoephedrine in clandestine laboratories.[5]

New Zealand also has a problem with volatile solvent abuse, known as "huffing". The inhalation of adhesives, solvents, fuels and propellant or flammable gases is a recognised cause of sudden death, particularly in young people, with one study reporting the majority of users being between 11 and 20 years of age.[6]

In this chapter we will review the main psychoactives available in NZ and discuss their involvement in forensic toxicology casework.

7.1.1 Novel Psychoactives

The introduction of novel psychoactives to NZ began in the year 1999 with legal stimulants being sold as party pills (Section 7.2). Another psychoactive drug, known as kava (Section 7.3), which is used mainly for ceremonial purposes by Pacific communities in NZ, was first encountered in forensic toxicology in 2003.

In more recent years, the NZ drug scene has witnessed the introduction of cathinones (Section 7.4), synthetic cannabinoids (also see Chapter 8) and NBOMe drugs (Section 7.5). In 2013, the new drugs most commonly mentioned in the IDMS were synthetic hallucinogens (NBOMe and 2C drugs), 3,4-methylenedioxymethamphetamine (MDMA) powder and oxycodone.[4] Other common psychoactives such as phenazepam,[7] methoxetamine[8] and benzofury (6-APB)[9] have been detected in seizures in NZ, but not yet in forensic toxicology.[10-12] The new opioid AH-9271[13] has not yet been encountered in NZ.

In NZ, drugs are controlled by the Misuse of Drugs Act 1975. Novel psychoactives, especially the synthetic cannabinoids, were not controlled by this Act. As concern grew with the availability of psychoactive substances in dairies (corner stores), the NZ government introduced the concept of a Temporary Class notice. This notice prohibited the sale of specified substances. It was meant to last for only one year, but was extended to two years while the Psychoactive Substances Act was put into place.

In 2013, the NZ government took a unique approach to dealing with new psychoactives, with the introduction of the Psychoactive Substances Act 2013 (see Chapter 17).[14] The purpose of this Act was to regulate the availability of novel psychoactive substances in NZ not controlled by other legislation. During the establishment phase of the

Act, all researchers, manufacturers, retailers and importers had to apply for interim licenses, and some substances were given interim approval. In 2014 the Psychoactive Substances Amendment Bill[15] was introduced and all interim licenses and approvals were revoked. The amendment stated that the psychoactive substances could not be sold legally without proof that they posed minimal risk to users. It was at this time that campaigning New Zealanders forced the Government to ensure that the safety and quality of these products would not be demonstrated by animal trials.

As of January 2015, the most recent drug seizures at the NZ border included analogues of methcathinone, phencyclidine (4-methoxy-phencyclidine, 4-MEO-PCP) and α-pyrrolidinovalerophenone (α-PVP). Nootropic substances are also increasingly being seen. A nootropic or memory enhancer is a drug used to treat cognitive/neurological disorders such as Alzheimer's disease.[16] These products are aimed not at recreational drug users but at the growing number of people looking to enhance their body and mind. Nootropics are sold as supplements with claims that they will increase energy, focus and memory.[17]

7.1.2   Forensic Toxicology in New Zealand

In NZ, the Institute of Environmental Science and Research (ESR) provides independent analytical and consulting services, including illicit drugs analysis and forensic toxicology. The forensic toxicology laboratory, based in Wellington, receives body fluid samples from deaths referred to the coroner, criminal cases (*e.g.*, homicide, drug-facilitated sexual assault) and drivers suspected of being impaired by drugs. The ESR drugs laboratory is based in Auckland and identifies the drugs seized by police and customs officials. Most new psychoactives in forensic toxicology cases are detected at ESR by a screen involving liquid chromatography with time-of-flight mass spectrometric detection (LC-TOFMS). This technique was introduced in December 2012, and almost all forensic toxicology cases received at ESR ($\sim$2200 per year) undergo this screen.

Figure 7.2 shows the forensic toxicology cases analysed at ESR since 2002 involving novel psychoactives. Of note, is the low number of 1-benzylpiperazine (BZP) cases between 2010 and 2013. This gap probably reflects a decline in the use of BZP following its control in 2008 (see Section 7.2 for more details). During the period 2010–2013, either former BZP users refrained from using other psychoactives, or chose substances that were not detected by ESR's forensic toxicology screening methods.

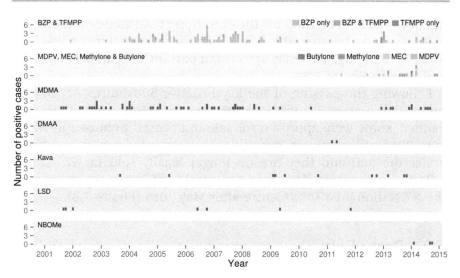

Figure 7.2 Forensic toxicology cases involving novel psychoactives in NZ from 2001 to 2015 (see text for the definitions of acronyms).

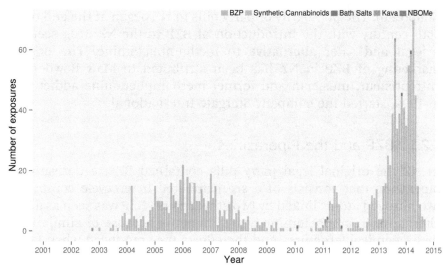

Figure 7.3 Adverse reactions due to exposures to novel psychoactives recorded by the NZ National Poisons Centre from 2001 to 2015.

Figure 7.3 shows the number of adverse reactions due to exposures to psychoactives recorded by the NZ National Poisons Centre, based in Dunedin, during the same period. It can be seen that between 2011 and 2014, synthetic cannabinoid use was a common reason to contact the NZ National Poisons Centre. ESR's forensic toxicology laboratory did not validate a method for the detection of synthetic cannabinoids in body fluids because the type of synthetic cannabinoid was continually changing. This may account in part for the gap between 2010 and 2013 in Figure 7.2.

Following the passing of the Psychoactive Substances Act 2013 in July of that year, whilst most synthetic cannabinoid compounds were banned, some were approved for sale in licensed premises in NZ. In May 2014, all synthetic cannabinoid compounds were controlled under the Act, and they are no longer legally sold in NZ. This is reflected in the steep decline in the number of exposures recorded by the NZ National Poisons Centre after May 2014 (Figure 7.3).

7.2 Party Pills

Party pills, also known as "legal highs", "herbal highs" and "dance pills", are pills sold for recreational purposes and taken for the induction of euphoria and an increased sense of energy (also see Chapter 6). The popularity of party pills in NZ began at the end of the 20th century with the introduction of BZP to the NZ drug scene as a legal and safer alternative to methamphetamine. The original marketing of BZP in NZ has been attributed to Matt Bowden, an entrepreneur, musician and former methamphetamine addict who, in 1999, started the company Stargate International.

7.2.1 BZP and the Piperazines

In NZ the original legal party pills contained BZP, a derivative of piperazine that consists of a six-membered heterocycle comprising two nitrogen atoms linked by two ethyl chains. BZP was seen as a legal alternative to amphetamine-type stimulants because of similarity in their chemical structures and the effects they produced when taken. Although BZP mimics the effects of amphetamine, it is reported to have only 10% of the potency of the drug.[18]

As the use of BZP increased, it was combined in a pill form with 1-(3-trifluoromethylphenyl)piperazine (TFMPP), which caused effects similar to those seen with Ecstasy use.[19,20] TFMPP alone has little or no psychoactive effect.

BZP is a synthetic sympathomimetic compound normally manufactured from piperazine, a compound used as an anthelmintic agent (a medicine that rids the body of parasitic worms).[21] BZP was evaluated as an antidepressant in the 1970s by the Wellcome Research

Laboratories in the UK, after animal trials suggested the drug had antidepressant activity. However, it caused euphoric and CNS stimulatory effects and had addictive properties, similar to those seen after amphetamine use, preventing the clinical use of BZP in humans.[21,22] During the period it was legally available, the use of BZP in NZ far exceeded that of any other illicit drug, except cannabis.[20]

In 2004, the Expert Advisory Committee on Drugs (EACD) reported to the NZ Government that although there was insufficient evidence at that time to ban the use of BZP, they recommended that it be re-classified and made available only to persons over the age of 18 years. The committee also recommended that there be a restriction on advertising, packaging and labelling, and that the packaging displayed any health warnings. Further research studies into the safety or harm of these drugs to humans were commissioned at this time.

In 2005, Gee *et al.*[23] reported results from a study they had carried out on patients admitted to the Christchurch Emergency Department (ED) after BZP party pill use. This study highlighted the serious potential side-effects of BZP, with 15 of the 80 party-pill-related admissions involving non-epileptic seizures. Those patients with seizures were reported to have taken no more tablets than non-seizing patients. Prior to this study, users had reported a series of adverse effects, including palpitations, vomiting, anxiety, insomnia, confusion, tremors[24] and aggression (see Case Study 7.1 below).[25]

In November 2006, the EACD reconsidered its previous position on BZP and related substances after receiving further advice and the results of additional research carried out since 2004.[26] The committee made the following recommendations:

1. BZP be classified under Schedule 3 Part 1 (Class C1) of the Misuse of Drugs Act 1975.[27]
2. The classification as a Class C1 drug should cover all known analogues and derivatives of BZP that have no known therapeutic use.

Case Study 7.1   BZP.

The NZ Herald reported a case involving BZP in November 2011. It involved six schoolgirls, the youngest being 13 years old, who took a number of pink pills that one of the girls had stolen from her mother's handbag. The girls all ended up in the ED at Waikato Hospital. The girl had told her friends that they were Ecstasy. The girls presented out-of-character aggressive behaviour in the ED and BZP was detected.

3. BZP be removed from Schedule 4 of the Misuse of Drugs Amendment Act 2005,[28] in order that it no longer be a restricted substance.

4. Work should continue to further develop the regulatory framework and enforcement capacity that would support the restricted substances provisions of the Misuse of Drugs Amendment Act 2005.

In April 2008 the NZ Government placed BZP and five related substances in the Misuse of Drugs Act 1975 and classified them as Class C drugs, making it illegal to manufacture, sell or use the drugs. The related substances were TFMPP, *meta*-chlorophenylpiperazine (*m*CPP), *para*-fluorophenylpiperazine (*p*FPP), *para*-methoxyphenyl-piperazine (MeOPP) and 1-benzyl-4-methylpiperazine (MBZP) (see Figure 7.4). The isomers of these substances were also listed.

Marketers of these party pills claimed to be concerned that prohibition of these drugs would lead to an increase in more dangerous illicit drug use. A study carried out in 2009 surveyed drug users to see whether the ban had changed their behaviour in relation to drug

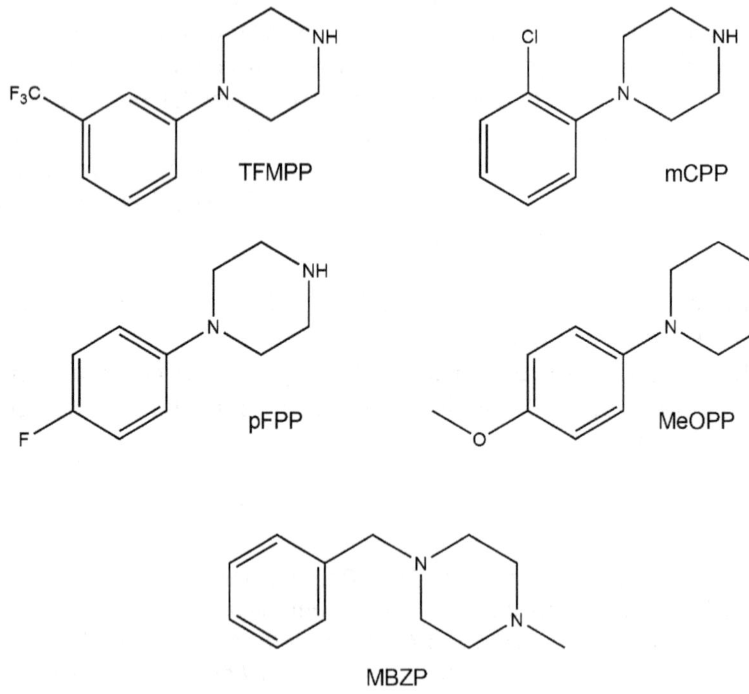

Figure 7.4 Structures of five piperazines (see text for the definitions of acronyms).

use.[29] The study found that fewer people were using BZP-type drugs, and that previous users had not moved onto more dangerous drugs like methamphetamine. The decline in BZP use was attributed to its lack of availability and the change in its legal status. This is also reflected in the drop-off in the number of exposures to these party pills recorded by the NZ National Poisons Centre after 2008 (Figure 7.3).

Prior to the introduction of the LC-TOFMS at ESR, BZP and related compounds were detected by screening with gas chromatography with mass spectrometric and nitrogen/phosphorus detection (GC-MS/NPD). Levels of BZP are reported following quantitation by liquid chromatography with tandem mass spectrometric detection (LC-MS/MS). Since 2002 there have been 102 forensic toxicology cases positive for BZP, with the first case appearing in that year. Figure 7.2 shows that most forensic toxicology cases involving BZP and/or TFMPP were reported between 2004 and 2009 (peaking in 2006), although these drugs continue to be observed in forensic toxicology casework in 2015 due to their availability in illegally produced pills. Figure 7.3 shows that most exposures to BZP were also recorded between 2004 and 2009 (also peaking in 2006).

The popularity of BZP stemmed from it being a legal alternative to the illegal dance party drug MDMA.[30] Figure 7.2 shows that the number of forensic toxicology cases involving MDMA in NZ has remained relatively stable up to 2008.

7.2.2 DMAA

Once BZP was banned, party pill manufacturers had to find an alternative ingredient that caused the desired effects and was not restricted for use. Initially that alternative ingredient was DMAA. DMAA is also known as methylhexanamine, 1,3-dimethylamylamine or geranamine and was patented in 1944 as a nasal decongestant.[31] It first appeared on the NZ market as a dietary supplement.[24] The stimulant effect of the drug has been likened to that seen after high doses of caffeine, and for this purpose it was an ingredient of training supplements used in the body-building community. The NZ Drug Foundation reported additional effects from DMAA use on its website, including nausea, vomiting, aggression, panic attacks and paranoia.[24]

Originally, DMAA was primarily found in powder form, which was then added to liquid and consumed as a drink by the user, although it was also available in the form of a capsule. Some users have also been known to snort or smoke the powder.[24]

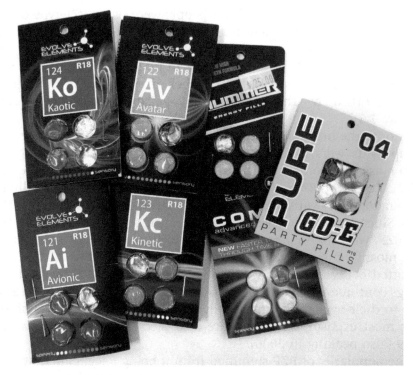

Figure 7.5   A range of party pills previously available in NZ.

Party pills containing DMAA became readily available in stores throughout NZ. They were marketed as "herbal highs", with the claim that all the contents were derived from natural products. DMAA itself is found in geranium oil. The pills were sold in bright attractive packaging (see Figure 7.5) and could be sold to anyone, at any dose. The pills themselves came in a variety of colours.

DMAA was banned under a Temporary Class Drug Notice in 2012, and since the introduction of the Psychoactive Substances Amendment Bill in 2014, manufacturers will have to prove the drug is safe and get it approved before it can be sold again legally.

DMAA was in use for four years in NZ prior to its ban. During this time there was a noticeable increase in the doses sold, from 50 mg pills to 5 g packets of powder. In the 18 months leading up to the ban, ED specialists reported at least five cases where DMAA was considered the cause of a cerebral haemorrhage (see Case Study 7.2).[32]

Figure 7.2 shows that there have been two forensic toxicology cases involving DMAA (both in 2011). These were an alleged drug-facilitated sexual assault, and a sudden death due to a brain bleed. Other drugs

---

Case Study 7.2 DMAA.

A 23-year-old woman took the recommended two-tablet dose of DMAA-containing Pure X-S® *party pills after ingesting a quantity of alcohol. Within 30 minutes she was vomiting, twitching and a brain scan showed bleeding across the frontal lobe. A plasma concentration of 1.09 mg L*$^{-1}$ *of DMAA was reported 100 minutes post-ingestion. After three days she showed improvement and was discharged.*

---

were detected in both cases. ESR analyses for DMAA using GC-MS/NPD in cases where its use is suspected, but levels are not reported. Recently, a related compound, 1,3-dimethylbutylamine (DMBA),[33] has appeared in pre-workout supplements, such as Frenzy, sold in NZ.

According to the most recent IDMS, legal party pills are now considered less popular than synthetic cannabinoids, with only 10% of frequent drug users having used them in the previous six months. The low strength of these products, most containing little more than caffeine, might explain their current low popularity.[4]

7.3 Kava

The *Piper methysticum* plant is a shrub that is native to a number of Pacific Islands, where it is also known as *yagona*. The drug known as kava is an aqueous preparation made from the peeled rhizome (rootstalk) of the plant, although over-the-counter preparations in the form of capsules are also available in NZ.[34] Kava is not controlled in NZ, and is used for medicinal purposes and during religious and cultural ceremonies for its ability to elicit an altered state of consciousness and quell anxiety.[34]

According to the NZ Ministry of Health, kava is the fifth most commonly used drug in NZ, with 6.3% of adults having used it in their lifetime.[1] To date, 18 different active ingredients, known as kavalactones, have been identified, with 96% of the pharmacological activity attributed to the presence of only six.[34] ESR analyses for the presence of one kavalactone, known as kavain (Figure 7.6) using GC-MS/NPD. Levels are not reported. Generally, cases are analysed for kava use when requested, or if its use is mentioned in the circumstances of the case.

Kava has a central nervous system (CNS) depressant action. It may also impair driving skills and reaction times, although a clear picture of its effect on driving has not yet emerged from the literature. Kava is often consumed concomitantly with alcohol.[34]

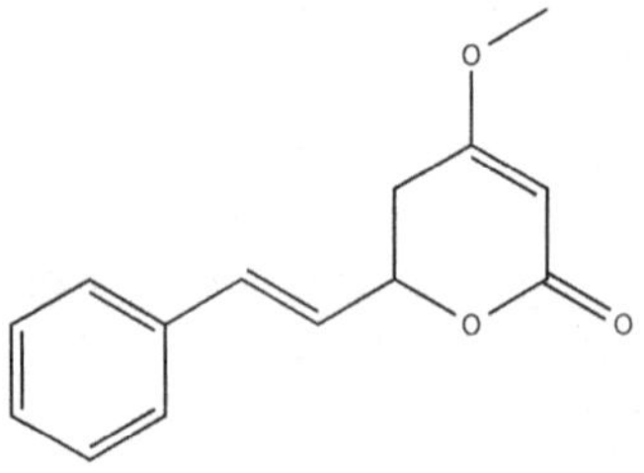

Figure 7.6   Chemical structure of kavain.

Since 2002 there have been 11 forensic toxicology cases positive for kavain, with the first appearing in 2003. These have included three sudden death cases where no other drugs (apart from alcohol) were detected, two fatal motor vehicle crashes (MVCs) and four cases of driving under the influence of drugs (DUID). Figure 7.3 shows that the NZ National Poisons Centre record very few exposures related to kava use.

7.4   Cathinones

Cathinones are analogues of cathinone, one of the psychoactive compounds that occurs naturally in the *Catha edulis* plant.[35] This plant is also known as khat (see Chapter 9). The khat plant is a Class C controlled substance, and cathinone and methcathinone are Class B controlled drugs under the Misuse of Drugs Act 1975, methcathinone having been added in 2003. Analogues of these compounds, such as mephedrone (see Chapter 6), are not listed separately in the Act, but are controlled under analogue legislation as Class C drugs. Cathinones are marketed as "bath salts" and "research chemicals" and have a stimulant effect on the CNS. Their effects include agitation, nausea and vomiting, hypertension and tachycardia.[36]

In NZ, cathinone analogues are sold as illicit party pills. From around 2008, with a decrease in the availability of MDMA, the pills marketed as Ecstasy now contained a range of compounds, including the cathinone analogues. Visually identical to MDMA tablets, the buyer did not know what was in the pills they were buying (see Figure 7.7 for an example).

Figure 7.2 shows the number of forensic toxicology cases involving cathinones since 2002, with the first case appearing in 2011. Prior to December 2012, the forensic toxicology laboratory of ESR did not screen routinely for cathinones. Since then, almost all forensic toxicology cases have been screened using LC-TOFMS, which is validated for eight cathinones. Prior to this, individual cathinones,

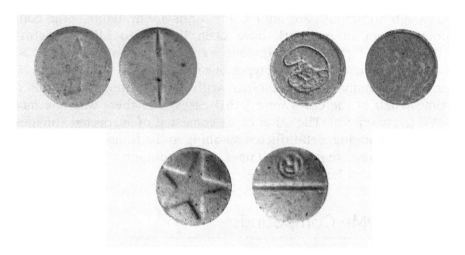

Figure 7.7 Party pills seized in NZ, shown without their packaging and illustrating how similar to MDMA tablets they appear.

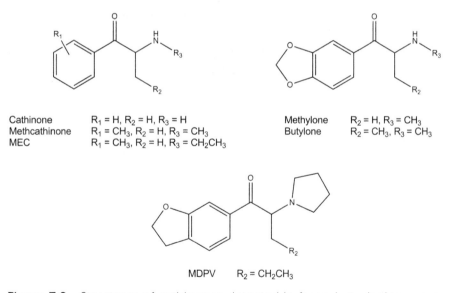

Cathinone	R_1 = H, R_2 = H, R_3 = H
Methcathinone	R_1 = CH$_3$, R_2 = H, R_3 = CH$_3$
MEC	R_1 = CH$_3$, R_2 = H, R_3 = CH$_2$CH$_3$

Methylone	R_2 = H, R_3 = CH$_3$
Butylone	R_2 = CH$_3$, R_3 = CH$_3$

MDPV R_2 = CH$_2$CH$_3$

Figure 7.8 Structures of cathinones detected in forensic toxicology case-work in NZ.

e.g. 3,4-methylenedioxy-α-pyrrolidinovalerophenone (MDPV, Figure 7.8) were targeted by LC-MS/MS in cases where their use was suspected.

As can be seen from Figure 7.2, cathinones do not have a large presence in NZ. However, in the most recent IDMS, an increasing proportion of frequent drug users reported using methylone and mephedrone for the first time in 2013.[4] Since 2002, 15 forensic toxicology cases involving cathinones have been reported by ESR.

Although numerous cathinone compounds are available,[37] the compounds seen in casework have been limited to MDPV, methylethcathinone (MEC), methylone and butylone (Figure 7.8). Concomitant use of different types of cathinones are often seen in NZ and occasionally use is combined with methamphetamine or BZP. Almost half of the cases were DUID cases, and there was one fatal MVC (a passenger). The other cases consisted of suspected suicides, including hanging, self-inflicted stabbing and a jump from a bridge. Figure 7.3 shows that bath salt use is not a common reason to contact the NZ National Poisons Centre.

7.5   NBOMe Compounds

NBOMe compounds, known as "synthetic" or "legal" LSD, are analogues of the 2C series of psychedelic phenethylamine drugs.[38] The terminology "2C" refers to an acronym created by Alexander Shulgin to describe the two carbons between the amino group and the benzene ring in the chemical structure.[39] The NBOMe analogues contain an *N*-methoxybenzyl group (the origin of the name NBOMe), which imparts significant pharmacological activity.[40] NBOMe compounds are typically sold as sheets of blotter paper that have been soaked in a solution and dried (analogous to LSD). Each dose is a small section of paper (6–8 mm^2) known as a ticket, tab or blotter that is administered sublingually. NBOMe compounds are also available in the form of liquids, pills, powders and capsules.[41–43] According to the IDMS, the use of LSD rose sharply in 2013, where previously it had been declining for a number of years (see Figure 7.2). This is almost certainly due to psychedelics such as NBOMe compounds being mis-sold as "LSD" in identical blotter tab formats.[4] There are numerous NBOMe compounds available,[44] but so far only three have been detected in forensic toxicology casework in NZ (Figure 7.9).

NBOMe compounds are not screened for as a matter of routine during forensic toxicology at ESR, but are targeted by LC-MS/MS when suspected. The effects of NBOMe include hallucinations, euphoria, mydriasis, agitation, tachycardia and hypertension,[39] and there are several reports in the literature of acute toxicity and sudden death associated with their use.[39,43,45–50]

Since February 2014, there have been three NBOMe-positive forensic toxicology cases at ESR. These were a death in custody (see Case Study 7.3),[51] a drowning and a DUID case. Other drugs were detected alongside the NBOMe compounds in all three cases.

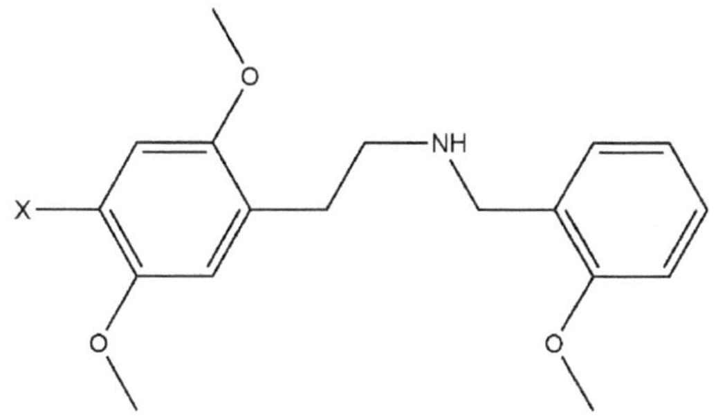

25I-NBOMe   X = I
25C-NBOMe   X = Cl
25B-NBOMe   X = Br

Figure 7.9 NBOMe compounds detected in forensic toxicology casework in NZ.

Case Study 7.3 NBOMe.

A 20-year-old man died in custody following his arrest for breach of the peace. Earlier in the evening he had become confused, aggressive and violent after suffering hallucinations. CCTV footage showed him falling and hitting his head on the concrete walls or floor of his police cell 83 times. Toxicology tests showed that he had consumed 25B-NBOMe, alcohol, cannabis and methamphetamine.

By the end of 2014, the drugs team at ESR had received seizures of over 15 000 tabs of NBOMe (see Figure 7.10 for an example), including 25H-NBOMe.[52] NBOMe compounds are currently unapproved under the Psychoactive Substances Act 2013. Figure 7.3 shows that NBOMe compounds are not yet a common reason to contact the NZ National Poisons Centre.

7.6 Conclusions

In this chapter we have reviewed some of the psychoactive substances available in NZ, including party pills, cathinones, kava and NBOMe compounds. Data from ESR show that BZP has been the main psychoactive involved in forensic toxicology casework in NZ, with most cases reported between 2004 and 2009 (peaking in 2006). Data from the NZ National Poisons Centre mirror this trend. Data from the NZ National Poisons Centre also show that synthetic cannabinoids had a large presence in NZ (with calls due to adverse reactions to exposures peaking in 2013–2014) until they were controlled in May

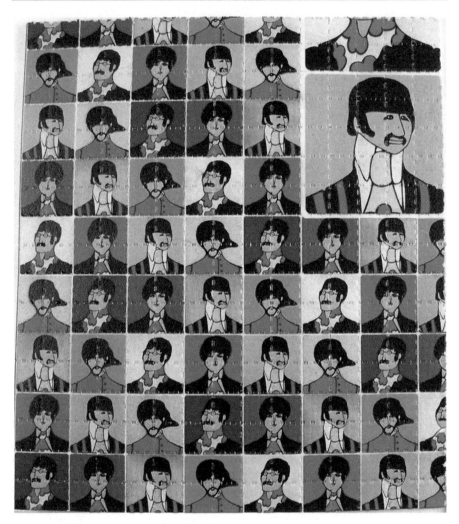

Figure 7.10   25C- and 25B-NBOMe tabs seized in NZ.

2014. The involvement of synthetic cannabinoids in forensic toxicology casework has not been recorded in NZ.

Other novel psychoactives, such as cathinones and kava, have had minimal impact. The most recent psychoactives to emerge in NZ are the NBOMe hallucinogens, with the first exposures and forensic toxicology cases being reported in late 2014. Some of the new psychoactives that are popular in Europe, *e.g.* methoxetamine, have not yet been encountered in forensic toxicology in NZ.

Analogues of methamphetamine, such as 2-fluoromethamphetamine and 3-fluoromethamphetamine, are becoming increasingly common in

New Zealand, and drug seizures of pills sold as "Ecstasy" containing multiple new psychoactives are also now being observed. However, strict biosecurity regulations in New Zealand are likely to limit the influence of Europe and the USA on future trends in new psychoactives.

Acknowledgements

The authors would like to thank Lucy Shieffelbien from the NZ National Poisons Centre for providing the exposure data and the ESR Drugs Group for seizure information. Figures 7.2 and 7.3 were prepared using R by Duncan Garmonsway of the New Zealand Police. The authors would also like to thank the following people from ESR: Dr Robyn Somerville, Vivienne Hassan and Kelly de Wit for the photo of the NBOMe tabs, Jenny Ralston for Figures 7.1, 7.5 and 7.7, and Dr Helen Poulsen and Dr Wendy Popplewell for reviewing the chapter.

References

1. K. Mason, A. Hewitt, N. Stefanogiannis, A. Bhattacharya, L. C. Yeh and M. Devlin, *Drug Use in New Zealand: Key Results of the 2007/08 New Zealand Alcohol and Drug Use Survey*, Ministry of Health, Wellington, 2010.
2. New Zealand Customs Service, *Contraband*, Wellington, 2012.
3. K. R. Bedford, S. L. Nolan, R. Onrust and J. D. Siegers, *Forensic Sci. Int.*, 1987, **34**, 197.
4. C. Wilkins, J. Prasad, K. Wong and M. Rychert, *Recent Trends in Illegal Drug Use in New Zealand, 2006-2013*, Massey University, Auckland, 2014.
5. Policy Advisory Group, *Tackling Methamphetamine: Progress Report (October 2014)*, Department of the Prime Minister and Cabinet, Wellington, 2014.
6. M. Beasley, L. Frampton and J. Fountain, *J. N. Z. Med. Assoc.*, 2006, **119**, 55.
7. P. D. Maskell, G. De Paoli, L. Nitin Seetohul and D. J. Pounder, *J. Forensic Leg. Med.*, 2012, **19**, 122.
8. J. Ward, S. Rhee and J. Plansky, *Clin. Toxicol.*, 2011, **49**, 874.
9. W. L. Chan, D. M. Wood, S. Hudson and P. I. Dargan, *J. Med. Toxicol.*, 2013, **9**, 278.
10. R. A. F. Couch and H. Madhavaram, *Drug Test. Anal.*, 2012, **4**, 409.

11. R. Somerville, personal communication, ESR Drugs, 2015.
12. J. Sibley, personal communication, ESR Drugs, 2015.
13. M. Katselou, I. Papoutsis, P. Nikolaou, C. Spiliopoulou and S. Athanaselis, *Forensic Toxicol.*, 2015, **33**, 195.
14. New Zealand Government, *Psychoactive Substances Act 2013*, Wellington. http://www.legislation.govt.nz/act/public/2013/0053/latest/DLM5042921.html.
15. New Zealand Government, *Psychoactive Substances Amendment Bill 2014*, Wellington. http://www.legislation.govt.nz/bill/government/2014/0206/latest/whole.html.
16. M. Russell, personal communication, ESR Drugs, 2015.
17. European Monitoring Centre for Drugs and Drug Driving, *New Psychoactive Substances in Europe*, European Union, Luxembourg, 2015.
18. Advisory Council on the Misuse of Drugs, *Control of 1-Benzylpiperazine (BZP) and Related Compounds*, ACMD, London, 2008.
19. M. H. Baumann, R. D. Clark, A. G. Budzynski, J. S. Partilla, B. E. Blough and R. B. Rothman, *Neuropsychopharmacology*, 2005, **30**, 550.
20. J. R. Kerr and L. S. Davis, *J. R. Soc. N. Z.*, 2011, **41**, 155.
21. D. G. Barceloux, *Medical Toxicology of Drugs of Abuse*, John Wiley & Sons, Hoboken, 2012.
22. H. Campbell, W. Cline, M. Evans, J. Lloyd and A. W. Peck, *Eur. J. Clin. Phamacol.*, 1973, **6**, 170.
23. P. Gee, S. Richardson, W. Woltersdorf and G. Moore, *J. N. Z. Med. Assoc.*, 2005, **118**, 35.
24. The New Zealand Drug Foundation, http://www.drugfoundation.org.nz/ (accessed 31/3/15).
25. A. Wade and J. Savage, *Ecstasy Puts Schoolgirls in Hospital*, The New Zealand Herald, 22nd November 2011.
26. Allen & Clarke, *Analysis of Submissions. Proposal to Classify BZP and Related Substances as Class C Controlled Drugs under the Misuse of Drugs Act 1975*, Allen & Clarke Policy and Regulatory Specialists Ltd, Wellington, 2007.
27. New Zealand Government, *Misuse of Drugs Act 1975*, Wellington. http://www.legislation.govt.nz/act/public/1975/0116/latest/DLM436101.html.
28. New Zealand Government, *Misuse of Drugs Amendment Act 2005*, Wellington. http://www.legislation.govt.nz/act/public/2005/0081/latest/DLM356224.html.

29. C. Wilkins and P. Sweetsur, *Drug Alcohol Depend.*, 2013, **127**, 72.
30. C. Wilkins, M. Girling and P. Sweetsur, *J. Subst. Use*, 2007, **12**, 213.
31. P. Gee, S. Jackson and J. Easton, *J. N. Z. Med. Assoc.*, 2010, **123**, 124.
32. P. Gee, C. Tallon, N. Long, G. Moore, R. Boet and S. Jackson, *Ann. Emerg. Med.*, 2012, **60**, 431.
33. P. A. Cohen, J. C. Travis and B. J. Venhuis, *Drug Test. Anal.*, 2015, **7**, 83.
34. J. Sarris, E. LaPorte and I. Schweitzer, *Aust. N. Z. J. Psychiatry*, 2011, **45**, 27.
35. A. L. Bretteville-Jensen and S. S. Tuv, *Forensic Sci. Rev.*, 2013, **25**, 7.
36. J. Jerry, G. Collins and D. Streem, *Cleveland Clin. J. Med.*, 2012, **79**, 258.
37. LGC Standards, *Know Your Cathinones (poster)*, LGC Ltd, Teddington, 2011.
38. A. Shulgin and A. Shulgin, *PIHKAL: A Chemical Love Story*, Transform Press, Berkeley, 2011.
39. S. l. Stellpflug, S. E. Kealey, C. B. Hegarty and G. C. Janis, *J. Med. Toxicol.*, 2014, **10**, 45.
40. D. G. E. Caldicott, S. J. Bright and M. J. Barratt, *Med. J. Aust.*, 2013, **199**, 322.
41. Expert Committee on Drug Dependence, *25I-NBOMe: Critical Review Report*, World Health Organisation, Geneva, 2014.
42. D. M. Wood, R. Sedefov, A. Cunningham and P. I. Dargan, *Clin. Toxicol.*, 2015, **53**, 85.
43. M. H. Y. Tang, C. K. Ching, M. S. H. Tsui, F. K. C. Chu and T. W. L. Mak, *Clin. Toxicol.*, 2014, **52**, 561.
44. J. F. Casale and P. A. Hays, *Microgram J.*, 2012, **9**, 84.
45. S. L. Hill, T. Doris, S. Gurung, S. Katebe, A. Lomas, M. Dunn, P. Blain and S. H. L. Thomas, *Clin. Toxicol.*, 2013, **51**, 487.
46. A. Kelly, B. Eisenga, B. Riley and B. Judge, Annual Meeting of the North American Congress of Clinical Toxicology (NACCT), Las Vegas, USA, 2012.
47. J. L. Poklis, K. G. Devers, E. F. Arbefeville, J. M. Pearson, E. Houston and A. Poklis, *Forensic Sci. Int.*, 2014, **234**, e14.
48. R. S. Rose, K. L. Cumpston, P. E. Stromberg and B. K. Wills, Annual Meeting of the North American Congress of Clinical Toxicology (NACCT), Las Vegas, USA, 2012.

49. S. Rutherfoord Rose, J. L. Poklis and A. Poklis, *Clin. Toxicol.*, 2013, **51**, 174.
50. J. P. Walterscheid, G. T. Phillips, A. E. Lopez, M. L. Gonsoulin, H.-H. Chen and L. A. Sanchez, *Am. J. Forensic. Med. Pathol.*, 2014, **35**, 20.
51. Independent Police Conduct Authority, *Death in Police Custody of Sentry Taitoko*, IPCA, Wellington, 2015.
52. R. Somerville, personal communication, ESR Drugs, 2014.

8 Cannabis and Synthetic Cannabinoids

Kim Wolff

King's College London, London, UK
Email: kim.wolff@kcl.ac.uk

8.1 Cannabis

8.1.1 Introduction and Prevalence

Several surveys demonstrate that cannabis is widely used across the world. In Europe it is the most commonly used illicit drug.[1,2] The Crime Survey for England and Wales, a household survey of adults aged 16 and over, estimated that 6.9% of adults aged 16–59 had used cannabis in 2011/2012, which extrapolates to around 2.3 million people nationally.[3] In the 2010 Global Burden of Disease study,[4] modelling was used to establish global prevalence of drug dependence and estimated that there were 13.1 million cannabis-dependent individuals (15.4 million opioid and 17.2 million amphetamine estimated cases); males formed the majority of cases (64%).

Generally, rates of use are higher among young people in high-income countries, but rates of recreational use may be increasing among young people in low- and middle-income countries. Daily cannabis users are more likely to be male, less well educated, and more likely to regularly use other drugs. Weekly or more frequent use in adolescence appears to carry significant risk for dependence in early adulthood.[5] It is thought that most cannabis use is irregular,

Forensic Toxicology: Drug Use and Misuse
Edited by Susannah Davies, Atholl Johnston and David Holt
© The Royal Society of Chemistry 2016
Published by the Royal Society of Chemistry, www.rsc.org

with very few users engaging in long-term daily use. In the US and Australia, about 10% of those who ever use cannabis become daily users.[6]

8.1.2 Different Patterns of Use

Naturally produced cannabis comes in the form of the dried leaves and the flowering heads of the marijuana plant, *Cannabis sativa*. Typically, cannabis is smoked in a water-cooled pipe (a "bong") (Figure 8.1) or as a small hand-rolled cigarette ("joint") or by using a vaporiser (Figure 8.2). The method of consumption of cannabis (inhalation or ingestion) is known to play a role in the length and intensity of the psychoactive effect, as does the quantity of cannabis used at any one time. Defining a typical dose of THC (Δ^9-tetra-hydrocannabinol), the main psychoactive constituent of cannabis, is difficult since consumption varies and different strains of the drug have different THC content. For instance, "non-skunk" strains of herbal cannabis are reported to contain only 3–4% THC, unchanged from a decade ago.[7] However, in 2004, the average THC content of Dutch home-grown marijuana (Nederwiet) was 20.4% and was significantly higher than that of imported marijuana (7.0% THC).[8] Other strains such as "Northern Lights" has a THC content of 15–20% and "Durban Poison", a South African strain, 8–15%.[9]

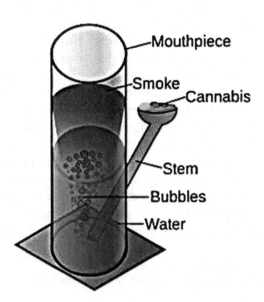

Figure 8.1 Scheme of a water bong used to smoke cannabis.[10]

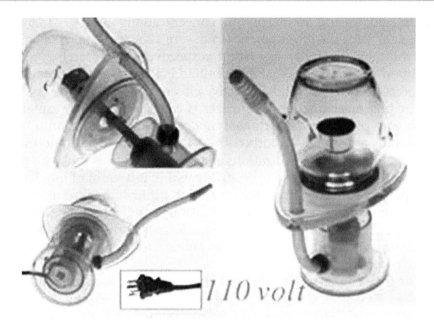

Figure 8.2 A vaporiser also used to vaporise and inhale cannabis.[10]

Cannabis is also widely ingested in foodstuffs (edibles), where quantities of THC can vary tremendously: brownies, 35 mg THC (25% extraction); pumpkin cake, 64 mg THC; a single chocolate bar, 45–60 mg THC;[11] although it is reported that one dose of an edible should contain about 20 mg THC.[9] A threshold for psychotropic effects of 0.2–0.3 mg THC per kg of body weight for a single oral dose in a lipophilic base, corresponding to 10–15 mg THC in an adult, has been suggested.[12]

8.1.3 Pharmacokinetics

8.1.3.1 Absorption

Smoking cannabis produces significant acute effects despite variability in the bioavailability of THC, ranging from 10–14% in recreational (infrequent/inexperienced) to 23–56% in regular users[13,14] and 18% in healthy volunteers.[15] Bioavailability varies according to depth of inhalation, puff duration and breath-hold. In a study of regular cannabis users, breath-hold (a common behaviour of cannabis smokers) duration (0, 10 and 20 s) and puff volume (30, 60 and 90 mL) were systematically varied. Post-smoking changes in CO exposure, plasma THC and subjective reports (especially acute intoxication) were significantly dose-related to puff volume: 45 min post-smoking for 1.75% THC and for 3.55% THC cigarettes and puff volume 30, 60 and 90 mL plasma concentrations were 5.7, 14.6 and 22.9 μg L^{-1} for the lower potency and were 12.8, 25.9 and 30.5 μg L^{-1} for the higher potency cigarettes, respectively.[16] Research suggests that puff volume rather than breath-hold is important in determining blood concentrations and subjective effects.

The pharmacokinetics of cannabis is complex and likely explained by a two- or three-compartment model. Following inhalation, cannabinoids are rapidly absorbed from the alveolar membrane in the lungs and plasma concentrations can be detected in 2 min, with peak plasma THC concentrations occurring within 3–15 min of inhalation. The plasma concentration time-course after inhalation resembles that after intravenous administration.[17] Unlike alcohol, however, and owing to its high lipid solubility and large volume of distribution, THC, regardless of the strain or the preparation consumed, is widely distributed into body tissues.[18,19]

The time-course of THC that has been ingested differs from inhaled THC as a result of the passage through the gut. Following ingestion, absorption of THC is slow and unpredictable, with maximal blood concentrations occurring between 1 and 7 h post-dose.[14,20] Bioavailability is low, with only 6% THC reaching the blood when orally administered,[14] and the onset of psychoactive effects occurs after a delay of 30–90 min, reaching their maximum after 2–3 h and lasting for about 4–12 h, depending on the dose.[21] In addition, when THC is taken orally, concentrations of the equipotent metabolite 11-OH-THC are higher (simply due to greater amounts of THC entering the liver), bringing about a synergistic effect.[22]

8.1.3.2 Distribution

In recreational users the difference in THC concentration between peak and trough plasma concentrations is greater than that observed in chronic users who, if smoking daily (or near daily), will achieve a steady-state condition: concentrations in the blood are maintained by the continual release of THC from the tissues into the general circulation. THC plasma concentrations in infrequent users have been reported to fall below the laboratory limits of quantitation within 8–12 h,[23,24] whereas a mean THC blood concentration was found to be 0.7 μg L^{-1} (SD 1.4 μg L^{-1}) after 24 h abstinence and after 7 days abstinence was 0.3 μg L^{-1} (SD 0.7 μg L^{-1}).[25] However, variable rates of release of THC from tissue stores have been reported.[26]

Maximal plasma THC concentrations after oral consumption of a 15 mg dose (2.7–6.3 μg L^{-1})[48] and a chocolate cookie (20 mg dose) were lower (4.4–11.0 μg L^{-1}) than that found after a single cannabis cigarette.[25,27] However, a meta-analysis of 21 studies investigating cannabis ingestion and driving performance revealed that a blood concentration of 3.7 µg L^{-1} THC (range 3.1–4.5 µg L^{-1}) impaired drivers to a level equivalent to a Blood Alcohol Concentration (BAC) of 50 mg alcohol per 100 mL blood.[8,27] Also see Chapter 13 on Drugs and Driving.

The plasma elimination half-life (the time taken for the concentration of a drug in blood to reduce by half), which is used to estimate how long a drug takes to leave the body, is multiphasic:[28] the distribution phase ($t_{1/2}\alpha$) is relatively short since THC is rapidly assimilated and distributed to adipose tissues and is marginally different for chronic (regular/frequent) users and recreational (infrequent) users of cannabis. For chronic users, $t_{1/2}\alpha$ is about 2 h and in infrequent users about 1.5 h, respectively.[29,30]

8.1.3.3 Metabolism

The major equipotent metabolite, 11-hydroxy-THC (11-OH-THC or THC-OH), is formed by hydroxylation after both inhalation and oral dosing, and further oxidation generates the inactive 11-nor-9-carboxy-THC (THC-COOH), which is eliminated *via* the faeces and urine.[31,32] 11-Hydroxy-THC has a half-life of 120 h for frequent users and 144 h for infrequent users of the drug.[29] Both this and the THC-COOH metabolite are detectable for a considerable time after a cannabis cigarette.[33] THC-COOH is detectable in plasma for up to 3 (range 2–7) days[21,34] and detected in urine for longer.[35,36]

8.1.3.4 Elimination

The transfer of THC to and from the tissues causes a relatively long terminal elimination half-life ($t_{1/2}\beta$) and after oral or intravenous administration is reported to be between 19 and 36 h.[37] Adipose tissue serves as a long-term storage site and is responsible for the lack of correlation between THC blood concentration and pharmacodynamic effect.[31] Unlike alcohol, there is no clear relationship between blood concentrations of THC and impairment,[21,23,38] with the time of maximum blood concentration preceding the time of maximum impairment of motor-based and cognitive abilities.[23,24] This makes it much harder to generate blood concentration time date *versus* impairment curves for cannabis than it is for alcohol.[39] However, a causal role of acute cannabis intoxication in RTC has been shown by the presence of measurable concentrations of THC.[38]

It has been reported that approximately 200 mg cannabis is typically smoked in an average rolled cigarette, about 5–30 mg active THC,[29] and one dose of a cannabis edible should contain about 20 mg of THC.[40] Much has been made of the differences between naïve (recreational/infrequent) and tolerant (chronic, dependent) use. The key points are discussed below.

8.1.4   Recreational Use

In doses given (often experimentally) to duplicate a single cannabis cigarette (18 mg THC or less), a maximal psychotropic effect was found 20–40 min after smoking, but effects had largely disappeared 2.5 h later.[41] Infrequent smoking of a single cannabis cigarette leads to higher plasma concentrations of THC in the body than that observed in regular users. Smoking single cannabis cigarettes containing 16, 27 or 34 mg THC, respectively, saw average peak plasma concentrations of 84.3 (range: 50.0–129.0), 21.5 (range: 3.2–53.3) and 162.2 µg L^{-1} (range: 76.0–267.0 µg L^{-1}), with concentrations rapidly decreasing, typically to 1–4 µg L^{-1} within 3–4 h;[42,43] for a 19 mg cigarette, a range of 33–77 µg L^{-1} THC was observed.[44,45]

The intoxicating effects of cannabis are typically short-lived, relatively benign and recovery is usually complete within a week of ceasing cannabis use (Table 8.1).

Table 8.1   Features of acute cannabis intoxication.[a]

Physical	Psychological	Mental
Markedly bloodshot eyes (conjunctivae)	Heightened nervousness, anxiety	Psychotic features such as suspiciousness, persecutory delusions
Sleepiness, sedation	Panic attacks: depersonalization and derealisation	Confusion and delirium: more common with high THC content
Slowed motor coordination	Perceptual distortions	Perceptual distortions: somatic or visual sensations
Bradycardia	Dysphoria: usually mild, brief and self-limiting	Hallucinations
Depressed respiratory rate	Cognitive impairment: altered attention, concentration, learning, and memory	Paranoia (usually transient)
	Emotional lability	Loss of insight

[a]Taken from Curran & Morgan, 2014; Watchel *et al.*, 2012; Macdonald *et al.*, 2008; Moore *et al.*, 2007; Wolff & Winstock, 2005; McCleod *et al.*, 2004.

8.1.5 Tolerance

Chronic, regular use of cannabis over many years is not unusual[46] and daily use is common in some users and almost always associated with dependent use.[47] Users may smoke five to ten cannabis cigarettes per day, thus tolerating daily doses of 100 mg THC or more. In those who use regularly, the maximal psychoactive effects of cannabis may persist for 4–6 h after use.[48] Harmful cannabis use refers to repeated use of cannabis causing physical or mental harm. This can include repeated episodes of clinically significant intoxication. A common example of physical harm is recurrent bronchitis and other chest infections.[6]

8.1.6 Cannabis Dependence

Chronic regular daily use may extend up to 14–16 h of continual smoking per day. The patient may be intoxicated for much of this period. A proportion of regular chronic users will exhibit features of dependence. The withdrawal syndrome occurs in some cannabis-dependent individuals when they cease cannabis use. The exact prevalence is not clear, but may be as high as 20% of regular heavy users. Symptoms of acute withdrawal start approximately 4 h after cessation of cannabis, peak at 4–7 days and last 2–3 weeks.[49] Protracted milder withdrawals symptoms may last several weeks. Cannabis withdrawal can be precipitated by the CB1 antagonist rimonabant.[50]

8.2 Medicinal Cannabinoids

Various analogues of cannabis have been manufactured commercially for medical purposes. They act on the CB receptors and can be detected in measureable quantities in blood, urine and oral fluid. Several therapeutic uses have been identified for medicinal cannabinoids, including:

- Chemotherapy or radiotherapy as an antiemetic
- In HIV and AIDS patients, as an appetite stimulant
- In the relief of glaucoma
- In neurological disorders, such as multiple sclerosis, spinal cord injuries and movement disorders for spasticity and pain

Those on the market include dronabinol (Marinol®), a pure isomer of THC, which is licensed in the US and Germany for anorexia in AIDS patients and in chemotherapy patients who have not responded to traditional antiemetics.[51] Experimental studies found that dronabinol (10–20 mg) caused impairment in on-the-road driving tests in a dose-dependent manner. In a single-dose, double-blind, placebo-controlled study of recreational and chronic cannabis users (dronabinol, 10 and 20 mg) a dose-dependent effect was observed on driving performance when under the influence of THC, regardless of the experience of the user.[52] Impairments were deemed bigger than the effects caused by a BAC of 50 mg alcohol per 100 mL blood, although effects were less pronounced after chronic dosing.[53,54]

Nabilone (Cesamet®), a synthetic analogue of THC,[25,26] is licensed in the UK, Canada and the USA for the nausea and vomiting associated with chemotherapy. Nabilone (usual dose 2–4 mg per day; 1 mg Cesamet® capsule contains 1 mg of nabilone) is well absorbed and, as with THC, there is a high first-pass effect. The $t_{1/2}$ of nabilone is about 2 h.[25] Oral administration of a 2 mg dose of radio-labelled nabilone achieved peak plasma concentrations of approximately 2 µg L^{-1} synthetic THC within 2.0 h (1 µg L^{-1} whole blood). Clinical trials have found that nabilone produces less tachycardia and less euphoria than THC for a similar antiemetic response,[55] and following 2 mg per day dosing reaction time, working memory, divided attention, psychomotor speed and mental flexibility did not deteriorate during a 4 week treatment period.[56]

Nabiximols (Sativex®), an oromucosal spray containing THC together with cannabidiol (CBD), is the first natural cannabis extract prescription medicine and is available in the UK and some European countries, Canada and New Zealand. Sativex spray is delivered in a fixed dose of 2.7 mg THC and 2.5 mg CBD and is indicated in the treatment of moderate to severe spasticity in multiple sclerosis[51,57] with reported mean C_{max} concentrations of <12 µg L^{-1} following single doses as multiple sprays, which is below those reported for cannabis cigarettes.

8.3   Synthetic Cannabinoids

8.3.1   Introduction and Prevalence

Over the last few years, synthetic cannabinoid receptor agonists have been detected in samples of smoking mixes such as "Spice", "Aroma", "K2" and "Silver" and are reported to have pharmacology similar to that of cannabis. One of the first compounds of this type was CP-47 497, which was developed by Pfizer and has a similar structure and properties to THC but is estimated to be 3–28 times more potent than THC.[58] Others include AM-2201, JWH-018, JWH-019, JWH-122, JWH-210, JWH-307, AM-2201-pMe (MAM-2201 or JWH-122 5-fluoropentyl derivative), AM-1220, AM-1220-azepane, UR-144, XLR-11, JWH-122-pentenyl, AM-2232 and STS-135.[59,60]

It is difficult to estimate the prevalence of "Spice" but household surveys report lifetime prevalence of 0.1% in adults (10.3% lifetime prevalence clubbers) in the UK[61] and 7% in Germany.[62] Surveys mostly reveal users to be single, young Caucasian males.[63] Synthetic cannabis use is also detected in prisoners and in the military service.[64,65]

8.3.2 Different Patterns of Use

Usually sprayed onto dried herbal tobacco, synthetic cannabinoids are marketed under a variety of names. "Annihilation" was found to contain synthetic cannabinoids (AM-2201, MAM-2201 and UR144) in two seizures from Scotland (Association of Chief Police Officers, 2008). Many of the mixtures available under different brand names contain the same compounds, one of which (AM-2201) has been identified in products traded as "Black Mamba", "Tai High Hawaiian Haze" and "Bombay Blue Extreme".[66] In the UK in 2011/2012 it was reported that 0.1% of 16–59 year olds had used "Spice" and other cannabinoids in the last year, with 0.4% of 16–24 year olds reporting use in 2010/2011.[67]

The synthetic cannabinoid HU-210 [the $(-)$-1,1-dimethylheptyl analogue of 11-hydroxy-Δ^8-tetrahydrocannabinol] is reported to be 100–800 times more potent than natural THC and has an extended duration of action.[68] It has been detected in three "Spice" products in the UK.[7] Another, JWH-018, has been reported to have intoxicating effects, with serum JWH-018 concentrations generally in the 1–10 μg L^{-1} range during the first few hours after recreational usage.[69] There is some evidence that the toxicity of the synthetic cannabinoids is worse than that of natural cannabis, owing to the higher potency, the difficulties of proper dosing and also the possibility of the presence of several different cannabinoids in one smoking mix.[70]

8.3.3 Tolerance and Dependence

There are growing reports of tolerance and physical withdrawal following regular use of synthetic cannabinoids.[71] Case reports document irritability, anxiety, mood swings, restlessness, tremor, palpitations, insomnia, myalgia, tachycardia, hypertension, hyperventilation, headache, nausea and vomiting as common withdrawal

symptoms.[72,73] There are also emerging reports of cardiac dysfunction and acute coronary syndromes[74] and periodontal and oral
damage (gingival enlargement) resulting from long term use.[75]

It is worth noting that smoking mixes do not cause a positive drug
test for cannabis or other illegal drugs using standard GC-MS drug-
screening with library search or multi-target screening by LC-MS/MS,
although bespoke methodology has enabled detection. Synthetic
cannabinoids are not detected by older immunoassay methods employed for detecting THC. However, JWH-018 usage is readily detected
in urine using "Spice" screening immunoassays from several manufacturers focused on both the parent drug and its ω-hydroxy and
carboxyl metabolites.[76]

8.4 Conclusion

Cannabis is undisputedly the most commonly used illicit drug across
the globe and estimates show that dependent use is increasingly common. THC, the main psychoactive constituent identified in the plant
Cannabis sativa L., causes impairment on a numbers of levels. There is a
significant dose-related decrement in cognitive performance following
cannabis use and raised concentrations of THC are significantly associated with increased traffic crash risk. Cannabis use behaviour is important in determining the impact on the individual. Smoking a single
cannabis cigarette on an infrequent basis (recreational use) will lead to
high concentrations of THC in the blood, causing acute intoxication. In
chronic, daily or near daily use over a prolonged period, steady-state
concentrations of THC are observed that are much lower than those
seen following infrequent use. Medicinal cannabinoids, however, have
demonstrated therapeutic potential. Prescribed in low doses, they
generally do not produce the debilitating effects of THC inhaled by
smoking. In recent years, synthetic cannabinoid receptor agonists have
been detected in samples of smoking mixes such as "Spice". Research is
emerging that suggests synthetic cannabinoids are of a higher potency
than natural cannabis and may thus produce more severe adverse effects. Further work to investigate the pharmacokinetic and pharmacodynamic properties of these compounds is required.

Further Reading

F. Grotenhmen, Reassessing the Drug Potential of Industrial Hemp,
    2002, pp. 0–48

European Monitoring Centre for Drugs and Drug Addiction (EMCDDA), 1999, Literature Review on the Relation between Drug Use, Impaired Driving and Traffic Accidents, (CT.97.EP.14), EMCDDA, Lisbon.

C. A. Hunt and R. T. Jones, Tolerance and disposition of tetrahydrocannabinol in man, *J. Pharmacol. Exp. Ther.*, 1980, **215**, 35–44.

G. Jager, Cannabis, in Drug Abuse and Addiction in Medical Illness, ed. J. C. Vester, K. Brady, M. Galanter and P. Conrod, Springer, New York, 2012, ISBN 987-1-4614-3374-3, pp. 151–162.

S. Harder and S. Rietbrock, Concentration-effect relationship of delta-9-tetrahydrocannabiol and prediction of psychotropic effects after smoking marijuana. *Int. J. Clin. Pharmacol. Ther.*, 1997, **35**, 155–159.

H. V. Curran and C. J. A. Morgan, Desired and undesired effects of cannabis on the human mind and psychological well-being, in ed. R. Pertwee, Handbook of Cannabis, Oxford, Oxford University Press, 2014, pp. 647–660.

J. G. Bramness, H. Z. Khiabani and J. Mørland, Impairment due to cannabis and oethanol: clinical signs and additive effects, *Addiction*, 2010, **105**, 1080–1087.

S. Agurell, M. Halldin and J. E. Lindgren, Cannabis and cannabinoids in the treatment of cancer, *Addiction*, 1996, **91**, 1585–1614.

Association of Chief Police Officers (ACMD), Written evidence presented to the ACMD, 2008.

M. S. Castaneto, D. A. Gorelick, N. A. Desrosiers, R. L. Hartman, S. Pirard and M. A. Huestis, Synthetic cannabinoids: epidemiology, pharmacodynamics, and clinical implications. *Drug Alcohol Depend.*, 2014, **144**, 12–41.

W. D. Darwin, R. S. Goodwin, S. Wright and M. A. Huestis, Plasma Cannabinoid Pharmacokinetics following Controlled Oral Δ9-Tetrahydrocannabinol and Oromucosal Cannabis Extract Administration, *Clin. Chem.*, 2011, **57**, 66–75.

R. Kronstrand and A. W. Jones. Drugs of abuse – analysis, in Encyclopedia of Forensic Sciences, ed. J. A. Siegel, P. J. Saukko and G. C. Knuffer, Academic Press, San Diego, CA, 2000, pp. 598–610.

J. Macleod, R. Oakes, A. Copello et al., Psychological and social sequelae of cannabis and other illicit drug use by young people: a systematic review of longitudinal, general population studies, *Lancet*, 2004, **363**, 1579–1588.

S. MacDonald, R. Mann, M. Chipman, B. Pakula, P. Erickson, A. Hathaway and P. MacIntyre, Driving behaviour under the influence of cannabis or cocaine, *Traffic Inj. Prev.*, 2008, **9**, 190–194.A.C

T. H. M. Moore, S. Zammit, A. Lingford-ughes et al., Cannabis use and risk of psychotic or affective mental health outcomes: a systematic review, *Lancet*, 2007, **370**, 319–328.

Office of National Drug Control Policy, 2011, Synthetic Marijuana, Accessed November 2015. http://www.whitehouse.gov/sites/default/files/ondcp/Blog/syntheticmarijuana.pdf.

S. W. Toennes, J. Röhrich and C. Wunder, Interpretation of blood analysis data found after passive exposure to cannabis, *Arch. Kriminol.*, 2010, **225**, 90–98.

United Nations Office on Drug and Crime, UN World Drug Report, 2011, pp. 11–18.

S. R. Watchel, M. A. ELSohly, S. A. Ross, J. Ambre and H. de Wit, Comparison of the subjective effects of delta-(9) tetra-hydrocannabinol and marijuana in humans, *Psychopharmacology*, 2012, **161**, 331–339.

K. Wolff, Biological Markers of Drug Use, in Addiction Psychiatry, Psychiatry, ed. E. J. Marshall and M. Farrell, The Medicine Pub Co Ltd, London, 2007.

References

1. M. F. Andreasen, J. B. Hasselstrom and I. Rosendal, Prevalence of Licit and Illicit drugs in whole blood *among Danish Drivers in Western Denmark*, *SOFT-TIAFT*, 2011, San Francisco.

2. T. Assum, M. P. Mathijssen, S. Houwing, S. C. Buttress, R. J. Sexton and R. J. Tunbridge, The prevalence of drug driving and relative risk estimations. A study conducted in the Netherlands, Norway and the United Kingdom, 22 June 2005, Report No.: D-R4.2.

3. CSEW, Drug Misuse Declared: Findings from the 2011/12 Crime Survey for England and Wales, 2nd edn, http://www.homeoffice.gov.uk/publications/science-research-statistics/research-statistics/crime-research/drugs-misuse-dec-1112/drugs-qmisuse-dec-1112-pdf.

4. S. S. Lim, T. Vos, A. D. Flaxman *et al.*, A comparative risk assessment of burden of disease and injury attributable to 67 risk factors and risk factor clusters in 21 regions, 1990–2010: a systematic analysis for the Global Burden of Disease Study 2010, *Lancet*, 2012, **380**, 2224–2260.

5. J. Macleod, R. Oakes, A. Copello, I. Crome, M. Egger, M. Hickman, T. Oppenkowski, H. Stokes-Lampard and G. Davey Smith,

Psychological and social sequelae of cannabis and other illicit drug use by young people: a systematic review of longitudinal, general population studies, *Lancet*, 2004, **15**(363), 1579–1588.

6. *Oxford Handbook of Addiction Medicine*, ed. J. B. Saunders, K. M. Conigrave, N. C. Latt, D. J. Nutt, E. J. Marshall, W. Ling and S. Higuchi, Oxford University Press, London, 2nd edn, 2016, ISBN 978-0-19-871475-0.

7. *European Monitoring Centre for Drugs and Drug Addiction (EMCDDA)*. 2010 Annual report on the state of the drugs problem in Europe. EMCDDA, Lisbon, November, 2010. www.emcdda. europa.eu/publications/annual-report/2010.

8. F. T. Pijlman, S. M. Rigter, J. Hoek, H. M. Goldschmidt and R. J. Niesink, Strong increase in total delta-THC in cannabis preparations sold in Dutch coffee shops, *Addict. Biol.*, 2005, **10**(2), 171–180.

9. http://www.marijuanagrowershq.com/how-best-to-use-medical-marijuana-smoking-vs-edibles-and-tinctures/ Accessed October 2015.

10. www.datuopinion.com.

11. http://www.bluelight.ru/vb/archive/index.php/t-549317.html.

12. F. Grotenhermen, Harm Reduction Associated with Inhalation and Oral Administration of Cannabis and THC, *J. Cannabis Ther.*, 2001, **1**, 133–152.

13. F. Musshoff and B. Madea, Review of Biologic Matrices (Urien, Blood, Hair, as Indicators of Recent or On-going Cannabis Use, *Ther. Drug Monit.*, 2006, **28**, 155–163.

14. S. Agurell and M. Halldin, Pharmacokinetics and metabolism of 9-tetrahydrocannabinol and other cannabinoids with emphasis on man, *Pharm. Rev.*, 1986, **38**, 21–43.

15. A. Ohlsson, J. E. Lindgren, A. Wahlen, S. Agurell, L. E. Hollister and H. K. Gillespie, Plasma delta-9-tetrahydrocannabinol concentrations and clinical effects after oral and intravenous administration and smoking, *Clin. Pharmacol. Ther.*, 1980, **28**, 409–416.

16. J. L. Azorlosa, M. K. Greenwald and M. L. Stitzer, Marijuana smoking: effects of varying puff volume and breathhold duration, *J. Pharmacol. Exp. Ther.*, 1995, **272**, 560–569.

17. M. Perez-Reyes, S. Di Guiseppi, K. H. Davis, V. H. Schindler and C. E. Cook, Comparison of effects of marihuana cigarettes to three different potencies, *Clin. Pharmacol. Ther.*, 1982, **31**, 617–624.

18. P. Sharma, P. Murthy and M. M. Bharath, Chemistry, metabolism, and toxicology of cannabis: clinical implications, *Iran. J. Psychiatry*, 2012, **7**(4), 149–156.

19. M. R. J. Baldock. Review of the literature on cannabis and crash risk (CASR010), Centre for Automotive Safety Research, Adelaide. Department for Transport, Energy and Infrastructure (SA), 2007, ISBN 9781920947095.
20. B. Law, P. A. Mason, A. C. Moffat, R. I. Gleadle and L. J. King, Forensic aspects of the metabolism and excretion of cannabinoids following oral ingestion of cannabis resin, *J. Pharm. Pharmacol.*, 1984, **36**, 289–294.
21. F. Gotenhermen, Pharmacokinetics and pharmacodynamics of cannabinoids, *Clin Pharmacol.*, 2003, **42**, 327–360.
22. C. Giroud, A. Menetrey and M. Augsburger, Delta(9)-THC, 11-OHDelta(9)-THC and Delta(9)-THCCOOH plasma or serum to whole blood concentrations distribution ratios in blood samples taken from living and dead people, *Forensic Sci. Int.*, 2001, **123**, 159–164.
23. F. J. Couper and B. K. Logan, Addicted to driving under the influence–a GHB/GBL case report, *J. Anal. Toxicol.*, 2004, **28**(6), 512–515.
24. O. H. Drummer, J. Gerostamoulos, H. Batziris, M. Chu, J. Caplehorn, M. D. Robertson *et al.*, The involvement of drugs in drivers of motor vehicles killed in Australian road traffic crashes, *Accid. Anal. Prev.*, 2004, **36**, 239–248.
25. E. L. Karschner, E. W. Schwilke, R. H. Lowe, W. D. Darwin, H. G. Pope, R. Herning *et al.*, Do Δ9-Tetrahydrocannabinol Concentrations Indicate Recent Use in Chronic Cannabis Users?, *Addiction*, 2009, **104**, 2041–2048.
26. E. L. Karschner, W. D. Darwin, R. S. Goodwin, S. Wright and M. A. Huestis, Plasma Cannabinoid Pharmacokinetics following Controlled Oral Δ^9-Tetrahydrocannabinol and Oromucosal Cannabis Extract Administration, *Clin. Chem.*, 2011, **57**(1), 66–75.
27. V. Vindenes, D. H. Strand, L. Kristoffersen, F. Boix and J. Mørland, Has the intake of THC by cannabis users changed over the last decade? Evidence of increased exposure by analysis of blood THC concentrations in impaired drivers, *Forensic Sci. Int.*, 2013, **226**, 197–201, DOI: 10.1016/j.forsciint.2013.01.017. Epub 2013 Feb 13.
28. J. P. Goullé, E. Saussereau and C. Lacroix, Delta-9-tetrahydrocannabinol pharmacokinetics, *Ann. Pharm. Fr.*, 2008, **66**(4), 232–244, DOI: 10.1016/j.pharma. 2008.07.006. Epub 2008 Aug 29. [English Abstract].
29. A. C. Moffat, M. D. Osselton and B. Widdop, *Clarke's Analysis of Drugs and Poisons*, Pharmaceutical Press, Bath, UK, 3rd edn, 2004, ISBN 0 85369 473 7.

30. M. D. Moffat Osselton and B. Widdop, *Clarke's Analysis of Drugs and Poisons*, 4th edn, 2011, ISBN 0 85369 711 4.
31. M. A. Huestis, J. E. Henningfield and E. J. Cone, Blood cannabinoids. II. Models for the prediction of time of marijuana exposure from plasma concentrations of delta 9-tetrahydrocannabinol (THC) and 11-nor-9-carboxy-delta 9-tetrahydrocannabinol (THCCOOH), *J. Anal. Toxicol.*, 1992, **16**(5), 283–290.
32. M. A. Huestis, J. E. Henningfield and E. J. Cone, Blood cannabinoids. II. Models for the prediction of time of marijuana exposure from plasma concentrations of delta 9-tetrahydrocannabinol (THC) and 11-nor-9-carboxy-delta 9-tetrahydrocannabinol (THCCOOH), *J. Anal. Toxicol.*, 1992, **16**(5), 283–290.
33. S. Frytak, C. G. Moertel and J. Rubin, Metabolic studies of delta-9-tetrahydro cannabinol in cancer patients, *Cancer Treat. Rep.*, 1984, **68**, 1427–1431.
34. A. Reiter, J. Hake, C. Meissner, J. Rohwer, H. J. Friedrich and M. Oehmichen, Time of drug elimination in chronic drug abusers. Case study of 52 patients in a "low-step" detoxification ward, *Forensic Sci. Int.*, 2001, **119**(2), 248–253.
35. G. Milman, D. M. Schwope, D. A. Gorelick and M. A. Huestis, Cannabinoids and metabolites in expectorated oral fluid following controlled smoked cannabis, *Clin. Chim. Acta*, 2012, **413**, 765–770, DOI: 10.1016/j.cca. Epub 2012 Jan 20.
36. P. Kelly and R. T. Jones, Metabolism of tetrahydrocannabinol in frequent and infrequent marijuana users, *J. Anal. Toxicol.*, 1992, **16**, 228–235.
37. M. E. Wall and B. M. Sadler, Metabolism, disposition, and kinetics of delta-9-tetrahydro cannabinol in men and women, *Clin. Pharmacol. Ther.*, 1983, **34**, 352–363.
38. H. Kalant, Adverse effects of cannabis on health: an update of the literature since 1996, *Prog. Neuro-Psychopharmacol. Biol. Psychiatry*, 2004, **28**(5), 849–863. Review.
39. R. A. Sewell, J. Poling and M. Sofuoglu, The effects of cannabis compared with alcohol on driving, *Am. J. Addict.*, 2009, **18**(3), 185–193, DOI: 10.1080/10550490902786934.
40. V. O. Von Leirer, J. A. Yesavage and D. G. Morrow, Marijuana carry-over effects on aircraft pilot performance, *Aviat., Space Environ. Med.*, 1991, **62**, 221–227.
41. F. Grotenhermen, Harm reduction associated with inhalation and oral administration of cannabis and THC, *J. Cannabis Ther.*, 2001, **1**, 133.

42. E. Manno, B. R. Manno, P. M. Kemp, D. D. Alford, I. K. Abukhalaf, M. E. McWilliams *et al.*, Temporal indication of marijuana use can be estimated from plasma and urine concentrations of delta-9-tetrahydrocannabinol, 11-hydroxy-delta-9-tetrahydro cannabinol, and 11-nor-delta-9-tetrahydrocannabinol-9-carboxylic acid, *J. Anal. Toxicol.*, 2001, **25**, 538–549.

43. M. A. Huestis, J. E. Henningfield and E. J. Cone, Blood cannabinoids: I. Absorption of THC and formation of 11-OH-THC and THC-COOH during and after marijuana smoking, *J. Anal. Toxicol.*, 1992, **16**, 276–282.

44. A. Ohlsson, J. E. Lindgren, A. Wahlen, S. Agurell, L. E. Hollister and H. K. Gillespie, Single dose kinetics of deuterium labelled delta-1-tetrahydrocannabinol in heavy and light cannabis users, *Biomed. Mass Spectrom.*, 1982, **9**, 6–10.

45. L. E. Hollister, H. K. Gillespie, A. Ohlsson, J. E. Lindgren, A. Wahlen and S. Agurell, Do plasma concentrations of delta 9-tetrahydrocannabinol reflect the degree of intoxication?, *J. Clin. Pharmacol.*, 1981, **21**(8–9 Suppl), 171S–177S.

46. S. E. Rooke, M. M. Norberg, J. Copeland and W. Swift, Health outcomes associated with long-term regular cannabis and tobacco smoking, *Addict. Behav.*, 2013, **38**, 2207–2213.

47. D. Reilly, R. Didcott, W. Swift *et al.* Long-term cannabis use: characteristics of users.

48. M. Kochanowski and M. Kała, Tetrahydrocannabinols in clinical and forensic toxicology, *Przegl. Lek.*, 2005, **62**, 576–580, English Abstract.

49. A. Budney and A. Hughes, The cannabis withdrawal syndrome, *Curr. Opin. Psychiatry*, 2006, **19**, 33–238.

50. D. J. Allsop, J. Copeland, N. Lintzeris *et al.*, Nabiximols as an agonist replacement therapy during cannabis withdrawal, *JAMA Psychiatry*, 2014, **71**(3), 281–291.

51. A. Green and K. de Vries, Issues for Pharmacists when patients use cannabis therapeutically at home, *Pharm. J.*, 2012, **289**, 531–532.

52. W. M. Bosker, E. L. Karschner, D. Lee, R. S. Goodwin, J. Hirvonen, R. B. Innis, E. L. Theunissen, K. P. Kuypers, M. A. Huestis and J. G. Ramaekers, Psychomotor function in chronic daily Cannabis smokers during sustained abstinence, *PLoS One*, 2013, **8**(1), e53127, DOI: 10.1371/journal.pone.0053127. Epub 2013 Jan 2.

53. DRUID, Prevalence of alcohol and other psychoactive substances in injured and killed drivers, Deliverable D3.2.2, http://www.west-info.eu/files/Druid-report-08-2011.pdf. Deliverable D3.2.2 2010.

54. DRUID, summary of main DRUID results: driving under the influence of drugs, alcohol, and medicines, TRB 91ST Annual meeting, USA Dutch Forensic Science.
55. I. J. McGilveray, Pharmacokinetics of cannabinoids, *Pain Res. Manage.*, 2005, **10**(Suppl A), 15A–22A.
56. I. Kurzthaler, T. Bodner, G. Kemmler, T. Entner, J. Wissel, T. Berger and W. W. Fleischhacker, The effect of nabilone on neuropsychological functions related to driving ability: an extended case series, *Hum. Psychopharmacol.*, 2005, **20**(4), 291–293.
57. C. G. Stott, L. White, S. Wright, D. Wilbraham and G. W. Guy, A phase I study to assess the single and multiple dose pharmacokinetics of THC/CBD oromucosal spray, *Eur. J. Clin. Pharmacol.*, 2013, **69**(5), 1135–1147, DOI: 10.1007/s00228-012-1441-0012-1441-0. Epub 2012 Nov 22.
58. A. Weissman, G. M. Milne and L. S. Melvin Jr., Cannabimimetic activity from CP-47,497, a derivative of 3-phenylcyclohexanol, *J. Pharmacol. Exp. Ther.*, 1982, **223**(2), 516–523.
59. F. Musshoff, L. Hottmann, C. Hess and B. Madea, "Legal highs" from the German internet-"bath salt drugs" on the rise, *Arch. Kriminol.*, 2013, **232**(3–4), 91–103.
60. J. Hillebrand, D. Olszewski and R. Sedefov, Legal highs on the Internet, *Subst. Use Misuse*, 2010, **45**(3), 330–340.
61. Office for National Statistics, *Drug Misuse Declared: Findings from the 2011/12 Crime Survey for England and Wales*, Home Office, London, 2012.
62. European Monitoring Center for Drugs and Drugs Addiction. (2014). Synthetic Cannabinoids in Europe, from http://www.emcdda.europa.eu/topics/pods/synthetic-cannabinoids.
63. L. D. Johnston, P. M. O'Malley, R. A. Miech, J. G. Bachman and J. E. Schulenberg, *Monitoring the Future National Survey Results on Drug Use:1975-2014: Overview, Key Findings on Adolescent Drug Use*, 2015.
64. C. S. Berry-Caban, J. Ee, V. Ingram, C. E. Berry and E. H. Kim, Synthetic cannabinoid overdose in a 20-year-old male US soldier, *Subst. Abuse*, 2013, **34**(1), 70–72.
65. D. Walker, C. Neighbors, T. Walton, A. Pierce, L. Mbilinyi, D. Kaysen and R. Roffman, Spicing up the military: Use and effects of synthetic cannabis in substance abusing army personnel, *Addict. Behav.*, 2014, **39**(7), 1139–1144.
66. www.talktofrank.com/drug/cannabis, 2015.
67. K. Smith and J. Flatley, *Drug Misuse Declared: Findings from the 2010/11 British Crime Survey England and Wales*, ISSN 1759 7005 ISBN 978 1 84987 482, 2011.

68. W. A. Devane *et al.*, A novel probe for the cannabinoid receptor, *J. Med. Chem.*, 1992, **35**(11), 2065–2069, DOI: 10.1021/jm00089a018. PMID 1317925.

69. A. A. Westin, W. R. Brede and K. A. Espnes, Synthetic cannabis, *Tidsskr Nor Laegeforen*, 2012, **132**(20), 2289, DOI: 10.4045/tidsskr.12.0661. English, Norwegian. No abstract.

70. R. Kronstrand, M. Roman, M. Andersson and A. Eklund, Toxicological findings of synthetic cannabinoids in recreational users, *J. Anal. Toxicol.*, 2013, **37**(8), 534–541, DOI: 10.1093/jat/bkt068. Epub 2013 Aug 22.

71. U. S. Zimmermann, P. R. Winkelmann, M. Pilhatsch, J. A. Nees, R. Spanagel and K. Schulz, Withdrawal phenomena and dependence syndrome after the consumption of "spice gold", *Dtsch. Arztebl. Int.*, 2009, **106**(27), 464–467.

72. N. Nacca, D. Vatti, R. Sullivan, P. Sud, M. Su and J. Marraffa, The synthetic cannabinoid withdrawal syndrome, *J. Addict. Med.*, 2013, 7(4), 296–298.

73. (a) V. Macfarlane and G. Christie, Synthetic cannabinoid withdrawal: a new demand on detoxification services, *Drug Alcohol Rev.*, 2015, **34**(2), 147–153; (b) M. S. Castaneto, D. A. Gorelick, N. A. Desrosiers, R. L. Hartman, S. Pirard and M. A. Huestis, Synthetic cannabinoids: epidemiology, pharmacodynamics, and clinical implications, *Drug Alcohol Depend.*, 2014, **144**, 12–41.

74. I. Casier, P. Vanduynhoven, S. Haine, C. Vrints and P. G. Jorens, Isrecent cannabis use associated with acute coronary syndromes? An illustrative case series, *Acta Cardiol.*, 2014, **69**(2), 131–136.

75. S. Y. Rawal, D. N. Tatakis and D. A. Tipton, Periodontal and oral manifestations of marijuana use, *J. Tenn. Dent. Assoc.*, 2012, **92**(2), 26–31, quiz 31-2.

76. EMCDDA (2009). Action on new drugs briefing paper: Understanding the 'spice' phenomenon. A report from an EMCDDA expert meeting, 6 March, 2009, Lisbon.

9 Khat – Chewing it Over: Continuing "Cultural Cement", Cardiac Challenge or Catalyst for Change?

John Martin Corkery

University of Hertfordshire, Hatfield, UK
Email: j.corkery@herts.ac.uk

9.1 Introduction

Khat is a herbal product consisting of the leaves and shoots of the shrub *Catha edulis* Forsk, a member of the evergreen Celastracae (moonseed or spindle-tree) family. It usually grows to heights of 2.7–3.7 m. On the moist slopes of the Ethiopian mountains it becomes a slender tree up to 6–8 m tall with a trunk over 50 cm in circumference at chest height.[1] In favourable conditions in the equatorial region it may attain 25 m with a trunk circumference of 60 cm.[2]

The leaves of this shrub resemble withered basil and when young are crimson-brown and shiny but become leathery and yellowy-green as they age. They vary greatly in size, ranging from 3 to 12.2 cm in length by 0.5 to 6.7 cm in breadth, the median veins and stalks being slightly reddish in colour. The leaves taste slightly bitter and astringent. Only the youngest leaves of the central twig, which are greenish-brown in colour, are used for consumption; the leaves of the lateral twigs are discarded.[2] Figure 9.1 shows khat leaves.

Forensic Toxicology: Drug Use and Misuse
Edited by Susannah Davies, Atholl Johnston and David Holt
© The Royal Society of Chemistry 2016
Published by the Royal Society of Chemistry, www.rsc.org

Figure 9.1 Khat leaves.

9.2   Natural Range and Cultivation

Under natural conditions, khat grows in and on the margins of dry evergreen forest and mist forest at elevations of 1500–2000 m, but is found at altitudes of 1200–2500 m. It is best cultivated at high elevations, with high rainfall in acidic, well-drained clay soils, but can survive long periods of drought. Khat's natural range extends throughout East Africa from Ethiopia to South Africa, as well as being found in central Southern Africa. *Catha edulis* is also found sporadically in Afghanistan, Israel, Saudi Arabia, Syria and Turkistan.[3,4] In Arabia, it is found principally in the Yemen, thought to have been introduced by the Abyssinians.

The primary areas for cultivation remain Eastern Africa – mainly Ethiopia, Kenya and to a lesser extent the Comoros, Madagascar and Tanzania[5] – and the Arabian Peninsula, especially Yemen. Emerging local markets, such as in Uganda and Rwanda, have been supplied by the harvesting of wild khat.[6] Apart from growing wild in the Horn of Africa and Arabia, *Catha edulis* has been successfully cultivated in Mumbai (Bombay), Sri Lanka, Algeria, Southern France, Portugal and the Southern USA.[1,2] There are reports of *Catha edulis* being cultivated in South Africa by the Xhosa and others.[7,8] There is even a report of the species growing freely in suburbs of Perth, Western Australia;[9] this is in keeping with khat use amongst Somalis, Ethiopians, Eritreans, Kenyans and possibly Yemenis in Melbourne, Victoria.[10]

A more recent development has been the sale on the internet of seeds and young plants for home cultivation.

9.3 History of Use and Diffusion

It is estimated that about 20 million people use khat in East Africa and the Arabian Peninsula.[11] Kalix[12] estimated that about 6 million individual portions are consumed daily worldwide. Prevalence of khat use varies considerably from one country to another, and within countries. Insufficient data are available to accurately estimate overall khat prevalence rates.[13]

Khat leaves have been used in traditional medicine for the treatment of depression, gastric ulcers, hunger, obesity and tiredness.[14] Its use as a medicine can be traced back to the 10th and 11th centuries C.E., when it was used to relieve vomiting and to "cool the stomach and liver".[15] An Arabic medical book written in 1237 C.E. described khat as a treatment for depression because it led to happiness and excitement.[16] Infusions of the leaf have been used in treating coughs, asthma and other respiratory diseases; there is now evidence that it could have proper medical applications in this respect.[17] Khat has also been used as an aphrodisiac.[18,19] Further details of its historic use can be found in UNODC,[2] McKee,[20] Al-Motarreb *et al.*,[11] Dhaifalah and Šantavý,[15] but especially in Kennedy.[21]

The diffusion of use of *Catha edulis* products appears capable of spreading to other communities and substance-using groups in society. In 1990 there did not appear to be any indication of chewing spreading outside these traditional using communities.[22] However, Grignon[23] suggests that *miraa* (a traditional name for khat) managed to penetrate the disco scene, being favoured especially by Jamaican Rastafarians, West Africans and South Africans, being consumed simultaneously with cannabis during reggae evenings. Other commentators believe that khat use may intensify amongst the second generation within immigrant communities; there is some anecdotal evidence of this amongst Somalis of school age in the London Borough of Greenwich.[24]

Even in the mid-1980s there was some evidence of use of khat by individuals without these communities in the UK.[25] There was some evidence of khat being tried by those involved in the UK club scene, but the bitter taste and amount of chewing needed did not have a lasting appeal. Even 25 mL bottles of khat tincture selling at £3 did not catch on,[6,26] although for a short while there was a brisk trade in small bottles containing cathine and cathinone at £3–4 each.[6]

Furthermore, there are other more potent stimulants that are widely available in countries such as the UK; the market is already saturated. The fact that English-language internet sites, some based in the UK, have advertised a range of khat-based products does suggest that some entrepreneurs consider there are other markets to exploit. The availability of khat in new urban markets in London in recent years suggests that other consumers may exist from outside of the traditional using communities who are keen to experiment with it.[6] The death of a Polish national from a drug overdose, including the presence of cathinone and (nor)pseudoephedrine, demonstrates that this is already occurring, although its extent is unknown.[27,28] Purchase of khat by Caucasians is also reported from East London.[29]

In the UK the khat session facilitates social cohesion and a sense of identity, being seen as part of the cultural tradition and to some extent underpinning religious practice.[30–33] These roles for khat play out particularly amongst long-settled and stable communities, such as Yemenis. Individuals who have come to the UK in order to avoid conflict in their native country, particularly Somalia, arrive with a variety of issues which make them more susceptible to self-medicate with khat in an attempt to forget their problems. Bashford *et al.*[34] suggest that cultural dislocation within the Somali refugee community may play a role in initiation into drug use, and "lead individuals to seek khat as a refuge", *i.e.* as a method of self-medication to mediate stress (although also prolonging and adding to existing stress). Alcohol and drugs play a role in 'killing time' for asylum seekers, and helping to forget past trauma.[35] Between 46% and 83% have tried khat by the time they are 20 years old. Many start using before coming to the UK, but a significant proportion after their arrival—perhaps to help compensate for the sense of isolation, uncertainty about the future, unemployment, *etc.*[30,32,33,36–39]

Griffiths *et al.*[36] found that 76% of those interviewed in London used more than they had done in Somalia. Two-thirds stated that chewing khat helped to maintain their cultural identity. Similar views have been expressed by East Africans in Australia,[10] Canada, Sweden and the USA.[40] This is underlined by the fact that seizures of khat have been increasing in Australia[5] and the USA.[13,41,42] The emergence of khat use as an issue in such places is part of the history of displacement at the individual and cultural levels.[10] It is important to note that differences in consumption patterns may also have their roots in the place of origin of the khat-using community.

9.4 Global Epidemiology

Catha edulis consumption has greatly increased in recent years due to improved transportation facilities.[21] This fact is reflected in figures contained in the *World Drug Report* which demonstrates a slower rate of increase but still growing number of countries reporting overall greater khat use: five in 2001, four in 2002, two in 2003 and one in 2004.[43,44] Although khat use and prevalence is at its most significant in East Africa and the Arabian Peninsula,[15] its spread to a global market is of most interest to Western toxicologists.

Khat use in the UK is mostly confined to ethnic communities traditionally accustomed to its use. These are located chiefly in Bristol, East London, the Somali communities in Liverpool[45,46] and Sheffield,[47] and the Yemeni community in South Wales (Barry, Cardiff and Newport).[48] It appears that khat use was only introduced to the UK in the early 1980s.[25,45,48,49] However, there were Somali communities in existence in the UK for several decades prior to this period. The UK Somali population grew dramatically during the 1990s because of the civil war and the collapse of the state. Over the period 1990–2005, many Ethiopian nationals applied for asylum.

There is no national estimate of the incidence, prevalence or population using khat in the UK. However, the Crime Survey for England and Wales, a general household survey, found that in both 2010/11 and 2011/12 about 0.2% of 16–59 year olds had used khat in the previous 12 months.[50] This equates to about 50 000 khat users in this age group in England and Wales. Given the nature of the survey, this is likely to be an underestimate; indeed, the London Borough of Hillingdon[51] suggested a figure of 88 000. Rates amongst those aged 16–24 were higher (0.3%) than for those aged 25–59 (0.1%) in 2010/11.[52]

Lifetime use of khat amongst Somalis living in four UK cities was reported to be 39%, with last year use at 34% and daily use at just under 4%.[32] A study of Somali, Ethiopian and Yemeni individuals found lifetime levels of 82%, current use at 60% and 26% used on a regular basis.[53] This is echoed by a survey of the 3000+ Somalis living in the London Borough of Lambeth, where 77% of those surveyed used khat on a regular basis.[54]

There is very little epidemiological data on khat use in North America, and few American researchers are involved in its study. Khat use was reported in New York City as early as 1904.[22] In the late 1980s its consumption was limited to immigrants from Somalia, Ethiopia and Yemen. Almost daily arrivals of khat are flown into New York from Europe and then distributed by minicabs to selling points. A study of about 50 Somali occasional users in 1990 found khat sessions were used to facilitate business deals, and that the venues served as foci for social networking and information. Khat leaves were chewed by students to keep them awake while studying, and by night guards to remain alert and vigilant.

9.5 Transportation, Trade and Price

Khat is normally packaged in plastic bags or foil, placed in burlap, wrapped in wet clothes or shawls or in banana leaves to retain moisture and keep it fresh. The cut end of the stem can be immersed in water for a minute or more to prolong freshness. During transportation, khat is often sprinkled with water to keep the leaves moist. Individual bundles/bunches of leaves are then packed into ventilated boxes lined with plastic, which also serves to reduce moisture loss. For instance, transhipments arriving in London were initially offloaded and treated with water and ice to prolong the khat's freshness, before being repackaged into smaller packages for concealment.[6] On reaching their final destination, bundles are often refrigerated or even frozen[48] by the end retailer or consumer. However, this treatment has disadvantages, in that when defrosted the leaves become mushy and disintegrate. There are reports from the USA of leaves being packed with dry-ice and freeze-dried on arrival to preserve their freshness.[55]

It is estimated that about 7 tons of khat from Kenya, Yemen and Ethiopia entered the UK through Heathrow Airport each week, with smaller amounts imported through other airports such as Birmingham, Liverpool and Manchester. For example, in the first half of 2005, 5–7 tons arrived daily from Kenya, 500 kg from Ethiopia and 175 kg from Yemen.[42] Anderson et al.[6] reckoned that whereas perhaps 500 tons of khat was flown to the UK each year in the late 1990s, by 2005 this amount may have grown to 10 000 tons. Some of this khat was re-exported to other European countries (e.g. Scandinavia), with Heathrow acting as a hub,[56] as well as to the USA and Canada (see below). Khat is also imported by air to other European countries, especially The Netherlands.[35]

The bulk of the khat landed at Heathrow in the past was in transit – sometimes via Dublin[57,58] – for the USA or Canada, where it is illegal; much of this is also thought to have originated in Kenya and Ethiopia.[13] Khat was also flown to Canada from Amsterdam. It is reported that 20 tons of khat originating in the Netherlands were confiscated in 2005–2007 by the Canadian authorities, compared to 15 tons coming from the UK.[59] The main means of exporting khat from Europe to the USA are: (a) human couriers who fly on commercial aircraft with baggage containing khat; and (b) express mail deliveries from various European cities. Khat-smuggling routes to Canada involve transit countries such as the USA, France, Germany and Italy.[60]

The normal place of sale is in street markets in the traditional centres of consumption, *e.g. mafrishi*, or in bars, grocery shops or restaurants catering for East African and Middle Eastern communities. This has been the case also in the UK, but some sales were also made from vehicles on the street in areas where these communities live. Somalis usually retail khat at establishments known as *mafrishi*, private houses where rooms are set aside for chewing. Khat is also sold in local shops and grocery stores run by Somalis, Ethiopians and Yemenis.[61] In East London, such vendors provided areas where male customers could gather and consume their purchases.[29] In recent years, the internet has become an alternative and anonymous source of supply. There are websites based in countries where khat is legal, as well as in countries where it is illicit. These sites offer other khat preparations, marketing them as "legal highs" or "herbal highs".

The UK price of khat appears to have ranged from £3 to £6 (US$6–12) a bundle, depending on the variety potency. The price remained very stable over time. For example, 30 years ago it was selling for £2.50 (US$5) in London,[49] and in the mid-1990s for £5 (US$10).[62] Part of this stability may be explained by the mixing of different types of khat from various producing areas.[6] There are two main varieties of khat sold in the UK: *kangeta* from Kenya costing £3 (US$6) per bundle and *harari* from Ethiopia at £5 (US$10) per bundle.[61] The latter appears to be preferred by older Somalis and other groups, whilst younger Somali users tend to take *kangeta*.[32,53] The price per bundle in Sweden ranges from 150 to 400 SKr (£13.20–44.40 or US$ 26.40–88.80) for 200 g; the price varies with the season.[63]

9.6 Legal and Religious Control

The growing international trade in khat has given rise to worries about its effect on consumers in the Diaspora countries, and its association with immigration, identity and cultural integration, as well as possible links with organised crime and violence. The United

Nations Office on Drugs and Crime (UNODC) even regards khat as "a particular NPS [New Psychoactive Substance] of concern".[13]

The issue of khat regulation has been examined several times over the years by international bodies. Although consumption of khat is associated with health risks and can have detrimental social consequences, the plant is only prohibited in some countries in Eastern Africa, such as Eritrea, Madagascar, Rwanda and Tanzania; in China, Jordan, Malaysia, New Zealand and Saudi Arabia; in some countries in Europe, including Belgium, Denmark, Finland, France, Germany, Greece, Ireland, Italy, Latvia, Lithuania, Norway, Poland, Slovenia, Sweden, Switzerland, the Netherlands and the UK (see below); and in Canada and the USA.[5,42] *Catha edulis* is not controlled in Cyprus, the Czech Republic, Malta or Portugal.[42,64] Khat was legalised in Kenya in 1977, but cathine and cathinone are controlled. In Australia, only licensed persons can import khat for personal use up to a maximum of 5 kg per month, and subject to permission from the local health authority. In Hong Kong a special prescription is needed before consuming khat. The position in South Africa appears to be unclear,[8] although the shrub has been given "protected species" status. Legal prohibition appears to follow consumption, which has been the product of migration.[6]

Khat is not under international control at present, but two substances that are usually present in khat, cathine and cathinone, are, since in the early 1980s all amphetamine-like substances were placed group-wise under international control.[65] Cathinone was included in Schedule I of the UN Convention on Psychotropic Substances in 1988, and cathine was then included in Schedule III of this Convention.

9.6.1 UK Legal Controls

Cathinone and cathine are controlled substances under Class C of the UK Misuse of Drugs Act 1971 by virtue of Schedules 1 and 3, respectively, of the Misuse of Drugs (Amendment) Regulations Act 1986. Thus an offence is committed if these substances are extracted from the plant; there appear to have been no successful prosecutions.[42,66] Khat was licensed under the Medicines Act 1968 as a medicinal product but was never imported in that way, being imported legally when declared as a vegetable. There is very little information on internal UK controls, but it appears that the London Borough of Brent banned the sale of khat to children less than 16 years.[67] In East London, sellers had imposed a voluntary ban on selling khat to children.[29]

The independent Advisory Council on the Misuse of Drugs (ACMD) has been asked to consider the issue of khat at the Government's request on several occasions.[42,68] In October 2010 the ACMD was again asked to undertake a review of khat, taking into consideration the most up-to-date evidence available. The ACMD's advice submitted in January 2013[64] was summarised thus: "khat has no direct causal link to adverse medical effects, other than a small number of reports of an association between khat use and significant liver toxicity. Some of the adverse outcomes are *associated* with khat use, *i.e.* a complex interaction of khat with other factors to produce the outcome, but not directly caused by khat use. It is apparent from the evidence on societal harms that it is often difficult to disentangle whether khat is the source of community problems or, to some extent, its prevalence and use is symptomatic of the problems for some individuals and groups within the community. On the basis of the available evidence, the overwhelming majority of Council members consider that khat should not be controlled under the Misuse of Drugs Act 1971. In summary the reason for this is that, save for the issue of liver toxicity, although there may be a correlation or association between the use of khat and various negative social indicators, it is not possible to conclude that there is any causal link. The ACMD considers that the evidence of harms associated with the use of khat is insufficient to justify control and it would be inappropriate and disproportionate to classify khat under the Misuse of Drugs Act 1971. In summary the ACMD considers that the harms of khat do not reach the level required for classification. Therefore, the ACMD recommend that the status of khat is not changed."

However, in a written statement to Parliament on 3 July 2013, the Home Secretary announced that the government was "concerned that we risk underestimating the actual harms of khat in our communities owing to the limitations of the evidence base available to the ACMD. To ensure a proportionate and robust policing response, the government will introduce an escalation framework for the possession of khat for personal use, similar to that in place for cannabis. The government will ban khat so that we can protect vulnerable members of our communities and send a clear message to our international partners and khat smugglers that the UK is serious about stopping the illegal trafficking of khat."[69]

On 24 June 2014, khat became a Class C drug under the provisions of the Misuse of Drugs Act 1971 (Amendment) Order 2014.[70] Khat warnings and penalty notices for disorder with associated fines of £60 or arrests for a third offence can be given to adults possessing khat, whereas those aged under 17 can be considered for a youth caution, youth conditional caution or prosecution. Offences of possession with intent to supply, supply and import or export of khat can attract unlimited fines and/or imprisonment of up to 14 years.

The effect of control may be to force up the price of illicit supplies in the UK, as happened in the USA.[71] There are some (unsourced) claims that the price of a bundle has increased by between three- and six-fold.[72,73] Higher prices probably will not deter hard-core chewers (*qodhadhi*) who will try and source it from wherever they can.[74] An immediate effect was that many local retail outlets stopped offering khat for sale,[150] as did online retailers (*e.g.* http://mykhat.co.uk/). Whilst most *mafrishi* have closed, some may go underground.[149] Short-term unavailability of fresh khat during the eruption of the Icelandic volcano Eyjafjallajökull in April 2010 meant that dried leaves were offered for sale.[29] There are initial indications of a drop in use following the UK khat ban: use fell from 0.2% in 2011/12 to 0.04% in 2014/15 according to the Crime Survey for England and Wales.[76]

9.7 Use

Khat is often used with other substances. A study of out-patients attending a Kenyan general hospital found that 29% chewed khat and two-thirds of these also consumed alcohol.[77] In Somalia, heroin and cocaine have been used by a high proportion of khat users.[78] Griffiths[56] found that 60% of Somali khat chewers in London also smoked cigarettes and 6% used cannabis. Smoking prevalence amongst university instructors in north-west Ethiopia was found to be 13.3% compared to 21% using khat.[79]

UK studies[30–33,36–39,47,53,54,80,81] indicate that most users are male, aged under 40, with relatively high rates of unemployment. From 46% to 83% had commenced using khat by the age of 20, usually being initiated by family or friends. Lifetime use for males ranges from 38% to 83%, female from 7% to 78%; khat is used three times a week on average. Males attended more sessions per week, as did the unemployed. Typically 2–3 bundles are consumed per session (range 0.25–6). Somalis tended to use more than other groups. Amounts increase with age and length of use: 0.5–8 h, average 5–6 h, but for some it could last up to 20 h. Length of sessions increased with age. The majority of all groups used socially, but some use on their own, especially women.

UK studies show that the use of alcohol varies from one ethnic khat-using community to another, with both very low rates and higher rates (up to 80%) being found amongst Somalis, and rates of 85% amongst Yemenis. Some groups reported use of other drugs: cannabis (6–27%), cocaine/crack (up to 15%) and heroin (up to 4%). Use of sedatives/tranquillisers was also reported.

Smoking cigarettes is common across all khat-using groups in the UK (range 35–83% for males, with much lower rates for women). Khat users exhibit even higher rates, up to 95% during khat sessions.[43] Not all groups in the UK report the use of other psychoactive substances. Where they do, cannabis is the most frequently cited (range 6–27%), as well as cocaine/crack and heroin.[30,36,37,39,47,56,80,81]

A bundle of khat typically consists of 15–35 twigs, 25–63.5 cm in length. Anderson *et al.*[6] report that according to an agreement between exporters and dealers in Djibouti, a bundle of khat has to be composed of two-thirds leaf, 28 cm long, weigh 90 g each and with two bundles packed in one plastic bag. In the UK the weight of a bundle is about 200–250 g,[32] and in Sweden about 200 g.[6]

The intoxicating effects of khat require a dosage of 25 g.[82] The amount of khat chewed per user is 100–300 g of leaves and stems over a period of 3–4 h.[83–88] Daily consumption is reported as 250–400 g.[89] The typical amount of khat consumed per day in the UK has been 2–3 bundles (range 0.25–6); Somalis tending to use more than other groups. The amount consumed increases with age[47] and length of khat-using career; there are no gender differences. Stevenson *et al.*[10] have suggested (using Kalix's analysis[12] of the wet weight content of fresh khat leaves) that an intake of 50 g of fresh leaves is equivalent to about 1.5 g of cathinone over a period of 3–4 h.

9.8 Social and Criminal Effects

Links between drugs, including khat, and criminal violence (including deaths) can be differentiated as follows: (a) psychopharmacological – violence due to the effects of the drug itself; (b) economic compulsive – acquisitive crime or "fund-raising" to finance drug purchases; and (c) systemic – violence to protect territory, enforce contracts, establish reputation, eliminate competition, *etc.*[90]

Another factor influencing the spread of khat use is that of legal restrictions. For example, during the period of British control in Kenya, tight restrictions were imposed by law so that the plant was not easily accessible to those outside the areas where it grew. A presidential decree of 1977 led to legal restrictions being placed on the plant and thus its availability in other parts of Kenya.[77]

Where khat is not controlled or regulated, acquisitive crime associated with the shrub tends not to occur or to be at a low level.[29,91] However, where khat is illegal, organised crime – often already involved in smuggling illicit substances – also trades in khat.[91] Alternatively, former legitimate khat traders can diversify into more lucrative drug markets.[73,92]

The banning of khat in the UK has led to a change in the type of khat smuggled into the UK; in addition to fresh leaves, dried leaves are being imported, misleadingly being described as "Chinese tea".[71,74,75,93,94] Another effect of the ban has been the diversion of khat destined for the UK market into the local markets in Somalia, Kenya and Ethiopia, the surpluses causing significant drops in price (to a half or less).[95–97] Similar experiences are reported with regard to *miraa* suppliers in Kenya, who are getting less for their produce in Somalia.[95,98]

9.9   Seizures of Khat

The United Nations Office on Drugs and Crime (UNODC) attempts to collect information on drug seizures or interdictions from every country and territory in the world so as to provide a comprehensive assessment of the state of the global illicit drug trade. However, many countries fail to comply with the annual requests for data, and the systems for gathering such information at national levels are very poor or non-existent in many countries. The statistics on drug seizures submitted to and collated by the UNODC allows an examination of data available online at http://data.unodc.org. Although the data are not comprehensive in terms of all countries and all seizures for all years, it is possible to make the following observations for the period 2010–2014, as at the end of March 2015.

During this period there were fluctuating numbers of countries reporting seizures, but overall 29 countries or territories reported seizures. These were primarily in Europe, Africa and North America, but there were also seizures in the Middle East, Hong Kong and New Zealand. Data submitted suggest that in 2012 the total quantity seized was about 134 115 kg: North America, 75 185 kg; Europe (including Turkey), 64 209 kg; and Africa, 7085 kg. The respective figures for 2011 were: global, 150 113 kg; North America, 58 791 kg; Europe (including Turkey), 90 358 kg; and Africa, 889 kg. The amount seized in particular countries can vary from year to year.

The UNODC[13] noted that the total quantity of khat seized in Europe and North America more than doubled between 2004 and 2012 (from

64 to 150 tons), whereas in Africa the quantity confiscated during this period varied between 1 and 10 tons, probably due to khat not being under national control in many East African countries. The market is also likely to have shifted from Africa to other parts of the world due to the Somali Diaspora, especially Western countries. The amount seized in Europe is likely to increase following the recent controlling of khat in the Netherlands (January 2013) and the UK (June 2014).

9.10 Chemistry

Fresh leaves of khat contain alkaloids of the phenylpropylamine type, of which the two psychoactive constituents are the stimulants cathinone [S-(−)-α-aminopropiophenone] and cathine [S,S-(+)-norpseudoephedrine] (see Figure 9.2). Cathinone (α-aminopropiophenone) has been isolated in variable amounts from fresh leaves. The two isomers of cathinone are roughly equally powerful in inducing release at peripheral noradrenergic nerve endings, but the (−)-isomer is about three times more potent than the (+)-isomer at dopamine terminals in the central nervous system (CNS).[99]

The psychoactive substances in khat act on two main neurochemical pathways: dopamine and noradrenalin. It has been suggested that cathinone, like amphetamine, releases serotonin into the CNS. Both these substances induce the release of dopamine from CNS dopamine terminals, thereby increasing the activity of dopaminergic pathways.[100] Cathinone affects release from storage sites of noradrenalin; this supports the notion that cathinone facilitates the transmission of noradrenalin. It has also been theorised that both cathinone and cathine inhibit the uptake of noradrenalin.[101]

Cathinone is a more powerful stimulant than cathine and is generally regarded as the most important element. However, cathinone is unstable in the presence of oxygen and decomposes within a few days of harvesting or if dried.[36] The stored product loses activity rapidly, becoming physiologically inactive after about 36 h. For maximum potency, khat must be picked in the morning and chewed that afternoon.[102] The red variety is considered to be superior by users and

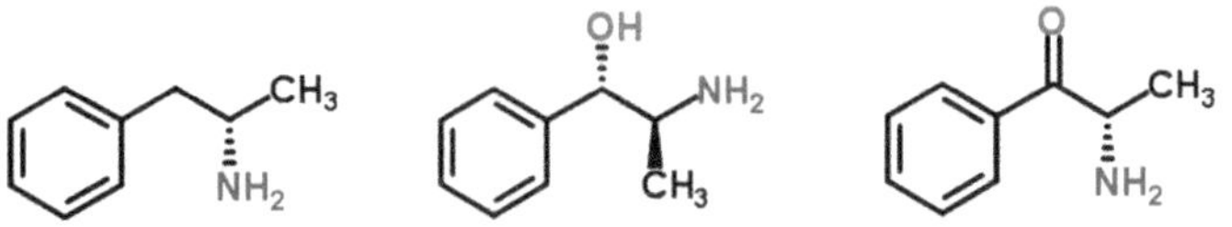

Figure 9.2   Chemical structures of amphetamine, cathine and cathinone.

contains more cathinone than the white variety. A 100 g of fresh khat leaves contains about 26 mg cathinone, 120 mg norpseudoephedrine and 8 mg norephedrine.[12,103] Toennes *et al.*[88] determined the content of khat leaves confiscated at Frankfurt airport as: cathinone 1.14, cathine 0.83 and norephedrine 0.44 mg g^{-1}. Whilst the amount of cathinone in fresh khat has been shown to range from 78 to 343 mg per 100 g,[16] it can range much higher. The study by Toennes *et al.*[88] isolated 1140 mg per 100 g of cathinone after a sample of khat had been stored at -20 °C before analysis. The cathine content of dried khat leaves ranges from 0.1 to 0.2%,[104] and has about one-tenth the stimulant activity of *d*-amphetamine.

9.11 Pharmacology

9.11.1 Psychological Effects

The psychotropic effects of khat are caused by the amphetamine-like compounds. Although amphetamine and cathinone act on different parts of the brain, they share common psychological effects, including an interaction with the dopaminergic pathways.[105] Central stimulation by khat is manifested by euphoria, increased alertness, garrulousness, hyperactivity, excitement, aggressiveness, anxiety, elevated blood pressure and manic behaviour. The simulation lasts for about 3 h.[12] A depressive phase, including insomnia, malaise and a lack of concentration, almost always follows.[106,107] For a detailed description of the different stages of intoxication associated with khat chewing, see Al-Motarreb and Broadley.[16] True psychotic reactions occur with much less frequency than with amphetamines. This is most likely because of the self-limiting dose of khat, which does not permit blood levels of the active compounds to rise high enough for toxic psychosis to occur. However, paranoid delusions, usually persecutory, have been seen.[108]

It has been postulated that khat use can exacerbate psychotic symptoms in individuals with pre-existing conditions, and precipitate psychiatric disorders in vulnerable subjects.[62] Some commentators suggest that it is not clear whether khat use may cause a psychotic disorder in an otherwise healthy individual or trigger the onset of schizophrenia in an individual with high vulnerability to the condition,[109,110] There have been suggestions that khat chewing exacerbates symptoms in patients with pre-existing psychiatric conditions, possibly caused by the sympathomimetic actions of cathinone on the CNS.[107] Odenwald *et al.*[111] propose that it is not khat consumption *per se* but specific patterns of its use are associated with the development of psychotic symptoms: early onset in life, excessive chewing (more than two bundles per day) and use as self-medication for war trauma-related symptoms. Other studies have found morbidity increased when daily consumption exceeds two bundles[112] or are dose related.[113,114]

Two main forms of psychoses associated with khat have been described in the literature: paranoid or schizophreniform and mania. The former is typically associated with a recent increase or heavy consumption of khat, the latter with unaccustomed heavy or prolonged consumption. Thus, for example, reviewing 16 cases (including 12 previously reported by Pantelis *et al.*[115]), Yousef *et al.*[62] noted that the majority (94%) were male, and 69% were in their 20s. A past personal psychiatric history was reported in 12.5% of cases, and a family psychiatric history in 19%. The majority (94%) of psychotic episodes were preceded by heavy consumption of khat, rapid resolution in 87.5%, but re-occurrence caused by khat use in 62.5%. Clinical features included manic symptoms (56%), paranoid delusions (69%) and auditory hallucinations (44%). There was a significant increase (94%) in psychiatric morbidity when consumption was excessive, and nearly two-thirds (62.5%) reported recurrence of symptoms when khat use was resumed.

Mental disturbances after khat use (typically anxiety, depression and mood swings) are commonly reported in UK studies, suggesting that there is potential for an increased incidence in the country.[27,28] Depression has also been related to khat chewing, and on occasion with self-harm and suicide, both attempted and completed.[113] Such cases have occurred during both chewing sessions and the subsequent intoxication state.[114,116]

9.11.2 Dependence and Withdrawal

Rural khat chewers in Ethiopia, who are mostly devout Muslims, make a distinction between khat "users" and khat "abusers". The latter are those who use alcoholic drinks to counteract the stimulant effects of the plant, and goes against the norms and customs associated with khat chewing.[117] In the mid-1990s it was suggested that the Somali community in London considered the chewing of more than two bundles a day as an indulgence, which is acceptable on occasions, but such an amount consumed on a regular basis would be regarded as excessive.[62] Some writers have suggested that the withdrawal symptoms should be interpreted as a "rebound phenomenon rather than a specific abstinence syndrome",[115] as a type of "cultural drug dependence" or as a "drug-facilitated sociability dependence".[21]

There is some debate as to whether or not *Catha edulis* can actually cause dependence. Physical dependence to khat does not occur. The available literature suggests that it is likely that khat use has the potential to develop into dependence.[104] However, it seems this is much less likely than dependence on stimulants such as amphetamine and more like the type of dependence seen with caffeine. The vast majority of people who chew khat do not use it in a dependent fashion and there is no evidence of more widespread drug misuse amongst khat users.

Tolerance to *Catha edulis* does not often occur; if it does, the doses are only very slowly increased. This may be due to the properties of the plant itself or to the physical limits on the amount that can be consumed.[104,118] Development of tolerance to the effects of cathinone is more rapid than to that of amphetamine, and there is cross-tolerance between cathinone and amphetamine.[119,120]

9.11.3 Medical Effects

Severe adverse effects have been associated with khat use: migraine, cerebral haemorrhage, myocardial infarction and pulmonary oedema, especially in older and predisposed individuals. Cardio-vascular effects occur within 15–30 min after the ingestion of khat, suggesting the absorption of the active ingredients through the oral mucosa. Effects include tachycardia and palpitations. In normal non-khat using volunteers, chewing khat leads to a significant increase in systolic and diastolic blood pressures, persisting for between 3 and 4 hours after the onset of chewing[88,121] and not returning to baseline an hour after chewing.[122] These effects may result from cathinone acting as an indirect sympathomimetic agent, facilitating the release of catecholamines from sympathetic nerve terminals and leading to significant increases in heart rate. There is evidence that tolerance develops to these sympathomimetic effects in chronic users.

Cases involving cardiovascular and cerebrovascular complications have been reported.[123,124] Khat use has important effects on acute cerebral infarction (ACI). Khat increases blood pressure more in ACI patients when compared to a control group.[125] The plant is more likely to affect older users, and more males (74.5%) than females (25.4%). A coronary spasm caused by khat is thought to have triggered ventricular fibrillation, leading to a fatal cardiac arrest, in one patient.[126]

Chronic khat use may lead to a marked deterioration of psycho-physical function, although noticeable symptoms of impairment (such as ability to drive a vehicle) are not visible after a single administration of khat.[127] However, there has been at least one instance of an individual whose driving was considered to be impaired through the use of khat.[128] Impaired motor co-ordination and fine tremor have been noted.[129] Vision can also be affected. A staring look is characteristic of many individuals under the influence of khat, and often mydriasis has been noted.

The effects of khat consumption on the reproductive system, pregnancy and neonates have also been noted. Khat is genotoxic, having teratogenic effects on the foetus if regularly consumed by pregnant mothers. Neonates have low birth weights – a risk factor for perinatal and young infant death.[130–132] Currently chewing lactating women have been found to excrete norpseudoephedrine in their breast milk, and traces were found in the urine of a breast-fed infant.[133] Cardiovascular complications amongst pregnant khat-chewing females have been reported: chest pains, tachycardia and hyperthermia.[134]

9.12   Toxicology

Administration of *Catha edulis* extracts has demonstrated a capacity to induce cytotoxic effects in liver and kidney cells, as well as cell death in various human leukaemia cell lines and peripheral human blood leukocytes.[135] Although studies on animals have shown that khat is hepatotoxic, with increases in liver enzymes, and there has been histopathological evidence of acute hepato-cellular degeneration,[136] it has only been recently that clinical human case reports have been published.[137,138] One case report suggested that hepatitis was probably due to right heart failure, which in turn was possibly due to direct khat toxicity as it did not respond to withdrawal of statins or proton pump inhibitors.[138]

Chapman *et al.*[139] report on six cases (two female) of khat chewers with severe, acute hepatitis that resulted in death or liver transplantation. As no other cause for their liver disease was found, and since re-exposure to khat-chewing resulted in further liver damage, a high concentration of cathinone was found in a sample of damaged liver tissue 3 weeks after the patient's last khat use, and there were no significant features to suggest an autoimmune or immunoallergic process in these cases, it was concluded that khat consumption was the aetiological cause of their liver injuries.

A study of adults in Yemen with acute sporadic hepatitis and jaundice found no viral cause in 51% of cases. Whilst there may be an unknown virus responsible or environmental toxins such as pesticides in khat leaves,[140] when khat chewing is stopped liver function test abnormalities resolve. Young Somali males who chew khat appear to be particularly at risk of autoimmune hepatitis[141,142] and liver disease.[142,143] Five Somali and one Yemeni male khat users (mean age 42.8) presenting to Sheffield hospitals between 2005 and 2010 with acute hepatitis were assessed with autoimmune hepatitis.[144] Two Somali males presented to UK hospitals with severe khat-related hepatitis that resolved on initial cessation of khat consumption, but resumption resulted in relapse.[145]

The dangers of synthetic forms of the active ingredients of khat should be noted, especially given the use of such substances amongst new populations who use it recreationally, *e.g.* in Israel. Bentur *et al.*[146] reported cases of poisoning with "hagigat" (Hebrew for "party") amongst patients aged 16–54 years of age.

The growing use of the internet as means of obtaining drugs of abuse,[147,148] including ordering khat products, has led to the widespread use of novel psychoactive products, especially those chemically related to khat, *i.e.* synthetic cathinones such as mephedrone, methylone, *etc.*[149,150] (see Chapter 6 for a further detailed discussion of these molecules).

9.13   Deaths

A paper designed to generate awareness of the range of potential contexts in which khat cultivation, transportation, trade and consumption can lead to fatalities was published.[27] This paper was not just about pharmacology, toxicology, *etc.*, but also about the wider phenomenon of deaths linked with a particular substance: khat. Its specific aims were to: (a) create a framework for discussion around the topic of khat-related death by generating a broad-based definition; (b) suggest dimensions of phenomena/aspects to consider without *a priori* theoretical paradigms/mindsets; (c) elucidate such dimensions by means of theoretical and actual examples. Apart from one paper looking at a series of khat-related fatalities in the UK reported to the National Programme on Substance Related Deaths (NPSAD),[28] there are still no known published reviews of, or statistics on, khat-related mortality howsoever defined. This is and continues to be an important gap in the knowledge base.

Following the journey of khat from its start to finish, from cultivation to consumption, provides a useful vehicle to identify such themes and to derive a taxonomy, based on context(s) in which death occurred or was caused, *e.g.* its aetiology and manner of occurrence, based on the observation of reported incidences/events and *post hoc* abstraction from observed phenomena (Table 9.1). The limitations of this approach are described in Corkery *et al.*[27,28]

Apart from poisoning from insecticide used on the khat crop, traumatic deaths can be a feature of the cultivation, transportation and trading activities undertaken by distributors and sellers. Violence can also occur in other settings. For example, psychopathological effects: (i) impaired judgement leading to accidents and violence; (ii) causing or exacerbating psychoses or causing depression leading to suicide and even homicide.[27,28]

9.13.1 Consumption-related Deaths

Table 9.2 summarises[27] some examples from a case-series of 15 UK deaths associated with the use of khat. Some of the data cannot be fully displayed in this publication. Please contact the author for full and further information.

Physiological effects encompass (a) mechanical problems, *e.g.* choking on pieces of the plant, and (b) toxicity, *e.g.* (i) causing heart problems leading to fatal heart attacks or (ii) liver failure. There is now a fairly well-developed literature on the cardiac and circulatory problems caused by khat, especially with regard to myocardial infarction. There is now evidence that such consequences can be triggered by high levels of the principal psychoactive substances in khat, *i.e.* left ventricular failure and acute pulmonary oedema. The physiological effects of khat consumption are quite clear for a number of cases. Khat chewing is a recognised risk factor for myocardial infarction.[11,106,151–153]

Until relatively recently, there have been few documented cases of khat causing liver failure. In three NPSAD cases, khat consumption ultimately led to liver failure and death. Khat toxicity was responsible in two cases for hepatic necrosis and sub-fulminant liver failure; and in the third case was due to sub-acute liver failure, in the presence of autoimmune hepatitis. These presentations of "khat-induced" liver

Table 9.1 Taxonomy of "khat-related" mortality.[a]

Directness	Type of association	Mechanism	Example
Trade related	Cultivation/ production	Poisoning	Fertiliser/pesticide not washed before consumption
		Disputes between actors	Disagreement over wages → homicide; possibly aggravated effects of khat chewing clouding judgement and triggering aggression
	Transportation	Fatigue	Long hours and driving at high speed → loss of concentration (can be in association with khat use to keep awake), leading to accidents
		Loss of concentration	Distraction whilst preparing khat for chewing whilst driving → accidents
		External factors	Bad weather conditions/mechanical failure → plane crashes
		High speed	High speed trying to escape from law enforcement officers → collision → fatal injuries
		High speed and mechanical failure	High speed, over-loaded vehicle, faulty brakes → failing to negotiate bend in road → fatal accident
	Distribution/ marketing/ wholesale	Disputes between actors	Fighting over 'turf' → violence and homicide
		Disputes between actors	Refusal to do business → violence and homicide
		Disputes between actors/theft	Possible fighting over 'turf' or theft going wrong → homicide
	Retail	Disputes over price	Homicide; possibly aggravated effects of khat chewing clouding judgement and triggering aggression

Category	Subcategory	Type	Description
	Violence in khat markets/ terrorism		Detonation of hand grenade in khat market; militia firing guns indiscriminately into crowds; suicide bomber exploding explosives immediately after chewing khat
	Law enforcement agencies	Execution	Beheading for possession and use of khat
		Hot pursuit	Smuggler trying to escape border guards
	Drug control	Opposition to imposition of ban on khat importation	Protestors against ban are fired on by militia enforcing ban
		Breach of drug controls	Trader killed by militia enforcing khat-selling restrictions
Consumption-related (medical)	Physiopathological	Mechanical	Choking on leaves/twigs or airway obstruction →asphyxia or cardiac arrest
		Toxicity	Myocardial infarction →fatal heart attack
			Liver failure
		Cancer related to khat use	Oral and other cancers caused by khat chewing→death
		Endocarditis related to khat chewing	Streptococcal bacterium *Gemella morbillorum* being ingested while khat chewing → endocarditis→death
		Reproductive health issues	Teratogenicity/low birth weight →perinatal and young infant mortality
	Neurological	Lack of motor co-ordination, shaking	Reduces control→accidents
		Eyesight problems	Impairs sight and focus →accidents
	Psychopathological	Causing and/or exacerbating psychosis and/or depression	Suicide and/or homicide, accidental overdose

Table 9.1 (*Continued*)

Directness	Type of association	Mechanism	Example
		Increases aggression and notions of personal supremacy	Violence in Somalian civil war → homicide; fighting over khat air-freight deliveries
		Impaired judgement/assessment of risk → accidents	Road traffic accident
		Impaired judgement/intoxicating effects	Misunderstanding/misinterpretation of others' motives/actions → offending others → retaliation involving fatal assault
			Fall from height
	Other medical contributory factors	Used with other psychoactive substances and/or positive toxicology	Role may not be clear
		Cerebral haemorrhage	
		Ischaemic stroke	

[a]This is a theoretical framework based on the findings of this research, and is subject to revision as new information is reported. Its aim is to provide a starting point for debate and refinement. Reproduced with permission.[27]

Table 9.2 Main characteristics of NPSAD 'khat-related' deaths, UK, 2004–2009.[a]

Characteristic	NPSAD no.					
	2	5	9	10	13	15
Year of death	2004	2005	2006	2008	2008	2009
Age (years)	32.5	27.3	37.8	29.9	19.3	31.9
Gender	Male	Male	Male	Male	Male	Female
Ethnicity	Somali	Somali	Somali	White (Polish)	Somali	Somali
Marital status	Divorced	Single	Not known	Not known	Single	Married
Living arrangements	With sibling	Alone	Alone	With friends	Not known	With partner
Employment status	Invalidity/ sickness benefits	Unemployed	Mini-cab driver	Employed (manual)	Student	Unemployed
UK residency (years)	15	3	Not known	Not known	Not known	4.5
Significant medical history	Yes	No	Not known	Not known	No	None
Known psychiatric history	Yes	No	Not known	Not known	No	Depression, mental health issues, on prescribed medications

Table 9.2 (*Continued*)

Characteristic	NPSAD no.					
	2	5	9	10	13	15
Known khat using history	Yes	Yes	Not known	Not known	Not known	but not being used Alcohol and drug abuse; chewed two bundles of khat, 2–4 times per week
Evidence of using khat prior to death	Yes	Yes	Yes	Yes	Yes	Yes
Place of death	Car park	At home	Road	At home	Hospital	Outside block of flats
Role of khat in death	Possible suicide/accidental fall whilst judgement impaired (found in body)	Ingestion of khat → high norephedrine levels → left ventricular failure → pulmonary oedema	Alcohol and khat in system → impaired judgement/co-ordination → loss of control of vehicle → traumatic road traffic accident (driver)	Overdose of injected heroin, but khat also in system	Had consumed alcohol, cannabis and khat which may have contributed to aggressive behaviour, leading to his assault and death	Its psychotic effects or hallucinations may have led her to jump or its effects could have led to an error of judgement

Cause(s) of death	1a Multiple injuries [severe multiple injuries, impact onto a firm surface, fall from a height]	1a Acute pulmonary oedema [left ventricular failure, high norephedrine levels] 1b Ingestion of khat	1a Multiple injuries	1a Morphine intoxication	1a Severe head injury [consistent with impacts from heavy blunt object]	1a Multiple injuries consistent with a fall whilst under the influence of khat
Verdict	Open	Narrative	Accident	Non-dependent abuse of drugs	Unlawful killing [attackers sentenced to life imprisonment]	Accidental
Coroner's area	Inner West London	North London	North London	North London	West London	Northants

aThese cases illustrate a range of deaths linked to khat consumption, from purely circumstantial to clear and unambiguous physiological causes. Reproduced with permission.27

failure (the Corkery khat liver condition) are consistent with findings in respect of hepatitis[137,139,141,154] and liver disease.[142] Such cases highlight that khat is a potential hepatotoxic agent and that this phenomenon needs further investigation and documentation.

There are probably more deaths that have occurred in the past and which are currently happening but that are not identified, recorded or reported in the literature as being related to the use of khat. This has meant that there is a demonstrable lack of documented cases in the literature, as well as nothing by way of quantitative data. These gaps in knowledge need to be filled, employing systematic and scientifically based approaches. This will lead to a much better understanding of the potential risks of death associated with the use of khat, based on empirical observation.

9.14   Toxicological Findings

Toxicology was not undertaken in all the NPSAD cases since the cause of death was known for three individuals who died in hospital. In two further cases there were negative results for khat. The results for the illustrated cases from Table 9.2 are given in Table 9.3; levels are only available for these six cases. There are only a few reports in the

Table 9.3   Summary of measured toxicological findings for the psychoactive constituents of khat (*Catha edulis* Forsk) in deceased humans.[28]

NPSAD case no.	Findings
2	Blood: cathinone 104 µg L^{-1}; norephedrine/norpseudoephedrine (total) 154 µg L^{-1}; zuclopenthixol 123.8 µg L^{-1}
5	Blood: cathinone <50 µg L^{-1}; norephedrine/norpseudoephedrine (total) 1000 µg L^{-1} Urine: norephedrine/norpseudoephedrine detected
9	Vitreous humour: alcohol 210 mg per 100 mL; cathinone 110 µg L^{-1}; cathine 310 µg L^{-1}
10	Blood: cathinone <25 µg L^{-1}; norephedrine/norpseudoephedrine (total) <50 µg L^{-1} Urine: cathinone detected; norephedrine/norpseudoephedrine detected
13	Blood: cathinone 19 µg L^{-1}
15	Peripheral blood: cathine 1447 µg L^{-1}; cathinones 122 µg L^{-1}

literature for levels in living humans of the psychoactive substances found in khat, and none for fatalities. There are three studies involving a total of 17 healthy adults, typically naïve volunteer users and one of 19 forensic non-fatal road traffic accidents. This severely limits interpretation and application to toxic/lethal/fatal doses. The previously published levels for cathinone, norpseudoephedrine/cathine and norephedrine in living persons are summarised in Table 9.4. Thus the data for deceased individuals given in Tables 9.2 and 9.3 will be of great interest to toxicologists and physicians alike.

How do the findings for the UK fatalities compare to the previously published results? The cathinone level for case 2 falls within the range given for the healthy adults and, as with those volunteers, well above the range for the forensic cases; the cathine level is higher than that given by Widler et al.[121] for healthy volunteers but in the middle of the range for the forensic cases.[127] For case 5 the cathinone level

Table 9.4 Summary of previously published toxicological findings for the psychoactive constituents of khat (*Catha edulis* Forsk) in living humans.[a]

Ref.	Study population	Findings
Brenneisen et al. (1990)[159]	Six healthy adult males	Maximum plasma concentration for cathinone after 1.2 h $= 110$ μg L^{-1}
Widler et al. (1994)[121]	Six healthy adult males	Peak plasma concentrations after 2.1 h: cathinone $= 74$–180 μg L^{-1} (max $= 127$ μg L^{-1}); norpseudoephedrine/ cathine $= 40$–138 μg L^{-1}; norephedrine $= 59$–161 μg L^{-1}
Halket et al. (1995)[155]	Five healthy adults (two female, three male), naïve users	Peak plasma levels for cathinone after 1.5–3.5 h $= 41$–141 μg L^{-1} (mean 83 μg L^{-1})
Toennes and Kauert (2004)[127]	19 forensic non-fatal road traffic accidents, aged 24–54 years	Urine: cathinone $= 0.1$–28.8 mg L^{-1} (median 8 mg L^{-1}); cathine $= 10.2$–> 300 mg L^{-1} (median 38.6 mg L^{-1}); norephedrine $= 10.2$–> 300 mg L^{-1} (median 43.0 mg L^{-1}) Serum: cathinone $= 9.2$–73.0 μg L^{-1} (median 33.2 μg L^{-1}); cathine $= 16.5$–314.7 μg L^{-1} (median 128.9 μg L^{-1})

[a]Reproduced with permission.[28]

was well within the range for healthy volunteers and the forensic cases; however, the norephedrine/norpseudoephedrine level far exceeds those given by Widler *et al.*[121] for healthy volunteers and the range for the forensic cases.[127] The results from vitreous humour for case 9 indicate that for cathinone the level is within the range for healthy adult volunteers but higher than that for forensic cases; for cathine the results are far higher than for the volunteers[121] but similar to the top of the range for forensic cases.[127] Case 10 and 13's blood results for cathinone are well within the ranges seen for healthy adults, healthy volunteers, naïve users[155] and forensic cases; the norephedrine/norpseudoephedrine levels are within the ranges for healthy adults and forensic cases. Blood cathinone levels for case 15 are within the ranges for healthy adults and healthy volunteers, but higher than in forensic sciences; the cathine level is far higher than that seen in both healthy adults and forensic cases. The cathine levels in cases 2, 9 and 13 exceed those in volunteers and in case 2 for forensic cases. As with other substances, there appears to be some considerable overlap between non-fatal and lethal levels of the major psychoactive substances in khat.

All but one of the decedents were male, aged between 22 and 47 (mean 35) years at death and thus comparatively young. All but one were of East African descent (one was born in Russia, one in Kuwait), 13 being Somali nationals and one Eritrean. However, one of the deceased was Polish; the correct ascertainment of this case is assured since the presence of both cathinone and norephedrine/norpseudoephedrine confirmed through analyses employing GM-CS and LC-MS by the one of the authors[27,28] demonstrated that khat had indeed been found rather than some other substance, such as a cathinone derivative. These findings are not in error; they clearly indicate the consumption of khat. This case suggests the possibility of the spread of khat use outside the traditional ethnic khat-using populations to other communities and substance-using groups in society.

9.15 Conclusions

Khat has been important socially and culturally in the Horn of Africa and Arabia for many centuries. Its use is continuing to spread amongst those in the Diaspora from the war-torn regions of East Africa where its consumption is part of the cultural tradition. Both in the countries of origin and destination it serves a number of beneficial roles: from providing cultural identity to self-medication. Whilst there are commonalities in approach to khat consumption in both settings, customs and practice are constantly evolving. New patterns of use are emerging in younger cohorts of users, there is increased use reported amongst women, as well as transmission to new ethnic communities. Whilst concomitant tobacco smoking continues as a feature of chewing sessions, users are consuming alcohol and other psychoactive substances. Synthetic derivatives of cathinone are becoming an issue amongst recreational drug users and are now available over the internet. Monitoring of such developments is vital to understanding the dynamics of change in this arena.

Khat use brings with it detrimental results in terms of damage to the community, families and individuals. This is evidenced through high unemployment rates, lack of financial resources for families, broken relationships, and impaired health – and even death. On the other hand, benefits can accrue through the provision of employment opportunities, generation of income and tax revenues, helping to create identity and cohesion within the community, especially when peoples are displaced.

The health risks associated with khat use and its psychoactive constituents appear to be greater than previously realised. Khat research is still in its infancy and there is little robust information on the subject.[156] Furthermore, the lack of negative health results for khat in the literature should not lead to complacency or an assumption that khat use is free from toxic consequences.[157] The NPSAD cases referred to above are thought to be the first reported (as distinct from recorded) cases of death due to khat toxicity since the case reported by Heisch.[158] This absence of negative reports is due to a lack of population-based studies,[111] particularly in respect of the toxicology of *Catha edulis*. Whilst anecdotal reports are informative, systematic investigations are needed to determine the incidence and prevalence of ill-effects of khat use. Only then can the best methods of supplying preventative and therapeutic interventions be considered in an informed way.[27,28]

These dangers, including serious adverse effects that can lead to death, need to be brought to the attention of those in producing/ growing countries, and those countries/regions that have become hosts to ever-increasing communities from these countries. There is also a need to consider the dissemination of such information to non-traditional groups that appear to be emerging as consumers of khat, *e.g.* those turning from the use of synthetic substances to "natural" herbal substances. This will need to be specifically focused on target groups using a variety of media and platforms, including a variety of languages.

Khat continues to act as a "cultural cement" for communities in both traditional and Disapora regions, but it has the potential to cause harm (mental, physiological and social), and its chemical constituents are acting as a catalyst for the development of novel psychoactive substances. The roles of *Catha edulis* continue to change.

References

1. P. J. Greenway, Khat, *East Afr. Agric. J.*, 1947, **13**, 98–102.
2. UNODC, Khat, *Bull. Narc.*, 1956, **4**(3), 6–12.
3. C. Brooke, Khat (Catha Edulis): Its production and trade in the Middle East, *Geogr. J.*, 1960, **126**(1), 52–59.
4. G. A. Balint and E. E. Balint, On the medico-social aspects of khat (catha edulis) chewing habit, *Hum. Psychopharmacol.*, 1994, **9**(2), 125–128.
5. INCB, *Report of the International Narcotics Control Board for 2006*, United Nations, New York, 2007, Available at: https://www.incb.org/documents/Publications/AnnualReports/AR2006/AR_06_English.pdf, Accessed on 10 July 2015.
6. D. Anderson, S. Beckerleg, D. Hailu and A. Klein, *The Khat Controversy: Stimulating the Debate on Drugs*, Berg Publishers, Oxford, 2007.
7. B. Manning, Home remedy or dangerous drug? *Daily Despatch.* 6 October 1999. Available at: http://www.dispatch.co.za/1999/10/26/features/DRUGORNO.HTM. Accessed on 2 August 2008.
8. Chewing the khat is illegal, says Sanab, *Saturday Star* 28th April 2000. Quoted at: http://www.iol.co.za/news/south-africa/chewing-the-khat-is-illegal-says-sanab-1.35917, Accessed on 10 July 2015.
9. J. Stefan and B. Mathew, Khat chewing: an emerging drug concern in Australia?, *Aust. N. Z. J. Psychiatry*, 2005, **39**(9), 842–843.
10. M. Stevenson, J. Fitzgerald and C. Banwell, Chewing as a social act: cultural displacement and khat consumption in the East African communities of Melbourne, *Drug Alcohol Rev.*, 1996, **15**(1), 73–82.
11. A. Al-Motarreb, M. Al-Kebai, B. Al-Adhi and K. J. Broadley, Khat chewing and acute myocardial infarction, *Heart*, 2002, **87**(3), 279–280.
12. P. Kalix, Catha edulis, a plant that has amphetamine effects, *Pharm. World Sci.*, 1996, **18**(2), 69–73.

13. UNODC, *Global Synthetic Drugs Assessment 2014: Amphetamine-type Stimulants and New Psychoactive Substances*, United Nations Office on Drugs and Crime, Vienna, 2014, Available at: https://www.unodc.org/documents/scientific/2014_Global_Synthetic_Drugs_Assessment_web.pdf. Accessed on 10 July 2015.

14. J. G. Kennedy, J. Teague, W. Rokaw and E. Cooney, A medical evaluation of the use of qat in North Yemen, *Soc. Sci. Med.*, 1983, **17**(12), 783–793.

15. I. Dhaifalah and J. Šantavý, Khat habit and its health effect. A natural amphetamine, *Biomed. Pap.*, 2004, **148**(1), 11–15.

16. A. L. Al-Motarreb and K. J. Broadley, Coronary and aortic vaso-constriction by cathinone, the active constituent of khat, *Auton. Autaccid. Pharmacol.*, 2003, **23**(5–6), 319–326.

17. V. C. Freund-Michel, M. A. Birrell, H. J. Patel, I. M. Murray-Lyon and M. G. Belvisi, Modulation of cholinergic contractions of airway smooth muscle by cathinone: potential beneficial effects in airway diseases, *Eur. Respir. J.*, 2008, **32**(3), 579–584.

18. E. L. Margetts, Miraa and myrrh in East Africa-clinical notes about Catha edulis, *Econ. Bot.*, 1967, **21**, 358–362.

19. A. D. Krikorian, Kat and its use: an historical perspective, *J. Ethnopharmacol.*, 1984, **12**(2), 115–178.

20. C. M. McKee, Medical and social aspects of qat in Yemen: a review, *J. R. Soc. Med.*, 1987, **80**, 762–765.

21. J. G. Kennedy, *The Flower of Paradise – The Institutionalized Use of the Drug Qat In North Yemen*, D. Reidel, Dordrecht, 1987.

22. D. L. Browne, Qat use in New York City, in *NIDA Research Monograph 105. Problems of Drug Dependence 1990* – Proceeding of the 52nd Annual Scientific Meeting, National Institute on Drug Abuse, 1991, pp. 464–465. Available at: http://archives.drugabuse.gov/pdf/monographs/105.pdf, Accessed on 10 July 2015.

23. F. Grignon, "Touche pas à mon khat!" Rivaltés Meru-somali autour d'un commerce en pleine expansion. ["Don't touch my khat!" Meru-Somali rivalries concerning a trade in full expansion], *Politique Africaine*, 1999, **73**(31), 177–185.

24. M. Barko and A. Saleh, *Report of the drugs misuse needs assessment carried out by African Health for Empowerment and Development (AHEAD) amongst the Sierra Leoneans and Ugandans community in Greenwich and Bexley Borough*, AHEAD, London, 2001. Available at: http://www.uclan.ac.uk/facs/health/ethnicity/communityengagement/documents/African%20Health%20for%20Empowerment%20&%20Development%20(AHEAD,%20London).pdf. Accessed on 18 May 2007.

25. A. Hogg and H. Rogers, Arab danger drug on sale legally in Britain, *The Sunday Times*, 20th January 1985, pp. 1–2.
26. E. R. S. Brown, D. R. Jarvie and D. Simpson, Use of drugs at 'raves', *Scott. Med. J.*, 1995, **40**, 168–171.
27. J. M. Corkery, F. Schifano, A. Oyefeso, A. H. Ghodse, T. Tonia, V. Naidoo and J. Button, Review of literature and information on 'khat-related' mortality: a call for recognition of the issue and further research, *Ann. Ist. Super. Sanità*, 2011, **47**(4), 445–464.
28. J. M. Corkery, F. Schifano, A. Oyefeso, A. H. Ghodse, T. Tonia, V. Naidoo and J. Button, 'Bundle of fun' or 'bunch of problems'? Case series of khat-related deaths in the UK, *Drugs: Educ., Prev. Policy*, 2011, **18**(6), 408–425.
29. S. Kassim, A. Dalsania, J. Nordgren, A. Klein and J. Hulbert, Before the ban – an exploratory study of a local khat market in East London, U.K, *Harm Reduct. J.*, 2015, **12**(1), 19.
30. G. Tewolde, B. Zerihun and W. Ekinu, *Report of the Drugs Misuse Needs Assessment Carried Out by Eritrean Community in Greenwich and Lewisham (ECGL) Amongst the Somali and Eritrean Communities in Greenwich*, ECGL, London, 2001, Available at: http://www.uclan.ac.uk/facs/health/ethnicity/communityengagement/documents/Eritrean%20Community%20in%20Greenwich%20(London).pdf, Accessed on 18 May 2007.
31. A. Klein, *Khat in Streatham: formulating a Community Response*, A DrugScope Report for the Streatham Town Centre Office and the Lambeth Drug and Alcohol Action Team, DrugScope, London, 2004.
32. S. L. Patel, S. Wright and A. Gammampila, *Khat use among Somalis in four English cities*, Home Office Online Report 47/05, Home Office Research Development and Statistics Directorate, London, 2005, Available at: http://webarchive.nationalarchives.gov.uk/20110218135832/http:/rds.homeoffice.gov.uk/rds/pdfs05/r266.pdf, Accessed on 10 July 2015.
33. S. Kassim and J. Croucher, Khat chewing amongst UK resident male Yemeni adults: an exploratory study, *Int. Dent. J.*, 2006, **56**(2), 97–101.
34. J. Bashford, J. Buffin and K. Patel, *The Department of Health's Black and Minority Ethnic Drug Misuse Needs Assessment Project: Part 2: The Findings*, Centre for Ethnicity and Health, University of Central Lancashire, Preston, 2003, Available at: http://clok.uclan.ac.uk/2591/1/Buffin_rep2comeng2.pdf, Accessed on 12 July 2015.

35. H. J. B. H. M. Dupont, C. D. Kaplan, H. T. Verbraeck, R. V. Braam and G. V. van de Wijngaart, Killing time: drug and alcohol problems among asylum seekers in the Netherlands, *Int. J. Drug Policy*, 2005, **16**(1), 27–36.

36. P. Griffiths, M. Gossop, S. Wickenden, J. Dunworth, K. Harris and C. Lloyd, A transcultural patterned drug use: qat (khat) in the UK, *Br. J. Psychiatry*, 1997, **170**, 281–284.

37. Y. Gatiso and L. W. S. Jembere, *Report of the Drugs Misuse Needs Assessment Carried out by Ethiopian Community in Lambeth (ECL) Amongst the Ethiopian Community in Lambeth, South London*, ECL, London, 2001, Available at: http://www.uclan.ac.uk/facs/health/ethnicity/communityengagement/documents/Ethiopian%20Community%20in%20Lambeth%20(London).pdf, Accessed on 18 May 2007.

38. A. Y. H. Al-Osaimi, N. Al-Kash, H. Yafai and M. Yafai, *Report of the Drugs Misuse Needs Assessment Carried Out by the Yemeni Community Association (YCA) Amongst the Yemeni Community in Sandwell*, YCA, Sandwell, 2001, Available at: http://www.uclan.ac.uk/facs/health/ethnicity/communityengagement/documents/Yemeni%20Community%20Association%20(Sandwell).pdf, Accessed on 18 May 2007.

39. M. A. Mela and A. J. McBride, 'What's in the banana leaves? Khat uses poses health threats to Somalis in Cardiff', Paper to Annual Meeting: *Psychiatry Today*, 24–27 June 2002. Cardiff.

40. N. Carrier, The need for speed: contrasting timeframes in the social life of Kenyan *miraa*, *Africa*, 2005, **75**(4), 539–558.

41. NDIC, *Intelligence Bulletin – khat (Catha edulis)*, National Drug Intelligence Centre, US Department of Justice, Johnstown, PA, 2003, Available at: http://www.justice.gov/archive/ndic/pubs3/3920/3920t.htm, Accessed on 10 July 2015.

42. ACMD, *Khat (Qat): Assessment of Risk to the Individual and Communities in the UK*, Advisory Council on the Misuse of Drugs, Home Office, London, 2006, Available at: https://www.gov.uk/government/uploads/system/uploads/attachment_data/file/119095/Khat_Report_.pdf, Accessed on 10 July 2015.

43. UNODC, *2004 World Drug Report-Volume 2: Statistics*, United Nations Office on Drugs and Crime, Vienna, 2004.

44. UNODC, *2006 World Drug Report-Volume 2: Statistics*, United Nations Office on Drugs and Crime, Vienna, 2006.

45. S. P. Gough and I. B. Cookson, Khat-induced schizophreniform psychosis in UK, *Lancet*, 1984, **323**(8374), 455.

46. S. P. Gough and I. B. Cookson, Khat-induced paranoid psychosis, *Br. J. Psychiatry*, 1987, **150**, 875–876.
47. D. Nabuzoka and F. A. Badhadbe, Use and perception of khat among young Somalis in a UK city, *Addict. Res.*, 2000, **8**(1), 5–26.
48. J. Mayberry, G. Morgan and E. Perkin, Khat-induced schizophreniform psychosis in UK, *Lancet*, 1984, **323**(8374), 455.
49. A. Busby, On sale: leaf drug that led to death, *The Observer*, 18 October 1987, p. 5.
50. Home Office, *Drug Misuse Declared: Findings From the 2011/12 Crime Survey for England and Wales*, Home Office, London, 2nd edn, 2012, Available at: https://www.gov.uk/government/uploads/system/uploads/attachment_data/file/147938/drugs-misuse-dec-1112-pdf.pdf, Accessed on 10 July 2015.
51. London Borough of Hillingdon, *The Hillingdon Khat Report – A Call for Action*, 2011, Available at: http://www.hillingdon.gov.uk/media/pdf/2/Khat_Final_Report_-_nationwide_11.pdf, Accessed on 29 March 2015.
52. K. Smith and J. Flatley, *Drug Misuse Declared: Findings From the 2010/11 British Crime Survey*, Home Office, London, 2011, Available at: https://www.gov.uk/government/uploads/system/uploads/attachment_data/file/116333/hosb1211.pdf, Accessed on 10 July 2015.
53. C. Havell, *Khat Use in Somali, Ethiopian and Yemeni Communities in England: Issues and Solutions. A Report by Turning Point*, Home Office and Turning Point, London, 2004, Available at: http://www.12steptreatmentcentres.com/Articles/khat%20report05.pdf, Accessed on 10 July 2015.
54. A. Yussuf, P. Asquith, and S. Ali, *Chewing It Over – Khat Use in Lambeth's Somali Community*, London Borough of Lambeth Drug and Alcohol Team, 2007, Available at: http://www.lambeth.gov.uk/sites/default/files/lambeth-khat-report.pdf, Accessed on 12 July 2015.
55. G. Emerling, D.C. seeks tougher penalties for khat, *The Washington Times*, 13th October 2008, Available at: http://washingtontimes.com/news/2008/oct/13/dc-seeks-tougher-penalties-for-khat/?page=2, Accessed on 10 July 2015.
56. P. Griffiths, *Qat Use in London: A Study of Khat Use Among a Sample of Somalis Living in London*, Drugs Prevention Initiative Paper 26, Home Office Central Drugs Prevention Unit, London, 1998.
57. B. Lavery, World Briefing Europe: Ireland: Huge shipment of khat is seized at airport, *New York Times*, 21 February 2003.

Available at: http://www.nytimes.com/2003/02/21/world/world-briefing-europe-ireland-huge-shipment-of-khat-is-seized-at-airport.html, Accessed on 10 July 2015.

58. Irish Revenue Service, 'Khat seizure at Dublin Airport.' Press release, 29 January 2004, Available at: http://www.revenue.ie/en/press/archive/2004/pr_290104khat.html, Accessed on 10 July 2015.

59. E. J. Pennings, A. Opperhuizen and J. G. van Amsterdam, Risk assessment of khat use in the Netherlands: A review based on adverse health effects, prevalence, criminal involvement and public order, *Regul. Toxicol. Pharmacol.*, 2008, **52**(3), 199–207, DOI: 10.1016/j.yrtph.2008.08.005.

60. INCB, *Report of the International Narcotics Control Board for 2008*, United Nations, New York, 2009, Available at: https://www.incb.org/documents/Publications/AnnualReports/AR2008/AR_08_English.pdf, Accessed on 10 July 2015.

61. N. Carrier, Under any other name: the trade and use of khat in the UK, *Drugs Alcohol Today*, 2005, **5**(3), 14–16.

62. G. Yousef, Z. Huq and T. Lambert, Khat chewing as a cause of psychosis, *Br. J. Hosp. Med.*, 1995, **54**(7), 322–326.

63. F. E. Khan, Khat: Sweden's forgotten immigrant drug, *The Local: Sweden's news in English*, 28 May 2008, Available at: http://www.thelocal.se/20080528/12066, Accessed on 10 July 2015.

64. ACMD, *Khat: A Review of Its Potential Harms to the Individual and Communities in the UK*, Advisory Council on the Misuse of Drugs, Home Office, London, 2013, Available at: https://www.gov.uk/government/uploads/system/uploads/attachment_data/file/144120/report-2013.pdf, Accessed on 10 July 2015.

65. ECCD, WHO Expert Committee on Drug Dependence. 22nd meeting, *World Health Organ Tech. Rep. Ser*, 1985, 729.

66. L. A. King, *The Misuse of Drugs Act: A Guide for Forensic Scientists*, Royal Society of Chemistry, London, 2003.

67. A. Hunter, E. Baker, M. H. Gladbaum, K. Hirani, P. Meshari and M. McLennon, *The Health and Social Impact of Khat Use in Brent*, Brent Council, London, 2012, Available at: http://www.brent.gov.uk/media/2025436/0702012-khat1.pdf, Accessed on 7 July 2015.

68. House of Commons, HC Deb 21 November 2000, vol. 357 c156W, Available at: http://hansard.millbanksystems.com/written_answers/2000/nov/21/khat-1, Accessed on 20 July 2015.

69. Home Office, Written statement to Parliament – Khat, 3 July, 2013, Available at: https://www.gov.uk/government/speeches/khat, Accessed on 10 July 2015.

70. UKSI, The Misuse of Drugs Act 1971 (Amendment) Order 2014, Statutory Instrument 2014 No 1352, Available at: http://www.legislation.gov.uk/uksi/2014/1352/pdfs/uksi_20141352_en.pdf, Accessed on 10 July 2015.

71. GDPO situation analysis: The UK khat ban: Likely adverse consequences, January 2014, Swansea University, Available at: http://www.swansea.ac.uk/media/GDPO%20Situation%20Analysis%20Khat.pdf, Accessed on 10 July 2015.

72. Latest update on implementing the khat ban, Council of Somali Organisations, 30 September 2014, Available at: http://www.councilofsomaliorgs.com/content/latest-update-implementing-khat-ban, Accessed on 10 July 2015.

73. Cocaine Route Programme, The Criminalisation of Khat, Undated, 2015, Available at: http://www.cocaineroute.eu/flows/criminalisation-khat/, Accessed on 10 July 2015.

74. N. Carrier, UK Khat ban likely to create innovative black market for drug, *Afr. Arguments*, 2014, Available at: http://africanarguments.org/2014/06/18/uk-khat-ban-likely-to-create-innovative-black-market-for-drug-by-neil-carrier/, Accessed on 10 July 2015.

75. R. Gardner, Bristol police claim biggest ever seizure of drug since khat made illegal, *The Bristol Post*, 15 September 2014, Available at: http://www.bristolpost.co.uk/Crime-Police-claim-biggest-seizure-Khat-illegal/story-22925377-detail/story.html, Accessed on 11 July 2015.

76. D. Lader, *Drug Misuse: Findings from the 2014/15 Crime Survey for England and Wales*, Statistical Bulletin 03/15, Home Office, London, 2015, Available at: https://www.gov.uk/government/uploads/system/uploads/attachment_data/file/447546/drug-misuse-1415.pdf, Accessed on 23 July 2015.

77. O. E. Omolo and M. Dhadphale, Prevalence of *khat* chewers among primary health clinic attendees in Kenya, *Acta Psychiatr. Scand.*, 1987, **75**(3), 318–320.

78. G. Cox and H. Rampes, Adverse effects of khat: a review, *Adv. Psychiatr. Treat.*, 2003, **9**, 456–463.

79. Y. Kebede, Cigarette smoking and khat chewing among university instructors in Ethiopia, *East Afr. Med. J.*, 2002, **79**(5), 274–278.

80. A. G. Ahmed and E. Salib, The khat users: a study of khat chewing in Liverpool's Somali men, *Med., Sci, Law*, 1998, **38**(2), 165–169.

81. D. Linehan, O. Abdi, R. Aziz, B. Hyman and L. Wisdom, *Report of the Drugs Misuse Needs Assessment Carried out by Sheffield Black Drugs Service amongst the Afro-Caribbean, Pakistani,*

Somali, Yemeni & Bangladeshi Communities in Burngreave, 2001, Available at: http://www.uclan.ac.uk/facs/health/ethnicity/community engagement/documents/Sheffield%20Black%20Drug%20 Service%20(Sheffield).pdf, Accessed on 18 May 2007.

82. T. Bønes and Ø. Bye, Khat – rekreasjons middel eller narkotikum? [Khat – recreational use or narcotic?], *Tidsskr. Nor. Legeforen.*, 2007, **5**(127), 545.

83. J. Date, N. Tanida and T. Hobara, Khat chewing and pesticides: A study of adverse health effects in people of the mountainous areas of Yemen, *Int. J. Environ. Health Res.*, 2004, **14**(6), 405–414.

84. V. A. Galkin, A. V. Mironychev and S. S. Korsakova, Effect of the narcotic kat (*Catha edulis*) on certain functions of the human body, *Zh. Nevropatol. Psikhiatr. Im.*, 1962, **62**, 1396–1397.

85. R. Nencini and A. M. Ahmed, Khat consumption: a pharmacological review, *Drug Alcohol Depend.*, 1989, **23**(1), 19–29.

86. P. Kalix, Pharmacological properties of the stimulant khat, *Pharmacol. Ther.*, 1990, **48**(3), 397–416.

87. P. Kalix, Khat, an amphetamine-like stimulant, *J. Psychoact. Drugs*, 1994, **26**(1), 69–74.

88. S. W. Toennes, S. Harder, M. Schramm, C. Niess and G. F. Kauert, Pharmacokinetics of cathinone, cathine and nor-ephredine after the chewing of khat leaves, *Br. J. Clin. Pharmacol.*, 2003, **56**(1), 125–130.

89. A. S. Elmi, The chewing of khat in Somalia, *J. Ethnopharmacol.*, 1983, **8**(2), 163–176.

90. P. Goldstein, The drugs 'violence nexus': a tripartite conceptual framework, *J. Drug Issues*, 1985, **14**(4), 493–506.

91. S. Thomas and T. Williams, Khat (*Catha edulis*): A systematic view of evidence and literature pertaining to its harms to UK users and society, *Drug Sci., Policy Law*, 2014, **1**, 1–25.

92. A. Klein, P. Metaal and M. Jelsma, Chewing over khat prohibition, *The Globalisation of control and regulation of an ancient stimulant*, Transnational Institute Series on Legislative Reform of Drug Policies, No. 17, Transnational Institute, Amsterdam, 2012, Available at: http://www.tni.org/files/download/dlr17.pdf, Accessed on 10 July 2015.

93. U. Amako, Khat smuggler is first to be jailed since drug outlawed, *Daily Star*, 1 October 2014, Available at: http://www.dailystar.co.uk/news/latest-news/402718/Khat-smuggler-jailed-drug-outlawed-Ernestas-Sidlauskas, Accessed on 10 July 2015.

94. D. Lemanski, Exclusive: fears grow over new killer strain of Khat, *Daily Star*, 2 December 2014, Available at: http://www.dailystar. co.uk/news/latest-news/413289/Fears-grow-over-killer-strain-Khat, Accessed on 10 July 2015.

95. Z. Hoke, British ban may hurt Kenya's khat business, *Voice of America*, 4 September 2014, Available at: http://www.voanews. com/content/british-ban-may-hurt-kenya-khat-business/ 2438085.html, Accessed on 10 July 2015.

96. Kenya: Miraa Khat prices fall as UK ban leads to surplus, *Djibouti Nation*, 24th July 2014, Available at: http://djiboutination.com/ kenya-miraa-khat-prices-fall-as-uk-ban-leads-to-surplus/, Accessed on 10 July 2015.

97. A. Khalif, Khat chewers pay less after UK bans herb, *Nation*, 8 July 2014, Available at: http://mobile.nation.co.ke/news/Khat-chewers-pay-less-after-UK-bans-herb—/-/1950946/2376594/-/ format/xhtml/-/b7q2q6z/-/index.html, Accessed on 10 July 2015.

98. L. Okulo and S. Astariko, Kenya: miraa khat prices fall as UK ban leads to surplus, *The Star*, 17 July 2014, Available at: http:// horseedmedia.net/2014/07/17/miraa-khat-prices-fall-uk-ban-leads-surplus/, Accessed on 10 July 2015.

99. P. Kalix, The releasing effect of the isomers of the alkaloid cathinone at central and peripheral catecholamine storage sites, *Neuropharmacology*, 1986, **25**(5), 499–501.

100. P. Kalix and O. Braenden, Pharmacological aspects of the chewing of khat leaves, *Pharmacol. Rev.*, 1985, **37**(2), 149–164.

101. P. H. Drake, Khat chewing in the near east, *Lancet*, 1988, **331**(8584), 532–533.

102. D. N. Baron, A memorable experience: the qat party, *Br. Med. J.*, 1999, **319**(7208), 500.

103. M. R. Paris and H. Moyes, Abyssinian tea (Catha edulis Forssk, Celastraceae), *Bull. Narc.*, 1958, **10**, 29.

104. H. Halbach, Medical aspects of the chewing of khat leaves, *Bull. W. H. O.*, 1972, **47**, 21–29.

105. E. A. Pehek, M. D. Schechter and B. K. Yamamoto, Effects of cathinone and amphetamine on the neurochemistry of dopamine in vivo, *Neuropharmacology*, 1990, **29**(12), 1171–1176.

106. A. Al-Motarreb, K. Baker and K. J. Broadley, Khat: Pharmaco-logical and medical aspects and its social use in Yemen, *Phytother. Res.*, 2002a, **16**(5), 403–413.

107. N. A. Hassan, A. A. Gunaid, F. M. El-Khally and I. M. Murray-Lyon, The effect of chewing Khat leaves on human mood, *Saudi Med. J.*, 2002, **23**(7), 850–853.

108. A. D. Jager and L. Sireling, Natural history of khat psychosis, *Aust. N. Z. J. Psychiatry*, 1994, **28**(2), 331–332.

109. R. Poole and G. Brabbins, Drug induced psychosis, *Br. J. Psychiatry*, 1996, **168**(2), 135–138.

110. P. Phillips and S. Johnson, How does drug and alcohol misuse develop among people with psychotic illness? A literature review, *Soc. Psychiatry Psychiatr. Epidemiol.*, 2001, **36**(6), 269–276.

111. M. Odenwald, F. Neuner, M. Schauer, T. Elbert, C. Catani, B. Lingenfelder, H. Hinkel, H. Häfner and B. Rockstroh, Khat use as risk factor for psychotic disorders: a cross-sectional and case-control study in Somalia, *BMC Med.*, 2005, **3**, 5–12.

112. M. Dhadphale and O. E. Omolo, Psychiatric morbidity among khat chewers, *East Afr. Med. J.*, 1988, **65**(6), 355–359.

113. S. Critchlow and R. Seifert, Khat induced paranoid psychosis, *Br. J. Psychiatry*, 1987, **150**, 247–249.

114. A. Alem and T. Shibre, Khat induced psychosis and its medico-legal implication: a case report, *Ethiop. Med. J.*, 1997, **35**(2), 137–141.

115. C. Pantelis, C. G. Hindler and J. C. Taylor, Use and abuse of khat (Catha edulis): a review of the distribution, pharmacology, side effects and a description of psychosis attributed to khat chewing, *Psychol. Med.*, 1989, **19**(3), 657–668.

116. A. Alem, D. Kebede and G. Kullgren, The prevalence and socio-demographic correlates of khat chewing in Butajira, Ethiopia, *Acta Psychiatr. Scand., Suppl.*, 1999, **399**, 84–91.

117. E. Gebissa, Scourge of life or an economic lifeline? Public discourses on khat (catha edulis) in Ethiopia, *Subst. Use Misuse*, 2008, **43**(6), 784–802.

118. P. Kalix, Khat: a plant with amphetamine effect, *J. Subst. Abuse Treat.*, 1988, **5**(3), 163–169.

119. C. R. Schuster and C. E. Johanson, Behavioural studies of cathinone in monkeys and rats, in *The Problems of Drug Dependence, Proceedings of the 41st Annual Scientific Meeting*, National Institute of Drug Abuse, Research Monograph No. 27, US Government Printing Office, Washington, DC, 1979, pp. 324–325.

120. P. Kalix, Cathinone, a natural amphetamine, *Pharmacol. Toxicol.*, 1992, **70**(2), 77–86.

121. P. Widler, K. Mathys, R. Brenneisen, P. Kalix and H. U. Fisch, Pharmacodynamics and pharmacokinetics of khat, a controlled drug, *Clin. Pharmacol. Ther.*, 1994, **55**(5), 556–562.

122. N. A. Hassan, A. A. Gunaid, A. A. Abdo Rabbo *et al.*, The effect of Qat chewing on blood pressure and heart rate in healthy volunteers, *Trop. Doct.*, 2000, **30**(2), 107–108.

123. S. de Ridder, F. Eerens and L. Hofstra, Khat rings twice: Khat-induced thrombosis in two vascular territories, *Neth. Heart J.*, 2007, **15**(7–8), 269–270.

124. T. J. Meulman, J. Bakker and E. J. van den Bos, Ischemic cardiomyopathy and cerebral infarction in a young patient associated with khat chewing, *Case Rep. Radiol.*, 2015, 893176.

125. H. M. Mujlli, X. Bo and L. Zhang, The effect of khat (Catha edulis) on acute cerebral infarction, *Neurosciences*, 2005, **10**(30), 219–222.

126. A. S. McLean and M. B. Kot, Cardiac collapse associated with the ingestion of khat, *Intern. Med. J.*, 2011, **41**(7), 579–581.

127. S. W. Toennes and G. F. Kauert, Driving under the influence of khat – alkaloid concentrations and observations in forensic cases, *Forensic Sci. Int.*, 2004, **140**(1), 85–90.

128. H. P. Walter, Drink-drive, broken jaw, lager and khat ingestion, *J. Clin. Forensic Med.*, 1996, **3**(1), 51–53.

129. P. Kalix, The pharmacology of khat, *Gen. Pharmacol.*, 1984, **15**(3), 179–187.

130. N. A. Ghani, M. Eriksson, B. Kristiansson and A. Oirbi, The influence of khat-chewing on birth-weight in full-term infants, *Soc. Sci. Med.*, 1987, **24**(7), 625–627.

131. M. Eriksson, N. A. Ghani and B. Kristiansson, Khat-chewing during pregnancy – effect upon the off-spring and some characteristics of the chewers, *East Afr. Med. J.*, 1991, **68**(2), 106–111.

132. J. M. Mwenda, M. M. Arimi, M. C. Kyama and D. K. Langat, Effects of khat (Catha edulis) consumption on reproductive functions: a review, *East Afr. Med. J.*, 2003, **80**(6), 318–323.

133. B. Kristiansson, N. A. Ghani, M. Eriksson, M. Larle and A. Qirbi, Use of khat in lactating women: A pilot study on breast-milk secretion, *J. Ethnopharmacol.*, 1987, **34**(5), 338–340.

134. K. M. Kuczkowski, Herbal ecstasy: cardiovascular complications of khat chewing in pregnancy, *Acta Anaesthesiol. Belg.*, 2005, **56**(1), 19–21.

135. M. Al-Habori, The potential adverse effects of habitual use of *Catha edulis* (khat), *Expert Opin. Drug Saf.*, 2005, **4**(6), 1145–1154.
136. M. Al-Mamary, M. Al-Habori, A. M. Al-Aghbari and M. M. Baker, Investigation into the toxicological effects of *Catha edulis* leaves: a short term study in animals, *Phytother. Res.*, 2002, **16**(2), 127–132.
137. J. M. Brostoff, C. Plymen and J. Birns, Khat – a novel cause of drug-induced hepatitis, *Eur. J. Intern. Med.*, 2006, **17**(5), 383.
138. S. Saha and C. Dollery, Severe ischaemic cardiomyopathy associated with khat chewing, *J. R. Soc. Med.*, 2006, **99**(6), 316–318.
139. M. H. Chapman, M. Kajihara, G. Borges, J. O'Beirne, D. Patch, A. P. Dhillon, A. Crozier and M. Y. Morgan, Severe, acute liver injury and khat leaves, *N. Engl. J. Med.*, 2010, **362**(17), 1642–1644.
140. A. A. Gunaid, T. M. Nasher, A. M. El-Guneid, M. Hill, R. Drayton, A. Pal, S. J. Skidmore, J. C. Coleman and I. M. Murray-Lyon, Acute sporadic hepatitis in the Republic of Yemen, *J. Med. Virol.*, 1997, **51**(1), 64–66.
141. R. D'Souza, P. O. Sinnott, M. J. Glynn, C. A. Sabin and G. R. Foster, An unusual form of autoimmune hepatitis in young Somalian men, *Liver Int.*, 2006, **25**(2), 325–330.
142. C. A. McCune, M. Moorghan, E. H. Gordon and P. L. Collins, Liver disease and khat chewing in young Somalian men, *J. Hepatol.*, 2007, **46**(Suppl 1), S276.
143. V. A. Luyckx and S. Naicker, Acute kidney injury associated with the use of traditional medicines, *Nat. Clin. Pract. Nephrol.*, 2008, **4**(12), 664–671.
144. S. Riyaz, M. Imran and M. Karajeh, Khat as a possible cause of drug induced autoimmune hepatitis: a case series, *Gut*, 2012, **61**(Suppl. 2), A413.
145. M. G. Jenkins, R. Handslip, M. Kumar, U. Mahadeva, S. Lucas, T. Yamamoto, D. M. Wood, T. Wong and P. I. Dargan, Reversible khat-induced hepatitis: two case reports and review of the literature, *Frontline Gastroenterol.*, 2013, **4**(4), 278–281.
146. Y. Bentur, A. Bloom-Krasik and B. Raikhlin-Eisenkraft, Illicit cathinone ("Hagigat") poisoning, *Clin. Toxicol.*, 2008, **46**(3), 206–210.
147. R. F. Forman, D. B. Marlow and A. T. McLellan, The Internet as source of drugs of abuse, *Curr. Psychiatry Rep.*, 2006, **8**(5), 377–382.
148. F. Schifano, P. Deluca, A. Baldacchino, T. Peltoniemi, N. Scherbaum, M. Torrens, M. Farre, I. Flores, M. Rossi, D. Eastwood, C. Guionnet, S. Rawaf, L. Agosti, L. Di Furia, R. Brigada, A. Majava, H. Siemann, M. Leoni, A. Tomasin,

F. Rovetto and A. H. Ghodse, Drugs on the web; the Psychonaut 2002 EU project, *Prog. Neuropsychopharmacol. Biol. Psychiatry*, 2006, **30**(4), 640–646.

149. J. M. Corkery, F. Schifano and A. H. Ghodse, Mephedrone-related fatalities in the United Kingdom: contextual, clinical and practical issues, *Pharmacology*, ed. L. Gallelli, In Tech, Rijeka, Croatia, 2012, ch. 17, pp. 355, Available at: http://www.intechopen.com/books/pharmacology/mephedrone-related-fatalities-in-the-united-kingdom-contextual-clinical-and-practical-issues, Accessed on 10 July 2015.

150. F. Schifano, J. Corkery and A. H. Ghodse, Suspected and confirmed fatalities associated with mephedrone (4-methylmethcathinone; 'meow meow') in the UK, *J. Clin. Psychopharmacol.*, 2012, **32**(5), 710–714.

151. A. Al-Motarreb, A.-N. Munibari, B. Al-Adhi and M. Al-Kebsi, Khat and acute MI. *Proceedings of Second Yemeni Cardiac Meeting, Sana'a, Yemen*: 12, 1997.

152. F. N. Croles, B. P. Brassé, M. Duisenberg-van Essenberg, H. F. Baars and C. M. Schweitzer, Verband tussen hypertensie en myocardinfarct en het kauwen van qatbladeren [Connection between hypertension and myocardial infarction and chewing of khat leaves], *Ned. Tijdschr. Geneeskd.*, 2009, **153**(1–2), 38–43.

153. Health Canada, *Straight Facts About Drugs & Drug Abuse*, 2007, Available at: http://www.hc-sc.gc.ca/hl-vs/pubs/adp-apd/straight_facts-faits_mefaits/tables-tableaux-eng.php, Accessed on 22 March 2009.

154. C. G. Peevers, M. Moorghen, P. L. Collins, F. H. Gordon and C. A. McCune, Liver disease and cirrhosis because of Khat chewing in UK Somali men: a case series, *Liver Int.*, 2010, **30**(8), 1242–1243.

155. J. M. Halket, Z. Karasu and I. M. Murray-Lyon, Plasma cathinone levels following chewing khat leaves (*Catha edulis* Forsk.), *J. Ethnopharmacol.*, 1995, **49**(2), 111–113.

156. M. Odenwald, A. Klein and N. Warfa, Khat use in Europe: implications for European policy, Drugs in focus 21, 4 July 2011, European Monitoring Centre for Drugs and Drug Addiction, Lisbon, Portugal, 2011, Available at: http://www.emcdda.europa.eu/publications/drugs-in-focus/khat, Accessed on 10 July 2015.

157. F. Carvalho, The toxicological potential of khat, *J. Ethnopharmacol.*, 2003, **87**(1), 1–2.

158. R. B. Heisch, A case of poisoning by *Catha edulis*, *East Afr. Med. J.*, 1945, **22**, 7–10.

159. R. Brenneisen, H.-U. Fisch, U. Koelbing, S. Geisshüsler and P. Kalix, Amphetamine-like effects in humans of the khat alkaloid cathinone, *Br. J. Clin. Pharmacol.*, 1990, **30**(6), 825–828.

10 Role of Analytical Screening in the Management and Assessment of Acute Recreational Drug Toxicity

David M. Wood* and Paul I. Dargan

Guy's and St Thomas' NHS Foundation Trust and King's College London, London, UK
*Email: david.wood@gstt.nhs.uk

10.1 Introduction

There is an increasing range of new psychoactive substances (NPS, also known colloquially and incorrectly by users as "legal highs"). In 2014 there were 101 new substances detected in Europe and reported to the European Monitoring Centre for Drugs and Drug Addiction (EMCDDA) through their Early Warning System (EWS).[1] These substances often involve small structural modifications to existing substances that are already controlled under national or international legislation. Many of the names of NPS sound very similar to existing substances. For example, when mephedrone (4-methylmethcathinone), the synthetic cathinone, was first seen, many clinicians incorrectly thought patients were reporting to have used methadone, an opioid used in opioid substation therapy.[2]

For clinicians who treat individuals who present to healthcare facilities with acute toxicity related to the suspected use of a

Forensic Toxicology: Drug Use and Misuse
Edited by Susannah Davies, Atholl Johnston and David Holt
© The Royal Society of Chemistry 2016
Published by the Royal Society of Chemistry, www.rsc.org

NPS or classical recreational drug, they may not be aware of the substance(s) reported to have been used. In this chapter we will discuss how clinicians make a diagnosis of acute toxicity related to the use of recreational drugs and NPS, and the role of analytical screening in managing acute recreational drug and NPS related toxicity.

10.2 Patterns of Acute Toxicity

There are broadly three patterns of acute toxicity seen with the use of classical recreational drugs:

(i) Stimulant toxicity: this includes hypertension, tachycardia, hyperpyrexia, agitation/aggression, sweating, convulsions, chest pain and arrhythmias, as seen with the use of cocaine, 3,4-methylenedioxymethamphetamine (MDMA) and other amphetamine-type stimulants (ATS).

(ii) Depressant toxicity: this includes neurological and associated respiratory depression, as seen with the use of opioids, benzodiazepines and γ-hydroxybutyrate (GHB) and its related analogues γ-butyrolactone and butane-1,4-diol.

(iii) Hallucinogenic toxicity: this includes hallucinations, mild aggression/anxiety, mild hypertension and tachycardia, as seen with the use of ketamine and LSD.

Typically, when an individual NPS or class of NPS enter the recreational drug scene, there are limited data available from animal or human studies on the potential for acute toxicity. There is no systematic collection of data on acute drug/NPS toxicity[3] and therefore collation of data on the pattern of toxicity of these drugs needs to be undertaken using data triangulation from a variety of sources.[4,5] These sources include: (i) user reports on internet discussion forums; (ii) subpopulation user surveys of acute harms; (iii) case reports and case series; (iv) information from regional or national poisons centres and poisons information services for accesses/contacts for advice on managing patients with acute toxicity. Each of these information sources has their own limitations. Therefore combining data from these different sources, through a process known as data triangulation, can provide a more robust pattern of acute toxicity related to the use of an NPS.[4–6] This process has demonstrated that NPS broadly are associated with the same patterns of acute toxicity as seen with the classical recreational drugs described above, although for some of the NPS there is more overlap between the different patterns of acute toxicity.[4–6]

10.3 Studies Assessing Toxicological Screening in the Night-time Economy Setting

Outside of the UK, there has been interest in screening recreational drugs and NPS in the night-time economy setting. The two main examples of this are the Drug Information and Monitoring System (DIMS) project in the Netherlands and the ChEck iT! Project in Vienna, Austria.[7,8]

10.3.1 ChEck iT! Project

The ChEck iT! project analyses samples at larger rave-type events in Austria, and uses high-pressure liquid chromatography (HPLC) to analyse a small sample of tablets/powders for individuals at the event.[8] Results are reported under a number rather than a picture of the tablet in an open tent area, which means that only the individual who asked for sampling knows the results. However, where appropriate the team will post alerts in this area if particularly high dose pills are identified or if there are new or unexpected substances detected. The study team have highlighted the risk of this type of approach, in that dealers can use clubbers to quality control/check products they are selling and/or confirm the strength, which then helps them market products to potential users within the rave event.

10.3.2 Drug Information and Monitoring System Project

The DIMS project is a network of analytical facilities at drug prevention institutions across the Netherlands.[7] Drug users engaged with services at these institutions can provide samples of the drugs that they are using (in any form). When tablet(s) are provided, at some institutions they are able to offer immediate comparison of the tablet(s) with the DIMS database to provide a visual identification and probable contents back to the drug user. Approximately 30% of tablets are identified in this manner. Where visual identification of a tablet is not possible, and for all non-tablet samples, they are analysed using a combination of thin-layer chromatography (TLC), gas chromatography-nitrogen/phosphorus detection (GC-NPD), gas chromatography-mass spectrometry (GC-MS) and nuclear magnetic resonance (NMR) spectroscopy. Initial studies have shown that users who provide samples to the DIMS project are comparable to the wider drug using population in the Netherlands, and therefore there is the potential to extrapolate the findings to be semi-representative of the overall drug availability and use in the Netherlands.[7] The DIMS project has been running since the early 1990s, and therefore is able to provide information on the trends in recreational drugs and NPS being used over this time. For example, this trend analysis demonstrated the appearance and subsequent disappearance of 1,3-benzodioxolyl-*N*-methylbutanamine (MBDB), 4-methylthioamphetamine (4-MTA) and mephedrone in "ecstasy" tablets submitted to the DIMS project for analysis in 1993–1997, 1997–2001 and 2008–2010, respectively.

More recently there has been comparison of the results from screening to patterns of toxicity described by users and information from the Dutch Poisons Information Service.[9,10] When 4-fluoro-amphetamine (4-FA) was detected in samples submitted to the DIMS project, the project undertook a questionnaire survey of 474 self-identified 4-FA users recruited from a variety of internet sites.[9] Although initial samples containing 4-FA were from misrepresented samples (contained in "ecstasy" tablets sold as containing MDMA), since 2010 there was an increase in samples submitted that were thought by users to contain 4-FA. Users reported a range of desired and unwanted effects, with the majority of unwanted effects similar to those reported for other sympathomimetic drugs such as MDMA, amphetamine and/or cocaine.

The DIMS project has also compared the range of drugs detected in samples submitted for analysis to the recreational drugs/NPS involved in cases discussed with the Dutch Poisons Information Centre (PIC).[10] In 2013, of the 431 samples demonstrated to contain one or more NPS, the four mostly frequently detected NPS were 4-bromo-2,5-dimethoxyphenethylamine (2C-B, 67 samples), 4-FA (105 samples), methoxetamine (70 samples) and 5-(2-aminopropyl)benzofuran/6-(2-aminopropyl)benzofuran (benzofury, 5-APB/6-APB, 38 samples). There were 35 NPS exposures discussed with the Dutch PIC; the most common NPS involved were 4-FA (10 cases), the cathinones (8 cases), methoxetamine (5 cases), 2C-type of drugs (4 cases) and 5-APB/6-APB (3 cases). This demonstrates that there is similarity in the range and frequency of the NPS being detected in samples submitted to the DIMS project to that being reported by clinicians in cases of acute toxicity that they are discussing with the Dutch PIC.

10.4 Role of Toxicological Screening in the Emergency Department

Currently in the UK and elsewhere in Europe there is no recommendation to routinely undertake detailed analytical screening in patients presenting with acute toxicity related to the use of classical recreational drugs or NPS. Over 70% of patients presenting with acute toxicity are discharged directly from the Emergency Department (ED).[11] Overall the median length of stay in hospital for those presenting with acute recreational drug/NPS toxicity is 4–5 h.[11] This means that the results of detailed analytical screening are not available within a timeframe that will alter the management of that individual patient. Furthermore, as discussed earlier, the clinical pattern of toxicity of both classical drugs and NPS can be divided into broad categories of stimulant, depressant and hallucinogenic toxicity. The management of individual drugs within each of these categories is the same. Therefore, initial assessment and management of patients with acute recreational drug/NPS toxicity can be guided by the clinical pattern of toxicity and analytical screening does not add significantly to guiding clinicians in the routine management of these patients.[12]

10.4.1   Urine Dipsticks

Near patient (or "bed-side") urine dipsticks for recreational drug screening generally use an enzyme multiplied immunoassay test (EMIT). Whilst these are simple to perform and have a rapid turnaround time, they have significant limitations because of the high false-positive rate due to cross-reaction with a variety of prescription and over-the-counter drugs and even food substances.[13–15] A further limitation is that generally these rely on the detection of metabolites rather than the parent drug and therefore a "true positive" may represent drug use hours or even days prior to presentation that is not of relevance to the clinical features seen in the current presentation.[16]

US guidelines from the National Academy of Clinical Biochemistry Laboratory Medicine caution against routine use of urine immunoassays in the assessment of patients with acute recreational drug toxicity because of the poor correlation with clinical effects and problems with sensitivity and specificity.[17]

10.4.2   Case Reports/Case Series

There is the potential for specialist centres to undertake analytical screening to identify what recreational drugs and/or novel psychoactive substances have been involved in the presentation to the ED. This can help with the characterisation of the acute toxicity of an NPS. In addition, analytical screening in a subset of patients where samples can be collected can allow comparison with larger cohorts where the drug(s) used is based on self-reported substances used.

In 2009, a synthetic cathinone, mephedrone (4-methylmethcathinone), was reported to be available and was being used as a recreational drug.[2] We reported the first case of acute mephedrone toxicity.[18] The patient, a 22-year-old male, presented with anxiety and agitation following the self-reported use of mephedrone; on review in the ED he was tachycardic (105 bpm) and hypertensive (177/111 mmHg) with no other features of sympathomimetic toxicity (no sweating, hyper-reflexia, clonus or bruxism). Subsequent analysis of serum and urine samples was undertaken by a combination of GC/MS detection (qualitative screening) and LC/tandem MS detection (quantitative screening). Urine and serum samples were both positive for mephedrone, with an estimated serum concentration of mephedrone estimated to be 0.15 mg L^{-1}.

Following this, we reported a case series of 15 individuals presenting to our ED following the self-reported use of mephedrone alone or in combination with other drugs/alcohol.[19] On presentation to the ED, 20% had clinically significant hypertension (defined as a systolic blood pressure >160 mmHg) and 40% of had a tachycardia (defined as a heart rate of >100 bpm). The majority of patients (60%) had a Glasgow Coma Score (GCS) of 15 on presentation to the ED; of the four who had a GCS of <8, all self-reported to have concomitantly used a CNS depressant (GHB/GBL in three presentations and opium in one presentation). In terms of other symptoms, these included: agitation (53.3%), seizure (20.0%), palpitations (13.3%), vomiting (13.3%), excessive sweating (13.3%), hyper-reflexia (6.7%) and bruxism (6.7%). The frequency of these clinical features in a larger case series of 72 self-reported mephedrone presentations to our ED was included in the European Monitoring Centre for Drugs and Drug Addiction (EMCDDA) Risk Assessment Report and is summarised in Table 10.1.[2]

Subsequently we reported the pattern of acute toxicity in a subset of patients presenting to our ED with analytically confirmed mephedrone use.[20] Of nine patients whose serum underwent analytical screening, seven had serum samples positive for mephedrone; the remaining two presented more than 24 h following use, which may explain why the serum samples were negative for mephedrone. The pattern of acute toxicity seen in the analytically confirmed mephedrone cases included tachycardia (71%), agitation (57%), hypertension (43%), palpitations (29%), chest pain (29%) and seizures (14%). As illustrated in Table 10.1, the frequency of clinical features is

Table 10.1   Comparison of clinical features in self-reported and analytically confirmed mephedrone-related acute toxicity presentations to our ED.

Clinical feature	Self-reported case series $(n = 72)$	Analytically confirmed case series $(n = 7)$
Tachycardia	36%	71%
Agitation	39%	57%
Hypertension	14%	43%
Palpitations	25%	29%
Chest pain	13%	29%
Seizures	7%	14%
Headaches	7%	14%
Vomiting	14%	0%

comparable in the self-reported cases and the analytically confirmed cases. This demonstrates the values from a research perspective of analytical confirmation in helping to confirm the pattern of toxicity of new NPS. However, it also demonstrates that clinical assessment is sufficient in determining the broad pattern of toxicity to guide clinicians in the routine management of these patients.

10.4.3   STRIDA Project

The STRIDA project is a collaboration between the Karolinska University Laboratory and the Swedish Poisons Information Centre (PIC) which attempts to link the clinical effects reported to the Swedish PIC to analytical findings from biological samples obtained from these individuals.[21] When a potential case is identified during a call to the Swedish PIC, the treating clinician is asked to obtain appropriate biological samples (blood/urine), which are then submitted for subsequent toxicological analysis. The case and the biological samples are then allocated a STRIDA project reference number, which allows pseudo-anonymisation of the sample so that data linkage between the analytical findings and the clinical features can be undertaken. Using this data linkage system it has been possible to enable the STRIDA project to undertake analysis of the pattern of toxicity of a range of NPS as they have been detected and/or reported to the project; examples of this include 5-(2-aminopropyl)indole (5-IT), 1-cyclohexyl-4-(1,2-diphenylethyl)piperazine (MT-45), 3-methoxyphencyclidine (3-MeO-PCP), 4-methoxyphencyclidine (4-MeO-PCP) and 3-methylmethcathinone (3-MMC).[21–25] These enable the STRIDA project to describe the pattern of toxicity and report any unexpected clinical features in cases of analytically confirmed use. Another advantage of a project such as STRIDA is the potential to retrospectively review cases when a new NPS is detected. This does rely on using appropriate analytical techniques that enable retrospective re-analysis of the initial analytical findings. Without this, retrospective review is only possible based on the analytical library of drugs screened for at the time that the analysis was undertaken.

With the substituted phenylethylamine 5-IT, the pattern of toxicity described in 14 analytically confirmed cases (five in which there was lone 5-IT detection) was consistent with that seen with other sympathomimetic drugs such as amphetamine, cocaine or MDMA.[22] With the novel opioid MT-45, of the nine analytically confirmed cases reported, the majority of effects were "opioid-like"; however, three patients also reported hearing loss, which is atypical for an opioid.[25] However, as with many cases of acute toxicity, often individuals will have used more than one psychoactive substance. In the 59 cases of analytically confirmed 3-MeO-PCP and 4-MeO-PCP use detected by the STRIDA project, 52 (88%) had one or more other psychoactive substance detected.[24] Therefore whilst the pattern of toxicity described was of apparent sympathomimetic toxicity, it is not possible to determine what the significance of the co-detected substances were.

10.5 National and International Guidelines

The UK Association of Clinical Biochemists (ACB) and National Poisons Information Service issued guidelines for laboratory analyses for poisoned patients in the UK in 2002, with updated guidelines published in 2014.[26,27] The guidelines divide laboratory analyses for the poisoned patient into essential, common and specialist or infrequent.[27] Screening for recreational drugs/NPS is considered an infrequent assay that is not routinely required for the management of poisoned patients.[27] US guidelines from the National Academy of Clinical Biochemistry Laboratory Medicine also caution against routine use of these assays (particularly urine toxicology screens) because of the poor correlation with clinical effects and problems with sensitivity and specificity.[17]

10.6 Conclusions

Currently the role of analytical screening for classical recreational drugs and/or NPS in the assessment and management of patients presenting with acute toxicity is limited. There is a role for analytical screening as a research tool in confirming the pattern of toxicity in case reports and case series of new NPS, to provide more robust information to clinicians to inform them on the pattern of toxicity seen with the NPS and to legislative bodies to inform risk assessment. In addition, analytical screening in subsets of populations and/or comparison of analytical screening of substances with user self-reports or calls can also make self-reported cases more robust.

However, analytical screening is not required to inform routine clinical management of patients with acute recreational drug/NPS toxicity. First, screening results are generally not available in a timeframe that would inform clinical management. Second, clinical management can be guided by the broad pattern of toxicity – stimulant, depressant or hallucinogenic – and it is not necessary for clinicians to know the actual drug(s) responsible for toxicity as management is the same for drugs within these categories.

References

1. European Monitoring Centre for Drugs and Drug Addiction 2015. European Drug Report: Trends and Development 2015. Available from: www.emcdda.europa.eu/attachements.cfm/att_239505_EN_TDAT15001.pdf (last accessed 26th September 2015).
2. European Monitoring Centre for Drugs and Drug Addiction 2011. Report on the risk assessment of mephedrone in the framework of the Council Decision on new psychoactive substances. Available from: www.emcdda.europa.eu/attachements.cfm/att_116646_EN_TDAK11001_-OPTIMISED%20.pdf (last accessed 26th September 2015).
3. F. Heyerdahl, K. E. Hovda, I. Giraudon, C. Yates, A. M. Dines, R. Sedefov, D. M. Wood and P. I. Dargan, Current European data collection on emergency department presentations with acute recreational drug toxicity: gaps and national variations, *Clin. Toxicol. (Phila)*, 2014, **52**, 1005–1012.
4. D. M. Wood and P. I. Dargan, Novel psychoactive substances: how to understand the acute toxicity associated with the use of these substances, *Ther. Drug Monit.*, 2012, **34**, 363–367.
5. D. M. Wood and P. I. Dargan, Understanding how data triangulation identifies acute toxicity of novel psychoactive drugs, *J. Med. Toxicol.*, 2012, **8**, 300–333.

6. D. M. Wood and P. I. Dargan, Mephedrone (4-methylmethcathinone): what is new in our understanding of its use and toxicity, *Prog. Neuropsychopharmacol. Biol. Psychiatry*, 2012, **39**, 227–233.

7. T. M. Brunt and R. J. Niesink, The Drug Information and Monitoring System (DIMS) in the Netherlands: implementation, results, and international comparison, *Drug Test. Anal.*, 2011, **3**, 621–634.

8. H. Kriener and M. Schmid, Check your pills. Check your life. ChEck iT! High quality on site testing of illicit substances. Information, counselling, and safer use measures at raves in Austria. Available from: http://www.drugtext.org/pdf/Dance/party-drugs-clubbing/check-your-pills-check-your-life-check-it.pdf (last accessed 26th September 2015).

9. F. Linsen, R. P. Koning, M. van Laar, R. J. Niesink, M. W. Koeter and T. M. Brunt, 4-Fluoroamphetamine in the Netherlands: more than a one-night stand, *Addiction*, 2015, **110**, 1138–1143.

10. L. Hondebrink, J. J. Nugteren-van Lonkhuyzen, D. Van Der Gouwe and T. M. Brunt, Monitoring new psychoactive substances (NPS) in The Netherlands: data from the drug market and the Poisons Information Centre, *Drug Alcohol Depend.*, 2015, **147**, 109–115.

11. A. M. Dines, D. M. Wood, C. Yates, F. Heyerdahl, K. E. Hovda, I. Giraudon, R. Sedefov and Euro-DEN Research Group, Dargan PI. Acute recreational drug and new psychoactive substance toxicity in Europe: 12 months data collection from the European Drug Emergencies Network, *Clin. Toxicol. (Phila)*, 2015, **53**, 893–900.

12. M. Tenenbein, Do you really need that emergency drug screen?, *Clin. Toxicol.*, 2009, **47**, 286–291.

13. A. Bhalla, Bedside point of care toxicology screens in the ED: Utility and pitfalls, *Int. J. Crit. Illn. Inj. Sci.*, 2014, **4**, 257–260.

14. C. R. Hamlin, A rapid toxicology screen for emergency and routine care of patients, *Clin. Chem.*, 1988, **34**, 158–162.

15. M. D. Krasowski, A. F. Pizon, M. G. Siam, S. Giannoutsos, M. Iyer and S. Ekins, Using molecular similarity to highlight the challenges of routine immunoassay-based drug of abuse/toxicology screening in emergency medicine, *BMC Emerg. Med.*, 2009, **9**, 5.

16. C. A. Hammett-Stabler, A. J. Pesce and D. J. Cannon, Urine drug screening in the medical setting, *Clin. Chim. Acta*, 2002, **315**, 125–135.

17. A. H. Wu, C. McKay, L. A. Broussard, R. S. Hoffman, T. C. Kwong, T. P. Moyer, E. M. Otten, S. L. Welch and P. Wax, National academy of clinical biochemistry laboratory medicine practice guidelines: recommendations for the use of laboratory tests to

support poisoned patients who present to the emergency department. National Academy of Clinical Biochemistry Laboratory Medicine, *Clin. Chem.*, 2003, **49**, 357–379.

18. D. M. Wood, S. Davies, M. Puchnarewicz, J. Button, R. Archer, H. Ovaska, J. Ramsey, T. Lee, D. W. Holt and P. I. Dargan, Recreational use of mephedrone (4-methylmethcathinone, 4-MMC) with associated sympathomimetic toxicity, *J. Med. Toxicol.*, 2010, **6**, 327–330.

19. D. M. Wood, S. L. Greene and P. I. Dargan, Clinical pattern of toxicity associated with the novel synthetic cathinone mephedrone, *Emerg. Med. J.*, 2011, **28**, 280–282.

20. D. M. Wood, S. Davies, S. L. Greene, J. Button, D. W. Holt, J. Ramsey and P. I. Dargan, Case series of individuals with analytically confirmed acute mephedrone toxicity, *Clin. Toxicol. (Phila)*, 2010, **48**, 924–927.

21. A. Helander, M. Bäckberg, P. Hultén, Y. Al-Saffar and O. Beck, Detection of new psychoactive substance use among emergency room patients: results from the Swedish STRIDA project, *Forensic Sci. Int.*, 2014, **243**, 23–29.

22. M. Bäckberg, O. Beck, P. Hultén, J. Rosengren-Holmberg and A. Helander, Intoxications of the new psychoactive substance 5-(2-aminopropyl)indole (5-IT): a case series from the Swedish STRIDA project, *Clin. Toxicol. (Phila)*, 2014, **52**, 618–624.

23. M. Bäckberg, E. Lindeman, O. Beck and A. Helander, Characteristics of analytically confirmed 3-MMC-related intoxications from the Swedish STRIDA project, *Clin. Toxicol. (Phila)*, 2015, **53**, 46–53.

24. M. Bäckberg, O. Beck and A. Helander, Phencyclidine analog use in Sweden-intoxication cases involving 3-MeO-PCP and 4-MeO-PCP from the STRIDA project, *Clin. Toxicol. (Phila)*, 2015, **53**, 856–864.

25. A. Helander, M. Bäckberg and O. Beck, MT-45, a new psychoactive substance associated with hearing loss and unconsciousness, *Clin. Toxicol. (Phila)*, 2014, **52**, 901–904.

26. National Poisons Information Service and Association of Clinical Biochemists, Laboratory analyses for poisoned patients: joint position paper, *Ann. Clin. Biochem.*, 2002, **39**, 328–339.

27. J. P. Thompson, I. D. Watson, H. K. Thanacoody, S. Morley, S. H. Thomas, M. Eddleston, J. A. Vale, D. N. Bateman and C. V. Krishna, Guidelines for laboratory analyses for poisoned patients in the United Kingdom, *Ann. Clin. Biochem.*, 2014, **51**, 312–325.

11 Workplace Drug Testing

Simon Walker

Alere Toxicology Plc, Abingdon, UK
Email: simon.walker@alere.com

11.1 How Does Workplace Testing Differ from Other Sorts of Drug Testing?

For simplicity and brevity, the term "workplace drug testing" refers to the testing for drugs and alcohol.

As in many areas of life, the field of drug testing has its own distinct niche categories. Examples of these specialties are:

- Therapeutic drug monitoring – analysing specimens (typically blood) for medication to ensure that the patient is compliant with their therapy and that the dosage is correct.
- Drug treatment programmes – the analysis of specimens (typically urine or oral fluid) from patients in a drug treatment programme to ensure that they are not taking illicit substances and that they are compliant with their therapy.
- Forensic toxicology – the analysis of samples from a post-mortem with a view to determine whether drugs played a role in the cause of death.
- Sports drug testing – the detection of performance enhancing substances in samples from athletes and sportsmen and -women.
- Workplace drug testing – the detection of illicit substances in employees and potential employees.

Forensic Toxicology: Drug Use and Misuse
Edited by Susannah Davies, Atholl Johnston and David Holt
© The Royal Society of Chemistry 2016
Published by the Royal Society of Chemistry, www.rsc.org

Although, on the face of it, workplace drug testing may appear to be a straightforward process, it is an area in which employer/employee relations, employment legislation, health & safety legislation, data protection and human rights converge to present a potential minefield for the unwary. Any specialist laboratory providing workplace drug testing services should be accredited to the ISO17025 standard (specifically for Workplace Drug Testing) or an equivalent standard, and should be able to provide expert advice to an employer wishing to implement a drug testing programme as part of their measures against drugs and alcohol in the workplace.

The recommended reading list at the end of this chapter will provide a very good background for anyone wishing to apply themselves to drug testing in the workplace in a professional manner.

11.2   Why Test?

The reasons that employers drug test their staff, or potential recruits, can be broadly summarised as "risk mitigation". Employers have a duty of care to their staff and to members of the public to ensure that they provide a safe working environment (*e.g.* UK Health and Safety at Work Act 1974) and they are also required, by UK law, to take reasonable precautions to prevent drug dealing and other illicit activities in the workplace (UK Misuse of Drugs Act 1971).

Drug testing in the workplace has been performed in the UK and elsewhere for many years – a report from the USA in 1986 suggested that workplace drug testing was becoming widespread from as early as 1981[1] and the US Government first issued guidelines for mandatory drug testing in 1988.[2] Despite this, the application of workplace drug testing remains controversial and, in some cases, poorly implemented. The 2004 report of an independent enquiry into drug testing at work reported that, although drug testing may have a role in some industries, it is no substitute for good management and that good all-round management is more likely to improve safety, productivity and performance than drug testing at work.[3]

Notwithstanding the potential ethical issues and management practices, drug testing provides *one* weapon in the employer's armoury but will only be effective if there is a robust policy in place too. Some excellent guidance on appropriate contents of drug and alcohol policies has been provided by Hadfield.[4] The approach adopted by the employer must be clearly stated in their policy. The policy needs to clearly define:

- What substances will be tested for, and the circumstances under which testing will be done.
- What action will be taken in the event of a positive test, a refusal to undertake a test, provision of an adulterated/fraudulent sample or any other abnormality associated with the test.
- Who "owns" the policy and who will see the results of the tests.
- What the employee's rights and safeguards are, especially if the results of the test are disputed.
- The policy also needs to allow for an arbitration process and an appropriate medical review of the test results to ensure that the employee is not unfairly discriminated against.

Typical reasons for testing are given in Table 11.1.

Table 11.1 Typical reasons for applying workplace drug testing.

Pre-employment	The employer has decided that they will test all potential employees to ensure that regular drug users are not recruited
Post-incident/post-accident for cause	The employer has decided that staff involved in an accident or incident at the workplace will be tested to identify whether drug use might have been a contributory factor
Random testing	The testing of staff in a truly random manner (that is, the employees to be tested are selected at random). This may mean that the same individual may be tested on more than one occasion (for example, the employer may decide that 10% of their workforce will be tested every 3 months; the same member of staff may get picked, at random, two or three times in the same year). For random testing to be seen as "truly random", many employers will choose to use a third party to manage the process for selecting the staff to be tested. In some industry sectors, staff may be given notice that they have been selected for testing
Unannounced testing	Samples are collected from employees with minimal notice. The benefit of this approach is that the individual does not have any opportunity to cheat the system

11.3   Confidentiality Issues

Most people would probably have no doubt that the results of a drug test should be treated as sensitive information. However, in many organisations, the fact that an individual has been tested is not, of itself, regarded as "sensitive". For example, in an environment where all staff are subject to pre-employment or random tests, then it would be reasonable to expect that everyone has been subjected to at least one test. Issues of confidentiality may arise inadvertently, as a consequence of some business practices. For example, it is not uncommon for finance departments to require details of which individual a specific invoice for drug testing applies to (so that the costs can be apportioned to, or approved by, the relevant head of department). If the charge also includes additional costs for confirmation testing, or medical review services, then the accounts payable clerk "knows" that there must have been something present in the specimen to trigger a positive response on an initial immunoassay screening test, even if the final outcome was a "pass/negative".

A laboratory performing the analysis of samples for workplace drug testing does not need to have any information identifying the individual being tested, provided procedures for ensuring that the sample which was tested can be traced back to the individual that provided the sample. Some (usually larger) employers or organisations have rigorous procedures in place to ensure this degree of traceability exists. In many smaller environments, employers cannot afford the administrative overhead associated with implementing the cross-referencing of individual to sample reference number that would be required to enforce complete anonymity to the laboratory. This can prove challenging, in some cases, because there is always the risk that the individual being tested may be known to one of the people associated with the sample analysis in the laboratory. For this reason, laboratory information management systems should be configured so that samples are referenced only by their reference number, with individual's names unavailable to anyone who does not specifically need that level of detail. The UK Information Commissioners Office (ICO) has published guidance for employers, including a section addressing workplace drug testing within the Data Protection Act 1998.[5]

11.4   How To Test?

Several factors need to be considered when determining the most appropriate analytical methodology to apply to workplace drug testing:

1. Speed; employers wishing to take on new staff, or who have staff
   suspended following an incident, need to have their test results
   returned promptly to avoid excessive costs and delays.
2. Robust/reliable test results; the outcome of the drug test is very
   likely to be a significant factor in determining whether an individual is given a job, or dismissed from a job. The employer
   needs to be confident that the results of the tests will withstand
   legal scrutiny if tested in court.
3. Which substances to test for.
4. Cost; employers are always trying to keep costs to a minimum
   and will often compromise on what can be achieved with point 3
   above to keep cost and turn-around time down.

Alcohol testing should always be done using an evidential breath-
alyser. The recommended approach for drugs[6,7] is for samples to be
"screened" using immunoassay and for any non-negative samples to
be "confirmed" by gas chromatography-mass spectrometry (GC-MS)
or liquid chromatography-mass spectrometry (LC-MS). This approach
has the advantage of combining the rapid, easily automated capabilities of immunoassays with the analytical accuracy of GC-MS or
LC-MS. The downside to this approach is that immunoassay reagent
manufacturers tend not to invest in the development of reagents for
novel substances until there is proven commercial demand. An additional concern with the introduction of novel substances on the drug
scene is that very little may be known about the substances, their
effects and the way in which they are metabolized, and reference
materials (for use as calibrators or quality control materials during
the analytical process) for the substances and their metabolites may
be unavailable for many months after the substances "hit the street".
This means, in practice, that by the time reagents are available the
drug users have moved to some new substance and there is limited
opportunity to offer a commercially viable service for new substances.
    The initial immunoassay screening test could be in the form of
a "POCT" (point-of-care test – usually a lateral-flow immunoassay
testing device incorporating tests for several substances in a single

device and intended for use outside the laboratory). The pros and cons of such devices are discussed elsewhere in this publication.

11.4.1 Matrices

The sample type used for the test will also need to be agreed.

11.4.1.1 Urine

Urine has the most history, is the best documented and the most tried/tested in court. Because urine is the major route of elimination of most metabolic products, the detection of metabolites provides strong evidence of use (as opposed to accidental exposure) of a substance. However, if urine testing is selected, there needs to be suitable provision for appropriate toilet/sample collection facilities. Urine sampling is frequently used for pre-employment and random/unannounced testing.

11.4.1.2 Oral Fluid (Saliva)

The presence of substance in oral fluid indicates very recent use/exposure. In most cases, using current analytical methodology, it is the parent compound and not metabolites that are detected. It can, therefore, be difficult to exclude passive exposure as a potential cause of the test result (especially for substances that are usually smoked). The use of oral fluid samples is potentially most useful for "for cause" testing because:

(a) It is a relatively easy sample to collect (no special facilities required).
(b) It may provide evidence of very recent use or exposure to the substance.

See Chapter 21 for more discussions around oral fluid.

11.4.1.3 Hair

Testing a sample of hair provides history of substance use. No special facilities are required for sample collection, but the laboratory processes are more laborious and time-consuming due to the need to extract drugs and their metabolites from the hair prior to analysis. In the workplace, the use of hair sampling is probably most benefit in pre-employment testing for law enforcement agencies, or for following up on a previous urine or oral fluid sample where the individual claims that a failed test was due to "one-off" use of a substance.

11.4.1.4 Other Matrices

These may also be tested for the presence of substances (*e.g.* sweat,[8,9] exhaled air[10]) although, at the time of writing, with the exception of breath-testing for the presence of alcohol, these have not established a significant presence in the field of workplace drug testing. See also Chapters 19 and 22 for other alternative matrices.

11.5 What Substances To Test For?

The list of substances that most employers test for is usually restricted to alcohol and "illicit" substances. There are significant ethical considerations when attempting to extend testing to include, for example, prescription medication, and this would only be sanctioned by trades unions or workers representative committees where there are compelling (*e.g.* safety related) grounds for doing so.

Analytical service providers can advise on the substances that can be tested for, but employers' drug and alcohol policies should allow for the facility to add new substances as drug use patterns change without recourse to rewriting the policy (for example, include the list of substances in an appendix that is easily updated).

11.6 What Cut-offs To Apply?

The "cut-off" concentration is the concentration of substance in a sample at or above[†] which a test would be reported as "positive" or "failed". Some organisations or industry sectors may adopt their own cut-off concentrations (for example, the UK Rail Safety & Standards Board[11,12]), but many will follow the guidelines established by the European Workplace Drug Testing Society (EWDTS), the US Substance Abuse and Mental Health Services Administration (SAMHSA)[13] or the Australian/New Zealand Standards bodies.[14] Different cut-offs will apply to different sample matrices and, for oral fluid, should refer to undiluted sample (many oral fluid sampling devices include a buffer solution to stabilise the sample; this has the effect of diluting the sample that was collected). Example cut-off concentrations taken

---

[†]In some organisations, a test is failed/positive if the measured result is greater than or equal to the cut-off concentration, whereas in others the test is failed/positive only if the measured result is greater than the cut-off. The laboratory must ensure that it has a comprehensive understanding of customer requirements to ensure that these distinctions are recognised and applied to their customers' samples.

from the EWDTS guidelines for urine,[7] oral fluid[15] and hair[16] are given in Tables 11.2–11.4.

Table 11.2   Example urine cut-off concentrations (from EWDTS guidelines[7]).

Group	Substance	Immunoassay ("screening") cut-off (ng mL^{-1})	Chromatographic/ mass spectrometric ("confirmation") cut-off (ng mL^{-1})
Amphetamines		500	
	Amphetamine		200
	Methamphetamine		200
	MDA		200
	MDMA		200
	Other members of the amphetamine group		200
Benzodiazepines		200	
	Alprazolam		100
	Bromazepam		100
	Clonazepam		100
	Diazepam		100
	Flunitrazepam		100
	Flurazepam		100
	Lorazepam		100
	Lormetazepam		100
	Midazolam		100
	Nitrazepam		100
	Nordiazepam		100
	Oxazepam		100
	Phenazepam		100
	Temazepam		100
Cannabis metabolites		50	
	Cannabis metabolite (THC-COOH)		15
Cocaine metabolites		150	
	Cocaine metabolite (benzoyl-ecgonine)		150

Table 11.2 *(Continued)*

Group	Substance	Immunoassay ("screening") cut-off (ng mL^{-1})	Chromatographic/ mass spectrometric ("confirmation") cut-off (ng mL^{-1})
Opiates		300	
	Morphine		300
	Codeine		300
	6-Monoacetyl- morphine		10
Methadone (or EDDP)		300 (100)	
	Methadone		250
	EDDP		75

Table 11.3 Example oral fluid cut-off concentrations (from EWDTS guidelines[15]).

Group	Substance	Immunoassay ("screening") cut-off (ng mL^{-1})	Chromatographic/ mass spectrometric ("confirmation") cut-off (ng mL^{-1})
Amphetamines		40	
	Amphetamine		30
	Methamphetamine		30
	MDA		30
	MDMA		30
	Other members of the amphetamine group		Not defined
Benzodiazepines		10	
	7-Aminofluni- trazepam		10
	7-Aminoclo- nazepam		10
	7-Amino- nitrazepam		10
	Alprazolam		10
	Bromazepam		10
	Clonazepam		10

Table 11.3 *(Continued)*

Group	Substance	Immunoassay ("screening") cut-off (ng mL^{-1})	Chromatographic/ mass spectrometric ("confirmation") cut-off (ng mL^{-1})
	Diazepam		10
	Flunitrazepam		10
	Flurazepam		10
	Lorazepam		10
	Lormetazepam		10
	Midazolam		10
	Nitrazepam		10
	Nordiazepam		10
	Oxazepam		10
	Phenazepam		10
	Temazepam		10
Cannabis		10	
	Cannabis (THC)		2
Cocaine metabolites		30	
	Cocaine metabolite (benzoyl-ecgonine)		8
	Cocaine		8
Opiates		40 (6-MAM 4)	
	Morphine		2
	Codeine		2
	Norcodeine		2
	6-Acetylcodeine		2
	Dihydrocodeine		2
	6-Monoacetyl-morphine		4
Methadone		50	
	Methadone		20
	EDDP		Not defined

It should be noted that "universal" agreement of cut-off concentrations is difficult to achieve, in particular for:

(a) Novel substances; cut-off concentrations tend to reflect analytical limits of detection or quantitation and may vary from one

Table 11.4 Example hair cut-off concentrations (from EWDTS guidelines[16]).

Group	Substance	Immunoassay ("screening") cut-off ($ng\,mg^{-1}$)	Chromatographic/mass spectrometric ("confirmation") cut-off ($ng\,mg^{-1}$)
Amphetamines		0.2	
	Amphetamine		0.2
	Methamphetamine		0.2
	MDA		0.2
	MDMA		0.2
	Other members of the amphetamine group		0.2
Benzodiazepines		0.05	
	Alprazolam		0.05
	Bromazepam		0.05
	Diazepam		0.05
	Flunitrazepam		0.05
	Lorazepam		0.05
	Nordiazepam		0.05
	Oxazepam		0.05
Cannabinoids		0.05	
	Cannabis (THC)		0.05
	Carboxy-THC (THC-COOH)		$0.2\ pg\,mg^{-1}$
Cocaine metabolites		0.5	
	Cocaine metabolite (benzoyl-ecgonine)		0.05
	Cocaethylene		0.05
	Norcocaine		0.05
	Ecgonine methyl ester		0.05
	Cocaine		0.5
Opiates		0.2	
	Morphine		0.2
	Codeine		0.2
	6-Acetylcodeine		0.2
	Dihydrocodeine		0.2
	6-Monoacetyl-morphine		0.4

provider to another, depending on the instrumentation they have in their laboratories. (Cut-off concentrations for more established substances were originally defined in a similar way, using instrumentation available at the time and have since been enshrined in Standards, *etc.*, and have proven resistant to change even though modern laboratory instruments are considerably more sensitive than their predecessors.) Also see Chapter 6 for NPS discussions.

(b) The reasons for the tests may vary. For example, when testing oral fluid for evidence of "impairment" (*e.g.* "drug-driving" scenarios) it would be preferable to apply relatively high cut-off concentrations reflective of the blood levels of substances when an individual is impaired, whereas many employers may prefer oral fluid testing to give comparable results and detection windows as seen with urine testing. In these cases, much lower cut-off concentrations would need to be adopted. Also see Chapters 1 and 13 for drug-driving discussions.

11.7   The "Testing Process"

The practical aspects of workplace drug testing break down on to a small number of distinct phases: sample collection, analysis and "review and interpretation" of results. These are described in more detail below.

11.7.1   Sample Collection

Regardless of the type of sample collected, and the reason for the test, any samples for workplace drug testing need to be collected under "chain-of-custody". This will be the start of an evidence chain that will ultimately provide confidence that the reported result of the test relates to the sample provided by that individual and that there has been no opportunity for the sample to have been interfered with between the point of collection and the completion of the analysis.

- The identity of the individual providing the sample must be verified (*e.g.* by photo ID or confirmation of ID by a manager).
- The individual must provide informed consent to the collection and analysis of the sample.
- The sample collection process (including the facilities available for collecting the sample) must maintain the dignity of the individual.
- The sample must, wherever possible, be divided into at least two representative portions (an "A" portion and a "B" portion),

each of which is placed into a container and sealed in the presence of the individual. There must be a unique reference number for "this sample" that appears on the consent form signed by the individual and on each of the sealed sample containers.

11.7.2 Analysis

As discussed above, the preferred mechanism[‡] for the analytical process is to adopt a two-stage approach. An initial immunoassay should be used to rapidly identify "negative" samples. Any sample that fails an immunoassay test must be analysed by a technique that combines chromatography and mass spectrometry (*e.g.* GC-MS or LC-MS) to confirm the presence of any of the target substances. Only the "A" portion of sample is analysed by the laboratory; the "B" portion *must* remain sealed and be stored in appropriate conditions to allow the individual an opportunity to have this portion tested by an alternative approved laboratory in the event that there is a dispute relating to the result reported following analysis of the "A" portion.

Certain sample types may also require additional validity tests. For example, the concentration of creatinine in a urine sample could be used to indicate that a sample has been artificially diluted. George[17] provides an excellent review of the effects of specimen adulterants on immunoassay and GC-MS/LC-MS "confirmation" tests. In all cases, the most effective means to reduce adulteration is to remove the opportunity for adulteration to occur (including, for example, arranging for drug tests to be done with minimal notice to the individual).

The analytical approach described is intended to minimize the risks of "false positive" or "false negative" results. In this context, false positive results occur:

(a) When there is an aberrant high reading on an immunoassay test, but there are no substances present in the sample to account for the analytical response (a "true" false positive).

(b) When there is a substance present that reacts with the immunoassay test to lead to a positive response, but that substance is not one of the target compounds of interest ("cross-reactivity").

[‡]This two-stage approach is recommended by UK and European guidelines for workplace drug testing and mandated by the US Substance Abuse and Mental Health Services Administration, US Dept of Transport, the US Coastguard, the Australian & New Zealand Standards Board and others.

(c) When there is a substance present that is converted to, or misinterpreted as, a compound of interest (a "true" false positive). Hornbeck *et al.*[18] reported just such an example where samples containing high concentrations of ephedrine or pseudoephedrine were mistakenly reported as containing methylamphetamine.[18]

(d) When there is a substance present that is metabolised to one of the target compounds of interest (see next section on interpretation of results); strictly speaking, this is not a false positive since the compound of interest is truly present in the sample but, from the lay person's perspective, it would not be a "true" result.

"False negative" results may occur when:

(e) A target substance is truly present in the sample, but an adulterant has been added to the sample to prevent that substance from being detected (for example, the addition of an oxidising agent to a sample may chemically modify the target substance[17]).

(f) Several compounds of interest, together in the sample, may lead to an immunoassay response above the cut-off but, when individually quantitated, each is below the cut-off for the test; strictly speaking, this is not a true "false negative" since the analytical findings indicate that substances are below the cut-off for the test but, from a lay perspective, this may be difficult to understand.

11.7.3 Result Review and Interpretation

Any non-negative result reported by the laboratory must be subjected to review by an experienced toxicologist or medical practitioner. The purpose of this review is to ensure that the individual is not falsely accused of substance misuse where a positive result could be accounted for by the legitimate use of over-the-counter or prescription medication (or by dietary sources). In 2006, the Faculty of Occupational Medicine at the Royal College of Physicians published guidance on alcohol and drug misuse in the workplace[19] and required that samples which test positive at the laboratory are reviewed by a Medical Review Officer ("MRO"). The purpose of the review process is to ascertain whether there is a legitimate reason to account for the laboratory findings and to determine whether the final outcome of

Table 11.5 Example outcomes of workplace drug testing.[a]

Outcome	Comment
Pass/ negative	Sample has tested negative (or below cut-off concentration) for all substances and all sample integrity tests have passed
Pass/ positive	At least one substance was detected above the cut-off concentration but the result can be accounted for by declared medication
Fail/positive	At least one substance was detected above the cut-off concentration and/or the sample failed one or more integrity checks
Inconclusive	It has not been possible to provide a definitive pass or fail outcome. This may be because, for example, there was insufficient specimen to conduct all the tests to the satisfaction of the laboratory analysts or the Medical Review Officer was not satisfied with responses given in explanation of an analytical positive result

[a]The use of the terms "Pass" and "Fail" allows clear distinction between, for example, samples that are "analytically positive" but for administrative reasons shoud be treated as "negative".

the test should be treated as a pass or as a fail. The person conducting the medical review must be suitably qualified: there are many examples of legitimate prescription medicines that, when metabolized in the human body, lead to detectable concentrations of illicit substances (for example, in 1993, Cody published an article citing 14 metabolic precursors to amphetamine or methylamphetamine[20]). In addition, it is widely documented that the dietary intake of poppy seeds can account for the presence of opiates in a urine sample.[21–23] Examples of outcomes of the review process are given in Table 11.5. The terminology used in the output of the review can be very helpful to a lay person when understanding the outcome. For example, it is not uncommon for an employer's drug testing policy to refer to some form of disciplinary action to be taken in the event of a "positive test". The use of the terms "pass" and "fail" can make it very clear that the outcome of the test is (or is not) in compliance with a policy, regardless of the analytical findings themselves.

11.7.4 Challenges to Drug Test Results

The individual who provided a sample for analysis may dispute the interpretation of the results, in which case the analytical data from the laboratory could be reviewed by an expert third party and an alternative interpretation put forward.

In the event that the analytical findings reported by the laboratory are themselves in dispute, the "B" portion of the sample will have been retained, sealed and kept in appropriate storage conditions in case of such an eventuality. This portion can be forwarded, along with the reported result from the original analysis and under chain-of-custody, to an alternative (accredited) laboratory. The alternative laboratory should then analyse the "B" portion and report their findings as either *consistent* or *not consistent* with the original report. This terminology allows for minor discrepancies in measured concentrations that might arise from differences in analytical methodology or deterioration of the sample since the original analysis was conducted.

11.8 Incidence of Drug Use at Work

In 2012, Alere Toxicology Plc (formerly Concateno Plc) published a report summarising the findings of some 1.6 million workplace drug tests conducted over the preceding 5 years.[24] The published figures indicated an increase positivity rate from 2.26% of samples collected in 2007 testing positive to 3.23% of samples collected in 2011 testing positive. Additional information, not published until now (Figures 11.1–11.3), supports the argument that drugs may play more of a role in workplace accidents than previously argued by factions who object to drug testing in the workplace. Of particular interest is

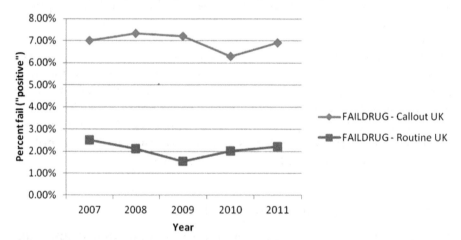

Figure 11.1 Incidence of passed/failed drug tests in the UK. These data are for drugs only (alcohol is excluded) and compares *illicit drug use* in the workforce for samples collected "routinely" and "post-incident in the workplace".

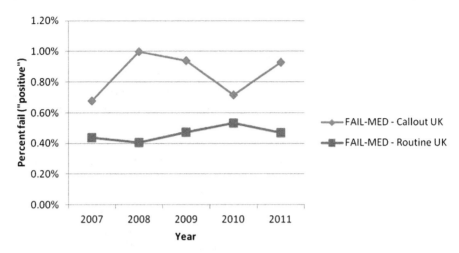

Figure 11.2 Incidence of passed/failed drug tests in the UK. These data are for drugs only (alcohol is excluded) and compares *unauthorised use of medication* in the workforce for samples collected "routinely" and "post-incident in the workplace".

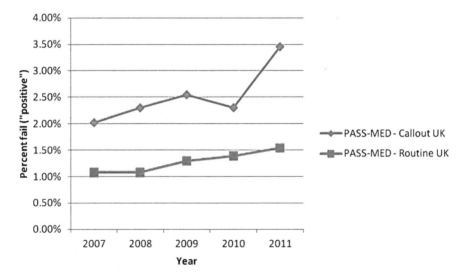

Figure 11.3 Incidence of passed/failed drug tests in the UK. These data are for drugs only (alcohol is excluded) and compares *authorised use of medication* in the workforce for samples collected "routinely" and "post-incident in the workplace".

the observation that even legitimate use of medication can be associated with an increased likelihood of being involved in an incident at work (Figure 11.3).

11.9 Conclusion

Workplace drug testing *per se* provides little benefit when conducted in isolation, but when applied properly, in conjunction with a comprehensive policy, it can be a very valuable tool. The correct use of drug testing *must* encompass rigorous procedures for maintaining chain-of-custody and specimen integrity and *must* be conducted by a laboratory accredited to ISO170256 (specifically for Workplace Drug Testing) to ensure that the results of the tests will withstand legal scrutiny.

Users of drug testing services should take appropriate measures to ensure that their service providers work within accepted guidelines and to appropriate standards. For their part, providers of drug testing services must remember that their work can directly affect the quality of life of the individuals being tested. Failure to adhere to professional standards of work will adversely affect the individual, the employer and the reputation of the service provider.

Recommended Reading

For more information relating to tackling issues related to drug and alcohol misuse in the workplace, including ethical considerations and policy recommendations, please refer to the following publications and sources: *Addiction at Work*;[25] *Workplace Drug Testing*;[26] TUC website.[27]

References

1. *National Institute on Drug Abuse Research Monograph 73: Urine Testing for Drugs of Abuse*, ed. R. L. Hawks and C. N. Chang, US Department of Health and Human Services, Rockville, 1986.
2. D. Bush, Review of Significant Changes in the Revised Mandatory Guidelines for Federal Workplace Drug Testing Programs. *TIAFT 50th Anniversary Conference*. London: April 26–27, 2013.
3. Ruth Evans (chair), *The Independent Inquiry into Drug Testing at Work*, York: Joseph Rowntree Foundation, 2004.
4. L. Hadfield, Policies for drugs and alcohol, *Workplace Drug Testing*, ed. A. Verstraete, Pharmaceutical Press, London, 2011, pp. 147–174.

5. ICO, The employment practices code, https://ico.org.uk/media/for-organisations/documents/1064/the_employment_practices_code.pdf. [Online] [Cited: September 20, 2015.] https://ico.org.uk/media/for-organisations/documents/1064/the_employment_practices_code.pdf.

6. *Mandatory Guidelines for Federal Workplace Drug Testing Programs*, Dept of Health and Human Services. 228, Rockville: Substance Abuse and Mental Health Services Administration, November 25, 2008, Federal Register, vol. 73, pp. 71858–71907, effective May 2010.

7. European Workplace Drug Testing Society, *European Guidelines for Workplace Drug Testing in Urine*, EWDTS: European Workplace Drug Testing Society, 2015, pp. 1–41.

8. N. De Giovanni and N. Fucci, The current status of sweat testing for drugs of abuse: a review, *Curr. Med. Chem.*, 2013, **20**(4), 545–561.

9. Y. H. Caplan and B. A. Goldbergerm, Alternative Specimens for Workplace Drug Testing, *J. Anal. Toxicol.*, 2001, **25**, 396–399.

10. N. Stephanson *et al.*, Method validation and application of a liquid chromatography–tandem mass spectrometry method for drugs of abuse testing in exhaled breath, *J. Chromatogr. B*, 2012, **985**, 189–196.

11. RSSB, *Guidance on the Management of Drugs and Alcohol GE/GN8570 Issue 2*, London: Rail Safety and Standards Board Ltd, 2012.

12. RSSB, Testing Railway Safety Critical Workers for Drugs and Alcohol GE/RT8070 Issue 3, London: Rail Safety and Standards Board Ltd, 2012.

13. Department of Health and Human Services Substance Abuse and Mental Health Services Administration, *Mandatory Guidelines for Federal Workplace Drug Testing Programs*, Rockville: Federal Register, 1994.

14. *Procedures for specimen collection and the detection and quanitita-tion of drugs of abuse in urine*, Standards Australia/Standards New Zealand. s.l.: Standards Australia/Standards New Zealand, 2008, AS/NZS 4308:2008.

15. European Workplace Drug Testing Society, *European Guidelines for Workplace Drug Testing in Oral Fluid*, European Workplace Drug Testing Society: EWDTS, 2015, pp. 1–40.

16. P. Kintz and R. Agius, Guidelines for European workplace drug and alcohol testing in hair, *Drug Test. Anal.*, 2010, **2**, 367–376, John Wiley & Sons Ltd.

17. C. George, Specimen adulteration, in *Workplace Drug Testing*, ed. A. Verstraete, Pharmaceutical Press, London, 2011, pp. 249–291.

18. C. L. Hornbeck, J. E. Carrig and R. J. Czarny, Detection of a GC/MS Artifact Peak as Methamphetamine, *J. Anal. Toxicol.*, 1993, **17**, 257–263.

19. Faculty of Occupational Medicine, *Guidance on Alcohol and Drug Misuse in the Workplace*, London: Faculty of Occupational Medicine of the Royal Collee of Physicians, 2006.

20. J. T. Cody, A. F. B. Lackland, Metabolic Precursors to Amphetamine and Methamphetamine, *Forensic Sci. Rev.*, 1993, **5**, 109.

21. C. M. Selavka, Poppy seed ingestion as a contributing factor to opiate-positive urinalysis results: the Pacific perspective, London: *J. Forensic Sci.*, 1991, **36**, 3, 685–696.

22. C. Meadway, S. George and R. Braithwaite, Opiate concentrations following the ingestion of poppy seed products-evidence for 'the poppy seed defence', London: *Forensic Sci. Int.*, 1998, **96**, 1, 29–38.

23. M. Thevis, G. Opfermann and W. Schanzer, Urinary Concentrations of Morphine and Codeine After Consumption of Poppy Seeds, *J. Anal. Toxicol.*, 2003, **27**, 53–56.

24. Alere Toxicology Plc. http://www.aleretoxicology.co.uk/highsociety/ (registration required), *Alere Toxicology Plc.* [Online] July 2012. [Cited: September 20, 2015.] MES0070 Ed.001. http://www.aleretoxicology.co.uk/highsociety/.

25. *Addiction at Work*, ed. H. Ghodse, Gower Publishing Limited, Aldershot, 2005.

26. *Workplace Drug Testing*, ed. A. Verstraete, Pharmaceutical Press, London, 2011.

27. Trades Union Congress, Drugs and Alcohol, *Trades Union Congress (TUC)*, [Online] [Cited: September 12, 2015.] https://www.tuc.org.uk/workplace-issues/health-and-safety/drugs-and-alcohol.

12 Current Issues in Human Sport Drug Testing: Clenbuterol, Erythropoietin and Xenon

A. T. Kicman,* D. A. Cowan and I. Gavrilović

King's College London, London, UK
*Email: andrew.kicman@kcl.ac.uk

12.1 Introduction

This chapter aims to give the reader an insight concerning three topical issues in anti-doping in human sport: clenbuterol, erythropoietin and the very recent issue of xenon inhalation. The physicochemical properties, pharmacology and analytical chemistry of the substances discussed are very different, exemplifying the complexity and diversity of this facet of toxicology.

12.2 Clenbuterol

The toxicologist is often asked to give an opinion on the relevance of a concentration of a drug or metabolite reported to be present in a biological matrix. Even when qualitative analysis is applied, an estimate of the concentration may still be highly relevant to the case. Depending on the analyte, a trace or small finding can be vitally important in some cases, but for a different analyte may be of no evidential significance whatsoever. Many sports have signed-up to the

Forensic Toxicology: Drug Use and Misuse
Edited by Susannah Davies, Atholl Johnston and David Holt
© The Royal Society of Chemistry 2016
Published by the Royal Society of Chemistry, www.rsc.org

rules of the World Anti-Doping Agency (WADA), this agency being established in 1999 under the initiative of the International Olympic Committttee (IOC) (the IOC rules on anti-doping then became in accordance with the World Anti-Doping Code). In sports that have adopted the World Anti-Doping Code, on occasion, the detection of a performance-enhancing drug in a urine sample within a low $ng\,mL^{-1}$ to $pg\,mL^{-1}$ concentration can raise problematic issues. A predominant factor is that the principle of strict liability applies even when there is a possibility that the drug can be unintentionally consumed in pharmacologically irrelevant amounts. Such adverse finding sometimes cause widespread controversy, with time-consuming investigations to consider whether a sanction should be applied and then sometimes appeals involving legal representatives on both sides and the time and expense that entails. In this respect, nandrolone is one well-known example, widely publicised in the past, but arguably the issue of clenbuterol has continued for longer, the latter starting around the time of the Barcelona Olympic Games in 1992 and continuing into the second decade of this century, such as the under-17 Soccer World Cup in Mexico in 2011. To date, there is no indication that the declarations by laboratories regarding findings for clenbuterol are diminishing (Table 12.1).

Bronchodilators, such as clenbuterol, are used in the treatment of asthma and other respiratory ailments. They are also considered as

Table 12.1   Number and % of clenbuterol findings declared by WADA-accredited laboratories.[a,b]

Year	Occurrences	Samples analysed	%
2014	251	283 304	0.09
2013	183	269 878	0.07
2012	98	267 645	0.04
2011	129	243 193	0.05
2010	116	258 267	0.04
2009	67	277 928	0.02
2008	73	274 615	0.03
2007	53	223 898	0.02
2006	53	198 143	0.03
2005	52	183 337	0.03
2004	46	169 187	0.03
2003	31	151 210	0.02

[a]Taken from the WADA anti-doping testing figures.
[b]Note: these statistics are not to be confused with adjudicated or sanctioned Anti-Doping Rule Violations.

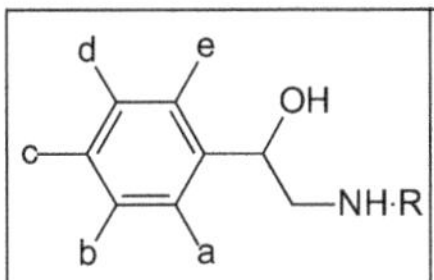

Substance	a	b	c	d	e	R
salbutamol	H	OH	H	CH$_2$OH	H	C(CH$_3$)$_3$
clenbuterol	H	Cl	NH$_2$	Cl	H	C(CH$_3$)$_3$
terbutaline	H	OH	H	OH	H	C(CH$_3$)$_3$
salmeterol	H	OH	H	CH$_2$OH	H	(CH$_2$)$_6$-O-(CH$_2$)$_4$-C$_6$H$_5$
formoterol	H	H	OH	HCONH	H	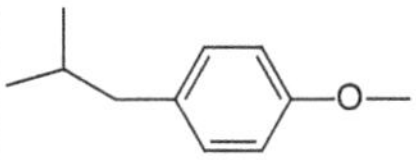 (structure)
fenoterol	H	OH	H	OH	H	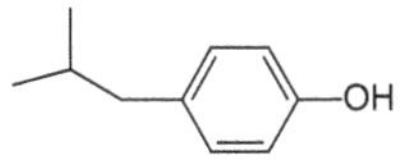 (structure)
isoprenaline	H	H	OH	OH	H	C(CH$_3$)$_3$

Figure 12.1 Examples of structures of β$_2$-agonists. (Reproduced from Van Eenoo and Delbecke,[1] with permission from Springer.)

potential ergogenic aids in sport, as beta-2-adrenergic receptor agonists (β$_2$-agonists), being structurally related to ephedrine (Figure 12.1).[1] The β$_2$-agonists fall under the category of sympathomimetic amines and therefore they were originally included within the doping class of stimulants,[2] but subsequently they were separated into their own class (Table 12.2). For therapeutic purposes, clenbuterol is usually adminstered as tablets, although it has also been given as inhalation, the usual oral dose being 20 µg to be administered twice daily.[3] Such tablets are likely to be obtained for doping purposes despite controls on medicines, with veterinary preparations offering another source, in granules, as a syrup and in an injectable form,[4] as well as non-medicinal sources including powders. With the exception as inhalation therapy for athletes with pulmonary disorders, usually asthma, the use of β$_2$-agonists were and are prohibited in sport under the WADA rules. By their bronchodilator action, the acute use of β$_2$-agonists anti-asthma medications seemed, nonetheless, unlikely to provide an improvement in performance beyond that to control asthma, based on a review of the literature published in 2008,[5] but a

Table 12.2   The rule concerning the β_2-adrenergic agonists as specified in the 2016 Prohibited List of the World Anti-Doping Code. Clenbuterol belongs in this class, but importantly it is the only β_2-agonist specifically named under "S1. Anabolic Agents".

S3. β-2 Agonists

All *β-2 agonists*, including all *optical isomers*, *e.g.* D- and L-where relevant, are prohibited.

Except:

- Inhaled *salbutamol* (maximum 1600 µg over 24 h)
- Inhaled *formoterol* (maximum delivered dose 54 µg over 24 h)
- Inhaled *salmeterol* in accordance with the manufacturers' recommended therapeutic regimen

The presence in urine of salbutamol in excess of 1000 $ng\,mL^{-1}$ or of formoterol in excess of 40 $ng\,mL^{-1}$ is presumed not to be an intended therapeutic use of the substance and will be considered as an *Adverse Analytical Finding (AAF)* unless the *Athlete* proves, through a controlled pharmacokinetic study, that the abnormal result was the consequence of the use of the therapeutic inhaled dose up to the maximum indicated above.

---

transient increase in performance could not be discounted when the rule was implemented, and nor can it be now.

In the past, there was no infringement of the rules if β_2-agonists were detected in urine samples collected out-of-competition, as it was not considered that they could provide anything but a transient advantage in performance following their administration. This view began to be questioned in the mid to late 1980s and early 1990s, with reports of oral β_2-agonists acting as growth promoters in animals (farm, show and laboratory). These adrenergic agonists were considered to be "repartitioning agents", a term used to encompass an increase in lean muscle mass to fat ratio. Attention particulary focused on clenbuterol, which was considered the most potent in stimulating protein deposition, with catabolism of fat also occurring due its thermogenic properties. For an excellent summary of what was known at that time regarding studies of the growth promoting effects of clenbuterol in animals, the interested reader is referred an editorial in *The Lancet*.[6] The issue that some athletes were using clenbuterol for the same purpose came to a head with the results of drug screening of athletes chosen to represent their countries at the Summer Olympics in Barcelona in 1992. It was decided that resultant positive findings of clenbuterol in samples collected from several athletes outside of competition was a doping infraction because clenbuterol was a

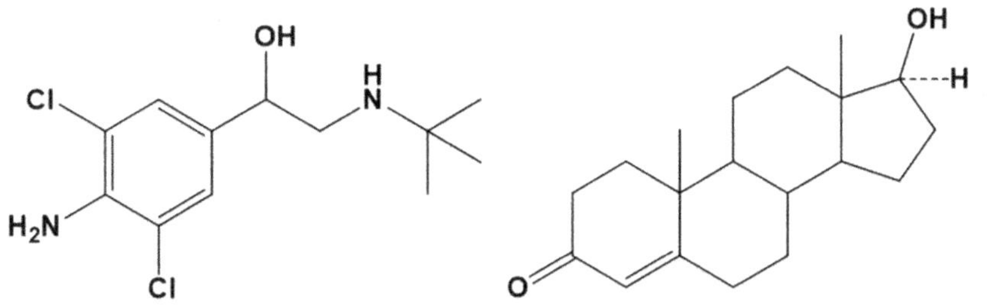

Figure 12.2 Structure of clenbuterol (*left*) in comparison with the anabolic steroid testosterone (*right*).

related compound within the banned class of anabolic-androgenic steroids, the use of anabolic steroids being prohibited at all times. As clenbuterol is completely different in structure to the anabolic steroids (Figure 12.2), the decision by the IOC Medical Commission (which incorporated its "Doping and Biochemistry of Sport Sub-Commission") was based on its assessment of the potential of clenbuterol to stimulate anabolism in the human. Even though some athletes were in all likelihood using clenbuterol chronically outside of competition for anabolic purposes, this decision was a very controversial one at the time. Professor Arnold Beckett, the established UK member of the IOC Medical Commission, openly disagreed with the decision that a doping violation had been committed and consequently he had to leave this scientific body. His eloquent letter to *The Lancet*,[7] published in November 1992, in response to the editorial in August of that year regarding clenbuterol as a repartitioning agent, makes a number of points, including that the rules should clearly state what is banned and what is not. The following year, the class of "anabolic steroids" was renamed by the IOC to "anabolic agents", with two subclasses being introduced: (i) anabolic-androgenic steroids and (ii) other anabolic agents (namely to include clenbuterol at that time). The anabolic effects of the β_2-agonists, in fact, are largely due to an inhibition in protein degradation of skeletal muscle protein,[5] *i.e.* an anti-catabolic effect, with "anabolic" being used as the encompassing term in the capacity of anti-doping.

Based on published evidence, it remains questionable as to whether the anabolic, bronchodilator and psychotropic effects of the β_2-agonists translate into improvements in sporting performance,[5] but there are several ongoing WADA-funded studies that may help to conclusively resolve this issue (see WADA web site). As an adjunct, bodybuilding web sites place much emphasis on clenbuterol as a "fat burner" and it is also marketed mainstream for slimming purposes.

For users seeking anabolism, achieving a systemic concentration that may be effective over a sustained period of weeks has to be balanced against the risk of experiencing unpleasant and potentially life-threatening adverse effects, chiefly tachycardia and cardiac arrhthymia.

With increasing analytical sensitivity, the question arises as to what concentration of a prohibited substance or metabolite should be reported? This first came to the fore in 1995, when out of 6700 samples tested by the anti-doping laboratory in Cologne, 116 adverse findings were reported, of which 41 were identified by the use of quadrupole gas chromatography–mass spectrometry, the typical analytical approach adopted internationally. The remaining 75 samples were detected only because of the application of a more sensitive technique, this being high-resolution mass spectrometry (HRMS), operating in the selected ion monitoring (SIM) mode, subsequent to an earlier report showing its potential.[8] Because of these "high sensitivity" findings, the IOC Medical Commission decided to place much greater emphasis on high sensitivity detection of a subset of anabolic agents, these being clenbuterol and the metabolites of particular anabolic steroids (of nandrolone, methandienone, 17α-methyltestosterone and stanozolol). The first GC-HRMS system was installed for dope testing at an Olympic Winter Games in 1994 held in Lillehammer, Norway. A mandatory minimum required performance level (MRPL) of 2 ng mL^{-1} was also introduced for these compounds, meaning that all accredited laboratories had to be capable of detecting and then confirming these target analytes at this concentration, whether by HRMS or tandem MS systems (importantly, this is not the same as a cut-off concentration/reporting threshold). Of note, it was permissible for laboratories to report smaller concentrations than 2 ng mL^{-1}, with the one subsequent exception being the targetted diagnostic metabolite of nandolone (19-norandrosterone). By contrast to the selected steroids, the presence of clenbuterol in urine was relatively easy to detect at 1 ng mL^{-1} and less, as electron ionization in positive mode of the bis-trimethylsilyl derivative of clenbuterol provides an intense fragment ion (*tert*-butylmethylamine cation).[9] The opportunity to report these substances, including clenbuterol below the MRPL of 2 ng mL^{-1}, was nevethetheless generally viewed with reticence, not least because of the analytical challenges and resources required regarding confirmatory analysis at such small concentrations. There was also apprehension over the rationale of targeting drugs and their metabolites at ever lower concentrations, not least because of the concern of athletes eating meat possibly tainted with clenbuterol.

It was known from a small research study in 1995 that following the eating of meat from calves that had received two doses of clenbuterol of 5 $\mu g\, kg^{-1}$ day^{-1}, urinary concentrations of clenbuterol could be measured (30–850 $pg\, mL^{-1}$).[10]

Such concern regarding tainted meat was not unfounded and, from a public health perspective, a major concern is the ingestion of toxic amounts of this highly potent agent. For example, several hundreds of Chinese people became ill after eating clenbuterol tainted pork in March 2011,[11] which led to the government banning the production, use and sale of clenbuterol tablets later that year. Even legislation by governments making it illegal to give clenbuterol to livestock for repartioning purposes is no guarantee that the practice will stop immediately or altogether. Within the EU countries, for example, the use of β-agonists has been banned since 1996 (Directive no. 96/22/CE, 1996), but there were cases of poisoning reported afterwards due to the eating of meat tainted with clenbuterol, such as in Portugal in 2005 involving 50 people,[12] illustrating that there is a considerable refractory period following legislation. On the other hand, eating meat containing clenbuterol far below that which may cause a noticeable adverse effect is a major issue for athletes, as this can result in them failing a drug test. Just before the Beijing Olympics in 2008, German athletes were officially warned about the possibility that meat in China may be contaminated with doping-relevant substances. A subsequent research investigation in 2010 demonstrated that this warning was well-founded, as analysis of urine specimens collected from each of 28 volunteers after they had visited various regions in China showed 22 were clenbuterol positive (79%), with the maximum concentration being a fraction under 51 $pg\, mL^{-1}$.[13] The risk of inadvertently producing a clenbuterol-positive urine sample thus depends on the legal controls in place in the country of origin of the meat, and the level of enforcement. Two notable and contrasting clenbuterol cases are described below.

In 2010, the reporting of a sub-ng mL^{-1} concentration of clenbuterol drew significant attention, this being an unusual case concerning Mr Alberto Contador, who was exonerated because of suspected meat contamination, but then subsequently sanctioned. Mr Contador, a Spanish professional cyclist, failed a test for the drug in the Tour de France. The A-sample contained a concentration of clenbuterol of only 50 $pg\, L^{-1}$, twenty-fold less than the MRPL of 2 $ng\, mL^{-1}$. A disciplinary committee within the Real Federación Española de Ciclismo (RFEC), the governing body of cycling in Spain and a member of the Union Cycliste Internationale (UCI), concluded that with great probability the positive test was a consequence of eating contaminated food, and this fact cannot be considered as negligent behaviour (as documented by the Court of Arbitration for Sport Report; CAS). The UCI and WADA then filed an appeal with CAS against the cyclist and the RFEC. CAS partially upheld the appeal, with Mr Contador being sanctioned with a "two-year period of in-eligibility" (in other words, he was banned from competing) and also being retrospectively disqualified from the 2010 Tour de France in which he had competed.[14] The CAS panel took into account that the probability of a piece of meat being contaminated in the EU by that time was very low. Furthermore, the provenance of the meat that this cyclist claimed that he had ingested the day before and on the day of collection of his sample had been investigated, including consider-ation of a report made by a Health Inspector of the Public Health Department of the Basque Government. The Court concluded that "the athlete's positive finding for clenbuterol is more likely to have been caused by the ingestion of a contaminated food supplement than by a blood transfusion or the ingestion of contaminated meat".

By contrast to the above, the ingestion of contaminated meat ac-counted for clenbuterol findings in 109 out of 208 urine samples collected during the 2011 Under-17 Soccer World Cup tournament in Mexico, with an average concentration of 105 $pg\,mL^{-1}$ (range: 1.3–1556 $pg\,mL^{-1}$).[15] It was discovered that of 47 food samples collected from meals served in the restaurants catering the teams, 14 (30%) contained clenbuterol at concentrations ranging between 60 $ng\,kg^{-1}$ and 11 $\mu g\,kg^{-1}$, with an average of 2.5 $\mu g\,kg^{-1}$. In these cases, the evidence of meat contamination being the cause was compelling and FIFA and WADA decided there was no infringement of the anti-doping rules.

In 2013, WADA lowered the MRPL for clenbuterol from 2 to 0.2 $ng\,mL^{-1}$ (200 $pg\,mL^{-1}$), but laboratories may continue to perfectly legitimately declare findings with smaller concentrations, as no cut-off has been set. The number of research articles regarding urine elimination profiles of clenbuterol is limited;[16,17] in the authors' opinion it is desirable to investigate where the MRPL of 200 $pg\,mL^{-1}$ might sit in relation to a multi-dose study since drug accumulation of this lipophilic agent is likely to occur and a starting dose of at least 20 $\mu g\ day^{-1}$ is more representative of both use and misuse. A pragmatic question that may be asked is whether choosing a cut-off concentration would be of benefit, as opposed to action required following any finding of clenbuterol, no matter how small? At present there appears to be no plan by WADA to introduce a urinary cut-off concentration for clenbuterol. Whatever the reason for WADA's decision, introducing a cut-off generally opens up fresh avenues of legal challenge based on the precision of measurement, when it may be viewed that qualitative analysis is sufficient for substances that only exists exogenously. Conversely, without implementation of a cut-off concentration, declaration of a finding can be theoretically extrapolated to $fg\,mL^{-1}$ of clenbuterol when, in the future, more sensitive mass spectrometers are no doubt developed. The current position may be considered to be an uncomfortable one and much more so for athletes present in countries where the eating of clenbuterol-tainted meat still remains a possibility or the risk is not known.

When recreational and performance enhancing drugs are also present in the food chain or as contaminants in the environment, ideally chemical analysis should be available to help discriminate the source of the drug present in a biological matrix, especially so when there is no reporting threshold. Assays for urinary clenbuterol do not currently have the irrefutable discrimination required to differentiate between accidental and purposeful administration, although stereospecific analysis may be helpful in evaluation.[18] Even so, if clenbuterol is administered close to the time of slaughter of an animal, as opposed to cessation sometime before slaughter, then the stereochemical composition is unlikely to differ sufficiently from the drug racemate. Segmental analysis of hair may also be helpful,[19] but under the current World Anti-Doping Code International Standard for Labaoratories, "Any testing results from hair, nails, oral fluid or other biological material shall not be used to counter *Adverse Analytical Findings* or *Atypical Findings* from urine". Notwithstanding this stipulation, such evidence can be considered in the review and appeal processes, albeit that WADA apply the principle of strict liability regardless of inadvertent or intentional administration.

It is difficult to make a firm conclusion regarding clenbuterol, as it depends on what perspective is taken. Assays with high analytical sensitivity are the norm in drug control in sport and limits of detection continue to advance because WADA-accredited laboratories usually adopt new mass spectrometer systems shortly after they are launched. This is desirable because long detection times help make an effective deterrent to drug misuse. Declaring findings of clenbuterol at $sub\text{-}ng\,mL^{-1}$ concentrations, nevertheless, comes at a cost to innocent athletes affected, with the likely worry and stigma experienced in the period between a laboratory reporting an adverse finding and subsequent acquittal (hopefully in all such cases). Assuming possible access to case histories held by WADA and other sports governing bodies, might it be possible for a cut-off concentration to be chosen that reduces the probability of innocent athletes failing the test without greatly increasing the probability of cheats evading detection? Is such a compromise worthy of consideration? Or should all findings of clenbuterol be declared, no matter what concentration, and then reliance be placed on inquiries to ascertain what is the likely source, doping or contamination?

12.3 Erythropoietin and Micro-dosing

Erythropoietin (EPO) is a glycoprotein hormone, with a molecular weight of 30 400 Da; it is the main regulator of red blood cell production (RBC) production. Haemoglobin is the main component of RBCs and its function is to transport oxygen from the high concentrations in the lungs to parts of the body where the oxygen is being consumed, *e.g.* muscle and brain. The EPO protein backbone consists of 165 amino acids. The post-translational glycosylation of EPO comprises a variable number of sialic acids at three N-linked and one O-linked carbohydrate chains, resulting in up to 14 sialic acids in total.[20] This glycosylation is considered to be unnecessary for biological activity but is important for its stability in the circulation; once the sialic acids are removed, the molecule is rapidly metabolised.[21]

Hypoxia stimulates EPO production, which in turn stimulates erythropoiesis, thereby reducing the hypoxia. The role of the hypoxia inducible factors 1α (HIF-1α) and 2α (HIF-2α) in affecting erythropoietin production is described elsewhere in this article. Some athletes have the benefit of having homes that are situated at high altitude. Simulated hypoxia in the form of high altitude training, the use of hypobaric chambers, or breathing nitrogen with reduced oxygen concentration, all of which are permitted under sports rules, have also been used by athletes. The use of EPO is prohibited in sport both because of its performance enhancing properties and concern about the harmful side effects of the polycythaemia produced by misuse and concomitant increase in blood viscosity, which could prove fatal. Elliott has published a review of EPO and other erythrocyte stimulating agents, including EPO, and other methods that enhance oxygen transport.[22]

In the mid-1980s, advances in recombinant biotechnology enabled the large-scale production of recombinant human EPO (rHuEPO). As glycosylation is required, the mammalian Chinese hamster ovary cell was used to produce this needed post-translational modification that bacterial cell lines cannot form. Unfortunately, this ready availability of rHuEPO meant that misuse by athletes soon followed. Athletes may administer iron in parallel with EPO a key component of haem, and one of the first references to EPO misuse was by Dine *et al.* when they described hyperferritinaemia in racing cyclists.[23] The detection of this misuse was the result of a breakthrough by French scientists.[24] Their method of analysis used isoelectric focusing (IEF) over a relatively narrow pH gradient to separate the different isoforms of EPO excreted in urine samples. rHuEPO is sulfated in about 3% of the carbohydrate chains whereas endogenous EPO has considerably more and is thus more acidic, enabling administration of rHuEPO to be detected by distinguishing endogenous from recombinant. The method was first employed at the Olympic Games in Sydney in 2000, but without any reported finding of EPO doping.

The biological half-life of EPO is very short (4–12 h) and the original preparation of rHuEPO (epoietin alpha) was administered subcutaneously for therapeutic use in daily dosing intervals. This short half-life contrasts markedly with the relatively long lifetime of the RBC (more than 100 days), which enables cheating athletes to go, for example, to a remote training location where drug testing was unlikely, administer EPO and thereby avoid detection.

A further concern is that, by tapering the dose of EPO, one could maintain an elevated concentration of haemoglobin in the circulation while reducing the detection time of the EPO use. In one study,[25] large doses (260 IU kg^{-1} daily for 6 days) of rHuEPO were given to two volunteers together with a single i.v. dose of an iron treatment (100 mg) to raise their haemoglobin to 170 g L^{-1}. Over the next 3 weeks, so called micro-doses of less than 10% of the original amount of rHuEPO were given every 2–3 days, which maintained the volunteers' peak haemoglobin concentrations. At the same time, they showed that the EPO isoform (IEF) method in use at that time did not detect these doses of EPO administration after about 18 hours after cessation.

More recently, Dehnes and colleagues[26] undertook a rHuEPO intravenous micro-dosing study with doses of 7.5 IU kg^{-1} NeoRecormon (epoetin beta) to nine healthy volunteers. The doses were given twice a week for 3 weeks. In this study the higher loading dose of rHuEPO to raise haemoglobin to a supraphysiological level, as used in the study by Ashenden *et al.*,[25] was not used. The haemoglobin concentration in these volunteers was not raised significantly. Nevertheless, the authors showed that their new EPO WGA MAIIA

(EPO Wheat Germ Agglutinin Membrane Assisted Isoform Immuno-Assay)[27] screening technique, which uses ultrasensitive lateral flow on a simple dipstick, was effective in detecting rHuEPO in either plasma or urine samples collected 18 hours after the previous dose. Furthermore, the approach recently adopted by many WADA accredited laboratories, using sarcosyl-polyacrylamide gel electrophoresis (SAR-PAGE), was able to confirm all the rHuEPO suspicious screening findings.[28,29]

Around the same time that Françoise Lasne and co-investigators were developing a method to distinguish the difference in the isoforms of endogenous EPO and those of rHuEPO, an Australian group was developing an indirect approach based on the effect of EPO administration. Both the IEF methodology and the indirect approach were employed at the Sydney Olympic Games. The indirect approach relies on a so-called "on-score", effective when EPO has been administered, and an "off-score" method that is effective soon after administration has ceased.[30] The Australian group modelled the measurements of reticulocyte haematocrit, serum EPO, soluble transferrin receptor, haematocrit and percent macrocytes. The ON-model detected use during the final two weeks of a 25 day administration study of 50 U per kilogram rHuEPO injected three times a week together with iron administration. The OFF-model used reticulocyte haematocrit, EPO and haematocrit and identified approximately 67% of users for 12–21 days after the last rHuEPO administration. In a follow-up study they refined their model and showed that the method could be applied successfully to both Caucasian and Asiatic subjects.[31] Russell and colleagues[32] investigated the ability of low-dose (20 $IU\,kg^{-1}$) rHuEPO administration three times a week for 5 weeks on maintaining an elevated haematocrit and maximal oxygen uptake (VO_{2max}). Gore and co-workers developed a so-called second-generation blood test,[33] which they claimed had better sensitivity for detecting rHuEPO administration than the former methods. The simplest of the models, the so-called OFF-hr score, is based on the haemoglobin concentration minus sixty times the square root of the reticulocyte concentration (in percentage), *i.e.* OFF-hr = $Hb - 60\sqrt{(Ret\,\%)}$. This is the formula used in the Athlete Biological Passport (ABP). The robustness of the blood variables used to determine the OFF-hr is considered in a paper by Robinson and co-workers[34] and confirms the reliability of the ABP system that has been adopted by WADA, with values being entered into the WADA Anti-Doping Administration & Management System (ADAMS). This system provides a confidential database of ABP entries for individual athletes that are then evaluated using an adaptive model, which uses a Bayesian statistical approach to report samples that go outside a confidence interval dynamically established for that sample against previous samples. Unfortunately, Ashenden and co-workers showed that the ADAMS ABP model would not detect the administration of a gradually increasing dosage regime of rHuEPO from 10 $IU\,kg^{-1}$ for 4 weeks up to 40 $IU\,kg^{-1}$ for the final three injections.[35]

Diamanti-Kandarakis and co-workers published a review of EPO misuse including EPO gene doping with detection strategies for the "genomic era".[36] A further review was published by Pitsiladis and colleagues[37] regarding the possibilities of using genomics-based approaches to evidence both EPO misuse and blood doping. The effectiveness of such approaches is still to be proven.

Table 12.3 lists the occurrences of EPO reported by WADA accredited laboratories showing a very small number of findings even when considered as a percentage of specific testing for EPO. Clearly the effectiveness of low-dose rHuEPO administration, its short half-life and the relatively long residence time of the RBCs produced by the rHuEPO administration justifies further research to provide more sensitive analytical methodologies. At the same time, the use of

Table 12.3 List of reported EPO cases from WADA accredited laboratories.[a]

Year	Occurrences	Samples analysed	% of all samples	Samples tested for EPOs	% Occurrences of EPO alone
2014	57 + 7 + 3	283 304	0.020	30 563	0.19
2013	56 + 5 + 2	269 878	0.021	25 623	0.22
2012	45 + 4 + 0	267 645	0.017	25 450	0.13
2011	43 + 5 + 0	243 193	0.018	n.a.[b]	n.a.[b]
2010	36 + 8 + 1	258 267	0.014	n.a.[b]	n.a.[b]
2009	56 + 4 + 8	277 928	0.020	n.a.[b]	n.a.[b]
2008	51 + 2 + 5	274 615	0.019	n.a.[b]	n.a.[b]
2007	22 + 2 + 0	223 898	0.010	n.a.[b]	n.a.[b]
2006	17 + 1 + 0	198 143	0.009	n.a.[b]	n.a.[b]
2005	15 + 1 + 0	183 337	0.008	n.a.[b]	n.a.[b]
2004	38 + 0 + 0	169 187	0.022	n.a.[b]	n.a.[b]
2003	51 + 7 + 0	151 210	0.034	n.a.[b]	n.a.[b]

[a]The *Occurrences* column gives three numbers: the first is for EPO, the second is for the synthetic more heavily glycosylated EPO darbepoetin and the third is for methoxy polyethylene glycol-epoetin beta (CERA, detectable only since 2008).
[b]n.a. = not available.

intelligence information to ensure that samples are collected at appropriate times will help better control EPO misuse in sport.

12.4 Xenon and Sport

Xenon is widely used in lasers, flash lamps, X-ray tubes and especially in medicine as an anaesthetic gas.[38,39] The application of xenon as an anaesthetic gas is supported by the results of studies examining the narcotic effects of noble gases and xenon partitioning in the blood and myocardium.[40–43] The anaesthetic effects of noble gases are correlated with their liposolubility and blood–gas partition co-efficient.[44,45] Xenon's liposolubility and a low blood–gas partition coefficient to quickly penetrate the blood–brain barrier make it a suitable candidate for gas anaesthesia.

Xenon, denser and more viscous than air, is considered safe, non-toxic, non-teratogenic, environmentally friendly, sedative in minor doses and features beneficial analgesic properties.[44,46,47] It was first introduced in Russia as an anaesthetic agent in 2000.[45] The anaesthetic effect of xenon relies on its antagonist effects on the excitatory N-methyl-D-aspartate (NMDA) receptors, a member of the glutamate receptor family.[48,49] However, the possibility of xenon binding to other types of glutamate receptors (kainic acid and/or α-amino-3-hydroxy-5-methyl-4-isoxazolepropionic acid receptors) are being considered.[50] Applying quantum mechanics and molecular dynamics models, Andrijchenko *et al.* described three credible xenon trapping sites at the NMDA receptor ligand binding domain.[51] When binding to the NMDA receptor, xenon competes with glycine, the receptor's natural agonist, and interacts with amino acids with aromatic rings (*e.g.* phenylalanine and to form π-type complexes, which are characteristic for each of the three trapping sites. These findings are further supported by the authors' studies using xenon–phenol complexes as demonstrative models. Nevertheless, the wider application of xenon in anaesthesia is prevented by its high cost.

A number of animal studies have shown that xenon has an organ protection effect, which may be independent of its application in anaesthesia. Ma *et al.* suggested that administration of xenon prior to renal ischemia can prevent renal failure in adult mice.[52] Valleggi *et al.* studied the neuroprotective effect of xenon on the rat brain after exposing animals to a 25% air/75% xenon mixture for two hours.[53] Xenon treatment appeared to be beneficial in preventing ischemia-reperfusion injury and chronic allograft nephropathy in rats undergoing allogenic kidney transplantation.[54] Tissue preconditioning with xenon and xenon's subsequent organ protective effect is thought to be due to the gas's ability to induce HIF-1α gene transcription and HIF-1α protein accumulation in the target tissues.[55] It is also believed that xenon tissue preconditioning induces gene expression of vascular endothelial growth factor (VEGF) and insulin-like growth factor I (IGF-I), which are involved in *de novo* tissue formation.[55,56] Male Wistar rats were exposed either to xenon/oxygen (70%/30%) or to nitrogen/oxygen (70%/30%) before or during gentamicin administration to model gentamicin-induced nephrotoxicity. Renal tubular damage, apoptosis and oxidative stress, but not tissue inflammation, were reduced in animals treated with xenon, perhaps by inducing HIF-2α and VEGF rather than HIF-1α.[57] Apart from an organ protecting effect, it was found that xenon appears to stimulate the release of noradrenaline in the brain, increases being found in the prefrontal cortex and in cerebrocortical slices in male Wister rats.[58]

Xenon use in sport became topical during the Olympic Winter Games in Sochi in February 2014, when the magazine *The Economist* published an article about xenon's ability to enhance athletic performance, although good evidence is lacking. The anonymous author(s) comment(s) about the protective role of xenon (*e.g.* against low temperatures, physical trauma, lack of oxygen) and its role in boosting erythropoietin levels, but also emphasise that xenon is used in Russia to improve athletic performance.[59] Nonetheless, it was indicated that Russian athletes were using xenon earlier than 2014. Koh and de Neef, supporting *The Economist* article, presented the acknowledgement letter from the president of the Russian Olympic Committee to the director of the medical centre where athletes were exposed to xenon as part of their government sponsored preparation programme for the 2006 Turin Winter Olympic Games.[60]

WADA, being aware of the publication in *The Economist*, took the unusual step of issuing a revision to the annual Prohibited List in September 2014. The revised list introduced xenon and argon as examples of the new category "HIF activators" together with "HIF stabilizers", the latter which were already prohibited. This article focuses on xenon as even less is known about argon. As described above, xenon induces HIF-1α, which has organ protective properties.[55] It is suspected that HIF-1α may induce erythropoietin gene transcription.[61] However, the literature suggests erythropoietin gene transcription appears to require HIF-2α up-regulation.[62,63] Erythropoietin controls erythropoiesis and hence oxygen transport by the RBCs in humans, thereby improving athletic endurance. The effect of xenon on erythropoietin levels in humans is not known[61] and research in this area is warranted.

(a)

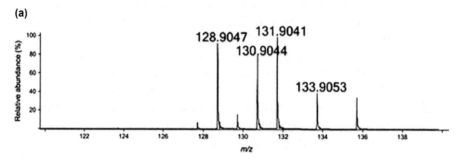

(b)

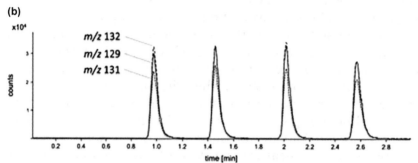

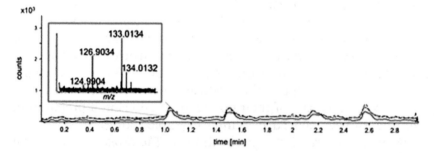

(c)

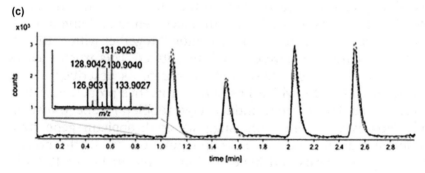

Two investigations have been reported describing the detection of xenon use by athletes in blood (and plasma) and urine.[64,65] Plasma samples were obtained from healthy volunteers and spiked with xenon at a range of concentrations. The presence of xenon in the samples was identified by accurate mass detection of three major xenon isotopes ($m/z = 128.9048$, $m/z = 130.9045$ and $m/z = 131.9042$) (Figure 12.3), to provide good legal defensibility, by the use of gas chromatography–time of flight mass spectrometry with headspace injection. However, low-resolution gas chromatography–mass spectrometers with headspace injection may also be utilized for xenon detection, since the isotope pattern of xenon is very characteristic and quadrupole mass analysers are commonly used in anti-doping laboratories. The method developed was applied for the analysis of xenon from pre-, during and post-operative blood samples collected from a patient anesthetised with xenon. According to the authors, detection of xenon in blood plasma up to 30 hours post-operation was a substantial proof of concept for the developed analytical assay.[64] Of note, the preliminary work regarding xenon detection in blood in our laboratory (Drug Control Centre, King's College London) confirms it is possible to use GC-MS with an ordinary heated auto-sampler (37 °C) by adjustment of the needle position to enable headspace injection.

Following their research on detecting xenon in blood and plasma, Thevis *et al.* investigated the detection of xenon in urine for doping control purposes. The limit of detection in xenon-enriched urine samples was 0.5 nmol mL^{-1} by GC–triple quadrupole MS with headspace injection, and xenon was detected up to 40 h in urine samples of xenon-anesthetized patients.[65]

At the time of writing this chapter, WADA is establishing the capability of its accredited laboratories to be able to implement testing for xenon in sports samples.

Figure 12.3   (a) Electron ionization mass spectrum of xenon. (b) Extracted ion chromatograms (m/z 131.904, 130.904 and 128.904 ± 50 ppm) of xenon measured from a fortified plasma sample containing 50 nmol mL^{-1} (*top*) and a blank plasma sample (*bottom*). (c) Extracted ion chromatograms of a blood sample collected 24 h after xenon-assisted anaesthesia. While the ion (isotope) ratios match in the case of fortified plasma and post-administration blood samples, the signals found in the blank specimen are attributed to a general raise of background noise with each injection (see EI mass spectrum inset). (Reproduced from Thevis *et al.*,[64] with permission from Wiley.)

12.5   Conclusion

This chapter has attempted to provide the reader with three topical issues currently confronting WADA and the anti-doping framework. The issue of environmental contamination is likely to increase in importance as analytical methods become even more sensitive, as has already been illustrated in the case of clenbuterol. On the other hand, improved analytical approaches are needed to provide greater sensitivity to detect EPO misuse. For agents such as xenon, more research to evidence efficacy or otherwise would support inclusion in the WADA Prohibited List. As in analytical toxicology generally, especially since the advent of modern LC-MS instruments, increases in the sensitivity of the analytical methods must be accompanied with careful consideration of the context and relevance of the data.

References

1. P. Van Eenoo and F. T. Delbeke, Beta-adrenergic stimulation, in *Handbook of experimental pharmacology*, ed. D. Thieme and P. Hemmersbach, Springer-Verlag, Berlin Heidelberg, 2010, pp. 227–249.
2. A. H. Beckett, in *International Athletic Foundation World Symposium On Doping in Sport (Florence, 10–12 May, 1987)*, ed. P. Bellotti, G. Benzi and A. L. Ljungqvist, Art Grafiche Danesi, Italy, Florence, 1987, pp. 1–11.
3. Clenbuterol Hydrochloride, in *Martindale: The Complete Drug Reference (Online)*. (last accessed September 2015).
4. Current authorised veterinary medicines containing controlled drugs. U.K. Government, vol. 2015, https://www.gov.uk/government/uploads/system/uploads/attachment_data/file/427845/controlled_drugs.pdf (last accessed September 2015).
5. E. Davis, R. Loiacono and R. J. Summers, The rush to adrenaline: drugs in sport acting on the beta-adrenergic system, *Br. J. Pharmacol.*, 2008, **154**, 584–597.
6. Muscling in on clenbuterol, *Lancet*, 1992, **340**, 403.
7. A. H. Beckett, Clenbuterol and sport, *Lancet*, 1992, **340**, 1165.
8. S. Horning and M. Donike, in *Recent Advances in Doping Analysis; Proceedings of the 11th Cologne Workshop on Dope Analysis 7th to 12th March 1993*, ed. M. Donike, H. Geyer, A. Gotzmann, U. Mareck-Engelke and S. Rauth, Sport und Buch Strauss, Köln, 1994, pp. 155–161.

9. *Mass Spectrometry in Sports Drug Testing: Characterization of Prohibited Substances and Doping Control Analytical Assays*, ed. M. Thevis, John Wiley and Sons Inc., Hoboken, New Jersey, 2010.

10. P. Hemmersbach, S. Tomten, S. Nilsson, H. Oftebro, O. Havrevoll, B. Oen and K. Birkeland, Illegal use of anabolic agents in animal fattening – consequences for doping analysis, in *Recent Advances in Doping Analysis*, ed. M. Donike, H. Geyer, A. Gotzmann and U. Mareck-Engelke, Sport und Buch Strauss, Cologne, 1995, vol. 2, pp. 185–191.

11. Clenbuterol – contaminated meat hits China, Mexico and sports. http://www.sgs.com/ ∼ /media/Global/Documents/ Technical%20Documents/Technical%20Bulletins/Scoop/sgs-cts-cc-food-clenbuterol-feb-2012-en-p4.pdf. (last accessed September 2015).

12. J. Barbosa, C. Cruz, J. Martins, J. M. Silva, C. Neves, C. Alves, F. Ramos and M. I. Da Silveira, Food poisoning by clenbuterol in Portugal, *Food Addit. Contam.*, 2005, **22**, 563–566.

13. S. Guddat, G. Fussholler, H. Geyer, A. Thomas, H. Braun, N. Haenelt, A. Schwenke, C. Klose, M. Thevis and W. Schanzer, Clenbuterol – regional food contamination a possible source for inadvertent doping in sports, *Drug Test. Anal.*, 2012, **4**, 534–538.

14. CAS 20111A12384 UCI v. Alberto Contador Velasco & RFEC; CAS 2011/A12386 WADA v. Alberto Contador Velasco & RFEC. ARBITRAL AWARD delivered by COURT OF ARBITRATION FOR SPORT. https://wada-main-prod.s3.amazonaws.com/resources/files/cas-2011-a-2384-contador.pdf (last accessed September 2015).

15. M. Thevis, L. Geyer, H. Geyer, S. Guddat, J. Dvorak, A. Butch, S. S. Sterk and W. Schanzer, Adverse analytical findings with clenbuterol among U-17 soccer players attributed to food contamination issues, *Drug Test. Anal.*, 2013, **5**, 372–376.

16. R. Nicoli, M. Petrou, F. Badoud, J. Dvorak, M. Saugy and N. Baume, Quantification of clenbuterol at trace level in human urine by ultra-high pressure liquid chromatography-tandem mass spectrometry, *J. Chromatogr. A*, 2013, **1292**, 142–150.

17. I. Yamamoto, K. Iwata and M. Nakashima, Pharmacokinetics of plasma and urine clenbuterol in man, rat, and rabbit, *J. Pharmacobio-Dyn.*, 1985, **8**, 385–391.

18. M. Thevis, A. Thomas, S. Beuck, A. Butch, J. Dvorak and W. Schanzer, Does the analysis of the enantiomeric composition of clenbuterol in human urine enable the differentiation of illicit clenbuterol administration from food contamination in sports drug testing?, *Rapid Commun. Mass Spectrom.*, 2013, **27**, 507–512.

19. A. Krumbholz, P. Anielski, L. Gfrerer, M. Graw, H. Geyer, W. Schanzer, J. Dvorak and D. Thieme, Statistical significance of hair analysis of clenbuterol to discriminate therapeutic use from contamination, *Drug Test. Anal.*, 2014, **6**, 1108–1116.

20. J. C. Egrie and J. K. Browne, Development and characterization of darbepoetin alfa, *Br. J. Cancer*, 2001, **84**, 3–10.

21. I. C. Macdougall, D. E. Roberts, G. A. Coles and J. D. Williams, Clinical pharmacokinetics of epoetin (recombinant-human-erythropoietin), *Clin. Pharmacokinet.*, 1991, **20**, 99–113.

22. S. Elliott, Erythropoiesis-stimulating agents and other methods to enhance oxygen transport, *Br. J. Pharmacol.*, 2008, **154**, 529–541.

23. G. Dine, A. Megert, F. Lasne, M. Guinot, A. Garnier, Y. Deugnier, G. Peres and A. Najman, Hyperferritinaemia in top level cyclists: Iatrogenic secondary iron overload?, *Blood*, 1999, **94**, 3203.

24. F. Lasne and J. d. Ceaurriz, Recombinant erythropoietin in urine, *Nature*, 2000, **405**, 635.

25. M. Ashenden, E. Varlet-Marie, F. Lasne and M. Audran, The effects of microdose recombinant human erythropoietin regimens in athletes, *Haematologica*, 2006, **91**, 1143–1144.

26. Y. Dehnes, A. Shalina and L. Myrvold, Detection of recombinant EPO in blood and urine samples with EPO WGA MAIIA, IEF and SAR-PAGE after microdose injections, *Drug Test. Anal.*, 2013, **5**, 861–869.

27. M. Lonnberg, M. Andren, G. Birgegard, M. Drevin, M. Garle and J. Carlsson, Rapid detection of erythropoiesis-stimulating agents in urine and serum, *Anal. Biochem.*, 2012, **420**, 101–114.

28. C. Reichel, F. Abzieher and T. Geisendorfer, SARCOSYL-PAGE: a new method for the detection of MIRCERA- and EPO-doping in blood, *Drug Test. Anal.*, 2009, **1**, 494–504.

29. C. Reichel, SARCOSYL-PAGE: a new electrophoretic method for the separation and immunological detection of PEGylated proteins, *Methods Mol. Biol.*, 2012, **869**, 65–79.

30. R. Parisotto, C. J. Gore, K. R. Emslie, M. J. Ashenden, C. Brugnara, C. Howe, D. T. Martin, G. J. Trout and A. G. Hahn, A novel method utilizing markers of altered erythropoiesis for the detection of recombinant human erythropoietin abuse in athletes, *Haematologica*, 2000, **85**, 564–572.

31. R. Parisotto, M. T. Wu, M. J. Ashenden, K. R. Emslie, C. J. Gore, C. Howe, R. Kazlauskas, K. Sharpe, G. J. Trout, M. H. Xie and A. G. Hahn, Detection of recombinant human erythropoietin abuse in athletes utilizing markers of altered erythropoiesis, *Haematologica*, 2001, **86**, 128–137.

32. G. Russell, C. J. Gore, M. J. Ashenden, R. Parisotto and A. G. Hahn, Effects of prolonged low doses of recombinant human erythropoietin during submaximal and maximal exercise, *Eur. J. Appl. Physiol.*, 2002, **86**, 442–449.

33. C. J. Gore, R. Parisotto, M. J. Ashenden, J. Stray-Gundersen, K. Sharpe, W. Hopkins, K. R. Emslie, C. Howe, G. J. Trout, R. Kazlauskas and A. G. Hahn, Second-generation blood tests to detect erythropoietin abuse by athletes, *Haematologica*, 2003, **88**, 333–344.

34. N. Robinson, P. E. Sottas, T. Pottgiesser, Y. O. Schumacher and M. Saugy, Stability and robustness of blood variables in an anti-doping context, *Int. J. Lab. Hematol.*, 2011, **33**, 146–153.

35. M. Ashenden, C. E. Gough, A. Garnham, C. J. Gore and K. Sharpe, Current markers of the athlete blood passport do not flag microdose EPO doping, *Eur. J. Appl. Physiol.*, 2011, **111**, 2307–2314.

36. E. Diamanti-Kandarakis, P. A. Konstantinopoulos, J. Papailiou, S. A. Kandarakis, A. Andreopoulos and G. P. Sykiotis, Erythropoietin abuse and erythropoietin gene doping: detection strategies in the genomic era, *Sports Med.*, 2005, **35**, 831–840.

37. Y. P. Pitsiladis, J. Durussel and O. Rabin, An integrative 'Omics' solution to the detection of recombinant human erythropoietin and blood doping, *Br. J. Sports Med.*, 2014, **48**, 856–861.

38. B. D. Jordan and E. L. Wright, Xenon as an anesthetic agent, *AANA J.*, 2010, **78**, 387–392.

39. P. H. Tonner, Xenon: one small step for anaesthesia...?, *Curr. Opin. Anaesthesiol.*, 2006, **19**, 382–384.

40. A. R. Behnke and O. D. Yarbrough, Respiratory resistance, oil-water solubility, and mental effects of argon, compared with helium and nitrogen, *Am. J. Physiol.*, 1939, **126**, 409–415.

41. J. H. Lawrence, W. F. Loomis, C. A. Tobias and F. H. Turpin, Preliminary observations on the narcotic effect of xenon with a review of values for solubilities of gases in water and oils, *J. Physiol.*, 1946, **105**, 197–204.

42. R. Carlin and S. Chien, Partition of xenon and iodoantipyrine among erythrocytes, plasma, and myocardium, *Circ. Res.*, 1977, **40**, 497–504.

43. T. Goto, K. Suwa, S. Uezono, F. Ichinose, M. Uchiyama and S. Morita, The blood-gas partition coefficient of xenon may be lower than generally accepted, *Br. J. Anaesth.*, 1998, **80**, 255–256.

44. J. Ruzicka, J. Benes, L. Bolek and V. Markvartova, Biological effects of noble gases, *Physiol. Res.*, 2007, **56**, S39–S44.

45. E. Esencan, S. Yuksel, Y. B. Tosun, A. Robinot, I. Solaroglu and J. Zhang, Xenon in medical area: emphasis on neuroprotection in hypoxia and anesthesia, *Med. Gas Res.*, 2013, **3**, 4.

46. M. Giacalone, A. Abramo, F. Giunta and F. Forfori, Xenon-related analgesia: a new target for pain treatment, *Clin. J. Pain*, 2013, **29**, 639–643.

47. L. L. Kirkland, Protecting both heart and brain: a noble goal for a noble gas, *Crit. Care Med.*, 2013, **41**, 2228–2229.

48. F. S. Servin, Update on pharmacology of hypnotic drugs, *Curr. Opin. Anesthesiol.*, 2008, **21**, 473–477.

49. B. Preckel, N. C. Weber, R. D. Sanders, M. Maze and W. Schlack, Molecular mechanisms transducing the anesthetic, analgesic, and organ protective actions of xenon, *Anesthesiology*, 2006, **105**, 187–197.

50. M. Derwall, M. Coburn, S. Rex, M. Hein, R. Rossaint and M. Fries, Xenon: recent developments and future perspectives, *Minerva Anestesiol.*, 2009, **75**, 37–45.

51. N. N. Andrijchenko, A. Y. Ermilov, L. Khriachtchev, M. Räsänen and A. V. Nemukhin, Toward molecular mechanism of xenon anesthesia: a link to studies of xenon complexes with small aromatic molecules, *J. Phys. Chem. A*, 2015, **119**, 2517–2521.

52. D. Ma, T. Lim, J. Xu, H. Tang, Y. Wan, H. Zhao, M. Hossain, P. H. Maxwell and M. Maze, Xenon preconditioning protects against renal ischemic-reperfusion injury via HIF-1α activation, *J. Am. Soc. Nephrol.*, 2009, **20**, 713–720.

53. S. Valleggi, C. B. Patel, A. O. Cavazzana, D. Ma, F. Giunta and D. Cattano, Xenon upregulates hypoxia inducible factor 1 alpha in neonatal rat brain under normoxic conditions, *ISRN Anesthesiol.*, 2011, **2011**, 1–7.

54. H. Zhao, X. Luo, Z. Zhou, J. Liu, C. Tralau-Stewart, A. J. T. George and D. Ma, Early treatment with xenon protects against the cold ischemia associated with chronic allograft nephropathy in rats, *Kidney Int.*, 2014, **85**, 112–123.

55. S. Hieber, R. Huhn, M. W. Hollmann, N. C. Weber and B. Preckel, Hypoxia-inducible factor 1 and related gene products in anaesthetic-induced preconditioning, *Eur. J. Anaesthesiol.*, 2009, **26**, 201–206.

56. A. Goetzenich, N. Hatam, S. Preuss, A. Moza, C. Bleilevens, A. B. Roehl, R. Autschbach, J. Bernhagen and C. Stoppe, The role of hypoxia-inducible factor-1α and vascular endothelial growth factor in late-phase preconditioning with xenon, isoflurane and

levosimendan in rat cardiomyocytes, *Interact. Cardiovasc. Thorac. Surg.*, 2014, **18**, 321–328.

57. P. Jia, J. Teng, J. Zou, Y. Fang, S. Jiang, X. Yu, A. J. Kriegel, M. Liang and X. Ding, Intermittent exposure to xenon protects against gentamicin-induced nephrotoxicity, *PLoS One*, 2013, **8**, e64329.

58. H. Yoshida, T. Kushikata, R. Tose, M. Kudo, T. Kudo and K. Hirota, Nitrous oxide and xenon increase noradrenaline release in the cerebral cortex *in vivo* and *in vitro*, *Neurosci. Lett.*, 2010, **469**, 199–203.

59. Breathe it in, *The Economist*, February 8th, 2014.

60. B. Koh and M. de Neef, *Xenon gas as a performance enhancing drug: doping or just hot air?*, 2014 http://cyclingtips.com.au/2014/03/xenon-gas-as-a-performance-enhancing-drug-doping-or-just-hot-air/ (last accessed September 2015).

61. W. Jelkmann, Xenon misuse in sports-increase of hypoxia-inducible factors and erythropoietin, or nothing but "hot air"?, *Dtsch. Z. Sportmed.*, 2014, **65**, 267–271.

62. T. Yamashita, O. Ohneda, A. Sakiyama, F. Iwata, K. Ohneda and Y. Fujii-Kuriyama, The microenvironment for erythropoiesis is regulated by HIF-2α through VCAM-1 in endothelial cells, *Blood*, 2008, **112**, 1482–1492.

63. H. F. Bunn, Erythropoietin, *Cold Spring Harbor Perspect. Med.*, 2013, **3**, 1–21.

64. M. Thevis, T. Piper, H. Geyer, A. Thomas, M. S. Schaefer, P. Kienbaum and W. Schänzer, Measuring xenon in human plasma and blood by gas chromatography/mass spectrometry, *Rapid Commun. Mass Spectrom.*, 2014, **28**, 1501–1506.

65. M. Thevis, T. Piper, H. Geyer, M. S. Schaefer, J. Schneemann, P. Kienbaum and W. Schänzer, Urine analysis concerning xenon for doping control purposes, *Rapid Commun. Mass Spectrom.*, 2015, **29**, 61–66.

13 Drugs and Driving

Kim Wolff

King's College London, London, UK
Email: kim.wolff@kcl.ac.uk

13.1 Introduction

There is a considerable amount of information in the scientific literature regarding drugs and driving. It is well known that the use of alcohol or drugs can impact on a number of driving skills, such as attention, tracking, reaction time, information processing, perception, psychomotor skills, visual function and increased risk-taking.[1,2]

13.2 Determining Thresholds for Drug Driving

13.2.1 International Approaches

Several approaches have been taken to manage those who drive under the influence of psychoactive substances. Setting a concentration threshold for a psychoactive drug in relation to road traffic legislation has been implemented across Europe. Some countries have instigated a programme of zero tolerance, which equates to a complete ban on the use of a specified drug whilst driving. The impairment approach has also proved popular and the United

Forensic Toxicology: Drug Use and Misuse
Edited by Susannah Davies, Atholl Johnston and David Holt
© The Royal Society of Chemistry 2016
Published by the Royal Society of Chemistry, www.rsc.org

Kingdom, like other EU countries, has legislation to prosecute someone who is driving while unfit (impaired) through drink or drugs (Section 4, Road Traffic Act 1988). A third approach is often referred to as the *"per se"* approach and is based on the detection of a drug in a driver above a defined cut-off concentration, predominantly in whole blood.

There are several different approaches to the implementation of the *"per se"* threshold. A threshold can be analytical and can refer to a laboratory's limit-of-detection (LOD), commonly employed in the laboratory. This is the lowest concentration of the drug that the analytical procedure can reliably differentiate from a concentration of zero and can be positively identified according to predetermined criteria or levels of statistical confidence. A threshold can be also technical and can refer to the laboratory limit-of-quantification (LOQ). This is defined as the lowest measurable quantity of a drug that can be detected according to the technological limits of the equipment with an acceptable level of accuracy and precision and that guarantees a valid and reliable analytical determination of the drug of interest. Thirdly, a threshold can specifically relate to the effects of a drug and can be set to where an effect on driving ability has been shown to occur. A "lower effect threshold" is set at the lowest concentration where an effect on driving has been observed. Detection of psychoactive substances in blood below this concentration does not imply recent drug use or being under the influence. A "lower effect threshold" limit is usually equivalent to a blood alcohol concentration (BAC) of 0.2 g L^{-1} alcohol. Finally, a threshold can also relate to risk and refer to a drug concentration threshold set in whole blood indicating a certain crash risk associated with driving under the influence of a drug above that threshold. "Risk thresholds" for instance, have been determined showing the same level of accident risk as a BAC of 0.5 g L^{-1} alcohol blood.[3]

13.2.2 Biological Fluids

Although the standard procedure for large-scale laboratory-based screening for illicit drug use typically involves the collection of urine samples, this can only provide retrospective information about past drug use rather than provide information about the "here and now" – the current effect of the drug on the person. It is widely acknowledged that blood and, to a lesser degree, oral fluid are likely to give the most accurate measurement of drugs currently active in the body; urine provides a somewhat broader time frame (drug use over the last 2–3 days), but with less quantitative accuracy;[†] hair provides a substantially longer time frame.[4]

Although urine has the advantage of being fairly easy to collect in large volumes and is the biological fluid of choice for laboratory-based drug-testing programmes, the interpretation of urine tests is often complex, with great variability in regard to the excretion of drugs from the body; some knowledge of the pharmacokinetics of the drug is usually necessary to interpret findings.[5] There is a time-lag between the consumption of a drug and its appearance in urine, which makes the relationship between urinary drug concentrations and driving behaviour difficult to describe, particularly as the time-lag may be affected by a myriad of factors such as gender, age, weight, disease state, *etc.* Drug concentrations in urine are not relevant as a means to calculate the relationship between drug effect and driving behaviour. It is generally accepted that urinary drug concentrations are not useful as an indicator of the effects of a drug on immediate driving safety.

Over the last 10 years there has been growing interest in the use of oral fluid for drug testing as an alternative to urine.[6] The major advantage of oral fluid over urine is the easy, rapid and non-intrusive sampling procedure.[4] Oral fluid has been shown to be a suitable matrix for community-based drug screening purposes and Toennes *et al.*[7] compared findings in oral fluid, serum and urine, and concluded that oral fluid was superior to urine in correlating with blood drug concentrations and driving behaviour, whilst field test results obtained from oral fluid samples compared with collected oral fluid and urine samples found that oral fluid proved to be a more effective confirmatory specimen, with more drugs being confirmed in oral fluid than urine.[8] Attempts have been made to establish fixed ratios or conversion factors between the drug concentrations in blood and those in oral fluid for confirmation testing. However, there are large individual variations, which mean that ratios cannot be easily determined for most psychoactive drugs, although some correlation has been described.[9,10]

[†]Unless voidance of urine is observed the authenticity of the sample may be called into question since urine can easily be contaminated, with the probability of false negative results following adulteration of urine with chemicals or by dilution. Artificial dilution can be a problem both before (by using diuretic agents widely available on the internet) or after voiding (by adding water); many laboratories have criteria for "normally concentrated" or "dilute" urine specimens.

Currently, oral fluid tests cannot be used to give a precise prediction of the concentration of a drug in blood (or plasma or serum) for confirmation testing and therefore prediction of possible drug effect.[9,11] From the point of view of setting thresholds in a biological fluid, reference values (concentration/effect ratios) are more readily available for blood (plasma or serum), which remains the matrix of choice. For establishing thresholds in the context of drug-driving legislation, blood is the preferred bodily fluid since it is generally well described in the scientific literature and is best related to behavioural effects on driving. Blood is the "gold standard";[9] however it is well known that drug concentrations in blood, plasma and serum cannot be used synonymously with each other, since the concentration of a drug in plasma and serum may be higher than in whole blood.

13.3 Consideration of Sampling Time

There is often an unavoidable delay between the witnessed impairment or road traffic incident and the time of blood sampling, such that concerns have been raised about the difficulty of relating blood concentration to driving under the influence of drugs. It is widely accepted that specimens should be obtained as soon as possible after the road traffic event, given the relatively rapid decline of drugs such as THC, cocaine and heroin in blood. For alcohol, many countries employ back-calculation; for drugs, because of variable pharmacokinetics, back-calculation is much more difficult and has not been widely adopted.

13.4 Alcohol

Alcohol has been shown to affect driving performance unequivocally and highly increases incident risk. Alcohol affects driving behaviour by increasing reaction time and decreasing concentration, coordination and tracking. In addition, increasing alcohol consumption leads to risk-taking behaviour, since drivers overestimate their skills and underestimate the risk due to the effects of alcohol.[12] A clear relationship has been shown between increasing alcohol consumption and risk of a road traffic collision (RTC). A meta-analysis revealed that in controlled experiments of real driving performance and complex divided attention tasks, alcohol impairment occurred at BACs as low as 0.2 g L^{-1}.[1,13]

A hospital study of seriously injured or killed drivers reported that alcohol (>0.2 g L^{-1}) was the most common toxicological observation.[14] DRUID (Driving Under Influence of Drugs, Alcohol and Medicines) researchers reported that in Europe amongst the drivers that tested positive, most had a high BAC: 90.5% of injured drivers and 87% of killed drivers had a BAC of ≥ 0.5 g L^{-1}. The mean values for alcohol concentration in these drivers were 1.59 g L^{-1} and 1.67 g L^{-1}, respectively. Alcohol was the only substance amongst those tested that appeared more often alone than in combinations.[15] In the UK (and elsewhere) drink-drivers continue to drive despite very high BACs. In 2010, 23% of car driver fatalities had a BAC concentration above the UK legal limit of 0.80 g L^{-1} alcohol, and 6% had a BAC concentration ≥ 2.0 g L^{-1} alcohol (Reported Road Casualties Great Britain 2011 Annual Report: Drinking and Driving).[‡]

13.5   Risk Estimates

Estimating the risk of a driver's involvement in RTCs has often been used to determine the impact on road safety of those driving under the influence of alcohol or drugs. In drug-driving research, calculation of the odds ratio (OR) or the relative risk (RR) ratio involves the comparison of two groups of drivers (*e.g.* drug driver *versus* non-drug driver) and gives an indication of the likelihood of a RTC happening to the one group compared to the other:[16] two levels of road collision risk (being a fatality or being seriously injured) are usually considered.[17,18] The European study DRUID has classified ORs as "low risk" (OR < 2.0), "medium risk" (OR > 2.0–10.0) and "high risk" (OR > 10.0).[19]

In the DRUID case-control study (D2.3.5) the risk of being seriously injured or killed was calculated for alcohol use against control data from the roadside survey (D2.2.3) and case data from the hospital study on killed drivers. The risk estimates (odds ratios) were adjusted for age and gender; the controls were weighted with traffic distribution in eight time periods over a week and estimated at different BACs for European drivers, as shown in Table 13.1.

13.6   Illicit Drug Use

It is widely acknowledged that cannabis is the most widely used illegal drug and this prevalence carries over into drug-driving populations, with notable exceptions.[20–22] The blood-concentration–time profile of

---

[‡]https://www.gov.uk/government/uploads/system/uploads/attachment_data/file/9276/rrcgb2011-03.pdf.

Table 13.1 Overview of odds ratios of being seriously injured or killed based on alcohol concentration from aggregated data from DRUID studies.[a]

BAC of seriously injured drivers	Adjusted odds ratios (95% CI)	BAC of fatally injured drivers	Adjusted odds ratios (95% CI)
$0.1 \leq BAC < 50$ mg alcohol per 100 mL blood (0.5 g L^{-1})	1.18 (0.81–1.73)	$0.1 \leq BAC$ < 0.5 g L^{-1}	8.01 (5.22–12.29)
$0.5 \leq BAC < 80$ mg alcohol per 100 mL blood (0.8 g L^{-1})	3.64 (2.31–5.72)	$0.5 \leq BAC$ < 0.8 g L^{-1}	45.93 (23.02–91.66)
$0.8 \leq BAC < 120$ mg alcohol per 100 mL blood (1.2 g L^{-1})	13.95 (8.15–21.88)	$0.8 \leq BAC$ < 1.2 g L^{-1}	35.69 (15.68–81.22)
$BAC \geq 120$ mg alcohol per 100 mL blood (1.2 g L^{-1})	62.79 (44.51–8.58)	$BAC \geq 1.2$ g L^{-1}	500.04 (238.07–∞)

[a]BAC = blood alcohol concentration; CI = 95% confidence intervals.

THC (tetrahydrocannabinol, primary metabolite of cannabis) shows a significant dose effect for cannabis and driving performance. This relationship has been observed in experimental and real-life situations (simulator, laboratory and forensic), in which raised concentrations of THC were associated with increased traffic crash risk. More frequent users of marijuana show less impairment than infrequent users (unless used in conjunction with alcohol) at the same dose, either because of physiological tolerance or learned compensatory driving behaviour.[23–26] Nevertheless, cannabis (or more specifically THC) has been enshrined in drug-driving legislation (Table 13.2).

The acute use of cannabis and driving behaviour has been evaluated. A meta-analysis of nine research studies summatively including 49 411 participants[27] that examined observational studies of the effects of acute cannabis use on the risk of RTCs showed that the pooled risk of a RTC whilst driving under the influence of cannabis was significant and almost twice the risk compared to driving having not consumed this drug (OR 1.92, CI 1.35–2.73; $P = 0.0003$). The summary estimate of risk for cannabis use was an OR of 2.10 for fatal accidents and 1.74 for non-fatal accidents.

Cocaine is often reported as the second most frequently used illicit drug in Europe.[28] Researchers have investigated the effects of cocaine use on driving and found that drivers often overestimate their driving skills and drive in a reckless fashion.[29] Common physical effects for

Table 13.2 International drug thresholds (set in or recommended for legislation): THC.[a]

Country	Approach to threshold	THC threshold in blood ($\mu g\ L^{-1}$)	Ref.
Belgium	Analytical cut-off	1.0	45
Norway	Impairment limit	1.3	46
	Comparable to 0.5 g L^{-1} BAC	3.0	
	Comparable to 1.2 g L^{-1} BAC	9.0	
Portugal	Analytical cut-off	3.0	
Sweden	Zero tolerance	0.3	23
Switzerland	Threshold for prosecution	1.5	
England & Wales	Strict liability offence	2.0	47

[a]It is important to note that the limits set in other countries need to be considered alongside their specific legal system and the specific drug-driving legislation. Some countries have set very low limits, which are often referred to as a zero-tolerance approach, but they may use these limits in conjunction with an impairment-type drug-driving offence, where the limits apply only if impaired driving is also recorded.

cocaine in drivers were heightened nervousness and greater alertness, but this was in combination with poor decision making and increased risk taking during driving. The negative effects of cocaine are particularly prominent when the drug is used in combination with alcohol or another drug.[2]

Cocaine is a short acting drug and the detection of cocaine alone, though unlikely, would suggest immediacy of use. The detection of cocaine and its primary metabolite, benzoylecgonine (BZE), is much more common and suggests use within the last 12 hours. Considerable effort has been targeted at the interpretation of the relationship between cocaine and BZE and driving performance. It is agreed that detection of BZE alone would indicate cocaine use in the past and may be indicative of the drug-induced exhaustion phase often experienced after the consumption of cocaine. The detection of BZE on its own has been associated with driver sedation and attributed to the "come down" period after cocaine use.[23] Researchers have observed that when both cocaine and BZE are detected together, the BZE concentration in blood was uniformly higher than the cocaine concentration (mean cocaine concentration 836 $\mu g\ L^{-1}$), with a typical BZE to cocaine ratio being 14.2:1 (range 1:1–55:1).[23] Some drug-driving legislation has included cut-off limits for BZE as well as cocaine (Table 13.3).

Risk estimates for cocaine and driving vary. Meta-analysis of four studies analysing the presence of cocaine in drivers fatally injured in RTCs reported an OR of 2.96 ($P<0.05$, CI 1.18–7.38).[30]

The estimation of risk of traffic incident (fatal or serious injury) involvement for the effects of other psychoactive drugs on driving

Table 13.3 International drug thresholds set or recommended for legislation for cocaine and benzoylecgonine.

Country	Approach to threshold	Cocaine[a] ($\mu g\ L^{-1}$)	BZE[b] ($\mu g\ L^{-1}$)	Ref.
England & Wales	Strict liability offence	10 (B)	50	48
Finland	Zero tolerance	15 (Se)	10 (Se)	
Germany	Zero tolerance	10 (Se)	75 (Se)	45
Netherlands	Threshold	50 (B)	—	49
Portugal	Zero tolerance	5 (B)	5 (B)	
Norway	Impairment limit	24 (B)	No limits	46

[a]Biological fluids: B = blood; Se = serum.
[b]BZE (benzoylecgonine) is the main metabolite of cocaine.

performance and accident risk have also been conclusive.[25,27,30,31] For instance, studies of the effects and influence of stimulant drugs, their interaction with sleep deprivation and with alcohol on driving performance and accident risk have been conducted[32,33] and form the basis for Dutch legislation. A common threshold is often set for all stimulant drugs, based on the fact that they all act in the same way,[34] whereas England and Wales have chosen a supra-therapeutic threshold set above concentrations expected to be seen in individuals legitimately prescribed a specific stimulant drug such as amphetamine (Table 13.4).

13.7 Psychoactive Medicines

In most European countries, benzodiazepines were the most common medicines detected in drivers, but there was high national variability. Epidemiological studies[11,21,30,35–37] indicate a major increase in the consumption of antidepressants and drugs for addictive disorders in the general population in Europe within the last few years. Medicinal drugs in Europe were mainly detected among older female drivers during daytime hours (DRUID deliverable 2.2.3), and among killed drivers in the DRUID studies the presence of benzodiazepines was the second most frequent toxicological finding after alcohol.[14]

The risk of traffic incident (fatal or serious injury) involvement for those using psychoactive medicines has also been widely studied.[37,38] For instance, Meuleners *et al.*[39] investigated 616 individuals aged 60 and older hospitalized as the result of a RTC between 2002 and 2008. Crash risk was significantly greater in men prescribed a benzodiazepine (OR = 6.2, 95% CI = 3.2–12.2, P < 0.001) or an antidepressant (OR = 2.7, 95% CI = 1.1–6.9, P = 0.03)

Table 13.4   International drug thresholds (set in or recommended for legislation): amphetamine and methamphetamine.

Country	Approach to threshold	Amphetamine (in blood) ($\mu g\ L^{-1}$)	Methamphetamine (in blood) ($\mu g\ L^{-1}$)	Ref.
Netherlands	Threshold	50[a]	50[a]	49
France	Threshold	50[a]	50[a]	24
England and Wales	Therapeutic threshold Strict liability offence	250	10	
Norway	Impairment limit	41	45	46
	Comparable to 50 mg/100 mL BAC	Legal limits for graded sanctions not defined	Legal limits for graded sanctions not defined	
	Comparable to 120 mg/ 100 mL BAC			

[a]The sum of the concentration of amphetamine, plus methamphetamine, plus MDMA, plus MDEA, plus MDA must not exceed 50 $\mu g\ L^{-1}$.

The risk of driver impairment has also been shown to increase significantly with increasing benzodiazepine blood–drug concentrations, with ORs being assessed impaired of OR 1.61 for diazepam ($P = 0.001$), OR 3.65 for oxazepam ($P = 0.05$) and OR 4.11 for flunitrazepam ($P = 0.05$), respectively.[26]

Meta-analysis of the effects of psychoactive medicinal opioids (morphine, methadone and buprenorphine) on driving performance[42] and the effects of psychoactive medicinal drugs (analgesics, hypnotics and antipsychotics) on safe driving and incident risk[30] have confirmed these findings. The evidence suggested a greater risk of a RTC for drivers aged >65 years who are prescribed opioid drugs than the risk for those in the same age group not prescribed medicinal opioids.[40,41] It was highlighted that females prescribed opioid analgesics have a significantly greater crash risk (OR = 1.8, 95% CI = 1.1–3.0, $P = 0.03$) than males under the same conditions.

13.8   Poly-substance Use

Poly-substance use is the norm for drug use behaviour rather than the exception today and is a growing and significant problem in driving populations.[43] The European DRUID studies (deliverable 2.3.5)

Table 13.5 DRUID risk estimates for a driver being seriously injured or killed in an accident when testing positive for a combination of drugs or a combination of drugs and alcohol.

Populations compared	Odds ratio (OR) and 95% confidence interval (CI)	Ref.
Multiple drug use compared with no drug use	OR: 6.05 (95% CI: 2.60–14)	41
Drugs + alcohol compared with no drugs	OR: 112 (95% CI: 14–893)	41

showed evidence of significantly increased risk, reported as odds ratios, for a driver being seriously injured or killed in an accident when testing positive for a combination of drugs and alcohol (Table 13.5). The odds ratio estimate suggests individuals are 20 times more likely to be involved in a RTC when both drugs and alcohol are used compared to those who have not taken these drugs together at one time (DRUID deliverable 2.3.5).

Evidence[1,13,19,44] suggests that even a small amount of alcohol when combined with a psychoactive drug leads to a significantly increased risk of a RTC compared to drivers who do not use this combination of substances.

Drug-driving is a complicated subject matter influenced by a multiplicity of factors, not least the difficulties associated with the toxicological analysis of a growing number of psychoactive substances that may be used alone or in combination by drivers. The implementation of rigorous medico-legal procedures are complicated by the circumstances under which testing must occur. The screening test will usually be undertaken under variable conditions (at the roadside). For the confirmatory test, time is of the essence particularly for short acting drugs with regard to the relationship between the sample collection (once in custody) and the road traffic incident. Many countries have introduced drug-driving legislation for the purpose of enforcement measures to deter the use of illegal substances when intending to drive, and to improve road safety for drivers themselves and other road users. Evaluation of this legislation should be undertaken regularly in order that it remains fit-for-purpose in an ever changing field.

References

1. M. Asbridge, C. Poulin and A. Donato, Motor vehicle collision risk and driving under the influence of cannabis: evidence from

adolescents in Atlantic Canada, *Accid. Anal. Prev.*, 2005, **37**(6), 1025–1034.

2. F. Barbone, A. D. McMahon, P. G. Davey, A. D. Morris, I. C. Reid, D. G. McDevitt *et al.*, Association of road-traffic accidents with benzodiazepine use, *Lancet*, 1998, **352**(9137), 1331–1336.

3. G. Berghaus, G. Sticht and W. Grellne, 2010 Meta-analysis of empirical studies concerning the effects of medicines and illegal drugs including pharmacokinetics of safe driving (EU Project DRUID Deliverable 1.1.2b).

4. W. M. Bosker, K. P. Kuypers, S. Conen, G. F. Kauert, S. W. Toennes, G. Skopp and J. G. Ramaekers, MDMA (ecstasy) effects on actual driving performance before and after sleep deprivation, as function of dose and concentration in blood and oral fluid, *Psychopharmacology*, 2012, **222**(3), 367–376, DOI: 10.1007/s00213-011-2497-8.

5. J. G. Bramness, S. Skurtveit and J. Mørland, Clinical impairment of benzodiazepines-relation between benzodiazepine concentrations and impairment in apprehended drivers, *Drug Alcohol Depend.*, 2002, **68**(2), 131–141.

6. H. J. Burch, E. Clarke, A. M. Hubbard and M. Scott-Ham, Concentrations of drugs determined in blood samples collected from suspected drugged drivers in England and Wales, *J. Forensic Legal Med.*, 2012.

7. S. W. Toennes, G. F. Kauert, S. Steinmeyer and M. R. Moeller, Driving under the influence of drugs – evaluation of analytical data of drugs in oral fluid, serum and urine, and correlation with impairment symptoms, *Forensic. Sci. Int.*, 2005, **152**(2–3), 149–155.

8. E. J. Cone and W. W. Weddington, Prolonged occurrence of cocaine in human saliva and urine after chronic use, *J. Anal. Toxicol.*, 1989, **13**(2), 65–68.

9. J. Davey and J. Freeman, Screening for drugs in oral fluid: drug driving and illicit drug use in a sample of Queensland motorists, *Traffic Inj. Prev.*, 2009, **10**(3), 231–236, DOI: 10.1080/15389580902826817.

10. T. Dassanayake, P. Michie, G. Carter and A. Jones, Effects of benzodiazepines, antidepressants and opioids on driving: a systematic review and meta-analysis of epidemiological and experimental evidence, *Drug Saf.*, 2011, **34**(2), 125–156.

11. H. T. O. Davies, Assessing chance variability in treatment trials, *B. J. Hosp. Med.*, 1998, **59**(8), 650–652.

12. DRUID, summary of main DRUID results: driving under the influence of drugs, alcohol, and medicines, 2012, TRB 91ST Annual

meeting, USA Dutch Forensic http://www.drugscope.org.uk/ Resources/Drugscope/Documents/PDF/Good%20Practice/ cannabis_submission.pdf.

13. DRUID, Prevalence of alcohol and other psychoactive substances in injured and killed drivers, Deliverable D2.2.5, 2010: http://www.west-info.eu/files/Druid-report-08-2011.pdf.

14. DRUID, summary of main DRUID results: driving under the influence of drugs, alcohol, and medicines, 2012, TRB 91ST Annual meeting, USA Dutch Forensic Institute report, DRUID deliverable 3.2.2.

15. DRUID, Main results to be communicated to different target groups, Deliverable 7.3.2, 2011: http://www.druid-oject.eu/cln_031/nn_107548/ Druid/EN/deliverales-list/downloads/Deliverable__7__3__2, templateId = raw,property = publicationFile. pdf/Deliverable_7_3_2.pdf.

16. O. H. Drummer, J. Gerostamoulos, H. Batziris, M. Chu, J. R. Caplehorn, M. D. Robertson and P. Swann, The incidence of drugs in drivers killed in Australian road traffic crashes, *Forensic Sci. Int.*, 2003, **134**(2–3), 154–162.

17. S. W. Toennes, G. F. Kauert, S. Steinmeyer and M. R. Moeller, Driving under the influence of drugs – evaluation of analytical data of drugs in oral fluid, serum and urine, and correlation with impairment symptoms, *Forensic. Sci. Int.*, 2005, **152**(2–3), 149–155.

18. R. Elvik, Risk of road accident associated with the use of drugs: A systematic review and meta-analysis of evidence from epidemiological studies, *Accid. Anal. Prev.*, 2013, **60**, 254–267.

19. A. Engeland, S. Skurtveit and J. Morland, Risk of road traffic accidents associated with the prescription of drugs: a registry-based cohort study, *Ann. Epidemiol.*, 2007, **17**(8), 597–602.

20. H. Gjerde, A. S. Christophersen, P. T. Normann and J. Mørland, Toxicological investigations of drivers killed in road traffic accidents in Norway during 2006–2008, *Forensic Sci Int.*, 2011, **212**(1–3), 102.

21. H. Gjerde and A. Verstraete, Can the prevalence of high blood drug concentrations in a population be estimated by analysing oral fluid? A study of tetrahydrocannabinol and amphetamine, *Forensic Sci. Int.*, 2010, **195**, 153–159.

22. M. Hjalmdahl, A. Vadeby, A. Forsman, C. Fors, G. Ceder, P. Woxler *et al.*, Effects of d-amphetamine on simulated driving performance before and after sleep deprivation, 2012, *Psychopharmacology*, 2012, **222**, 401–411.

23. A. W. Jones, A. Holmgren and Kugelberg, Concentrations of cocaine and its major metabolite benzoylecgonine in blood samples from apprehended drivers in Sweden, *Forensic Sci. Int.*, 2008, **177**(2–3), 133–139.

24. T. M. Kelly, J. E. Donovan, J. R. Cornelius and T. R. Delbridge, Predictors of problem drinking among older adolescent emergency department patients, *J. Emerg. Med.*, 2004, **27**(3), 209–215.
25. B. Laumon, B. Gadegbeku, J. L. Martin and M. B. Biecheler, Cannabis intoxication and fatal road crashes in France: population based case-control study, *Br. Med. J.*, 2005, **10**; **331**(7529), 1371.
26. S. G. Leveille, D. M. Buchner, T. D. Koepsell, L. W. McCloskey, M. E. Wolf and E. H. Wagner, Psychoactive medications and injurious motor vehicle collisions involving older drivers, *Epidemiology*, 1994, **5**, 591–598.
27. B. K. Logan, A. L. Mohr and S. K. Talpins, Detection and prevalence of drug use in arrested drivers using the Dräger Drug Test 5000 and Affiniton DrugWipe oral fluid drug screening devices, *J. Anal. Toxicol.*, 2014, **38**(7), 444–450, DOI: 10.1093/jat/bku050.
28. S. MacDonald, R. Mann, M. Chipman, B. Pakula, P. Erickson, A. Hathaway and P. MacIntyre, Driving behavior under the influence of cannabis or cocaine, Traffic, *Injury Prevention*, 2008, **9**(3), 190–194.
29. L. B. Meuleners, J. Duke, A. H. Lee, P. Palamara, J. Hildebrand and J. Q. Ng, Psychoactive Medications and Crash Involvement Requiring Hospitalization for Older Drivers: A Population-Based Study, *J. Am. Geriatr. Soc.*, 2011, **59**(9), 1575–1580.
30. K. L. Movig, M. P. Mathijssen, P. H. Nagel, T. van Egmond, J. J. de Gier, H. G. Leufkens and A. C. Egberts, Psychoactive substance use and the risk of motor vehicle accidents, *Accid. Anal. Prev.*, 2004, **36**(4), 631–636.
31. P. Mura, P. Kintz, B. Ludes, J. M. Gaulier, P. Marquet, S. Martin-Dupont *et al.*, Comparison of the prevalence of alcohol, cannabis and other drugs between 900 injured drivers and 900 control subjects: results of a French collaborative study, *Forensic Sci. Int.*, 2003, **133**, 79–85.
32. Netherlands Forensic Institute (NFI), Toxicological investigation into drugs and medicines in traffic incidents, draft 0112209.
33. W. R. Nickel and J. J. de Gier, DRUID (Driving under the Influence of Alcohol and Drugs): A Survey on Zero Tolerance, Saliva Testing and Sanctions, 2009, In, Deliverable 1.4.2 Per se limits – Methods of de-fining cut-off values for zero tolerance, 2010.
34. E. J. Ogden and H. Moskowitz, Effects of alcohol and other drugs on driver performance, *Traffic Inj. Prev.*, 2004, **5**(3), 185–198.
35. L. Orriols, L. R. Salmi, P. Philip, N. Moore, B. Delorme, A. Castot and E. Lagarde, The impact of medicinal drugs on traffic

safety: a systematic review of epidemiological studies, *Pharmacoepidemiol. Drug Saf.*, 2009, **18**(8), 647–658.

36. R. Penning, J. L. Veldstra, A. P. Daamen, B. Olivier and J. C. Verster, Drugs of abuse, driving and traffic safety, *Curr. Drug Abuse Rev.*,3(1), 23–32.

37. M. J. Rapoport, K. L. Lanctot, D. L. Streiner, M. Bedard, E. Vingilis, B. Murray *et al.*, Benzodiazepine use and driving: a meta-analysis, *J. Clin. Psychiatry*, 2009, **70**(5), 663–673.

38. S. Ravera and J. J. de Gier, 2007, Prevalence of Psychoactive Substances in the General Population. DRUID deliverable 2.1.1, www.druid-project.eu.

39. L. B. Meuleners, J. Duke, A. H. Lee, P. Palamara, J. Hildebrand and J. Q. Ng, Psychoactive medications and crash involvement requiring hospitalization for older drivers: a population-based study, *J. Am. Geriatr. Soc.*, 2011, **59**(9), 1575–1580.

40. C. Stough, L. Downey, R. King, K. Papafotiou, P. Swann and E. Ogden, The acute effects of 3,4-methylenedioxymethamphetamine and methamphetamine on driving: A simulator study, *Accid. Anal. Prev.*, 2012, **45**, 493–497.

41. S. W. Toennes, J. Röhrich and C. Wunder, Interpretation of blood analysis data found after passive exposure to cannabis, *Archive fur Kriminologie*, 2010, **225**, 90–98.

42. V. Vindenes, D. Jordbru, A. B. Knapskog, E. Kvan, G. Mathisrud, L. Slørdal and J. Mørland, Impairment based legislative limits for driving under the influence of non-alcohol drugs in Norway, *Forensic Sci. Int.*, 2012, **219**(1–3), 1–11.

43. N. D. Volkow, G. J. Wang, M. W. Fischman, R. Foltin, J. S. Fowler, D. Franceschi *et al.*, Effects of route of administration on cocaine induced dopamine transporter blockade in the human brain, *Life Sci.*, 2000, **67**(12), 1507–1515.

44. S. M. Wille, E. Raes, P. Lillsunde, T. Gunnar, M. Laloup, N. Samyn, A. S. Christophersen, M. R. Moeller, K. P. Hammer and A. G. Verstraete, Relationship between oral fluid and blood concentrations of drugs of abuse for drivers suspected of driving under the influence of drugs, *Ther. Drug Monit.*, 2009, **31**(4), 511–519.

45. K. Wolff and A. Winstock, Ketamine: From medicine to misuse. Review Article, *CNS Drugs*, 2006, **20**, 199–218.

46. F. M. Wylie, H. Torrance, A. Seymour, S. Buttress and J. S. Oliver, Drugs in oral fluid Part II. Investigation of drugs in drivers, *Forensic Sci. Int*, 2005, **150**(2–3), 199–204.

14 Alcohol Technical Defences in Road Traffic Casework

Mike Scott-Ham

Principal Forensic Toxicology and Drugs Consultancy Ltd and Principal
Forensic Services Ltd, Bromley, UK
Email: michael.scottham@gmail.com

14.1 The Law (United Kingdom)

In the UK, traffic law distinguishes between drivers who are impaired
through alcohol or drugs and those who are in excess of prescribed
limits. The legislation is contained within the Road Traffic Act (RTA)
1988, Section 4 of which states:

1. A person who, when driving or attempting to drive a mechanically propelled vehicle on a road or other public place, is unfit to
   drive through drink or drugs is guilty of an offence.
2. Without prejudice to subsection (1) above, a person who, when
   in charge of a mechanically propelled vehicle which is on a road
   or other public place, is unfit to drive through drink or drugs is
   guilty of an offence.

Section 5 of the RTA 1988 states:
If a person

1. Drives or attempts to drive a motor vehicle on a road or other
   public place, or

Forensic Toxicology: Drug Use and Misuse
Edited by Susannah Davies, Atholl Johnston and David Holt
© The Royal Society of Chemistry 2016
Published by the Royal Society of Chemistry, www.rsc.org

2. is in charge of a motor vehicle on a road or other public place after consuming so much alcohol that the proportion of it in his breath, blood or urine exceeds the prescribed limit he is guilty of an offence.

Currently the prescribed limits in England and Wales are:

- Breath 35 micrograms per 100 millilitres (35 µg dL^{-1})
- Blood 80 milligrams per 100 millilitres (80 mg dL^{-1})
- Urine 107 milligrams per 100 millilitres (107 mg dL^{-1})

However, in December 2014 the Scottish Government introduced lower limits *via* the Road Traffic Act 1988 (Prescribed Limit) (Scotland) Regulations 2014 to lower the prescribed limits to:

- Breath 22 µg dL^{-1}
- Blood 50 mg dL^{-1}
- Urine 67 mg dL^{-1}

Northern Ireland is also in the process of adopting lower prescribed limits.

In a situation where a motorist is convicted of an offence where disqualification is obligatory, such as the drink and drug driving offences, it would be normal for the court to order that the offender would be disqualified from driving for not less than a 12 month period, although the court can disqualify for a shorter time, or not at all, if special reasons are found.

A special reason applies to the facts of a particular offence only. It is a mitigating or extenuating circumstance which is directly connected with the commission of the offence and which can properly be taken into consideration by the sentencing court.

A special reason can include inadvertent or unknown ingestion of alcohol and this is where expert opinion should be sought. The expert(s) would consider whether or not this unknown alcohol could have been sufficient to make a difference between the motorist being above the prescribed limit or below it. Scenarios where this could occur include addition of extra alcohol by another individual or possible ingestion of alcohol within food or liquid medication such as cough medicines, some of which may contain relatively high concentrations of alcohol, albeit usually a small amount in a standard dose.

Regarding possible defences to the above offences in England, Wales and Scotland, some legislation is contained within the Road

Traffic Offenders Act 1988, Section 15.2 and in Northern Ireland in Article 18(2) of the Road Traffic Offenders (Northern Ireland) Order 1996 which states that:

> "Evidence of the proportion of alcohol or any drug in a specimen of breath, blood or urine provided by or taken from the accused shall, in all cases be taken into account and, subject to subsection (3) below, it shall be assumed that the proportion of alcohol in the accused's breath, blood or urine at the time of the alleged offence was not less than in the specimen."

That assumption shall *not* be made if the accused proves:

(a) that he consumed alcohol before he provided the specimen and
   (i) in relation to an offence under Section 3A, after the time of the alleged offence, and
   (ii) otherwise, after he had ceased to drive, attempt to drive or be in charge of a vehicle on a road or other public place, and
(b) that had he not done so the proportion of alcohol in his breath, blood or urine would not have exceeded the prescribed limit and, if it is alleged that he was unfit to drive through drink, would not have been such as to impair his ability to drive properly.

This allows for a defence of post-driving alcohol consumption, the so-called "hip-flask" defence, to be put forward. The onus is on the Defence to prove their case and the standard of proof is on a "balance of probabilities" approach, *i.e.* >50% chance that it happened, rather than "beyond reasonable doubt" that the post-incident consumption caused the motorist to be above the prescribed limit [Section 15(3) Road Traffic Offenders Act 1988].

Another possible defence arises within the Road Traffic Act itself, Section 4, subsection 3:

> "A person shall be deemed not to have been in charge of a mechanically propelled vehicle if he proves that at the material time the circumstances were such that there was no likelihood of his driving it so long as he remained unfit to drive through drink or drugs."

Another similar possible defence arises to the prescribed limit offence within the Road Traffic Act Section 5, subsection 2:

> "It is a defence for a person charged with an offence under subsection (1)(b) above to prove that at the time he is alleged to have committed the offence the circumstances were such that there was no likelihood of his

driving the vehicle whilst the proportion of alcohol in his breath, blood or urine remained likely to exceed the prescribed limit."

This legislation allows for the motorist to claim that they were not in charge or were not going to drive until their alcohol concentration had decreased to below the prescribed limit.

14.2 Evidential Samples

For the purposes of alcohol measurement in driving cases, three main sample types will be encountered, these being breath, blood and urine.

Following consumption, alcohol is rapidly distributed throughout the body in body water, with alcohol concentrations reflecting the water content of the medium. For example, alcohol concentrations in urine, based on relative water content, would be expected to be approximately 20–25% higher than those in blood.

Breath is the most commonly encountered medium and its use for alcohol measurement was introduced in many countries in the 1980s as it was seen as a faster, and less intrusive, means of determining the concentration of alcohol in a motorist than by provision of either blood or urine samples. One of the main issues with breath alcohol measurement is ensuring that the sample of breath analysed contains alcohol at a concentration representative of that circulating within the bloodstream. Alcohol can remain within the mouth for a while after ingestion and a minimum of 20 minutes must be allowed to permit dissipation and avoid elevation of a breath alcohol reading. When the alcohol content of the air in the lungs is in equilibrium with the alcohol content of the blood in the alveolar capillaries, the concentration in the air will reflect that in the pulmonary arterial blood. Consequently, only a sample of deep lung air will contain an alcohol concentration representative of that in the blood. For this reason it is essential that any device measuring breath alcohol content performs such measurement on a sample of deep lung air.

Since there are differences between arterial and venous blood alcohol concentrations[1] it follows that a venous blood sample, as taken from a motorist, will only contain alcohol at a concentration equivalent to that in a deep lung breath sample at equilibrium. It is not safe to take arterial blood samples from motorists, of course!

In the UK, calculations to determine the contribution from a stated alcohol consumption to a measured breath alcohol concentration, so-called forward calculations, assume a blood to breath partition ratio of

2300 : 1. The theoretical figure is 2280 : 1, but this is rounded upwards for convenience. In practice, this may overestimate the contribution but as the calculation, based on Widmark's equation (see below), calculates a theoretical maximum value only which will never be reached in practice, any overestimation in such calculations will not disadvantage the motorist. No ranges should be applied since this calculation will specifically address the situation where the blood supply, carrying the alcohol, is in contact with the alveoli in the lungs. This ratio varies little between individuals and depends on the partition ratio of alcohol between blood and air at 34 °C (expired air temperature), but with allowance for alcohol lost and gained on inspiration and expiration.

In a large survey in the 1980s written up in the Paton Report,[2] only 0.5% of subjects were found to have a blood/breath ratio <2000 : 1 and 3.4% had a value >3000 : 1. Therefore it is suggested that calculations involving conversion from breath alcohol concentrations to blood alcohol concentration equivalents take into account a range of blood to breath *concentration* ratios from 2000 : 1 to 3000 : 1. The most-likely blood/breath concentration ratio is taken as 2300 : 1 and this conversion ratio is used in the UK.

Up until recently the UK offered a "statutory option" to motorists whereby if a breath alcohol sample was supplied which gave a result in the region 40–50 µg dL^{-1} an alternative sample, either blood or urine, could be taken to replace the breath as the evidential specimen. This option was removed in April 2015 and the number of blood, and urine, samples taken dropped dramatically. Consequently, most cases where calculations are required now involve breath.

Despite this, blood will of course always be the best sample scientifically for measurement of alcohol content as the alcohol concentration in the bloodstream will be that affecting the person's behaviour. Sampling is, however, intrusive and painful and there are health and safety risks associated with the use of needles and hence fewer blood samples are taken. In hospital, whole blood samples may have been taken for diagnostic purposes and be processed to separate them into red blood cells and serum, and occasionally such separated samples may be encountered in road traffic cases where a request has been made for analysis of a sample taken at hospital. If it is required to assess a serum, or plasma, alcohol concentration against a prescribed limit, the distribution of alcohol between the different media must be taken into account. Given the higher water content in serum and plasma compared to blood, higher alcohol concentrations therein are to be expected. A mean ratio of around 1.18 serum or plasma to blood concentration should be used, although this can vary from 1.09 to 1.35.[3]

Urine is not an ideal sample for measurement of alcohol concentration in a scenario where there is a prescribed limit. This is because blood/urine alcohol ratios can vary widely depending on many factors, including when alcohol was last consumed and when the bladder was last emptied. For the first 1–2 hours after bolus alcohol consumption the urine alcohol concentration is generally less than, or equal to, that for blood and reflects the absorption phase. Thereafter, in the post-absorptive phase, the urine alcohol concentration will exceed the blood concentration. Typical urine/blood alcohol concentration ratios are from 1.4 to 1.7 in this latter phase, but can increase significantly as low blood alcohol concentrations are reached.[4] At equilibrium a ratio of 1.33 urine to blood should be used, but any calculations undertaken from a urine specimen should be done with caution.

Under road traffic legislation in the UK a motorist must supply an initial urine specimen which is then discarded and then supply another sample, the evidential specimen, within an hour. Scientifically this should permit the measured urine alcohol concentration to most closely reflect the alcohol concentration circulating in the blood, assuming the bladder is completely emptied when the first specimen is supplied. If so, the urine alcohol concentration measured will be that at the mid-point between supplying both specimens. Emptying of the bladder is not a legal requirement and situations have occurred where two urine samples supplied within a minute of each other have been found to be legally admissible, even though the scientific argument behind the reason for the requirement has been neglected. In such a situation, case law has differentiated between when the motorist "has adjusted their clothing" between sampling and when two samples have been supplied without adjustment of dress in between, with the former having been ruled admissible, the latter inadmissible (Prosser v Dickeson [1982] RTR 96 and R v Musker [1985] RTR 84). Knowledge of the exact process occurring when urine samples have been taken is clearly important when considering prosecution and information contained within the completed paperwork should be scrutinised carefully.

14.3 Pre-analytical Considerations

Blood and urine samples must be preserved and stored in appropriate conditions to prevent changes in alcohol concentration between sampling and analysis.

Suitable forensic kits are available and must be used, and a specification for all forensic kits has recently been issued by the British

Standards Institute (PAS 377:2012) which relates to "Consumables used in the collection, preservation and processing of material for forensic analysis – Specification for performance, manufacturing and forensic kit assembly".

For optimal prevention of microbial alcohol production a sodium fluoride content of at least 1.5% weight/volume (w/v) is recommended and the widely used 1% w/v may not be sufficient in all circumstances. Blood kits should contain an anti-coagulant, normally potassium oxalate, to ensure the sample is in a suitable condition for analysis. Once taken, samples should be agitated, preferably by shaking, to ensure complete dissolution of the additives. Samples should be stored refrigerated prior to analysis, although many samples are sent *via* post to analytical laboratories, which can expose the sample to elevated temperatures, especially during summer months. Blood samples which have not been stored correctly may normally be identified by a change in colour and/or consistency and this should be noted by the forensic analyst and appropriate caution applied to the result.

Motorists must be offered their own sample, the second or "B" sample, which could be analysed privately if required. Differing results between the prosecution and defence analysis may be challenged at court with arguments over analytical quality becoming important. Laboratory accreditation, participation in quality assurance programs, such as proficiency testing schemes, are all important when considering analytical accuracy and possible reasons for differences in results.

It has been suggested that the presence of a salt can influence the alcohol concentration in the headspace above a blood or urine sample. Since most analytical methods involve headspace analysis, this could affect a measured blood alcohol concentration. The salt relevant to blood samples is sodium fluoride, which is added as a preservative to prevent microbial changes to alcohol content. All headspace GC analytical methods should employ an internal standard and, in practice, although a salting-out effect can be noted, it will inevitably result in a lowering of the alcohol concentration since an enhanced preferential salting-out effect occurs to the internal standard which has a higher molecular weight.[5] Therefore this is not a useful challenge for defence lawyers!

14.4 Alcohol Calculations

There are many situations where calculations are required, including where a motorist has consumed alcohol after driving, where additional alcohol may have been consumed without their knowledge ("spiking") or when a motorist has no intention of driving until they are sober even though they may be sitting or sleeping in their vehicle and technically "in charge" of it.

A so-called back-calculation is a calculation to extrapolate the measured alcohol concentration back to a previous time. This may be required when a motorist has left the scene of a collision or driving incident, and thereby delayed sampling, or has been injured and taken to hospital, again delaying sampling.

In order to perform such calculations it is sound practice to have a standard, agreed set of guidelines to ensure that all practitioners will obtain the same outcome given the same set of circumstances. To facilitate this the United Kingdom and Ireland Association of Forensic Toxicologists (UKIAFT) has recently produced such a set of guidelines.[6] These guidelines include details as to what range of elimination rates should be used, how long to allow for alcohol to be fully absorbed into the bloodstream and defining what calculations should be performed. Bearing in mind that there is now a legal requirement in England and Wales for all experts to comply with Part 33 of the Criminal Procedure Rules introduced in 2013, which states in subsection 33.2 that the expert's opinion must be objective and unbiased and that they stay within their areas of expertise. Importantly it also states that the expert's duty overrides any obligation to the person from whom the expert receives instructions or by whom the expert is paid. Consequently, and hopefully, the days of "rent an expert" are now long gone.

A calculation upon which an otherwise viable defence often fails is the "total" alcohol calculation used to assess whether or not a stated alcohol intake over a period of time could have accounted for the measured alcohol concentration. For example, many post-incident drinking calculations will show that had it not been for that alleged consumption the motorist would not have exceeded the prescribed limit, but further calculations often show that the stated total consumption could not possibly have accounted for the actual measured alcohol concentration. An account of post-incident drinking in such an instance may not be believed at court. Information required to satisfactorily perform such calculations includes the time of the driving incident, time of screening breath test and result, time of evidential breath analysis and the results and the time and analytical result of any other samples taken subsequently, *e.g.* blood.

Information specific to the motorist includes name, age, gender at birth, weight and height. If the weight of the subject is not available but the height and build is known, it may be possible to estimate the motorist's weight from tables available on the internet, but any assumptions made should be reported. Details concerning food and drink consumption should include any food consumed within the previous 24 hours as well as full details of alcohol consumption. It is important to include as much detail as possible – such as times of drinking (especially the start time), volumes consumed and any brand names. Any medication or medical conditions may be relevant in some situations.

Different countries may serve different volumes as small, medium and large glasses of wine, for example, and beers may be served in subunits of pints, litres or others. Cases not infrequently involve a "mouthful" of drink – how much is this? Experiments have shown that 50 mL would be sufficient for most such occurrences and would normally be in the motorist's favour for most calculations.

14.4.1 Alcohol Absorption

Alcohol may be absorbed into the bloodstream rapidly after ingestion, but the rate depends on a number of factors including the presence of food in the stomach and the alcohol strength of the beverage consumed. Food can significantly delay absorption of alcohol and result in a much lower peak blood alcohol concentration and there have been many studies investigating this.[7,8] The optimum concentration of alcohol for rapid absorption is around 20% w/v, as could be expected when drinking a strong spirit with a mixer. The presence of carbohydrates in beers can slow down absorption. A recent study by Mitchell *et al.*[9] investigated the absorption of alcohol and timing of peak blood alcohol concentrations following ingestion of the same amount of alcohol in three different drinks (vodka with mixer, white wine and beer) in fasted subjects. The study demonstrated that the peak blood alcohol concentration was higher, and the time to peak was much faster, for vodka with a mixer, with white wine producing the second highest concentration and being second fastest to be fully absorbed and beer producing the lowest peak concentration and being the slowest to reach maximum concentration. Mean peak blood alcohol concentrations attained were 77 mg dL^{-1} (vodka with mixer), 62 mg dL^{-1} (white wine) and 50 mg dL^{-1} (beer) and were reached for vodka at 36 ± 10 min, for wine at 54 ± 14 min and for beer at 62 ± 23 min. These findings are consistent with others[10] and clearly demonstrate that the rate of absorption is a key factor in determining the peak blood alcohol concentration likely to be reached.

The effect of food on absorption, and hence degree of subsequent intoxication, is well known with many binge-drinkers deliberately avoiding food intake before starting consumption; the term "eating is cheating" is widely used for those who eat a meal first!

When performing a back-calculation a minimum absorption time of an hour is generally considered a safe assumption for full absorption of alcohol to have occurred for many drinking drivers, if they have not eaten recently. Caution should be used if a meal has been consumed within the previous few hours. If it is known that a substantial meal has been consumed recently, 2 hours should be allowed after last drinking to allow for full absorption to have occurred. The drinking history is important and in the UK a "Drink/Drive Information Form" (MG DD/D) or equivalent should be supplied in any cases where a calculation is required and this should contain the required information.

14.4.2 Alcohol Elimination

Alcohol is eliminated from the body at a constant rate and metabolism follows zero-order kinetics once the body's enzyme systems are saturated. The major metabolic pathway is *via* hepatic alcohol dehydrogenase (ADH), with some alcohol also being broken down in the stomach *via* gastric ADH. The latter pathway, although relatively minor, can be important when only a small amount of alcohol is consumed, especially over a long period of time. Saturation of this gastric pathway typically occurs when more alcohol than is contained within a half a pint of normal strength beer, or a measure of spirit, is consumed in less than an hour. In addition to these enzymic pathways, small amounts of alcohol are also eliminated *via* the breath and in the urine.

Numerous studies have investigated alcohol elimination rates and a useful review is provided by Jones.[11] He concludes that within the general population a range of elimination rates from 10 to 35 mg dL^{-1} per hour (mg dL^{-1} h^{-1}) is applicable and that for moderate drinkers a good average is 15 mg dL^{-1} h^{-1}. Apprehended drivers typically have a higher average elimination rate, as many are thought to be binge drinkers or alcoholics and a more appropriate average was suggested to be 19 mg dL^{-1} h^{-1}. Within the UK this study[12] has been suggested as being most representative of the driving population, with a range of 9–29 mg dL^{-1} h^{-1} covering 95% of the drinking population, and an average of 19 mg dL^{-1} h^{-1}. This working range has been recommended for use in alcohol-related calculations.[6]

Once the main enzymic pathway for alcohol breakdown is no longer saturated, which occurs once a blood alcohol concentration falls below about 10–20 mg dL^{-1}, the elimination kinetics become first order. Any back-calculations from low blood alcohol concentrations should be undertaken with caution although, in practice, no large differences in predicted concentrations would occur; however, when working with a prescribed limit offence, as applicable to most road traffic casework, even a small difference in blood alcohol concentration can be highly significant if close to the limit.

Many analytical laboratories work with a limit of quantification of 10 mg dL^{-1} so back-calculations from concentrations lower than this would not be attempted anyway. The lowest breath alcohol concentration printed out on an evidential breath testing instrument is 4 µg dL^{-1}, which is equivalent to a blood alcohol concentration of 9 mg dL^{-1} at a blood/breath concentration ratio of 2300:1.

It should be noted that in chronic alcohol consumption, other metabolic processes are likely including enzymic breakdown by CYP2E1 located within liver hepatocytes. Such enzymes are induced following periods of prolonged drinking and reflect the body's desire to eliminate a toxic substance more quickly. This can lead to elevated elimination rates in heavy drinkers with a mean of 33 mg dL^{-1} h^{-1} reported and a range from 20 to 62 mg dL^{-1} h^{-1},[12] although the highest rate may be questionable. The mean initial blood alcohol concentration in the study was 405 mg dL^{-1}.

Elimination rates in breath are, not surprisingly, very similar to those in blood once relevant ratios are taken into consideration, with rates of 8.2 µg dL^{-1} h^{-1} (95% range from 5.0 to 11.4 µg dL^{-1} h^{-1}) being reported.[13] Alcohol calculations involving use of elimination rates where breath specimens have been provided are also therefore entirely feasible and safe to undertake, providing the consumed alcohol has been completely absorbed by the relevant time.

14.4.3 Calculations Required

A large number of calculations can be performed in drinking and driving casework and a standardised set has been suggested to answer all questions raised in such cases. These are:

1. Calculate the contribution of the alcohol contained within the alleged post-incident alcohol consumption, or the additional alcohol allegedly drunk unknowingly, to the measured concentration.
2. Calculate the estimated blood or breath or urine concentration in the absence of the post-incident, or additional alcohol in the laced drink, at the time of the evidential analysis.

3. Calculate the estimated concentration in the absence of the post-incident or laced drink contribution at the time of the incident, *e.g.* collision. This should be extrapolated ("back-calculated") from the time of test after allowance for the post-incident consumption has been deducted but must only be performed where this calculation is scientifically reliable.

In some situations it may be necessary to change the order and back-calculate from the time of test and then subtract the post-incident consumption unless the electronic method of calculation, if used, can deal with negative numbers.

4. Calculate the expected alcohol concentration at the time of supplying the evidential sample using the total alcohol consumption claimed by the motorist. The concentration is derived by calculation of the maximum contribution from all of the drinks consumed and then allowing for alcohol elimination between the start of drinking and the time of the evidential test.
5. An optional additional calculation is to estimate the motorist's concentration at the time of the incident, from the claimed pre-incident consumption. This involves the calculation of the maximum contribution that the claimed pre-incident drinking could have produced.

In addition, the stated drinking scenario should be thoroughly examined to ensure that the motorist's alcohol level would not have fallen to zero at any point in the time period under investigation (*e.g.* between drinks). If this could have occurred the calculation must be modified accordingly. It may be that a concentration could have fallen to zero at a fast elimination rate but not a slow one and this should be considered carefully to provide a full range of possible values.

14.4.3.1 *Widmark Equation*

Many alcohol technical defence (ATD) calculations require a factor to be determined which takes into account personal details, such as the motorist's weight, height and sometimes age, in order to perform forward calculations to estimate an expected blood or breath alcohol concentration from a stated intake. The basic Widmark equation[14] is shown in eqn (14.1):

$$\text{Dose} = \frac{C_{\max} \times \text{Weight} \times r}{100}\text{g} \qquad (14.1)$$

where Dose is the amount alcohol that has been consumed (g), C_{max} is the resulting theoretical maximum blood alcohol concentration (mg per 100 mL), Weight is the body weight (kg) and r is the proportion of body weight in which alcohol is distributed (related to total body water) and is commonly known as the Widmark factor. Rearranging this gives the equation shown in eqn (14.2):

$$C_{max} = \frac{100 \times \text{Dose}}{\text{Weight} \times r} \text{ mg dL}^{-1} \tag{14.2}$$

This equation originated from the pioneering work of Erik Widmark on the pharmacokinetics of alcohol in the early part of the 20th century.[14] It can still be used in calculations to determine the theoretical maximum blood alcohol concentration that can be produced following a stated alcohol intake even though it was proposed way back in 1932. However, nowadays it is commonplace to amend this to reflect personal information available. The resultant factor is known as the modified Widmark factor, which considers the body water available for alcohol to be distributed within. It will rarely be the factors proposed by Widmark (0.68 for males, 0.55 for females). The most commonly used methods for calculating the modified Widmark factor are those described by Forrest[15] and Watson et al.;[16,17] where both factors are calculated, the lowest of these factors will normally be used in all subsequent calculations in the UK as this will usually be in the motorist's favour for most of the calculations performed. If no age is available for a male subject, Forrest's method should be used.

Seidl[18] has described an alternative approach, but since this always leads to higher modified Widmark factors than those calculated using the other methods, which would not be in the motorist's favour for the majority of calculations, this approach has not been adopted in the UK.

Unfortunately, methods may not always allow for an ageing population and in such instances the factors may not be as accurate as may be required for legal purposes.[19,20] Detailed studies on the applicability of current methods of estimating modified Widmark factors in an increasingly obese population would be useful.

If insufficient detailed information has been supplied for calculating an accurate modified Widmark factor (*e.g.* weight, height), it may still be possible to perform the calculations although any assumptions made must be detailed in any subsequent report or statement so they can be challenged if necessary. Steps should be taken to ascertain the required information or provide a reasonable

estimate. Any estimates made should be clearly stated in the practitioner's statement/report to allow for challenge if the information is
incorrect.

Both Forrest[15] and Watson *et al.*[16,17] use a water distribution constant (factor) of 0.8, as in eqn (14.3):

$$r = \frac{\text{Total body water}}{\text{Weight} \times 0.8} \tag{14.3}$$

Where 0.8 (80%) is the supposed water content of whole blood.
However, this factor is a weight/weight or volume/volume value and
the calculation relies on weight/volume. The water content of whole
blood can vary between 78–82%.[18,21–23] The specific gravity of blood is
around 1.05.[24–26] Therefore the correct value for water content should
be 0.84–0.85 as w/v depending on reference and we should substitute
this value into the calculation from eqn (14.3) to make that shown in
eqn (14.4):

$$r = \frac{\text{Total body water}}{\text{Weight} \times 0.84} \tag{14.4}$$

In addition, the "Widmark" equation can be further simplified to that
shown in eqn (14.5). Thus there is no need to specifically calculate or
to use *r*. Combining eqn (14.2) and (14.4) we have:

$$C_{\text{max}} = \frac{100 \times \text{Dose} \times 0.84}{\text{Total body water}} \text{ mg dL}^{-1} \tag{14.5}$$

14.5 Report Content

It has been proposed in the UKIAFT Guidelines that the following
information should be included in all statements and reports to ensure that the court has all relevant information to assess an expert's
competence and to see what information their opinion was based
upon. It should include:

- Practitioner's qualifications and experience – why is this person
  qualified to carry out these calculations?
- Purpose of statement/report – what has the practitioner been
  requested to do?
- Information received – what information has been supplied?
  This allows interested parties to ascertain if all relevant information has been supplied to the practitioner.

- Receipt and results of examination of any items submitted.
- The scientific basis of the calculation (this could be included as a standardised appendix). The particular method used for calculation of the modified Widmark factor or total body water must be included in the practitioner's statement/report.
- The information/assumptions on which the calculations are based. This allows the evidence to be challenged should any information or assumptions be incorrect or invalid.
- Comments including calculations – lay them out logically and with appropriate headings to make them easy to follow.
- Conclusions – keep them short and simple to understand. Remember that most people reading the report are not scientists.

A list of suggested calculations was proposed and in order to maximise the clarity of the report/statement it is suggested that each calculation has a separate heading. A logical order has been suggested:

1. The contribution due to the additional alcohol consumed either post-incident or unknowingly, using the widest range of values.
2. The estimated breath or blood result in the absence of the post-incident, or laced drink, at the time of the test using the widest range of values.
3. The estimated concentration in the absence of the post-incident or laced drink contribution at the time of the incident.
4. The expected blood or breath alcohol concentration at the time of the evidential test based upon the total intake of alcohol as claimed by the motorist and a comparison between this and the actual measured value, with a comment stating whether or not the alleged total alcohol consumption is consistent with the results obtained, using the widest range of values. If the "total" alcohol calculation does not add up, do not forget to include a proviso in your statement that the inferences drawn from the calculations may be unreliable as clearly the motorist has not given a full and accurate account of their drinking pattern.

It is also suggested that if the result is only consistent with the claimed pattern if extreme factors are used, this should be stated and may be important to the court when assessing the likelihood or otherwise of evidence presented. If the calculations are not compatible with the results and claimed drinking pattern, a warning should be included that caution is required when considering the remainder of the practitioner's statement. This would normally only apply if the motorist's account details insufficient alcohol; an adverse comment should only be used in situations where extra alcohol is detailed in exceptional circumstances, *e.g.* where an unrealistically large amount of additional alcohol has been claimed.

If the lower end of the range in a breath calculation falls between the prescribed limit and charging threshold (*i.e.* prosecution limit), this must be clearly stated.

As it is not known which alcohol concentration will be used at court, calculations to the time of the incident, as well as to the time of the evidential test, are recommended to be included in a report/statement. All information is then available for the Prosecution to proceed as they wish. The approach may vary according to region.

The practitioner should always carefully consider the assumptions made in their calculations, particularly when they are aware of uncertainties surrounding case information such as the nature and volume of the alcohol consumed and possible inaccuracies in weight, height, *etc.* When a calculated range is close to the prescribed limit or the charging threshold, extra care should be taken in wording statements and any uncertainties clearly expressed to avoid possible miscarriages of justice.

In a situation where clearly more alcohol has been consumed than stated, as we cannot say when this extra alcohol had been consumed, the statement must clearly reflect that it is not possible to specify whether this additional alcohol consumption occurred before or after the incident, or both.

14.5.1 Alcohol from Food

Occasionally a defence of accidental ingestion of alcohol from food, or medication, may be put forward as a reason for being above the prescribed limit. In such instances it will be necessary to calculate the possible contribution from this additional alcohol and assess firstly whether it was sufficient to have taken the motorist from below to above the prescribed limit and also whether the motorist's account of known drinking, added to this additional alcohol, could account for the measured alcohol concentration.

Whereas the alcohol content of medication may be readily available, that possibly present in food may well not. Cooked/heated food is likely to lose a significant amount of alcohol by evaporation and a controlled experiment may be required to assess the alcohol content after food preparation.

14.5.2 Unknown Alcohol Consumption (Laced Drinks)

Occasions may arise where a motorist claims to have not realised that they had consumed alcohol or had consumed additional alcohol to that known. This may be a fruit cocktail which contained alcohol of which the motorist was not aware, additional alcohol may have been bought by a friend or someone may have added additional alcohol. The latter may or may not have had malicious intent. Sometimes an account may be plausible, sometimes not, but as scientists we may be asked to perform various calculations to determine any numerical effects this may have had. For a case to stand a chance of success the "lacer" must prepare a statement and attend court to give evidence of what happened, including details of the amount of alcohol added without the motorist's knowledge, or how a drink was prepared for example. Studies have been published around the likelihood or otherwise of alcohol being detected in a drink by subjects.[27]

14.6 Medication

Ranitidine (Zantac®) and cimetidine (Tagamet®) are prescribed to treat gastric and duodenal ulceration, reflux oesophagitis and other conditions where reduction in gastric acid secretion is beneficial. They are histamine H2-receptor antagonists and reduce gastric acid output. Ranitidine and cimetidine can reduce gastric ADH activity when they are present at a high enough concentration, but when taken at therapeutic dosage (150 mg twice a day for ranitidine, 400 mg twice a day for cimetidine) neither drug is present at a high enough concentration in the bloodstream to have any noticeable effect on gastric ADH activity; however, this concentration is of course sufficient to produce the desired effect of decreasing gastric acid secretion. In order for the drugs to have any effect on the gastric ADH activity they need to be present in the stomach and therefore this will only be for a short time after drug ingestion. Peak blood drug concentrations for both drugs are reached 1–2 h after oral ingestion. Studies giving alcohol intravenously showed no effect.[28]

A study published in 1992,[29] and some earlier and since, showed that patients taking ranitidine or cimetidine had higher blood alcohol concentrations than when not taking it, even though the alcohol consumption had been the same. In total there have been more than 100 papers published which have investigated this interaction.

Papers showing an increase in blood alcohol concentration after taking either drug used low doses of alcohol (1–2 units) and/or carried out testing within an hour or so of taking the drugs.[30] The DiPadova study[29] gave around 2 units of alcohol to volunteers and measured their peak blood alcohol concentrations. They found that peak blood alcohol concentrations after taking ranitidine were 34% higher and after taking cimetidine were 92% higher; no significant effects were found with famotidine. Most papers found no effect on blood alcohol concentration, including those which mirrored social drinking.[31–35] The various findings can be reconciled since gastric ADH only destroys a small amount of alcohol but, as a proportion of intake, this is significant when only a small amount of alcohol is consumed. When larger amounts of alcohol are consumed the proportion is much smaller and any effect cannot be distinguished from variations in expected blood alcohol concentrations.

The key point is that although these drugs can result in an increase in blood alcohol concentration in certain situations, they do not actually produce alcohol in the body. Since the calculations we perform calculate a theoretical maximum blood alcohol concentration based on personal factors, such calculations are not affected by medication and therefore the calculated theoretical maximum blood alcohol concentration from a stated alcohol intake will not be exceeded. The subject might get closer to their theoretical maximum than if they were not taking medication, but will not produce a blood or breath alcohol concentration exceeding it. A defence should of course fail in such a situation.

Additional Reading

R. G. Gullberg, Estimating the uncertainty associated with Widmark's equation as commonly applied in forensic toxicology, *Forensic Sci. Int.*. 2007, **172**(1), 33–39.

A. W. Jones and L. Andersson, Comparison of ethanol concentrations in venous blood and end-expired breath during a controlled drinking study, *Forensic Sci. Int.*, 2003, **132**(1), 18–25.

A. W. Jones and M. Fransson, Blood Analysis by Headspace Gas Chromatography: Does a deficient sample volume distort ethanol concentration? *Med., Sci. Law*, 2003, **43**(3), 241–247.

A. W. Jones, Biomarkers of recent drinking, retrograde extrapolation of blood-alcohol concentration and plasma-to-blood distribution ratio in a case of driving under the influence of alcohol, *J. Forensic Leg. Med.*, 2011, **18**(5), 213–216.

A. W. Jones, Letter to the editor: Body mass index and blood-alcohol calculations, *J. Anal. Toxicol.*, 2007, **31**(3), 177–178.

A. W. Jones, Ultra-rapid rate of ethanol elimination from blood in drunken drivers with extremely high blood-alcohol concentrations, *Int. J. Leg. Med.*, 2008, **122**(2), 129–134.

M. Lewis, Blood alcohol: the concentration-time curve and retrospective estimation of level, *J. Forensic Sci. Soc.*, 1986, **26**(2), 95–113.

References

1. A. W. Jones, L. Lindberg and S.-G. Olsson, Magnitude and time-course of arterio-venous differences in blood-alcohol concentration in healthy men, *Clin. Pharmacokinet.*, 2004, **43**(15), 1157–1166.
2. P. Cobb and M. Dabbs, Report on the performance of the Lion Intoximeter 3000 and the Camic Breath Analyser evidential breath alcohol measuring instruments during the period 16 April 1984 to 15 October 1984, 1985.
3. W. J. Frajola, Blood alcohol testing in the clinical laboratory: problems and suggested remedies, *Clin. Chem.*, 1993, **39**(3), 377–379.
4. A. W. Jones, Ethanol distribution ratios between urine and capillary blood in controlled experiments and in apprehended drinking drivers, *J. Forensic Sci.*, 1992, **37**(1), 21–34.
5. B. Miller, S. Day, T. Vasquez and F. Evans, Absence of salting out effects in forensic blood alcohol determination at various concentrations of sodium fluoride using semi-automated headspace gas chromatography, *Sci. Justice*, 2004, **44**(2), 73–76.
6. UKIAFT 2014. UKIAFT Guidelines for Performing Alcohol Technical Defence Calculations – version 2.1 [Online], Available: http://tinyurl.com/pom537x [Accessed July 7th 2015].
7. A. W. Jones and K.-A. Jonsson, Food-induced lowering of blood-ethanol profiles and increased rate of elimination immediately after a meal, *J. Forensic Sci.*, 1994, **39**(4), 1084–1093.
8. A. Jones and A. Neri, Evaluation of blood-ethanol profiles after consumption of alcohol together with a large meal, *J. Can. Soc. Forensic Sci.*, 1991, **24**(3), 165–173.
9. M. C. Mitchell Jr., E. L. Teigen and V. A. Ramchandani, Absorption and peak blood alcohol concentration after drinking beer, wine, or spirits, *Alcohol.: Clin. Exp. Res.*, 2014, **38**(5), 1200–1204.

10. A. W. Jones, K.-A. Jonsson and A. Neri, Peak blood-ethanol concentration and the time of its occurrence after rapid drinking on an empty stomach, *J. Forensic Sci.*, 1991.

11. A. W. Jones, Evidence-based survey of the elimination rates of ethanol from blood with applications in forensic casework, *Forensico Sci. Int.*, 2010, **200**(1–3), 1–20.

12. A. W. Jones, Ultra-rapid rate of ethanol elimination from blood in drunken drivers with extremely high blood-alcohol concentrations, *Int. J. Leg. Med.*, 2008, **122**(2), 129–134.

13. M. Pavlic, P. Grubwieser, K. Libiseller and W. Rabl, Elimination rates of breath alcohol, *Forensic Sci. Int.*, 2007, **171**(1), 16–21.

14. E. M. P. Widmark, *Die theoretischen Grundlagen und die praktische Verwendbarkeit der gerichtlich-medizinischen Alkoholbestimmung*, Urban & Schwarzenberg, 1932.

15. A. Forrest, The estimation of Widmark's factor, *J. Forensic Sci. Soc.*, 1986, **26**(4), 249–252.

16. P. E. Watson, I. D. Watson and R. D. Batt, Total body water volumes for adult males and females estimated from simple anthropometric measurements, *Am. J. Clin. Nutr.*, 1980, **33**(1), 27–39.

17. P. E. Watson, I. D. Watson and R. D. Batt, Prediction of blood alcohol concentrations in human subjects. Updating the Widmark Equation, *J. Stud. Alcohol*, 1981, **42**, 547–566.

18. S. Seidl, U. Jensen and A. Alt, The calculation of blood ethanol concentrations in males and females, *Int. J. Legal Med.*, 2000, **114**(1–2), 71–77.

19. A. Thierauf, J. Kempf, J. Eschbach, V. Auwarter, W. Weinmann and H. Gnann, A case of a distinct difference between the measured blood ethanol concentration and the concentration estimated by Widmark's equation, *Med. Sci. Law*, 2013, **53**(2), 96–99.

20. L. Bielefeld, V. Auwarter, S. Pollak and A. Thierauf-Emberger, Differences between the measured blood ethanol concentration and the estimated concentration by Widmark's equation in elderly persons, *Forensic Sci. Int.*, 2015, **247**, 23–27.

21. F. Davis, K. Kenyon and J. Kirk, A rapid titrimetric method for determining the water content of human blood, *Science*, 1953, **118**(3062), 276–277.

22. T. Lijnema, J. Huizenga, J. Jager, A. Mackor and C. Gips, Gravimetric determination of the water concentration in whole blood, plasma and erythrocytes and correlations with hematological and clinicochemical parameters, *Clin. Chim. Acta*, 1993, **214**(2), 129–138.

23. A. Jones, R. Hahn and H. Stalberg, Distribution of ethanol and water between plasma and whole blood; inter-and intra-individual

variations after administration of ethanol by intravenous infusion, *Scand. J. Clin. Lab. Invest.*, 1990, **50**(7), 775–780.

24. R. J. Trudnowski and R. C. Rico, Specific gravity of blood and plasma at 4 and 37 °C, *Clin. Chem.*, 1974, **20**(5), 615–616.
25. W. S. Snyder, M. J. Cook, E. S. Nasset, L. R. Karhausen, G. P. Howells and I. H. Tipton, Report of the task group on reference man, *Int. Commission Radiol. Prot.*, 1974, **23**, 112.
26. *Geigy Scientific Tables No. 3*, ed. C. Lentner, 1984.
27. N. Langford, T. Marshall and R. Ferner, The lacing defence: double blind study of thresholds for detecting addition of ethanol to drinks, *Br. Med. J.*, 1999, **319**(7225), 1610.
28. R. Hernandez-Munoz, J. Caballeria, E. Baraona, R. Uppal, R. Greenstein and C. S. Lieber, Human Gastric Alcohol Dehydrogenase: Its Inhibition by H2,ÄêReceptor Antagonists, and Its Effect on the Bioavailability of Ethanol, *Alcohol.: Clin. Exp. Res.*, 1990, **14**(6), 946–950.
29. C. DiPadova, R. Roine, M. Frezza, R. T. Gentry, E. Baraona and C. S. Lieber, Effects of ranitidine on blood alcohol levels after ethanol ingestion: comparison with other H2-receptor antagonists, *JAMA, J. Am. Med. Assoc.*, 1992, **267**(1), 83–86.
30. J. Caballeria, E. Baraona, M. Rodamilans and C. S. Lieber, Effects of cimetidine on gastric alcohol dehydrogenase activity and blood ethanol levels, *Gastroenterology*, 1989, **96**(2 Pt 1), 388–392.
31. A. G. Fraser, M. Hudson, A. M. Sawyerr, M. Smith, J. Sercombe, S. B. Rosalki *et al.*, Ranitidine has no effect on postbreakfast ethanol absorption, *Am. J. Gastroenterol.*, 1993, **88**(2), 217–221.
32. G. Dobrilla, G. De Pretis, L. Piazzi, F. Chilovi, M. Comberlato, M. Valentini *et al.*, Is ethanol metabolism affected by oral administration of cimetidine and ranitidine at therapeutic doses?, *Hepato-gastroenterology*, 1984, **31**(1), 35–37.
33. R. H. Palmer, W. O. Frank, P. Nambi, J. D. Wetherington and M. J. Fox, Effects of various concomitant medications on gastric alcohol dehydrogenase and the first-pass metabolism of ethanol, *Am. J. Gastroenterol.*, 1991, **86**(12), 1749–1755.
34. K.-A. Jonsson, A. Jones, H. Bostrom and T. Andersson, Lack of effect of omeprazole, cimetidine, and ranitidine on the pharmacokinetics of ethanol in fasting male volunteers, *Eur. J. Clin. Pharmacol.*, 1992, **42**(2), 209–212.
35. J. L. Holtzman, R. L. Gebhard, J. H. Eckfeldt, L. R. Mottonen, D. K. Finley and F. N. Eshelman, The effects of several weeks of ethanol consumption on ethanol kinetics in normal men and women, *Clin. Pharmacol. Ther.*, 1985, **38**(2), 157–163.

15 New Psychoactive Substances and the Criminal Law[†]

Rudi Fortson

Queen Mary University of London, London, UK
Email: rfortson@25bedfordrow.com

15.1 Regulating Psychoactive Substances that Fall Outside the UN Drug Conventions

15.1.1 Descriptors of Non-medicinal Psychoactive Substances

There has been growing concern (worldwide) over the prevalence and use of drug substances that are used recreationally, but which fall outside international drug-control treaties[1] and national drugs legislation.[2] To the extent that these substances are not listed within the schedules to the three main UN Drug Conventions,[3] they have sometimes been described as "new"[4] or "novel"[5] psychoactives,[6] but both adjectives are apt to mislead because some substances have a long history. However, for the sake of convenience, the abbreviation "NPS" is used in this paper for New Psychoactive Substances.

NPS have also been described as "emerging" or "designer drugs". When not listed as "controlled drugs", for the purposes of the Misuse of Drugs Act 1971, they have been popularly (but ill-advisedly) termed

---

[†]This chapter is *not* intended for Court use, and it is *not* to be taken as giving legal advice.

Forensic Toxicology: Drug Use and Misuse
Edited by Susannah Davies, Atholl Johnston and David Holt
© The Royal Society of Chemistry 2016
Published by the Royal Society of Chemistry, www.rsc.org

"legal highs", and sold through retail outlets ("headshops") or *via* the internet.

None of the above descriptors are suitable for inclusion in legislation, not least because they lack precision.

15.1.2   The Misuse of Drugs Act 1971

The primary drug control legislative instrument in the UK, which has given effect to the aforementioned UN drug treaties, is the Misuse of Drugs Act (MDA) 1971 (as amended). The MDA 1971 possesses four striking features. First, it does not impose a "blanket-ban" on all psychoactive substances, but usually specifies drugs by their botanical or chemical name.[7] Secondly, and related to the first, section 1(2) provides more than just a hint that the Act was intended to control those substances the misuse of which "is having or appears [to be] capable of having harmful effects *sufficient to constitute a social problem*".[8] Thirdly, the Act says something about the relative harm of the drugs which it controls [by way of the three Classes of "controlled drugs" (A, B and C) set out in schedule 2 to the Act]. The Classes are relevant to the statutory maximum penalties that may be imposed in respect of offences created under the Act. Fourthly, although the Act creates general prohibitions, the detailed Regulations made thereunder are largely permissive (creating exceptions and savings).

Given the aforementioned structure of the MDA, the process for bringing a drug under the control of that Act can be lengthy, and not apt to deal with an emerging substance that is claimed to have harmful effects. Accordingly, the MDA 1971 was amended[9] to treat as a drug of Class B (in effect)[10] a "Temporary Class Drug" ("TCD") "that is being, or is likely to be, misused and that the misuse is having, or is capable of having, harmful effects" [s.2A(4) MDA].[11] Most of the MDA offences apply to a TCD, save that it is neither unlawful nor a criminal offence to be in *simple* possession of it.[12] As at November 2015, only five Temporary Class Drug Orders were been made.[13] Each order survives for no longer than one year.

The MDA may have further relevance in relation to NPS because there can be cases where, in the production of a psychoactive substance, a controlled drug has been produced (either as a necessary precursor to the making of the non-controlled variant or accidentally). Furthermore, traders in psychoactive substances cannot be entirely confident that an MDA controlled substance (*e.g.* ketamine) has not been added to the product in question. Many packages do not particularise the chemical ingredients that they contain or (even if they purport to do so) the details may not be complete or accurate. Some NPS wholesalers and retailers submit their products for chemical analysis (or receive an analyst's report). Their reasons for doing so are to give them reassurance that the products do not contain a controlled drug. However, experience has shown that such analysis is not always complete or wholly reliable.

15.1.3 Medicinal Products Legislation

Attempts by the UK to regulate NPS under its medicines legislation have not been effective. The Medicines Act 1968,[14] largely superseded by the Human Medicines Regulations 2012,[15] regulates "medicinal products".

Before 1-benzylpiperazine (BZP) was controlled under the MDA 1971, there had been successful prosecutions under the 1968 Act for the unauthorised supply (by retail) of that substance on the basis that it was a "medicinal product".[16] Unhappily, this was not the legal profession's "finest hour" because BZP has no recognised medicinal value whatsoever.[17] The reasoning may have been (not unreasonably) that in the light of a determination made by the regulatory agency that BZP was a "medicinal product", this was to be taken as being conclusive of the matter.[18] In fact, that determination might have been open to challenge. One legal practitioner[19] has resolutely argued that "legal highs" contravene UK medicines legislation (at least when the Medicines Act 1968 applied). However – as we shall see – experience has shown this argument to be an oversimplification.

Regulation 2(1) of the Human Medicines Regulations 2012 defines a "medicinal product" as:

(a) any substance or combination of substances *presented* as having properties of preventing or treating disease in human beings; or
(b) any substance or combination of substances that *may be used by or administered to human beings with a view to*
  (i) *restoring, correcting or modifying a physiological function by exerting a pharmacological, immunological or metabolic action, or*
  (ii) *making a medical diagnosis*. [Emphasis has been added]

The above definition mirrors (almost to the letter) the wording of article 1(2) of Council Directive 2001/83/EC (as amended). It is a definition that has (now[20]) a consistent meaning for the purposes of: (a) the MDA 1971; (b) the MA 1968;[21] and (c) the Human Medicines Regulations 2012.

It is often said that there are two limbs to the statutory definition of the expression "medicinal product",[22] namely, a "presentation limb" [reg.2(1)(a)] *and* a "functional limb"[23] [reg.2(1)(b)]. Lyn Brown MP astutely commented (in respect of the above definition) that "whether a product is medicinal is subjective; it depends on how the drugs are presented or viewed, rather than what they actually do".[24] The actual position is more complex. A decision of the European Court of Justice held, in the *Cases of D and G* (2014),[25] that the two limbs cannot be regarded as being unconnected with each other. In other words, the manner in which a drug is presented and its functional properties are relevant to each limb (depending on the circumstances) of a given case. Suppliers of NPS appear to have been mindful of these considerations, marking their products "not for human consumption", "research chemicals" or "plant food".

The term "presentation of a product" must be interpreted broadly[26] and includes cases where the product is presented (*e.g.* by labelling, leaflets) as intended for treating or preventing disease, or "any averagely well informed consumer gains the impression" that this is the purpose of the product.

The functional limb of the definition of "medicinal product" is the subject of slowly developing case-law. Unfortunately, the courts (in different jurisdictions) have diverged on whether a "medicinal product" is one that must be shown to have therapeutic benefit. The courts of the United Kingdom have not addressed this issue directly, but in *Regina v Foster*[27] the Court of Appeal proceeded on the basis (implicitly rather than fully spelt out) that the word "medicinal", as it appeared in section 130(1) of the MA 1968,[28] enjoys its popular meaning, namely, the quality of being beneficial to human health. By contrast, in *Law Officers of the Crown v Le Billon*[29] the Court of Appeal of the Island of Guernsey concluded that *mephedrone* is a medicinal product by function within the meaning of the 2001 EC Directive even though it has no beneficial or therapeutic effect and does not have the function of treating and preventing disease. However, in the *Cases of D and G* the ECJ held that article 1(2)(b) [the second limb] did *not* cover substances such as synthetic cannabinoids (which produced effects that merely modified physiological functions and *which did not bring about any improvement in those functions and which were harmful to human health*).[30] Indeed, the Court stated that each limb alludes to the existence of a beneficial effect for human health [evident in limb (a) and implicit in (b)]. On that basis, the prosecutions in the UK in respect of BZP were not well founded (it is submitted). It should be noted that the UK Advisory Council on the Misuse of Drugs alluded to the effect of a line of decisions of the European Court of Justice that the second limb of the definition of a "medicinal product" is one that has a significant effect on human physiological functions through a pharmacological, immunological or metabolic action.[31] Given the status of the European Court of Justice, the judgment in *D and G* (and other decisions that is has handed down on this issue) cannot be lightly disregarded by the member states of the EU.

Rather than having to rely on case law, the ideal would be for statutory definitions of a "medicinal product" to make clear that such a product is one that has a therapeutic or medicinal value (*i.e.* as the word "medicinal" is popularly understood). As we shall see, this is not something which Parliament has done for the purposes of the Psychoactive Substances Act 2016.

15.1.3.1 Offences under the Medicines Legislation

If an NPS is determined to be not only a "medicinal product" but also an "unauthorised medicinal product" – that is to say, a "medicinal product" in respect of which there is *not* in force[32] a UK or an EU "marketing authorisation",[33] a "certificate of registration",[34] a "traditional herbal registration",[35] or an "Article 126a authorisation"[36] – a person may not sell, supply, or offer to sell or supply, that product to a person within the European Economic Area. A number of offences are created under the HM Regulations 2012 and the MA 1968, some which are merely summarised here:

(a) Regulation 46 makes it *unlawful* (the *offence* of acting contrary to Regulation 46 is created by Regulation 47[37]) to sell, supply, or offer to sell or supply, an *unauthorised* medicinal product to a person within the European Economic Area,[38] or to *possess* an *unauthorised* medicinal product if the person knows or has reasonable cause to believe that the product is intended to be sold or supplied to another person *within the European Economic Area* [reg.46(3)]; or to manufacture or assemble a medicinal product, or procure the sale, supply, manufacture or assembly of a medicinal product where a person knows or has reasonable cause to believe that the medicinal product has been or is intended to be sold or supplied contrary to Regulation 46(1).[39]

(b) A person may not [except in accordance with a licence (a "wholesale dealer's licence")] distribute a medicinal product by way of wholesale dealing; or possess a medicinal product for the purpose of such distribution.[40]

(c) Under section 62 of the MA 1968 (as amended) where, in the interests of safety, it appears to Ministers to be necessary to prohibit the sale or supply, or the importation, of medicinal products of any description (or falling within any class), they may do so by Order. The prohibition may be total or subject to exceptions specified in the Order. A contravention of section 62 is an offence under section 67(2)[41] of the MA 1968.[42]

(d) Under the MA 1968, it is an offence[43] for a person to be in unauthorised possession of a "medicinal product" (knowing or having reasonable cause to suspect) that it had been sold, supplied or imported in contravention of an order made under section 62 of the 1968 Act. The adulteration of a "medicinal product" is also an offence.[44]

The 2012 Regulations also restrict the importation, manufacture and distribution of an "active substance"[45] that is intended to be used in a "medicinal product" unless the actor is registered with the "licensing authority" in accordance with the Regulations.[46] A failure to comply with those requirements is an offence.[47] Accordingly, were a UK court to rule that a given NPS was a "medicinal product", it follows that the manufacture, importation or supply (as the case may be) of the "active substance" would also be subject to the terms of the 2012 Regulations.

15.1.4 Other Statutory Measures Deployed Against the Trade in NPS

In 2013 – following publication by the ACMD of its report in 2011 into NPS[48] – the Home Office published "Guidance for local authorities on taking action against 'head shops' selling new psychoactive substances".[49] Both the ACMD and the Home Office alluded to various statutory measures (other than the MDA and the medicines legislation) that might be deployed in respect of the production, importation and distribution by retail of NPS. Some of these are included in the list below (which is not to be taken as being exhaustive):

(a) Offences Against the Person Act 1861
(b) The Open General Import Licence[50]
(c) The Trade Descriptions Act 1968
(d) Poisons Act 1972

(e) The Criminal Law Act 1977 (conspiracy)
(f) The Criminal Attempts Act 1981
(g) Intoxicating Substances (Supply) Act 1985
(h) The Enterprise Act 2002
(i) The General Product Safety Regulations 2005
(j) The Fraud Act 2006
(k) The Serious Crime Act 2007
(l) The Consumer Protection from Unfair Trading Regulations 2008
(m) The Anti-social Behaviour, Crime and Policing Act 2014 (part 4); notably Public Space Protection Orders (PSPOs)
(n) The Consumer Rights Act 2015

Space in this chapter does not permit detailed consideration of the above measures. In any event, it is clear that the Psychoactive Substances Act 2016, together with the MDA 1971, will be the primary instruments directed against those who trade in, or otherwise handle, NPS (other than the simple possession or use of a psychoactive substance). However, some of the satellite measures [save (g) above] may continue to have relevance.

15.1.4.1 Import of Goods (Control) Order 1954[51]

Article 1 prohibits the importation of *all* goods into the UK *unless* permitted under Licence made under article 2.[52] The Open General Import Licence (OGIL) effectively neutralises the blanket ban imposed by Article 1, so that goods may be imported subject to exceptions set out in a schedule to the OGIL. On the 22nd July 2011, the Home Office used its powers under the 1954 Order to prohibit the importation of phenazepam[53] as well as any preparation or products containing that substance, by adding them to the schedule to the Open General Import Licence (OGIL). The practical effect was to remove phenazepam from the licence – pending its control as a Class C drug under the 1971 Act. This procedure was also used in relation to mephedrone and its related compounds before those substances became controlled drugs under the MDA 1971.

15.1.4.2 General Product Safety Regulations 2005[54]

The 2005 Regulations typically apply to durable goods such as electric heaters and chain saws. It is questionable whether the Regulations are apt to deal with the production and supply of psychoactive products. However, it is reported by Norfolk County Council (UK) that two Great Yarmouth traders were convicted in 2013 for contravening a Safety Notice issued under the 2005 Regulations[55] in respect of the sale of "legal highs".[56] More recently (June 2015), it was reported that Belfast City Council obtained a permanent injunction against three people which prohibited them from selling NPS in Northern Ireland.[57]

15.1.4.3 Consumer Protection from Unfair Trading Regulations 2008 (CPR)[58]

A trader[59] is not entitled to pursue an "unfair commercial practice"[60] (which may be a "misleading action" or a "misleading omission").[61] A "misleading action" is defined by Regulation 5 and includes the provision of "false information" in relation to (among other things) "the existence or nature of the product", "the main characteristics of the product" (including the "benefits" and "risks" of the product), or if the "overall presentation in any way deceives or is likely to deceive the average consumer" in relation to any of the aforementioned matters. A breach of Regulation 5 is a criminal offence. Central to the operation of the Regulations is the requirement/test that the practice *"causes or is likely to cause the average consumer to take a transactional decision[62] he would not have taken otherwise"*. However, it seems likely that the average consumer who buys a NPS will not be misled into making the decision to buy it (*i.e.* the "transactional decision") by reason of the marketing literature or product labelling. Accordingly, the 2008 Regulations provide limited scope for regulating NPS.

15.1.4.4 Trade Descriptions Act 1968[63]

Section 2 of the TDA 1968 defines a "trade description" as one that refers to any goods in terms of (among other specified matters) its quantity, size, "method of manufacture", "composition" or its "fitness for purpose, strength, performance, behaviour or accuracy". A "false trade description" is a "trade description" that is false "to a material degree" (s.3, TDA 1968). The TDA 1968 is of limited application given the primacy of the Consumer Protection from Unfair Trading Regulations 2008 (see above).[64] As originally drafted, section 1(1) of the TDA 1968 made it an offence for a person in the course of a trade or business to apply a false trade description to any goods, or to supply, or offer to supply any goods to which a false trade description is applied. However, section 1(1) was repealed by the 2008

Regulations. Nevertheless, by section 16 of the TDA 1968 there is a prohibition on the importation of goods from outside the UK, if a "false trade description" has been applied to them with regards to the place of manufacture, production, processing or reconditioning of the goods. In this connection, one should note that offences exist in relation to unlawful importations charged under the Customs and Excise Management Act 1979.

15.1.4.5   Serious Crime Act 2007

The common law offence of inciting the commission of a crime was abolished,[65] and three complex offences of encouraging or assisting the commission of an offence were created (contrary to part 2 of the Serious Crime Act 2007). Liability arises (in essence) where a person ("D") does an act, which is *capable* of encouraging or assisting another person (P), to commit a criminal offence (the "anticipated offence") – regardless of whether or not the anticipated offence was actually committed. Lability will often turn on what the defendant intended or believed, or whether his conduct was reckless. The offences, set out in ss.44, 45, and 46 of the SCA 2007, are:

> Cases where D *intended* to encourage or to assist another to commit an offence (s.44).
> Cases where D *believed* that the anticipated offence *would* be committed (not might be committed) and he believes that his act will encourage or assist its commission (s.45).
> Cases where D did an act that was capable of encouraging or assisting the commission of one or more of a number of offences and he *believed* (a) that one or more of those offences *would* be committed (but has no belief as to which), and (b) that his act *would* encourage or assist the commission of one or more of them (s.46).

15.1.4.6   Intoxicating Substances (Supply) Act 1985

The Act (now repealed by the PSA 2016) created a summary-only offence [s.1(1)] for a person to supply, or offer to supply, a substance (other than a controlled drug) in respect of which he knew (or had reasonable cause to believe) was likely to be inhaled (even by its fumes) by a person under the age of 18 for the purpose of causing intoxication. The offence was introduced to tackle (for the most part) "glue sniffing".

15.1.4.7 Offences Against the Person Act 1861

Section 23 and 24 of the 1861 Act enacted two offences concerning the *administration by* a person ["D"] of (or *causing* to be administered) a *"poison or other destructive or noxious thing"* to *another* person ["V"]. Whereas the section 23 offence is concerned with acts that would endanger V's life or would inflict upon him grievous bodily harm, the section 24 offence (a less serious offence) is concerned with acts done by D *with intent to* injure, or aggrieve, or annoy, V. The section 24 offence is unlikely to arise in relation to NPS because D is unlikely to *intend* to "injure, aggrieve, or annoy" V.

As for the section 23 offence, a number of NPS are likely to be viewed by the Courts as being "noxious" (*R. v. Marcus*[66]). It appears to be no defence to a charge under s.23 (and presumably s.24) that V consented to an act of "administration" by D.[67] Accordingly, D might commit an offence contrary to s.23 OAPA 1861 if he/she *directly administers* a NPS to another (*e.g.* by injecting V with it; or placing a pill *directly* into the mouth of V). However, this is an unlikely scenario. A charge based on D having *"caused"* the administration of the NPS to V is problematic. There is no offence of *self-administration* of a noxious thing [see the House of Lords decision in *R v Kennedy No.2*;[68] and see D. Ormerod and R. Fortson, *"Drug Suppliers as Manslaughterers (Again)]"*.[69]

15.2 The Introduction of the UK Psychoactive Substances Act (2016)

On May 28, 2015, the Government introduced (in the House of Lords) the Psychoactive Substances Bill (PSB). This would have been an ideal opportunity for the UK administration to have trialled a new approach – very different from the regime enacted under the MDA 1971 – perhaps along the lines of the New Zealand model [Psychoactive Substances Act 2013 (NZ)] under which a blanket ban is imposed other than on a psychoactive product that is approved for use by individuals and which poses no more than a low risk of harm to those who use it. Alternatively, the government might have piloted the scheme proposed by the European Commission under which the intensity of control over the production and distribution of an NPS varies depending on whether the substance is of "low", "moderate" or "severe" risk.[70] The New Zealand approach,[71] discussed in some detail in another chapter of this book, is one that the UK All-Party

Parliamentary Group for Drug Policy Reform had invited the Government to consider adopting.[72] Instead, the Government opted for a regime akin to that enacted in Eire, Poland and certain other countries that have imposed a "blanket ban" (subject to exceptions) on psychoactive substances supplied for human consumption.

The PSB was introduced following publication of the "New Psychoactive Substances Review" (September 2014) by an Expert Panel that had been set up by the Government in December 2013. The Panel had been tasked "to make a clear recommendation for an effective and sustainable legislative response to new psychoactive substances".[73] Although the Government asserted that the panel had "recommended that the Government develop proposals for a general prohibition on the supply of non-controlled psychoactive substances",[74] its actual recommendations were not clear-cut. Indeed, it identified a number of "key risks" associated with such an approach in the areas of supply, demand, enforcement, harms, forensic science, law, communications and cost. In relation to "supply" and "demand" the Panel foresaw that the closure of retail outlets of NPS was "likely to result" in the market moving into "criminal supply", and that demand was likely to shift towards other substances such as alcohol and traditional illicit substances.[75]

At Second Reading of the PSB, the Government indicated that the objective of the Bill was "to protect the public" and it made various assertions, some of which were highly contentious [for example, deaths attributed to NPS, and the extent to which the Irish legislation (the PSA 2013) had been effective].[76]

15.2.1 An Overview of the Psychoactive Substances Act 2016 (UK)

The PSA imposes a "blanket ban" on "psychoactive substances" (defined by s.2) save in respect of those substances, or categories of substances, listed in schedule 1 to the Act (for example, alcohol, tobacco products, caffeine and food). The Act creates a legal fiction that schedule 1 substances are not "psychoactive substances" within the terms of section 2.

It is important to note that despite the fact that every "controlled drug" is psychoactive, the PSA is a self-contained regime that consists of *civil* powers as well as *criminal* offences to enforce the activities which the Act prohibits. The PSA does not attempt to consolidate or to codify other statutory regimes (notably the MDA 1971, Temporary Class Drug Orders, and the Human Medicines Regulations 2012). The co-existence and operation of the PSA and the MDA, in respect of two sets of psychoactive substances, do not make for a coherent strategic approach (it is submitted) for regulating the production and distribution of such substances.

Given that a principal objective of the PSA is to reduce the availability of NPS by closing down retail outlets, it seems plausible that the objectives of the PSA will be achieved largely by the exercise of the civil powers rather than by way of criminal prosecutions. From a law enforcement perspective, the civil powers conferred by the Act do not possess all of the disadvantages of the criminal offences. Firstly, the latter are typically reactive to acts done, whereas the civil powers may be used to prevent any of the "prohibited activities" stated in the PSA. Secondly, whereas proving a substance to be capable of having "psychoactive effect" for the purposes of criminal offences will be problematic in the absence of a dependable scientific test (*e.g.* human clinical trials), the exercise of the civil powers does not necessarily entail proving that the substance in question exists or, if it does, that it is psychoactive. Thus, "prohibition notices" and "premises notices" may be issued by a senior police officer, or by a local authority, on the basis of a *reasonable belief* that a "prohibited activity" is being (or is likely to be) carried on by a person or on premises.

15.3 Meaning of "Psychoactive Substance" and "Exempted Substances"

15.3.1 "Psychoactive Substance"

A "psychoactive substance" is defined by section 2(1) as (emphasis added):

> ".... any substance which (a) is *capable* of producing a *psychoactive effect* in a person who *consumes* it, and (b) is not an *exempted substance*."

It follows that whether something is a "psychoactive substance" depends primarily on whether it is *capable* of producing a "psychoactive effect". The latter expression is defined by section 2(2) [emphasis added]:

> "..... a substance produces a psychoactive effect in a person if, by stimulating or depressing the person's central nervous system, it affects the person's mental functioning *or emotional state*; and references to a substance's psychoactive effects are to be read accordingly."

The above can be compared with the Irish Criminal Justice (Psychoactive Substances) Act 2010 – "CJ(PS)A" (on which the PSA is said to be based) that reads [section 1(1)] [emphasis added]:

> "Psychoactive substance" means a substance, product, preparation, plant, fungus or natural organism which has, when consumed by a person, the capacity to (a) produce *stimulation* or *depression* of the central nervous system of the person, resulting in *hallucinations* or a *significant disturbance* in, or *significant change* to, motor function, thinking, behaviour, perception, awareness or mood, or (b) *cause a state of dependence*, including physical or psychological addiction.

The PSA definition is considerably more open-ended. Rather than listing various mental states (such as perception, awareness, mood) the PSA uses the loose expression "emotional state" – which is not further defined in the Act. There are many substances which (arguably) can have that effect, including perfumes.

Both the PSA and the CJ(PS)A 2010 refer to the *capacity* of the substance to produce one of the effects stated. However, whereas the PSA merely refers to a result that "affects" a person, the CJ(PS)A is considerably more specific in setting result-thresholds, such as "*significant* disturbance" or causing "a state of *dependence*". Although the words "*capable* of producing a psychoactive effect" (PSA) recognise that different people may respond differently to the ingestion of the same dosage quantity of a psychoactive substance, the PSA makes no attempt to say something about the *degree* of potency necessary to satisfy the definition of a "psychoactive substance". Thus, the question arises whether any pharmacological effect is sufficient (*e.g.* on a key target-area of the brain) regardless of whether that effect would be noticeable or not to the user, or to an observer. That said, it is just possible that the courts might rule that the "psychoactive effect" of the substance on a person must be one that is more than "merely trivial".

In a letter by the ACMD to the Home Secretary (August 17, 2015), the Council proposed that the definition of a "psychoactive substance" should be modified (and made more specific) to read:

> ".... any compound, which is capable of producing a pharmacological response on the central nervous system or which produces a chemical response in vitro, identical or pharmacologically similar to substances controlled under the Misuse of Drugs Act 1971."

The ACMD also proposed limiting the definition to a substance *produced by synthesis*[77] being a "compound,[78] irrespective of chemical state....or metabolites thereof" [emphasis added].

In a subsequent letter (October 23, 2015), the ACMD proposed an alternative definition of a "psychoactive substance":

> "For the purposes of this Act a substance produces a psychoactive effect in a person if, by stimulating or depressing the person's central nervous system, it affects the person's mental functioning or emotional state; as measured by the production of a pharmacological response on the central nervous system or which produces a response in in vitro tests *qualitatively identical to substances controlled under the Misuse of Drugs Act 1971*, and references to a substance's psychoactive effects are to be read accordingly and…" [emphasis added].

By "qualitatively identical to", the ACMD meant that "the substance interacts with the same target as a known psychoactive drug controlled under the Misuse of Drugs Act 1971."

The first version of the ACMD's proposed definitions was moved in Parliament as an amendment to the Psychoactive Substances Bill,[79] but it was successfully resisted by the Government who were "adamant that a definition cannot be arrived at through 'synthetic'; it has to be a blanket ban".[80]

There was force in the points made by Lyn Brown MP that (in effect) the expression "identical or pharmacologically similar to substances controlled under the Misuse of Drugs Act 1971" was capable of two interpretations, namely, that the relevant "effect" of a given substance is, (i) of a kind produced by drugs controlled under the MDA 1971 [a pharmacological effect/response], or (ii) molecularly similar to a particular controlled drug [a chemical response].[81] As Lyn Brown pointed out, the first interpretation might be the one intended by Parliament, whereas if the second interpretation was correct, the Act could "capture a benign substance that had only a limited effect but happened to be similar to a drug controlled under the MDA. Many exclusions might then be needed".[82] In the event, the proposed amendment was withdrawn.

It is important to note that even if a substance is capable of causing a psychoactive effect, it does not mean that an offence under the PSA is necessarily committed. This is because each of the primary offences under the PSA (production, supply, *etc.*) requires proof that the specified acts were done "*for* their psychoactive effect". In other words, acts must be done in the knowledge that, or being reckless whether, the substance is likely to be consumed for its psychoactive effect. Some church leaders have been concerned that the use of incense in religious ceremonies might be fall foul of the legislation. However, the Government wrote to the Churches' Legislation Advisory Service in September 2015 to lay to rest their concerns, which were unfounded (it is submitted) for the reasons just given.

15.3.1.1   Psychoactive Effect – Regardless of Harm

Unlike the MDA 1971, there is no scale of relative harm set out in the PSA 2016 in respect of banned psychoactive substances (*e.g.* there is no drug classification of such substances depending on their relative harm (which might not easily be capable of assessment in any event). Accordingly, It is not a requirement that the substance should be demonstrated to be *prima facie* harmful to health or that it would be likely to give rise to a social problem. Even psychoactive substances that are shown to be beneficial to health (or low risk) are *prima facie* compassed within section 2 (unless exempted under section 3 and schedule 1). The latter point was raised by the ACMD to the Home Secretary in a letter dated the 2nd July 2015.[83]

The absence of criteria of harm is not an academic point. In the event of a conviction for an offence under the PSA, a factor that would typically be taken into account for a drug offence, namely, its potential for causing harm, may not be available (albeit that the maximum penalty is 7 years imprisonment on indictment). As this writer stated to the Home Affair Committee, "in the absence of drug classification, or an expert's opinion (if accepted) as to harm, the courts will have little option but to assume that all psychoactive substances are equally harmful".[84] None of those matters persuaded the Government to change the terms of the Bill, which asserted that although the Courts should take account of the harms and the type of offence, those were matters for the Sentencing Council.[85]

15.3.2   "Exempted Substances"

Section 2 takes subject to the combined effect of section 3 and schedule 1 (the latter lists "exempted substances"). Exempted substances currently include "controlled drugs" (under the MDA 1971), "medicinal products" (as defined by Regulation 2 of the Human Medicines Regulations 2012), alcohol or alcoholic products, nicotine and tobacco products,[86] caffeine or caffeine products, and food.

The wording of a number of exemptions may seem puzzling. Thus, to take alcohol as an example, para.3 of schedule 1 states:

> "In this paragraph 'alcohol' means ethyl alcohol, and 'alcoholic product' means any product which (a) contains alcohol, and (b) *does not contain any psychoactive substance*" [emphasis added].

Note the words in italics. On the one hand, the provision exempts "alcohol products" – *unless it contains a "psychoactive substance"*. Alcohol clearly has a "psychoactive effect" within the meaning of section 2. Thus, the draftsperson might have specified that an "alcohol product" is one which "does not contain any *other* psychoactive substance". However, as we have seen, the Act creates a fiction (s.2) that alcohol (being an exempted substance) is not a "psychoactive substance" – even though, in reality, it plainly is.

"Caffeine" and "caffeine products" are specifically exempted, but so is "food" (and drink). The description of the latter exemption is ambiguous:

> "Any substance which (a) is ordinarily consumed as food, and (b) does not contain a prohibited ingredient.
> .....'food' includes drink;
> 'prohibited ingredient', in relation to a substance, means any psychoactive substance (a) which is not naturally occurring in the substance, and (b) the use of which in or on food is not authorised by an EU instrument".

This provision appears to be intended to capture nitrous oxide when used other than in the preparation and presentation of food. Arguably, there would have been merit had the Act simply explicitly specified that substance in schedule 1.

Questions remain. Is coffee, tea, or an energy drink that contains caffeine, to be described as "food/drink" or are they "caffeine products"? Is an energy drink – that contains a psychoactive substance other than caffeine – "ordinarily consumed as food"? Are exotic foods, which have a psychoactive effect "ordinarily consumed as food" and, if so, does it matter that such foods are "ordinarily consumed" in places other than in the UK? It is not clear whether the Government has thought through all the foods and drinks in which a psychoactive substance does not naturally subsist, but which its use in, or on, foods is not authorised by an EU instrument. The Constitutional Committee raised similar concerns (2nd Report, 18th June 2015):

> "23. Most of the definitions in schedule 1 are relatively clear. However, some are not. For example, to count as 'food' or 'drink', the substance must be ordinarily consumed as food or drink and must not contain a prohibited ingredient. However, the boundary delineating things that are and are not 'ordinarily consumed' is not without difficulty. Meanwhile the circumstances in which food will be free from 'prohibited ingredients' is not wholly clear, a prohibited ingredient being defined as 'any psychoactive substance which is (a) not naturally occurring in the substance, and (b) the use of which in or on food is not authorised by an EU instrument'. It is also worth noting that the relationship between some exemptions is not clear. For instance, 'caffeine products' are an exempted substance to

which no qualification relating to 'prohibited ingredients' applies, yet many such products will also be 'drinks' to which that qualification does apply."

The Act exempts "medicinal products" – the definition of which applies Regulation 2 of the Human Medicines Regulations 2012. The Government believes that the definition of a "medicinal product" will be wide enough to include "investigational medical products", homoeopathic medicinal products and traditional herbal medicines.[87]

This chapter has briefly discussed the definition of a "medicinal product" (above). It may be asked whether the "medicinal product" exemption could be a loophole for retailers of NPS who represent that their products are "medicinal products". However, this would by a *volte face* by such retailers whose products have often been marked "not for human consumption". It is submitted that no such loophole exists *provided that* the definition of a "medicinal product" is limited to reg.2(1)(b) and it follows the reasoning of the European Court of Justice in the *Cases of D and G*, namely, that the product is one that has a [medically recognised] benefit for human health. The assumption is that NPS, when used recreationally, are not used medicinally as treatment.

15.3.3 Proving that a Substance is Capable of Having "Psychoactive Effect"

Crucial to the operation of the PSA is proof that the substance in question is capable of producing a "psychoactive effect" in a person who consumes it. For the purposes of the principal criminal offences under the Act (sections 4, 5, 7–9), the burden will be on the prosecution to prove this element (as an issue of fact) to the criminal standard of proof. However, this begs the question of *how* this is to be proved in the absence of human clinical trials (animal testing is unlikely to be approved). A similar difficulty has been said to have arisen in relation to the application of the Irish CJ(PS)A 2010, and which may or may not explain the low prosecution/conviction rate in respect of offences created under that Act.

It is conceivable that psychoactivity could be proved by an admission made by an accused person who had used the substance in question. However, it is unlikely that a user would testify against third parties (there is no guarantee that he/she would do so) and, in any event, the testimony would only be probative if it was referable to the precise drug substance in question. Substances may not only be mis-described, but there is also no guarantee that the chemical makeup of a "branded" product will be consistent and not change, even from batch to batch. An analytical test of psychoactivity must be reliable and capable of independent verification (notably by the defence or by a third party).

The Home Secretary, clearly aware of this fundamental weakness in the Psychoactive Substances Bill, wrote to the ACMD[88] two days before the Bill was introduced in the House of Lords, seeking the Council's views "on how best we can establish a comprehensive scientific approach for determining psychoactivity for evidential purposes". The Council responded that "[t]he only definitive way of determining psychoactivity is via human experience which is usually not documented".[89]

The ACMD reiterated the difficulties of proving psychoactivity in a further letter to the Home Secretary (13th July 2015):

> "….psychoactivity in humans cannot be definitively established in many cases in a way that would definitely stand up in a court of law where a high threshold of evidence is required. There is currently no way to define psychoactivity through a biochemical test, therefore there is no guarantee of proving psychoactivity in a court of law. The only definitive way of determining psychoactivity is via human experience, which is usually not documented."

In evidence to the Home Affairs Committee, the Chair of the ACMD (Professor Les Iversen) stated that "brain-targeting" could be considered:[90]

> "What we are suggesting is that there would be a limited number of targets in the brain that psychoactive drugs act upon and activate. It would be quite easy for the Government to set up a contract laboratory *to test any new substance against any one of these five—or not more than five or six key—targets in the brain and if drug X were to activate any one of these, it would be considered to be an active psychoactive drug for the purposes of the Bill* we would say. Maybe this could be argued." [Emphasis added]

However, this suggestion is not free of difficulty. Firstly, there is the question of how speedily *reliable* test results could be obtained in this way (between the moment that the substance is submitted for analysis and obtaining the final results). Secondly, it is not clear whether limits would (or should) be set on the quantity of a questioned drug (as an assumed dosage unit) in order to determine whether it has sufficient potency to activate a particular "target" in the brain. What degree of effect would be regarded as being sufficient to constitute a "psychoactive effect"? The point has added relevance given that the PSA is intended to include *emotional* states such as depression, happiness, *etc.* Thus:

1. Would it be sufficient that any psychoactive effect is detected (albeit using sophisticated laboratory equipment), or must the effect be significant (*e.g.* that there is significant/manifest/ externally visible impairment or change in respect of a given bodily or mental function)?
2. Would the test be satisfied notwithstanding that the intended dosage (*e.g.* a specified quantity of powder, or a tablet) would produce no discernible effect but a significantly larger quantity would do so (albeit not in the dosage form marketed/supplied)?
3. Is it possible to establish thresholds of potency/intensity of "psychoactive effect" for the purposes of satisfying the statutory definition of "a psychoactive effect" as it appears in section 2(2) of the PSA 2016?
4. Would it be possible to reach consensus within the expert scientific community in respect of (3) above; and if not, who would set the relevant thresholds?

Bearing in mind that criminal offences are created under the PSA (most of which may attract confiscation proceedings under the Proceeds of Crime Act 2002 in the event of conviction), what opportunities will be given to the defence to independently verify the analysis undertaken by the "contract laboratory"? How would this be organised (more than one "contract laboratory")? To date, it is not clear how the Government anticipates these issues being satisfactorily resolved.

15.4   Principal Offences Under the 2016 Act

15.4.1   Overview

Six principal offences are included within the terms of the Act:

1. Producing a psychoactive substance.
2. Supplying a psychoactive substance.
3. Offering to supply a psychoactive substance.
4. Possession of a psychoactive substance with intent to supply it to another person.

5. Unlawful importation/exportation of a psychoactive substance.
6. Possession of a psychoactive substance in a custodial institution.

The first five offences (essentially "trafficking offences") carry a maximum of seven years imprisonment on indictment (or, on summary conviction: 6 months in England, Wales and Scotland; and 6 months in Northern Ireland): s.10. Those offences are "criminal lifestyle" offences for the purposes of the Proceeds of Crime Act 2002 (confiscation of the proceeds of crime following conviction).[91]

The sixth offence (s.9), carries a maximum of two years imprisonment on indictment (or, on summary conviction: 12 months in England, Wales[92] and Scotland; and 6 months in Northern Ireland): s.10. The section 9 offence is not a "criminal lifestyle" offence for the purposes of POCA 2002.

At least two other significant offences are created under the PSA:

7. Failing to comply with a "prohibition"/"premises" order [the civil powers].
8. Failing to comply with an "access prohibition" (s.27).

Note that a person may be convicted of offences 1–6 (and perhaps 7–8 too) by acting as a *secondary party*, that is to say, by "aiding", "abetting", "counselling or procuring" an offence which is committed by another person.

An offence under s.5 (supply, offering to supply) is aggravated for the purposes of sentencing if the matters set out in section 6 of the Act are established (*i.e.* the offence was committed in the vicinity of a school, or the offender used a courier under the age of 18 years, or the offence was committed in a custodial institution).

Each of the offences mentioned in 1–6 above contains a number of mental elements, which it is necessary for the prosecution to prove in respect of the principal actor. However, in the case of a secondary party, the mental elements will not necessarily be the same. Thus, where (for example) P unlawfully produces a psychoactive substance (with the purpose of doing so), and D sells the glassware and equipment in the knowledge that P will use them for that purpose, D may be liable under the PSA as a secondary party even though D did not desire the substance to be produced at all but only wished to make a profit from the sale of the equipment.

Liability is likely to be further extended by *inchoate liability* (*e.g.* where D and P *conspired* to commit an offence of producing a psychoactive substance, or *attempted* to do so). There may also be inchoate liability under the sweeping but complex provisions of part 2 of the Serious Crime Act 2007 (see above) of assisting or encouraging the commission of an offence under the PSA. Furthermore, it will be an offence for a person who participates in the "criminal activities" of an "organised crime group" [three or more persons acting with a common purpose for gain or benefit], if he knows or reasonably suspects that the activities are criminal, or that this conduct "will help an organised crime group to carry on" such activities: section 45 of the Serious Crime Act 2015.

An offence may be committed by a director, manager or company secretary of a body corporate where that offence was committed "with the consent or connivance" of that person, or the commission of the offence was "attributable to any neglect on the part of such a person".[93]

15.4.1.1 Simple Possession and Use of an NPS

Although the Government has made much of the fact that the PSA does not create a specific offence of *possessing* a psychoactive substance (simple possession) or *using* a psychoactive substance, it cannot be said that personal use is wholly exempt. This is because the offences of producing (s.4) and importing/exporting (s.8) a psychoactive substance may be committed if the actor "*intends to consume [the substance] for its psychoactive effect*". The offence under section 9 (possession in a custodial institution) was added to the Bill at the Committee stage in the House of Commons in October 2015.

15.4.1.2 Premises

Note that there is no equivalent to section 8 of the MDA 1971 (permitting or suffering specified activities to take place on premises by an occupier or manager). Thus, it would seem that where an owner of premises (other than a custodial institution) knows that X, Y and Z are taking psychoactive substances on his premises, no offence is committed by him. This is because it is not an offence to possess or to use a psychoactive substance and therefore the owner/manager cannot "aid, abet" (*etc.*) a non-offence. However, if X, Y and Z unlawfully share prohibited psychoactive substances, and the owner/occupier assists or encourages them in that conduct, then he risks being convicted as a secondary party to that criminal supply. Alternatively, the owner/occupier may be at risk of prosecution under part 2 of the

Serious Crime Act 2007 [assisting or encouraging the commission of an offence (*i.e.* under the PSA)].

15.4.1.3   Exempted Activities

The above offences are not committed in respect of an "exempted activity" (s.11) of a kind specified in schedule 2 to the Act. These include activities by a "health care professional"[94] acting in the course of his or her profession, or an activity in respect of an "active substance"[95] performed by a person registered under regulation 45N of the Human Medicines Regulations 2012 [unless exempted from registration by reg.45M(2) or (3)]; or any activity carried on in connection with "approved scientific research".[96] The exempted activities had been foreshadowed in debates in the House of Lords and later in the House of Commons; the Bill was then amended to include those exemptions (October 2015).

15.4.2   Producing a Psychoactive Substance

It is an offence [section 4] for a person to produce a psychoactive substance. A number of elements must be proved:

(a) The person *intentionally* produces a psychoactive substance;
(b) The person *knows or suspects* that the substance is a psychoactive substance; and
(c) The person *either* (i) *intends* to consume the psychoactive substance for its psychoactive effects, *or* (ii) *knows*, or is *reckless* as to whether, the psychoactive substance is likely to be consumed by some other person for its psychoactive effects.

Regulations may be made under the PSA to provide that specified activities are not an offence under this provision. Exceptions are made in respect of an "exempted activity" (see s.11 and schedule 2). A number of observations are tentatively made in connection with this provision.

Firstly, it will be seen that the prosecution must prove the fact of production of the substance in question, and that it was a "psychoactive substance".

Secondly, the word "production" is partially defined by section 59(2)(a) to mean producing the substance "by manufacture, cultivation *or any other method*". Those last four words are unclear as to their scope. The law is underdeveloped in relation to the MDA 1971 as to the distinction to be drawn (if at any) between "production" and "preparation", or whether what must be produced is the chemical substance itself (*i.e.* its molecular structure) or whether it is sufficient to blend a drug substance with a bulking agent. The courts are likely to answer such questions by discerning the intention of Parliament having regard to the language employed in the relevant statutory provision.[97] It is submitted that there would be merit if the word "produce" (and its concomitant expressions) were construed in terms similar to the word "manufacture", as it appears in Regulation 8 of the Human Medicines Regulations 2012, namely:

> "[produce] in relation [to a psychoactive substance] includes any process carried out in the course of making the product, but does not include dissolving or dispersing the product in, or diluting or mixing it with, a substance used as a vehicle for the purpose of [human consumption]".

Regulation 8 might have added that the term "production" does not include "*preparing* the product for human consumption".

Thirdly, when the provision (section 4) is read as a whole, the words "intentionally produces a psychoactive substance" means that D intended to produce psychoactive substance: see the Government's Explanatory Notes (paragraph 51). This is because [noting (b) above] it is sufficient that D "*knows* or *suspects*" that the substance is psychoactive.

Fourthly, D must either *intend* to consume the product as a "psychoactive substance", or he *knows* or is *reckless* whether somebody else would do so (or was likely to do so). It is probable that the courts would hold that D must be *subjectively* reckless – *i.e.* that he or she was aware of the risk but went ahead regardless of it.

15.4.3 Supplying a Psychoactive Substance

By section 5(1) of the PSA, it is an offence to supply a "psychoactive substance" to another person. Exceptions are made in respect of an "exempted activity" (see s.11 and schedule 2). The elements of the offence would appear to be that:

(a) The person *intentionally* supplies a substance to another person;
(b) The substance is a psychoactive substance;
(c) The person *knows* or *suspects*, or *ought to know or suspect*, that the substance is a psychoactive substance; and
(d) The person *knows*, or is *reckless* as to whether, the psychoactive substance is likely to be consumed by the person to whom it is supplied, or by some other person, for its psychoactive effects.

The prosecution must prove that the substance was supplied and that it was "psychoactive".

A partial definition of "supply" is given by s.59(2)(b), namely that "any reference to supplying a substance includes a reference to distributing it". This broadly corresponds to the approach taken in the MDA 1971, which states that "'supplying' includes distributing".[98] It is likely that the word "supply" as it appears in the PSA, and in the MDA 1971, will be construed similarly. Not every transfer of goods is necessarily an act of "supply" for the purposes of the MDA 1971,[99] and this may hold true for the purposes of the PSA.

There are a number of mental elements of the offence which must be proved. For the purposes of (a) above, the words "intentionally supplies" appear to mean merely that D deliberately supplied *something* to another – not that he intended that the thing should be a psychoactive substance. This is because (c) states that it is sufficient that D knows, suspects – or "ought to know or suspect" – that the thing was a psychoactive substance.

Note that the words "ought to know or suspect" do *not* feature in respect of the offence of "producing" a psychoactive substance (s.4) or "possessing a psychoactive substance with intent to supply it to another" (s.7). It is not clear why this difference exists. The words "ought to know or suspect" imports an objective element, but it is submitted that D must at least known of the facts which ought to have put him on enquiry to "know or suspect" that the substance was psychoactive.

15.4.4   Offer to Supply a Psychoactive Substance

An offence will be committed [section 5(2), PSA] if:

(a) D "offers to supply" a psychoactive substance to another person; and

(b) D *knows*, or is *reckless* as to whether, that other (or someone else) would be likely to consume the substance for its psychoactive effects (if the substance was what it was represented by D to be).

Exceptions are made in respect of an "exempted activity" (see s.11 and schedule 2).

This provision is likely to be construed in terms similar to the offence of offering to supply a "controlled drug" to another (MDA 1971, section 4). Thus, the offence is in the making of the offer. By section

5(3) of the PSA, reference to a substance's "psychoactive effects" includes a reference to the psychoactive effects which the substance would have "if it were the substance which D had offered to supply to another". Thus, if D displays herbal material on the palm of his hand, and offers to supply it as "Formula A" (which, had the representation been true, is a "psychoactive substance"), the offence is in the making of that offer, even if the thing was actually harmless and not psychoactive at all.

Regulations may be made under the PSA to permit offers to be made in the circumstances specified in the regulations.

15.4.5 Possession of a Psychoactive Substance with Intent to Supply it

A person will commit an offence [section 7] if:

(a) The person is in possession of a psychoactive substance;
(b) The person *knows* or *suspects* that the substance is a psychoactive substance; and
(c) The person *intends* to supply the psychoactive substance to another person for its consumption, whether by any person to whom it is supplied or by some other person, for its psychoactive effects.

Exceptions are made in respect of an "exempted activity" (see s.11 and schedule 2).

A person must be in "possession" of the substance (*i.e.* in its legal, technical, sense). By section 59(3), possession includes items which are subject to that person's control, but they are in the custody of another person. Thus, where D stores a psychoactive substance with P for safe-keeping or warehousing, D retains possession of the substance. D must *intend* to supply the thing to another person. This merely means that D intended to hand *something* to another person: he need only *suspect* that the thing was a psychoactive substance.

15.4.6 Importing or Exporting a Psychoactive Substance

The elements of the offences of importing or exporting a "psychoactive substance" [section 8] are essentially the same. Exceptions are made in respect of an "exempted activity" (see s.11 and schedule 2).

(a) The person *intentionally* imports/exports a substance;
(b) The substance is a psychoactive substance;

(c) The person *knows* or *suspects*, or *ought to know or suspect*, that the substance is a psychoactive substance; and

(d) The person (i) intends to consume the psychoactive substance for its psychoactive effects, or (ii) knows, or is reckless as to whether, the psychoactive substance is likely to be consumed by some other person for its psychoactive effects.

Although the section states that D must "intentionally" import (export) a substance, this means only that D acted deliberately, knowing that the thing had been brought within the limits of a port/airport (which defines the moment of importation).[100] The prosecution must prove that the thing is a "psychoactive substance" but, once that is done, it is only necessary to prove that D knew, suspected – *or ought to have known or suspected* – that the substance was psychoactive. It is not clear why this objective element exists in respect of this offence. Note that D must either have intended to consume the substance himself (and thus personal use is criminalised to that extent by this provision), or that he knew or was reckless whether someone else would consume the substance for its psychoactive effects.

15.4.7   Possession of a Psychoactive Substance in a Custodial Institution

By section 9(1) PSA 2016 a person commits an offence if he is in simple of a psychoactive substance in a "custodial institution" (see s.6). No offence is committed if that act pertained to an "exempted activity" (as defined in s.11). The substance must be proved to be psychoactive, and he knows or suspects that to be the case. It must also be proved that the person intended to consume the psychoactive substance for its psychoactive effects.

15.4.8   Civil Powers

15.4.8.1   *"Prohibited Activities"*

Quite apart from the aforementioned *criminal* offences, the PSA makes provision for the exercise of *civil* powers by specified police officers and by a local authority in respect of a "prohibited activity" (that is to say, producing/supplying/offering to supply/importing or exporting, a psychoactive substance, or assisting or encouraging any one of those activities): see section 12. The standard of proof is on the balance of probabilities [s.32(2)]. A court that is exercising civil powers is not bound by rules of evidence applicable in criminal proceedings [s.32(3) in relation to ss.19 and 29].

Section 12 is ambiguously worded. By section 12(2) an activity is not carried on if to do so would not be an offence by virtue of section 11 (*i.e.* it is an "exempted activity"). However, section 12 of the PSA does not expressly limit the meaning of a "prohibited activity" to acts that are criminal under the PSA. In other words, whereas the *offences* require (among other ingredients) proof of *mens rea* (guilty mind), it is arguable that a "prohibited activity" requires merely proof of a *conduct element* (*e.g.* the fact of supply, importation, exportation, *etc.*).

15.4.8.2 *Prohibition Notices/Orders, and Premises Notices/Orders*

The civil powers include issuing "prohibition notices" (s.13) and/or "premises notices" (s.14), the effect of which is to require specified persons to cease an activity, or to prohibit or to prevent a "prohibited activity". A notice takes effect "when it is given" [s.16(2)]. Note that it is not a prerequisite that the approval of a court is required before a notice may issued. There is no mechanism under the PSA for the party affected by the notice to appeal against it (judicial review may be available).

The party affected by a notice might choose not to comply with it, but sections 18 and 20 provide that certain persons (specified in s.21) may apply to an "appropriate court"[101] for a "Prohibition Order" or a "Premises Order" directing compliance. A further power is granted by section 19 by which a court, having dealt with an offender convicted of a "relevant offence" [s.19(5)],[102] may make a "prohibition order". In these situations, a route of appeal is provided by section 30 of the Act, with further provision being made with respect to the variation or discharge of an Order.

The ability of the Court to make Orders under ss.18, 19 and 20 (or as varied) includes the power to impose "any prohibitions, restrictions or requirements that the court considers appropriate" in addition to other prohibitions or requirements specified in section 22 of the PSA. The prohibitions may include "prohibiting access to premises owned, occupied, leased, controlled or operated by the person for a specified period (an 'access prohibition')": section 22(6). "Access prohibitions" may be enforced by an "authorised person" (s.23), with exemption from liability in damages for anything done, or omitted to be done, by the authorised person in the exercise or purported exercise of the s.23 power – unless the act was done in bad faith (see s.25). Offences of "failing to comply with a prohibition order or premises order", or of "failing to comply with an access prohibition", are created by sections 26 and 27 of the PSA.

Note that there is a degree of overlap in relation to the civil powers enacted under the PSA and those enacted under part 4 of the Anti-social Behaviour, Crime and Policing Act 2014 (ABCPA 2014), which came into force on the 20 October 2014.[103] The ABCPA creates various "community protection powers" where persistent or continuing conduct is having a detrimental effect on the quality of life in a locality. Specified persons and legal entities are empowered (under the 2014 Act) to apply for, and to issue, "Community Protection Notices"[104] and/or "Public Spaces Protection Orders",[105] and to obtain orders for the "Closure of Premises Associated with Nuisance or Disorder".[106] Those powers have been deployed in respect of the distribution of NPS (*e.g.* from so-called "headshops"[107]) and/or the use of such substances in public places.

15.4.8.3   *Investigative Powers Under the PSA 2016*

Where a police officer or customs officer has "reasonable grounds to suspect" that a person has committed, or is likely to commit, an offence under the PSA, he may search that person for "relevant evidence" and "stop and detain" the person for the purposes of the search: see section 36.

A power to enter and search vehicles for "relevant evidence" is given by section 37, with a similar power to board and search vessels or aircraft (s.38). In respect of premises, a "search warrant" may be issued by a justice authorising a "relevant enforcement officer" to enter and to search premises for "relevant evidence" (s.39). In respect of searches carried out in relation to vehicles, vessels, aircraft and premises, power is vested in a "relevant enforcement officer" to examine "anything that is in or on premises", and he may "carry out any measurement or test of anything" which the officer has power under the PSA to examine, including taking a sample from any live plant: section 41.

A further power is given to such officers to "require any person in or on the premises to produce any document or record that is in the person's possession or control": section 42. At this stage (production of documents) there would appear to be no requirement that the power is exercisable only in relation to relevant documents (subject to the provision relating to legally privileged documents or other "excluded items": s.42). However, although an officer has a power to seize, detain or remove "any item found on the premises" (and/or to take copies), the section 12 power is exercisable only if the officer reasonably believes the item to be "relevant evidence" or "a psycho-active substance (whether or not it is relevant evidence)": section 43(5). Items seized may be retained as long as necessary for use as evidence at a trial or for forensic examination, or for an investigation into an offence under the PSA (section 49).

15.4.8.4 Power of Disposal of Seized Psychoactive Substances

Where, following a lawful search and seizure of a substance, an officer reasonably believes that the item is a psychoactive substance which, if it had not been seized, "was likely to be consumed by an individual for its psychoactive effects", *and* it was not evidence of an offence under the PSA, *and* the officer has no reason to believe that the item was being used in connection with an "exempted activity", he may "may dispose of the item in whatever way the officer thinks is suitable": section 50.

A court must forfeit a substance if it is satisfied that the item is a psychoactive substance which was likely to be consumed by an individual for its psychoactive effects, *and* the item was not being used in connection with an "exempted activity": section 51. However, there may be circumstances in which a substance – even if it is psychoactive – must be returned to the person entitled to it [for example, the substance was not likely to be consumed for its psychoactive effect: s.51(9)].

A psychoactive substance may be forfeited following a conviction of a PSA offence (or an "ancillary offence"[108]): section 54.

15.4.8.5 Internet Service Providers ("Information Society Services")

The Government could not ignore the extent to which NPS have been available *via* the internet. Concern has been expressed by many commentators that the closure of retail outlets of NPS (head-shops) – many of which charged (and paid to HMRC) the Value Added Tax on sales – will result in an increase in internet sales of such substances. As a result, section 57, together with schedule 4 of the PSA, make provision to address this issue.

A feature of an "information society service" is that it is a "service normally provided for remuneration, at a distance, by means of electronic equipment for the processing (including digital compression) and storage of data, and at the individual request of a recipient of a service" (see schedule 4, part 3, para.10). A "service provider" means a person providing an information society service.

The extent to which the PSA is effective in reaching internet retailers as opposed to those who provide and manage internet electronic data traffic, remains to be seen.

Part 1 to schedule 4 extends criminal liability for the offence of offering to supply a psychoactive substance [s.5(2), PSA] if the service provider is established in the UK *and* an act is done by it within in an EEA state (other than the UK) that would amount to a s.5(2) offence had it been done in the UK. However, limits are placed on instituting proceedings in respect of non-UK service providers, including a requirement to satisfy the so-called "derogation condition". This condition is that proceedings are (i) necessary for the purposes of the "public interest objective", (ii) relates to an information society service that prejudices that objective or presents a serious and grave risk of prejudice to that objective, and (iii) the proceedings are proportionate to that objective. There are exceptions from liability in respect of "mere conduits", "hosting" services and in respect of data that is automatically and temporarily cached: see schedule 4, part 1.

Prohibition notices and orders may be given and made in respect of service providers (non-compliance is an offence: schedule 4, part 2, para.6), but this is subject to the requirement that the terms of the order were necessary to pursue the "public interest objective", *etc.*

15.5   Other Models for Regulating NPS

15.5.1   Proposed EU Action in Respect of NPSs

On September 17, 2013, the European Commission proposed a revised set of measures to control NPS.[109] The aim is to "speed up the Union's ability to fight new psychoactive substances by providing for…a quicker procedure" by which the EU will be able to act within ten months rather than within two years. On the 17th April 2014 the European Parliament "gave its strong backing (507 in favour of the Regulation, 37 against and 33 abstentions)".[110]

Under the proposed system, and following a joint EMCDDA and Europol Report, the substance or product in question would typically be risk assessed *unless* there is an "immediate risk" to public health, in which event there is power to prohibit, temporarily, the making available of the substance on the consumer market (see draft Article 9). Where, however, the risk was assessed as being "low" (*i.e.* posing low health, social and safety risks), the Commission would introduce no restrictions (draft Article 11). A "moderate risk" would empower the Commission to prohibit the making available *on the consumer market* of the NPS (draft Article 12). A "severe risk" would empower the Commission to prohibit the production, manufacture, making available on the market, transport, importation or exportation of the NPS (draft Article 13).[111]

The current draft of Article 9 (immediate risk of harm) suggests the existence of a high threshold before temporary consumer market restrictions may be imposed (pending assessment) in respect of a NPS that has come to the attention of a Member State. The risk of "immediate harm" to public health is to be "evidenced by"[112] reported fatalities and severe health consequences associated with the consumption of the NPS, related to its "serious acute toxicity", and that the prevalence and patterns of use of the new psychoactive substance in the general population and in specific groups indicates that the "scale of risk is considerable".

15.5.2 The Approach in Eire

By section 1 of the Criminal Justice (Psychoactive Substances) Act 2010, the expression "psychoactive substance" is broadly construed to mean "a substance, product, preparation, plant, fungus or natural organism which has, when consumed by a person, the capacity to":

(a) Produce stimulation or depression of the central nervous system of the person, resulting in hallucinations or a significant disturbance in, or significant change to, motor function, thinking, behaviour, perception, awareness or mood; or
(b) Cause a state of dependence, including physical or psychological addiction.

The Act will not apply to a psychoactive substance specified by an Order made under section 2 of the Act by a Minister.[113]

It is an offence for a person:[114]

(a) To *sell* a psychoactive substance knowing or being reckless as to whether that substance is being acquired or supplied for human consumption [s.3(1)].
(b) To *import* or *export* a psychoactive substance knowing or being reckless as to whether that substance is being acquired or supplied for human consumption [s.3(2)].
(c) To *sell any object* knowing that it will be used to cultivate by hydroponic means any plant in contravention of section 17 of the Misuse of Drugs Act 1977 (s.4).

By section 10, a "closure order" may be made against a person who has been convicted of an offence under ss. 3 or 4 of the 2010 Act [or under section 5 or section 8(6)]. The order prohibits the person, at a specified place or places, from operating any business or engaging in any specified activities "which may reasonably be considered to be connected with the sale, importation or exportation or advertisement of psychoactive substances for human consumption or, as may be appropriate, the sale or advertisement of an object for use in cultivating by hydroponic means any plant in contravention of [s.17 of the Misuse of Drugs Act 1977 (Eire)]".

The "closure order" may be made in addition to, or as an alternative to, any other penalty. However, a BBC investigation has suggested that although the legislation has resulted in the closure of "head-shops", Ireland's drugs squad is "unable to act against a range of legal-high type drugs because of problems with the legislation", notably, proving that the substance under investigation is in fact psychoactive.[115] At the time that the Psychoactive Substances Bill (UK) was being debated in the House of Lords, Baroness Meacher called for an impact assessment of the Irish legislation.

15.5.3 The New Zealand Model: Psychoactive Substances Act 2013

In 2013, New Zealand enacted the Psychoactive Substances Act, consisting of 110 sections and two schedules (also see Chapter 17). The Act came into force on the 18th July 2013. It is proposed to say little about the NZ legislation in this chapter, given its detailed treatment by another author in this book. Suffice to say that – unlike the UK Psychoactive Substances Act – the NZ Act does not impose a blanket ban on psychoactive substances (subject to exempted substances) but enacts a structured regime for regulating the *"importation, manufacture,*[116] *sale,*[117] *supply,*[118] *or possession*[119] *of a psychoactive substance or approved product*[120]..." [see section 5 of the PSA 2013 (NZ)]. The stated aims of the NZ Act are "...*to protect the health of, and minimise harm to, individuals who use psychoactive substances*" (section 3). Three of the five statutory principles (mentioned in section 4) are that [emphasis added]:

(a) A psychoactive product that poses no more than a *low risk of harm* to individuals who use the product *should be approved.*

(b) A psychoactive product that poses *more than a low risk of harm* to individuals who use the product *should be prohibited.*

(c) A psychoactive *product that has not been approved* by the Authority *should be prohibited, on a precautionary basis, until it has been assessed* by the Authority and the Authority is satisfied that it poses no more than a low risk of harm to individuals who use it.

A "psychoactive substance" is defined by section 9 of the PSA 2013 to mean (unless the context otherwise requires), "*a substance, mixture, preparation, article, device, or thing that is capable of inducing a psychoactive effect*[121] *(by any means) in an individual who uses the psychoactive substance*". Excluded from the definition of a "psychoactive substance" are (among others) "controlled drugs",[122] or "medicines"[123] or "herbal remedies".[124]

As originally drafted, the PSA 2013 included transitional provisions that were enacted "in order to bridge the gap between the enactment of the legislation and the introduction of the regulations" so as "to minimise the risk of black market activity under the assumption that demand for psychoactive substances would continue irrespective of the existence of a legal market".[125] However, in April 2014, following disquiet about harms and other social issues, the NZ government introduced a Bill to revoke all interim product approvals, and all interim retail and wholesale licences that had been granted under the PSA 2013, and to require the recall and disposal of all products for which interim product approval were granted, as well as suspending, until regulations come into effect, the requirement that the Psychoactive Substances Regulatory Authority must consider product approvals. In effect, the amending legislation "placed a moratorium on processing applications for product approvals and for licences until regulations came into force".[126] On the 7th May 2014 the Psychoactive Substances Amendment Act 2014 received Royal Assent, and gives effect to the aforementioned modifications to the 2013 Act.

That is not to say that the PSA 2013 is a "dead letter". In a recent speech on the 12th June 2015, the Hon Peter Dunne, Associate Minister of Health, said:[127]

"The Government enacted the Psychoactive Substances Act 2013 to regulate psychoactive substances because illicit street drugs and untested psychoactive products were already widely available and long-term effective prohibition of all psychoactive products was proving impractical. Internationally, no other jurisdiction has found a more effective approach to dealing with new psychoactive substances. Some countries have introduced product bans, however all that has achieved has been to drive the market underground and out of sight of the Police and health authorities, to the detriment of those adversely affected by their use. The Government's view is that it is preferable to regulate such substances and to permit access to low-risk products rather than drive potentially more harmful products underground. The Act specifies that if a substance is psychoactive, it must be approved before it may be sold and that it pose no more than a low risk of harm to the user. The criteria for product approvals are very stringent, similar to the pre-market approval regime for medicines. Currently there are no approved psychoactive products and it is likely to be some time before any are approved, due mainly to the prohibition on animal testing for psychoactive substances……"

On the 5th June 2015, the Psychoactive Substances Regulatory Authority (NZ) posted on its website that The Psychoactive Substances Retail Regulations "that will allow for the licensing of retail premises for the sale of psychoactive substances, planned for June 2015, are now due to come into force in November 2015":[128]

"This will give territorial authorities more time to adopt Local Approved Product Policies (LAPPs). Territorial authorities are able to make policies that describe where products can be sold within their area. This includes highlighting specific areas that products must be sold in, excluding areas where they cannot be sold, and minimum distances between retailers, and between retailers and sensitive sites. It is preferable for as many LAPPS as possible to be in place prior to the retail regulations coming into force. A LAPP will provide direction to retail licence applicants and the Psychoactive Substances Regulatory Authority about the locations where psychoactive products may be sold in a particular area. If there is no LAPP in place then the Authority will not have the benefit of the community's view when considering retail licence applications. As of 1 June 2015, around 37 local authorities have adopted LAPPs for the sale of psychoactive substances. Retailers licensed by the Authority will only be able to sell psychoactive products approved by the Authority."

Notes and References

1. Notably, the three main UN Conventions, namely, the 1961 Single Convention on Narcotic Drugs, as amended by the 1972 Protocol; the Convention on Psychotropic Substances of 1971, and the United Nations Convention against Illicit Traffic in Narcotic Drugs and Psychotropic Substances of 1988.
2. Much has been written about this issue, but see (as examples), '*New Psychoactive Substances in England: A review of the evidence*',

Giles Stephenson and Anna Richardson, Home Office, Crime and Policing Analysis Unit (Home Office Science, October 2014); *'New Psychoactive Substances'*, Sarah Barber, Briefing Paper Number CBP07215 (4 June 2015, UK House of Commons Library); *'New Psychoactive Substances Review Report of the Expert Panel'* (September 2014); *'Review of the current Legal Framework available to Govern the Sale and Supply of New Psychoactive Substances'*, The New Psychoactive Substances Expert Review Group (Scottish Government, February 2015); *'Consideration of the Novel Psychoactive Substances ('Legal Highs')'*, Advisory Council on the Misuse of Drugs (October 2011); *'Global Synthetic Drugs Assessment'* United Nations Office on Drugs and Crime (2014); *'Towards a Safer Drug Policy: Challenges and Opportunities arising from "legal highs"'*, UK All-Party Parliamentary Group for Drug Policy Reform (2013); *'Law Reform Issues Regarding Synthetic Drugs'*, Legal Affairs Committee Report 1/55, Legislative Assembly of New South Wales (May 2013); *'Perspectives on Drugs. Controlling new psychoactive substances'*, European Monitoring Centre for Drugs and Addiction (2013); House of Commons, Home Affairs Committee, *'Psychoactive substances'*, First Report of Session 2015–16, October 2015.

3. The 1961 Single Convention on Narcotic Drugs, as amended by the 1972 Protocol; the Convention on Psychotropic Substances of 1971, and the United Nations Convention against Illicit Traffic in Narcotic Drugs and Psychotropic Substances of 1988.

4. The expression "New Psychoactive Substances" has been adopted by the Commission on Narcotic Drugs (resolution 55/1; 16 March 2012; see the World Drug Report (UNODC; 2013, p. 63); and by the European Monitoring Centre for Drugs and Drug Addiction (EMCDDA): and see 'Responding to new psychoactive substances', Drugs in Focus, No.22, 2011; 'Perspectives on Drugs: Controlling new psychoactive substances', EMCDDA, 2013, p. 3. See also L. A. King and R. Sedefov, Early-Warning System on New Psychoactive Substances: Operating Guidelines, EMCDDA (2007)). It is not an expression that has yet appeared in a UK statute.

5. 'Consideration of the Novel Psychoactive Substances ("Legal Highs")'; ACMD; and note the letter of the ACMD to the UK Home Secretary (July 13, 2015): https://www.gov.uk/government/uploads/system/uploads/attachment_data/file/444670/3-7-15_ACMD_letter_on_PSB.pdf.

6. The expressions "new psychoactive substance", "new narcotic drug", and "new psychotropic drug" have been defined for the

purposes of Council Decision 2005/387/JHA as follows: (a) 'new psychoactive substance' means a new narcotic drug or a new psychotropic drug in pure form or in a preparation; (b) 'new narcotic drug' means a substance in pure form or in a preparation, that has not been scheduled under the 1961 United Nations Single Convention on Narcotic Drugs, and that may pose a threat to public health comparable to the substances listed in Schedule I, II or IV; (c) 'new psychotropic drug' means a substance in pure form or in a preparation that has not been scheduled under the 1971 United Nations Convention on Psychotropic Substances, and that may pose a threat to public health comparable to the substances listed in Schedule I, II, III or IV.

7. Generic definitions also appear, but these are not readily understood by lay person, and their meaning can be contentious.

8. Emphasis added. Section 1(2) specified one duty of the Advisory Council on the Misuse of Drugs (created under the Act) when offering guidance to the Government.

9. From the 15th November 2011.

10. Introduced by the Police Reform and Social Responsibility Act 2011 (s.151, Schd.17). See R. Fortson '*Misuse of Drugs and Drug Trafficking Offences*' (para.1-090; Sweet & Maxwell, 2012, 6th ed).

11. There is no reference here to effects "sufficient to constitute a social problem" – contrast with s.1(2) MDA.

12. See section 5(2A) of the MDA, inserted by section 152, Sch.17 of the PRSRA.

13. The Misuse of Drugs Act 1971 (Temporary Class Drug) Orders SI 2012 No.980; SI 2013 No.1294, and SI 2015 No.1027, and SI 2015 No.1396. The latter treated the following as "controlled drugs" (other than simple possession): 3,4-Dichloromethylphenidate (3,4-DCMP); 4-Methylmethylphenidate; Ethylnaphthidate; Ethylphenidate; Isopropylphenidate (IPP or IPPD); Methyl-naphthidate (HDMP-28); Propylphenidate; as well as any stereoisomeric form of the aforementioned substances; or any salt of the foregoing; or any preparation (or other product) containing any of the aforementioned substances: see Misuse of Drugs Act 1971 (Temporary Class Drug) (No. 2) Order 2015 (SI 2015 No. 1396).

14. And the Medicines for Human Use (Marketing Authorisations Etc) Regulations 1994 (SI 1994/3144).

15. The 2012 Regulations came into force on the 14th August 2012. The regulations largely consolidate the pre-existing position.

The 1994 Regulations were revoked in their entirety (along with their amending, and related, regulations). Large parts of the Medicines Act 1968 were repealed. The Regulations 2012 do not provide a complete code in relation to "medicinal products" and a number of provisions enacted under the MA 1968 survive. The 2012 Regulations have been significantly amended.

16. As then defined by s.130 (as originally enacted) of the Medicines Act 1968. At Sheffield Crown Court on Monday 13 July 2009, Patrick Wilson pleaded guilty to the importation of an unlicensed medicinal product (BZP), contrary to s.8(2) of the Medicines Act 1968; wholesale dealing of an unlicensed medicinal product (BZP), contrary to s.8(2) of the Medicines Act 1968; placing unlicensed medicinal product (BZP) on the market contrary to para 1 sched 3, S.I. 3144/1994; possession of an unlicensed medicinal product (BZP) knowing it will be placed on the market contrary to para.2 sched.3 S.I. 3144/1994. Wilson was sentenced to a two year conditional discharge for each offence. An order was made for the forfeiture and destruction of all BZP seized by the Medicines and Healthcare products Regulatory Agency (MHRA).

17. R. Fortson, Are We Misusing The MDA 1971?, *Criminal Law Today*, 2011, (2). Sweet & Maxwell Ltd.

18. See, for example, *R v Henderson* [2011] EWCA Crim 2035. BZP was designated by the MHRA as a 'medicinal product' for the purposes of the Medicines Act 1968.

19. Mr Roger Birch: New Law Journal, 159 NLJ 1695.

20. From the 14th August 2012, section 130(1) is substituted by SI 2012/1916, reg 348, Sch 34, Pt 1, paras 1, 31(a).

21. The Human Medicines Regulations 2012 definition of "medicinal product" differs significantly from the original definition of that term as it appeared in the now repealed section 130(1), (2), MA 1968, which read: *"(1) Subject to the following provisions of this section, in this Act "medicinal product" means any substance or article (not being an instrument, apparatus or appliance) which is manufactured, sold, supplied, imported or exported for use wholly or mainly in either or both of the following ways, that is to say – (a) use by being administered to one or more human beings for a medicinal purpose; (b) use, in circumstances to which this paragraph applies, as an ingredient in the preparation of a substance or article which is to be administered to one or more human beings for a medicinal purpose. (2) In this Act "a medicinal purpose" means any one or more of the following purposes, that is to say – (a) treating or*

preventing disease; (b) diagnosing disease or ascertaining the existence, degree or extent of a physiological condition; (c) contraception; (d) inducing anaesthesia; (e) otherwise preventing or interfering with the normal operation of a physiological function, whether permanently or temporarily, and whether by way of terminating, reducing or postponing, or increasing or accelerating, the operation of that function or in any other way".

22. See regulation 2(1) of the 2012 Regulations.
23. See Hecht-Pharma GmbH v Staatliches Gewerbeaufiichtsamt Liineburg [2009] 2 Crim. L.R. 23.
24. Hansard, Public Bill Committee, Psychoactive Substances Bill, October 27, 2015, HC., col.33.
25. The *Cases of D and G* [2014] PTSR 1217, para.29.
26. See *Van Bennekom* [1983] ECR 3883, and *Cantoni v France* ('Ter Voort' case; Judgment 28 October 1992 (C-219/91 [1992] ECR 5485)).
27. [2010] EWCA Crim. 2247.
28. See, now, regulation 2 of the Human Medicines Regulations 2012.
29. 15th September 2011, Court of Appeal, Island of Guernsey (The Hon Michael J Beloff QC Presiding, Michael Scott Jones QC, and Clare Montgomery QC).
30. Judgment: paragraph 50.
31. Consideration of the Novel Psychoactive Substances, 'Legal Highs'; October 2011; para.7.19.
32. Human Medicines Regulations 2012: Regulation 46.
33. See "General interpretation": article 8(1) of the Human Medicines Regulations 2012.
34. Granted by the licensing authority under Part 6 of the Human Medicines Regulations 2012: see "General interpretation": article 8(1) of the Human Medicines Regulations 2012.
35. Granted by the licensing authority under the 2012 Regulations: see "General interpretation": article 8(1) of the Human Medicines Regulations 2012.
36. That is to say, authorisation granted by the licencing authority under Part 8 of the Human Medicines Regulations 2012: see "General interpretation": article 8(1) of the Human Medicines Regulations 2012.
37. By regulation 47(2), the penalties for a breach of regulation 46 is, (a) on summary conviction to a fine not exceeding the statutory maximum; or (b) on conviction on indictment to a fine, to imprisonment not exceeding two years or to both.

38. Note regulation 46(9). By regulation 46(2), a person may not sell or supply, or offer to sell or supply, a medicinal product to a person within the EEA, otherwise than in accordance with the terms of (a) a marketing authorisation; (b) a certificate of registration; (c) a traditional herbal registration; or (d) an Article 126a authorisation. By regulation 46(9), "Paragraphs (1) and (2) do not apply to the sale, supply, or offer for sale or supply, of a medicinal product to a person outside the European Economic Area".

39. Regulation 46(4).

40. See regulations 18(1) and 34(1).

41. A contravention of section 62 is an offence contrary to section 67(2) of the MA 1968. By section 67(4), the penalties are: (a) on summary conviction, to a fine not exceeding [the prescribed sum]; (b) on conviction on indictment, to a fine or to imprisonment for a term not exceeding two years or to both.

42. By section 67(4), MA 1968, the penalties are: (a) on summary conviction, to a fine not exceeding [the prescribed sum]; (b) on conviction on indictment, to a fine or to imprisonment for a term not exceeding two years or to both.

43. See section 67(3). By section 67(4), MA 1968, the penalties are: (a) on summary conviction, to a fine not exceeding [the prescribed sum]; (b) on conviction on indictment, to a fine or to imprisonment for a term not exceeding two years or to both.

44. Section 63 of the Medicines Act 1968. By section 67(4), the penalties are: (a) on summary conviction, to a fine not exceeding [the prescribed sum]; (b) on conviction on indictment, to a fine or to imprisonment for a term not exceeding two years or to both.

45. Regulation 8(1) of the Human Medicines Regulations 2012 defines an "active substance" to mean "any substance or mixture of substances intended to be used in the manufacture of a medicinal product and that, when used in its production, becomes an active ingredient of that product intended to exert a pharmacological, immunological or metabolic action with a view to re-storing, correcting or modifying physiological functions or to make a medical diagnosis".

46. Namely, Regulation 45N and the requirements in Regulation 45O of the Human Medicines Regulations 2012: see regulation 45M.

47. Regulation 45U of the Human Medicines Regulations 2012.

48. "Consideration of the Novel Psychoactive Substances ('Legal Highs')" (Advisory Council on the Misuse of Drugs: October 2011).

49. "Guidance for local authorities on taking action against 'head shops' selling new psychoactive substances" (Home Office); updated in 2015 ("Guidance for local authorities on taking action against 'head shops' selling new psychoactive substances – working with local partners").

50. The Import, Export and Customs (Defence) Act 1939, and the Import of Goods (Control) Order 1954.

51. Made pursuant to powers conferred by section 1 of the Import, Export and Customs Powers (Defence) Act, 1939).

52. SI 1954/23; "Subject to the provisions of this Order, all goods are prohibited to be imported into the United Kingdom".

53. 7-bromo-5-(2-chlorophenyl)-1,3-dihydro-2H-1,4-benzodiazepin-2-one.

54. The regulations were made pursuant to section 2(2) of the European Communities Act 1972, and came into force (with some qualification) in late 2005; as amended by SI 2015 No. 1630.

55. Presumably under regulation 20 and regulation 12 of the 2005 Regulations.

56. http://www.norfolk.gov.uk/Community_and_living/Consumer_advice_and_protection/Enforcement/index.htm.

57. http://www.belfastlive.co.uk/news/belfast-news/trio-banned-selling-legal-highs-9552381.

58. The 2008 Regulations were made pursuant to section 2(2) of the European Communities Act 1972, and came into force on the 26th May 2008. A number of provisions in the Trade Descriptions Act 1968 were repealed by Reg 30 and Schedule 2. The regulations have been amended by SI 2015 No. 1630.

59. The expression 'trader' is defined by reg.2, and it is to be broadly construed: see *R v Scottish & Southern Energy* [2012] EWCA Crim 539. The Regulations have been amended by SI SI 2014/870, and by the Consumer Rights Act 2015.

60. By reg.2, a 'commercial practice' is "....any act, omission, course of conduct, representation or commercial communication (including advertising and marketing) by a trader, which is directly connected with the promotion, sale or supply of a product to or from consumers, whether occurring before, during or after a commercial transaction (if any) in relation to a product..." In *R v X* [2013] EWCA Crim 818, the Court of Appeal (Criminal Division) held that a "commercial practice" can be derived from a single incident. It will depend on the circumstances.

61. See regulation 3.

62. Regulation 2 (as amended by SI 2014/870) defines a "transactional decision" as "any decision taken by a consumer, whether it is to act or to refrain from acting, concerning–(a) whether, how and on what terms to purchase, make payment in whole or in part for, retain or dispose of a product; or (b) whether, how and on what terms to exercise a contractual right in relation to a product (*but the application of this definition to regulations 5 and 7 as they apply for the purposes of Part 4A is subject to regulation 27B(2)* [italicised words inserted by SI 2014/870]).

63. A number of provisions in the Trade Descriptions Act 1968 were repealed by Reg 30 and Schedule 2, of the Consumer Protection from Unfair Trading Regulations 2008, namely, section 1(1) (prohibition of false trade descriptions); section 5 (trade descriptions used in advertisements); section 6 (offer to supply); sections 7-10 (power to define terms and to require display etc of information); sections 13 to 15 (false representations or statements concerning services etc); sections 21(1) and (2) (accessories to offences committed abroad); section 22 (restrictions on institution of proceedings and admission of evidence); section 32 (power to exempt goods sold for export, etc); section 37 (market research experiments); section 39(2) (interpretation). By para.9 of schedule 2, of the 2008 Regs, section 1(2) of the TDA 1968 was substituted so that sections 2-4 of the TDA shall have effect for the interpretation of expressions used in the TDA 1968. The TDA 1968 has been further amended by the Consumer Rights Act 2015.

64. There is an important limitation in the application of the TDA 1968 where the NPS is a "medicinal product" for the purposes of the Human Medicines Regulations 2012 and the Medicines Act 1968. This relates to requirements under the Human Medicines Regulations 2012, "for packaging and package leaflets relating to medicinal products" (Chapter 1 to Part 13). Regulations 268 and 269 of the HM Regs 2012 create offences relating to breaches of Chapter 1. Accordingly, the TDA 1968 is effectively disapplied in respect of such a breach: see section 2(5)(b) of the TDA 1968 (and see regulation 348 and schedule 34, Part 2, at para 37). The amendment came into force on the 14th August 2012. Presumably, the aim is to prevent unnecessary overlapping of enforcement measures in respect of the same activity.

65. Section 59 of the Serious Crime Act 2007.

66. 73 Cr.App.R. 49, CA.

67. See *Brown* [1994] A.C. 212; *Cato* (1976) 62 Cr. App. R. 41, CA; *McShane* (1977) 66 Cr. App. R. 97, but see the Criminal Law

Revision Committee, 14th Report Offences Against the Person, Cmnd.7844, para.190.

68. *R v. Kennedy* [2007] UKHL 38.

69. [2005] Crim. L.R. 819.

70. See the Press Release: http://europa.eu/rapid/press-release_IP-14-461_en.htm?locale=en; and see the Opinion of the European Economic and Social Committee on the Proposal for a Regulation of the European Parliament and of the Council on new psychoactive substances COM(2013) 619 final – 2013/0305 (COD) COM(2013) 618 final – 2013/0304 (COD); January 2014.

71. A useful paper in relation to the New Zealand legislation following the Psychoactive Substances Act 2013, has been published by C. Wilkins, '*A Critical first assessment of the new-market approval regime for new psychoactive substances (NPS) in New Zealand*'; Addiction, 2014, p. 1. However, as the author fairly concedes, a "pressing task for researchers is to provide empirical evaluations of [different] approaches to inform future policy responses". See also, '*Implementation of the 2013 Psychoactive Substances Act and Mental Health Harms from Synthetic Cannabinoids*': P. Glue, et al, NZMJ 15 May 2015; Vol.128 No.1414, p.13.

72. Report of an Inquiry into new psychoactive substances; All-Party Parliamentary Group for Drug Policy Reform; '*Towards a Safer Drug Policy: Challenges and Opportunities arising from "legal highs"*'; 2013, para.24.

73. See the speech of the Minister of State for the Home Office (Lord Bates); Hansard, 9 Jun 2015: Column 735, HL.

74. Hansard, 9 Jun 2015: Column 736, HL.

75. New Psychoactive Substances Review: Report of the Expert Panel; September 2014, page 36. See the 'Government response to New Psychoactive Substances Review Expert Panel Report'; 2014.

76. Hansard, 9 Jun 2015: Column 735, HL.

77. The ACMD proposed that this term should be defined to mean "the process of producing a compound by human instigation of at least one chemical reaction".

78. The ACMD proposed that this term should be defined to mean "any chemical species that is formed when two or more atoms join together chemically".

79. Amendment 43: see Hansard, October 27, 2015, Public Bill Committee, col.15.

80. Mike Penning MP (Minister for Policing, Crime and Criminal Justice); Hansard, Public Bill Committee, October 27, 2015; col.16.

81. Hansard, October 27, 2015, Public Bill Committee, col.18.
82. Hansard, October 27, 2015, Public Bill Committee, col.18.
83. Letter, ACMD to the Home Secretary (2 July 2015): "It is almost impossible to list all possible desirable exemptions under the Bill. As drafted, the Bill may now include substances that are benign or even helpful to people including evidence-based herbal remedies that are not included on the current exemption list".
84. Hansard, October 27, 2015, Public Bill Committee, col.51; House of Commons, Home Affairs Committee, ' Psychoactive Substances", First Report of Session 2015–16, October 2015, para.41.
85. Hansard, October 27, 2015, Public Bill Committee, col.53.
86. Within the meaning of the Tobacco Products Duty Act 1979.
87. The Government said, at the Committee stage of the PSB, that it would give the matter further attention when the PSB was considered on Report: Hansard, October 27, 2015, Public Bill Committee, col.34.
88. Letter, 26th May 2015.
89. Letter, ACMD to the Home Secretary (2nd July 2015): "The only definitive way of determining psychoactivity is via human experience, which is usually not documented. However, most psychoactive drugs share similar mechanisms of action. The ability of a Novel Psychoactive Substance to target one or other of these mechanisms can be determined by in vitro neurochemical tests, however, such proxy measures may not stand up in a court of law.".
90. House of Commons, Home Affairs Committee; Oral evidence, 15th September 2015, HAC, Q.126.
91. See schedule 5, para.2, of the Psychoactive Substances Act 2015.
92. Unless the offence was committed before the commencement of s.154(1) of the Criminal Justice Act 2003.
93. Clause 56 of the Psychoactive Substances Act.
94. As defined by the Human Medicines Regulations 2012 (S.I. 2012 No.1916) (regulation 8).
95. Defined by reg.8 of the Human Medicines Regulations 2012.
96. "Approved scientific research" means scientific research carried out by a person who has approval from a "relevant ethics review body" to carry out that research; and "relevant ethics review body" means (a) a research ethics committee recognised or established by the Health Research Authority under Chapter 2 of Part 3 of the Care Act 2014, or (b) a body appointed by any of the following for the purpose of assessing the ethics of research

involving individuals, namely, (i) the Secretary of State, the Scottish Ministers, the Welsh Ministers, or a Northern Ireland department; (ii) a relevant NHS body; (iii) a body that is a Research Council for the purposes of the Science and Technology Act 1965; (iv) an institution that is a research institution for the purposes of Chapter 4A of Part 7 of the Income Tax (Earnings and Pensions) Act 2003 (see section 457 of that Act); (v) a charity which has as its charitable purpose (or one of its charitable purposes) the advancement of health or the saving of lives". The expressions "charity" and "relevant NHS body" are also defined in schedule to the 2015 Act.

97. In Williams [2011] EWCA Crim 232, the Court of Appeal held as a correct statement of the law for the purposes of the MDA 1971 was that the addition of adulterants or bulking agents can amount to the production of a controlled drug. However, the definition of a "controlled drug" is wide.

98. Section 37(1) of the Misuse of Drugs Act 1971.

99. See R v. Maginnis [1987] A.C. 303, HL.

100. See section 8(4), PSA 2015; and s.5 of the Customs and Excise Management Act 1979.

101. See section 18(9), PSA 2015; and section 20(9), PSA 2015.

102. Despite the persistent pleas of the writer to the Government, the latter has retained in the Act "an offence of inciting a person to commit an offence under any of sections 4 to 8" and "an offence of aiding, abetting, counselling or procuring the commission of an offence under any of sections 4 to 8". The Act does not create an offence of "incitement" and Parliament abolished common law incitement under the Serious Crime Act 2007 (s.59). There is no offence of "aiding and abetting" (etc) as such. These concepts are ones that give rise to accessorial liability by persons acting as secondary parties to a crime committed.

103. See SI 2014 No. 2590.

104. Part 4, chapter 1 of the ABCPA 2014.

105. Part 4, chapter 2 of the ABCPA 2014.

106. Part 4, chapter 3 of the ABCPA 2014.

107. That is to say, retail outlets that sell psychoactive substances.

108. Again, despite the persistent pleas of the writer to the Government, these continue to include "an offence of inciting a person to commit an offence under any of sections 4 to 8" and "an offence of aiding, abetting, counselling or procuring the commission of an offence under any of sections 4 to 8". The Act does not create an offence of "incitement" and Parliament abolished

common law incitement under the Serious Crime Act 2007 (s.59). There is no offence of "aiding and abetting" (etc) as such.

109. See the Press Release: http://europa.eu/rapid/press-release_IP-13-837_en.htm.

110. See the Press Release: http://europa.eu/rapid/press-release_IP-14-461_en.htm?locale=en; and see the Opinion of the European Economic and Social Committee on the Proposal for a Regulation of the European Parliament and of the Council on new psychoactive substances COM(2013) 619 final – 2013/0305 (COD) COM(2013) 618 final – 2013/0304 (COD); January 2014.

111. See The European Commission's proposal for a "Regulation of the European Parliament and of the Council" (2013/0305), 17th September 2013.

112. It is submitted that "evidenced by" the matters stated, is not the same as imposing a closed set of mandatory conditions.

113. After consultation with the Minister for Health and Children and such other Minister of the Government as he or she considers appropriate.

114. A person who is guilty of an offence under this Act (other than an offence under section 15) shall be liable – (a) on summary conviction, to a fine not exceeding €5,000 or imprisonment for a term not exceeding 12 months or both, or (b) on conviction on indictment, to a fine or imprisonment for a term not exceeding 5 years or both.

115. See http://www.bbc.co.uk/news/uk-33226526; and http://www.bbc.co.uk/news/world-europe-33128818.

116. "Manufacture"... "(a) means to make up, prepare, produce, or process the substance or product for the purpose of sale; and (b) includes packaging the substance or product for the purpose of sale": section 8.

117. "Sell" is defined by section 8 of the PSA 2013 as "sell includes every method of disposition for valuable consideration, for example, (a) bartering: (b) offering or attempting to sell or having in possession for sale, or exposing, sending, or delivering for sale, or causing or allowing to be sold, offered, or exposed for sale: (c) retailing: (d) wholesaling".

118. "Supply"... "(a) includes distribute or give; but (b) does not include sell": section 8.

119. "Possess" includes a psychoactive substance "that is subject to a person' s control but that is in the custody of another person": section 8.

120. An "approved product" is a "psychoactive product approved by the Authority under section 37" of the Act: section 8.
121. "Psychoactive effect" is defined by section 8 of the 2013 Act to mean "in relation to an individual who is using or has used a psychoactive substance, means the effect of the substance on the individual' s mind".
122. That is to say, controlled under the Misuse of Drugs Act 1975 (NZ).
123. That is to say, "...within the meaning of section 3 of the Medicines Act 1981 [NZ] or a related product within the meaning of section 94 of that Act" (section 9(3)(c)).
124. That is to say, "within the meaning of section 2(1) of the Medicines Act 1981 [NZ]" (section 9(3)(d)).
125. Regulatory Impact Statement (c. May 2014).
126. See http://psychoactives.health.govt.nz/psychoactive-substances-act-2013/background-act-and-regime posted by the Psychoactive Substances Regulatory Authority.
127. NZMA General Practitioners Conference, The Rotorua Energy Events Centre, New Zealand, http://www.nznewsuk.co.uk/news/?id=61843&story=NZMA-General-Practitioners-Conference-Speech.
128. http://psychoactives.health.govt.nz/psychoactive-substances-retail-regulations.

16 Scheduling of Drugs in the United States

Jeffery Hackett

Office of Chief Medical Examiner, San Francisco, CA, USA
Email: hackett.jeffery@gmail.com

16.1 Scheduling Drugs and Pharmaceuticals in the USA

The scheduling of drugs in the USA was enacted by the Congress of the USA as part of the nation's public health service laws. This scheduling was brought about for the prevention of drug/pharmaceutical abuse and dependence, to assist with treatment and rehabilitation of drug abusers and drug dependent individuals. These schedules also help and assist with the strengthening of existing law enforcement authority in the important area of drug abuse.

The US Congress, by enacting legislation, created five classifications (better known as Schedules I through V) which contain varying qualifications for a particular substance to be included. In the USA there are two main agencies appointed at the federal level to determine which substances are added to or removed from the various schedules. These agencies are the Drug Enforcement Agency (DEA) and the United States Food and Drug Agency (USDA). The initial statute passed by the US Congress created the original listing of substances; it has placed other substances in the schedules through additional US legislation such as the Hillory J. Farias and Samantha Reid Date-Rape Prevention Act of

Forensic Toxicology: Drug Use and Misuse
Edited by Susannah Davies, Atholl Johnston and David Holt
© The Royal Society of Chemistry 2016
Published by the Royal Society of Chemistry, www.rsc.org

2000, which placed γ-hydroxybutyrate (GHB) in Schedule I. Classification decisions are required to be made on criteria including potential for abuse, currently accepted medical use in treatment in the US, and international treaties.

Historically, the US first outlawed addictive drugs in the early 1900s and the International Opium Convention was employed to lead international agreements regulating trade in pharmaceuticals. The Food and Drugs Act (enacted in 1906) was the first of over 200 laws concerning public health and consumer protections. Others were the Federal Food, Drug, and Cosmetic Act (1938), and the Kefauver Harris Amendment of 1962. In 1969, President Richard Nixon announced that Attorney General Mitchell was preparing a comprehensive new measure to more effectively meet the narcotic and dangerous drug problems at the federal level by combining all existing federal laws into a single new statute. His initiative was the Controlled Substance Act (CSA). The CSA worked to amalgamate the existing federal drug laws but it also changed the nature of federal drug law policies, expanded the scope of federal drug laws and expanded federal law enforcement as pertaining to controlled substances. Part F of the Comprehensive Drug Abuse Prevention and Control Act of 1970 established the National Commission on Marijuana and Drug Abuse. This is commonly referred to as the Shafer Commission after its chairman (Raymond P. Shafer). The remit of this commission was to study cannabis abuse in the US. During his presentation of the commission's first report to Congress, the Shafer Commission recommended the decriminalization of marijuana in small amounts, indicating that criminal law is too harsh a tool to apply to personal possession even in the effort to discourage use. It implies an overwhelming indictment of the behavior which we believe is not appropriate. The actual and potential harm of use of the drug is not great enough to justify intrusion by the criminal law into private behavior.

The CSA is broken down into two sub-chapters. The first sub-chapter I defines Schedules I–V. It also lists chemicals used in the manufacture of controlled substances, and differentiates lawful from unlawful manufacturing, distribution, and possession of controlled substances. It defines the possession of Schedule I drugs for personal use. Sub-chapter I also specifies the dollar amounts of fines and durations of prison terms for violations. Sub-chapter 2 describes the laws for exportation and importation of controlled substances, again specifying penalties that can be applied by courts for violations of the act.

Since its enactment in 1970, the CSA, which covers the scheduling of pharmaceuticals in the USA, has been amended several times including:

- The Medical Device Amendments of 1976.
- The Psychotropic Substances Act of 1978 added provisions implementing the Convention on Psychotropic Substances.
- The Controlled Substances Penalties Amendments Act of 1984.
- The Chemical Diversion and Trafficking Act of 1988 (implemented August 1, 1989 as Article 12) added provisions implementing the United Nations Convention Against Illicit Traffic in Narcotic Drugs and Psychotropic Substances that went into force on November 11, 1990.
- The Anabolic Steroids Act, passed as part of the Crime Control Act of 1990, which placed anabolic steroids into Schedule III[12]:30.
- The Domestic Chemical Diversion and Control Act of 1993 (effective on April 16, 1994) in response to the methamphetamine trafficking.
- The Federal Analog Act (which was brought in to cover the emergence of synthetic cannabinoids such as JWH 018, 073, AM 2201 and the synthetic cathinones such mephedrone, butylone, *etc.*).
- The Ryan Haight Online Pharmacy Consumer Protection Act of 2008.
- The Electronic Prescriptions for Controlled Substances (EPCS) 2010.

Placing a drug or other substance in a certain schedule category or removing it from a certain schedule category is primarily based on the provisions laid out in Title 21 of the United States Code (USC) Controlled Substances Act sections: §§ 801, 801a, 802, 811, 812, 813, and 814. The particular schedule requires finding and specifying the "potential for abuse" before a substance can be placed within that particular schedule category.

Schedule I drugs fulfill the following conditions. The drug or other substance has a high potential for abuse. The drug or other substance has no currently accepted medical use in treatment in the US. There is a lack of accepted safety for use of the drug or other substance under medical supervision.

16.2   Drugs that are Categorized in Schedule I

- α-Methyltryptamine, an anti-depressant from the tryptamine group of compounds. It was originally developed in the former Soviet Union and marketed under the brand name Indopan.

- Dimethyltryptamines, naturally occurring psychedelic drugs that are widespread throughout the plant kingdom and endogenous to the human body; they are the main psychoactive constituent in the psychedelic South American brewed liquor ayahuasca, for which some shamanic practitioners are granted exemption from Schedule I status on the grounds of religious freedom.
- Mescaline, a naturally occurring psychedelic drug and the main psychoactive constituent of peyote (*L. williamsii*), San Pedro cactus (*E. pachanoi*), and Peruvian torch cactus (*E. peruviana*).
- LSD (lysergic acid diethylamide), a semi-synthetic psychedelic drug famous for its involvement in the counterculture of the 1960s.
- Peyote *(L. williamsii)*, a cactus growing naturally mainly located in northeastern Mexico; this is one of the few plants specifically scheduled, with a narrow exception to its legal status for religious use by members of the Native American Church.
- Psilocybin and psilocin, naturally occurring psychedelic drugs and the main psychoactive constituents of psilocybin mushrooms.
- Benzylpiperazines, a group of synthetic stimulants once sold as potential designer drugs. They have since been shown to be associated with an increase in seizures if taken alone Although the effects of benzylpiperazines are not as potent as the psychoactive drug MDMA, they can produce neuroadaptions which can cause an increase in the potential for abuse of this drug.
- MDMA (also referred to as "ecstasy"), a stimulant, psychedelic, and entactogenic drug which was initially gained attention in psychedelic therapy as a treatment for post-traumatic stress disorder. The term entactogen gives rise to the commonly known term "loved up". The medical community originally agreed upon placing it as a Schedule III substance, but the government denied this proposal, despite US court rulings by the DEA's administrative law judge that placing MDMA in Schedule I was illegal. It was temporarily unscheduled after the first administrative hearing from December 22, 1987 to July 1, 1988.
- Etorphine, a semi-synthetic opioid possessing an analgesic potency approximately 1000–3000 times the potency of morphine.
- Heroin (diacetylmorphine), which is used in some European countries as a potent pain reliever in terminal cancer patients, and as a second option, after morphine (it is twice as potent as morphine).
- GHB, a general anesthetic and treatment for narcolepsy-cataplexy and alcohol withdrawal with minimal side-effects and controlled

action, but a limited safe dosage range. It was placed in Schedule I in March 2000 after widespread recreational use led to increased emergency room visits, hospitalizations, and deaths. This drug is also listed in Schedule III for limited uses, under the trademark Xyrem.

- Marijuana and related cannabinoids. Pure $(-)$-*trans*-Δ^9-tetrahydrocannabinol (THC) is also listed in Schedule III for limited uses, under the trademark Marinol. Following several ballot measures, certain states in the USA, such as Colorado, Washington, and Oregon, permitted allowances for recreational and medical use of marijuana and/or have decriminalized possession of small amounts of marijuana. These measures operate only on state laws, and have no effect on the existing federal law. Despite such ballot measures, and multiple studies showing medicinal benefits, marijuana nevertheless remains on Schedule I, effective across all US states and its territories.
- Methaqualone, a sedative that was previously used for similar purposes as barbiturates, until it was rescheduled. It was marketed under such names as Quaalude, Sopor, and Mandrax. This drug has now been removed from the general legitimate distribution process in the US.

No prescriptions are allowed to be written for Schedule I substances, and such substances are subject to production quotas which the DEA imposes. Moreover, it is illegal, and indeed a Class 1 federal felony, even to conduct any otherwise legitimate scientific research of any kind on Schedule I substances.

When a drug or pharmaceutical is listed in Schedule I, and it is undisputed that such drug has no currently accepted medical use in treatment in the USA and a lack of accepted safety for use under medical supervision, and it is further undisputed that the drug has at least some potential for abuse sufficient to warrant control under the CSA, the drug must remain in Schedule I. In such circumstances, placement of the drug in Schedules II through V would conflict with the CSA since such drug would not meet the criterion of a currently accepted medical use in treatment in the US.

16.3 Schedule II

Drugs listed in Schedule II are those that fulfill the following conditions. The drug or other substances have a high potential for abuse.

The drug or other substances have currently accepted medical use in treatment in the US, or currently accepted medical use with severe restrictions. Abuse of the drug or other substances may lead to severe psychological or physical dependence. These drugs vary in their potency, *e.g.* fentanyl is approximately 80 times as potent as morphine (heroin has a potency approximately four times that of morphine). More significantly, they vary in nature, pharmacology, and CSA scheduling and have a weak relationship. Because these compounds are in the main prescription drugs, and refills of prescriptions for Schedule II substances are not allowed, it can be burdensome to both the practitioner and the patient if the substances are to be used on a long-term basis. To provide relief, in 2007, Section 21 of the Code of Federal regulations (C.F.R) 1306.12 was amended (at 72 FR 64921) to allow practitioners to write up to three prescriptions at once, to provide up to a 90 day supply, specifying on each of the prescriptions the earliest date on which it may be filled.

16.3.1   Drugs in Schedule II

- Morphine, hydromorphone (semi-synthetic opioid; active ingredient in Dilaudid), palladone, pure codeine and any drug for non-parenteral administration containing the equivalent of more than 90 mg of codeine per dosage unit.
- Oxycodone (semi-synthetic opioid; active ingredient in Percocet, OxyContin, and Percodan).
- Oxymorphone (semi-synthetic opioid, active ingredient in Opana), hydrocodone in any formulation as of October 2014 (examples include Vicodin, Norco, Tussionex). Prior to this date, formulations containing hydrocodone and over-the-counter non-steroidal anti-inflammatory drugs (better known as NSAIDs) such as acetaminophen (paracetamol outside of USA) and ibuprofen were placed into Schedule III.
- Opium, opium tinctures (laudanum): treatment as a potent antidiarrheal.
- Pethidine (also marketed as Meperidine, Demerol).
- Fentanyl and most other strong pure opioid agonists, *i.e.* levorphanol, opium.
- Tapentadol (Nucynta) A drug with mixed opioid agonist and norepinephrine re-uptake inhibitor activity.
- Methadone: treatment of heroin addiction, extreme chronic pain such as back pain.

- Nabilone (Cesamet), a synthetic cannabinoid. This compound is an analogue of dronabinol (Marinol) which is listed under Schedule III.
- Amphetamine (originally placed in Schedule III, but reassigned to Schedule II in 1971), adderall, dextroamphetamine (Dexedrine), lisdexamfetamine (Vyvanse), methamphetamine for the treatment of ADHD, narcolepsy.
- Methylphenidate (Ritalin, Concerta), dexmethylphenidate (Focalin): treatment of ADHD, narcolepsy.
- Cocaine (used as a topical anesthetic). It also has had legitimate use in ocular surgery but this is now being replaced.
- Secobarbital (Seconal) and short-acting barbiturates, such as pentobarbital, nembutal. phencyclidine (PCP).
- Pure diphenoxylate.

Except when dispensed directly by a practitioner, other than a pharmacist, to an ultimate user, no controlled substance in Schedule II, which is a prescription drug as determined under the Federal Food, Drug, and Cosmetic Act (21 USC 301 *et seq.*), may be dispensed without the written prescription of a practitioner, except in emergency situations.

16.4 Schedule III

Compounds that are categorized as Schedule III drugs are those that meet the following conditions. The drug or other substance has a potential for abuse less than the drugs or other substances in Schedules I and II. The drug or other substance has a currently accepted medical use in treatment in the US. The potential for abuse of the drug or other substance may lead to moderate or low physical dependence or high psychological dependence.

16.4.1 Drugs in Schedule III

- Buprenorphine (semi-synthetic opioid: active in Suboxone, Subutex), dihydrocodeine when compounded with other substances such as acetaminophen (paracetamol) at prescribed dosages and concentrations (for liquid formulations). Paregoric, an antidiarrheal and anti-tussive, which contains opium combined with camphor (which makes it less addiction-prone than laudanum, which is in Schedule II).

- Phendimetrazine tartrate and benzphetamine [as the hydrochloride (Didrex)], which are both classified as stimulants designed for use in the treatment of anorexia.
- Xyrem, a preparation of GHB used to treat narcolepsy; is in Schedule III but with a restricted distribution system.
- Marinol, synthetically prepared tetrahydrocannabinol (officially referred to by the name, dronabinol), used to treat nausea and vomiting caused by chemotherapy, as well as appetite loss caused by AIDS.
- Anabolic steroids (including prohormones such as androstenedione); the specific end molecule testosterone in many of its forms (Androderm, AndroGel, Testosterone Cypionate and Testosterone Enanthate) are categorized as Schedule III drugs, while low-dose testosterone when compounded with estrogen derivatives have been exempted from scheduling restrictions by the FDA.
- Intermediate-acting barbiturates, such as talbutal or butalbital; fast-acting barbiturates such as secobarbital (Seconal) and pentobarbital (Nembutal), when combined with one or more additional active ingredient(s) not in Schedule II [*e.g.* Carbrital (no longer marketed), a combination of pentobarbital and carbromal].
- Ergine (lysergic acid amide), listed as a sedative but considered by some to be psychedelic. An inefficient precursor to its *N,N*-diethyl analog, LSD, ergine occurs naturally in the seeds of the common garden flowers *Turbina corymbosa*, *Ipomoea tricolor*, and *Argyreia nervosa*.

Conditions for use include those in which, when dispensed directly by a practitioner, other than a pharmacist, to an ultimate user, no controlled substance in Schedule III or IV, which is a prescription drug as determined under the Federal Food, Drug, and Cosmetic Act (21 USC 301 *et seq.*), may be dispensed without a written or oral prescription in conformity with section 503(b) of that Act [21 USC 353 (b)]. Such prescriptions may not be filled or refilled more than six months after the date thereof or be refilled more than five times after the date of the prescription unless renewed by the registered practitioner. A prescription for controlled substances in Schedules III, IV, and V issued by a practitioner, may be communicated either orally, in writing, or by facsimile to the pharmacist, and may be refilled if so authorized on the prescription or personal visit to the appropriate dispensary.

The control measures applied in Schedule III of wholesale distribution is somewhat less stringent than Schedule II drugs. Provisions for emergency situations are less restrictive within the "closed system" of the Controlled Substances Act than for Schedule II, though no schedule has provisions to address circumstances where the closed system is unavailable, nonfunctioning, or otherwise inadequate.

16.5 Schedule IV

Drugs that are categorized in Schedule IV fulfill the following conditions. The drug or other substance has a low potential for abuse relative to the drugs or other substances in Schedule III. The drug or other substance has a currently accepted medical use in treatment in the USA. Abuse of the drug or other substance may lead to limited physical dependence or psychological dependence relative to the drugs or other substances in Schedule III.

16.5.1 Drugs in Schedule IV

- Benzodiazepines, such as alprazolam (Xanax), chlordiazepoxide (Librium), clonazepam (Klonopin), diazepam (Valium), lorazepam (Ativan), temazepam (Restoril) (note that some States require specially coded prescriptions for temazepam). Note also that flunitrazepam (also known as rohypnol) falls into Schedule IV but is not used medically in the USA. Rohypnol is not approved or available for medical use in the US, but it is temporarily controlled in Schedule IV pursuant to a treaty obligation under the 1971 Convention on Psychotropic Substances. At the time flunitrazepam was placed temporarily in Schedule IV (November 5, 1984), there was no evidence of abuse or trafficking of the drug in the US. The benzodiazepine-like so-called Z-drugs, zolpidem (Ambien), zopiclone (Imovane), eszopiclone (Lunesta), and zaleplon (Sonata), also fall into Schedule IV but note zopiclone is not currently commercially available in the USA as a prescribed medication.
- Chloral hydrate, a sedative-hypnotic.
- Long-acting barbiturates such as phenobarbital.
- Some partial agonist opioid analgesics, such as pentazocine (Talwin); tramadol (Ultram).
- The stimulant-like drug modafinil (sold in the US as Provigil) as well as its (R)-enantiomer armodafinil (marketed in the US as Nuvigil).

- Difenoxin, an antidiarrheal drug, such as when combined with atropine (Motofen); difenoxin is 2–3 times more potent than diphenoxylate, the active ingredient in lomotil, which is in Schedule V.
- Carisoprodol (Soma) has become a Schedule IV medication as of January 2012.
- Rohypnol (flunitrazepam) is not approved or available for medical use in the US, but it is temporarily controlled in Schedule IV pursuant to a treaty obligation under the 1971 Convention on Psychotropic Substances.
- Flunitrazepam was placed temporarily in Schedule IV (November 5, 1984); at that time there was no evidence of abuse or trafficking of the drug in the US, but this drug has since been implicated in drug facilitated sexual assaults (DFSA) and crimes (DFC).

The control measures applied to Schedule IV are similar to Schedule III. Prescriptions for Schedule IV drugs may be refilled up to five times within a six-month period. A prescription for controlled substances in Schedules III, IV, and V, issued by a practitioner, may be communicated either orally, in writing, or by facsimile to the pharmacist, and may be refilled if so authorized on the prescription or by personal visit to the appropriate dispensary.

16.6   Schedule V

Drugs that are categorized in Schedule V fulfill the following conditions. The drug or other substance has a low potential for abuse relative to the drugs or other substances in Schedule IV. The drug or other substance has a currently accepted medical use in treatment in the US. The potential for abuse of the drug or other substance may lead to limited physical dependence or psychological dependence relative to the drugs or other substances in Schedule IV.

16.6.1   Drugs in Schedule V

- Cough suppressants containing small amounts of codeine (*e.g.* promethazine + codeine).
- Preparations containing small amounts of opium or diphenoxylate (used to treat diarrhea).
- Some anticonvulsants, such as pregabalin (Lyrica), lacosamide (Vimpat), and retigabine (Ezogabine) (Potiga/Trobalt).

- Pyrovalerone (used to treat chronic fatigue and as an appetite suppressant for weight loss). This compound has now been found to be a drug of abuse under the Bathsalts or Synthetic Cathinones along with its methylenedioxy analogue MDPV (methylenedioxypyravalerone).
- Some centrally-acting antidiarrheal medications, such as diphenoxylate (Lomotil) when mixed with atropine (to make it unpleasant for subjects to self-administer *via* grinding the medication, cooking with appropriate liquids, and injecting the filtered liquor).
- Difenoxin with atropine (Motofen) has been reassigned to Schedule IV. Without containing atropine, these drugs are categorized in Schedule II.
- Federal regulation of pseudoephedrine and ephedrine due to them both being widely employed in the illicit synthesis of methamphetamine, the US Congress passed the Methamphetamine Precursor Control Act which places restrictions on the sale of any medicine containing pseudoephedrine. The bill was then superseded by the Combat Methamphetamine Epidemic Act of 2005, which was passed as an amendment to the Patriot Act renewal and included wider and more comprehensive restrictions on the sale of pseudoephedrine containing products. This law requires customer signature of a "log-book" and presentation of valid photo ID such as a valid driving license in order to purchase pseudoephedrine containing products from any permitted retailer.

Additionally, the law restricts an individual to the retail purchase of no more than three packages or 3.6 g of such product per day per purchase – and no more than 9 g in a single month. A violation of this statute constitutes a misdemeanor. Retailers now commonly require pseudoephedrine containing products to be sold behind the pharmacy or service counter. This affects many preparations which were previously available over-the-counter without restriction, such as Actifed and its generic equivalents.

No controlled substance in Schedule V which is a drug may be distributed or dispensed other than for a medical purpose. A prescription for controlled substances in Schedules III, IV, and V, issued by a practitioner, may be communicated either orally, in writing, or by facsimile to the pharmacist, and may be refilled if so authorized on the prescription or by personal visit to the appropriate dispensary.

16.7 Conclusion

In the US, like other countries such as the UK, there has been much vocal criticism against the schedule classifications of the listed drugs which has tended to be reactive rather than proactive; an excellent example is the newer psychoactive substances (NPS which includes the synthetic cannabinoids, cathinones, and phenylethylamines) in the CSA. Citation in the schedules (typically Schedule I) for these particular substances has been performed without extensive clinical trials, and is considered a "knee jerk reaction" in some quarters after some fatalities have been reported either in the general press or scientific literature.

When extensive research is performed, several substances on the list of Schedule I substances have been found to have an actual accepted medical use and low abuse potential, despite the requirement for a Schedule I listing mandating that any substance so scheduled have both a high potential for abuse and no accepted medical use; the most recent example is the legalization of medical marijuana in over 20 states in the US, and several including California are moving to legalize recreational use of marijuana.

Further Reading

The Federal Food, Drug, and Cosmetic Act 1906 (US); http://www.gpo.gov/fdsys/pkg/USCODE-2011-title21/pdf/USCODE-2011-title21-chap9.pdf.

Comprehensive Drug Abuse Prevention and Control act 1970 (US); http://www.fda.gov/regulatoryinformation/legislation/ucm148726.htm.

Misuse of Drugs Act 1971 (UK); http://www.legislation.gov.uk/ukpga/1971/38/pdfs/ukpga_19710038_en.pdf.

The Chemical Diversion and Trafficking Act of 1988 (US); http://www.deadiversion.usdoj.gov/chem_prog/34chems.htm.

The Federal Food, Drug, and Cosmetic Act (1938) (US); http://www.fda.gov/AboutFDA/WhatWeDo/History/ProductRegulation/ucm132818.htm.

Kefauver Harris Amendment of 1962 (US); http://www.fda.gov/AboutFDA/WhatWeDo/History/CentennialofFDA/CentennialEditionofFDAConsumer/ucm093787.htm.

The Federal Analog Act (US); http://moritzlaw.osu.edu/students/groups/osjcl/amici-blog/the-federal-controlled-substances-analogue-act-an-antiquated-solution-meets-an-evolving-problem/.

The Anabolic Steroids Act (US); http://www.gpo.gov/fdsys/pkg/BILLS-108s2195enr/pdf/BILLS-108s2195enr.pdf; https://www.federalregister.gov/articles/2008/04/25/E8-8842/classification-of-three-steroids-as-schedule-iii-anabolic-steroids-under-the-controlled-substances.

Title 21 United States Code (USC) Controlled Substances Act (US); http://www.deadiversion.usdoj.gov/21cfr/21usc/.

Section 21 C.F.R 1306.12 (US); http://www.deadiversion.usdoj.gov/21cfr/cfr/1306/1306_12.htm.

17 Drug Legislation in New Zealand

Keith Bedford

Institute of Environmental Science and Research Limited, Auckland, New Zealand
Email: keith.bedford@esr.cri.nz

17.1 Historical Background and New Zealand Drug Policy Framework

In New Zealand (NZ) until recently, the legislative framework used to control and regulate the availability of medicines, drugs and other substances that may be abused or misused has been provided through the:

- Medicines Act (1981)[1] and Regulations; and the
- Misuse of Misuse of Drugs Act (1975)[2] and Regulations.

Substances not listed in the appropriate schedules of the Medicines Regulations cannot be marketed as medicines. The Misuse of Drugs Act (MoDA) applies to certain therapeutic medicines for which an increased level of control is considered necessary, as well as to substances that are prohibited because of the history or risk of abuse or misuse. Substances not listed in the schedules to the Act are not controlled, although there is a provision covering

Forensic Toxicology: Drug Use and Misuse
Edited by Susannah Davies, Atholl Johnston and David Holt
© The Royal Society of Chemistry 2016
Published by the Royal Society of Chemistry, www.rsc.org

"controlled drug analogues", which is explained later in this chapter. Also see Chapter 7 for full discussions on general New Zealand drug use.

Both Acts have been extensively amended and added to over time. It has been stated by the New Zealand Law Commission when reviewing the state of drug legislation in New Zealand,[3] particularly in reference to the MoDA, that:

> "This Act is now 35 years old. Its main components were developed in the 1970s, when the 'hippie' counterculture was at its height and the illegal drugs of choice were cannabis, cocaine, opiates and psychedelics like LSD. Since that time, a great amount of research has been undertaken into the effects of different drugs. We now know much more about the harms of drug use, and what can be done to reduce that harm. While cannabis use remains relatively high, new drugs have appeared. In the 2000s, party pills like BZP and more harmful drugs like methamphetamine have joined cannabis at the forefront of New Zealand's drug scene. New Zealand's drug landscape is vastly different from that which the Act contemplated in 1975. Over the years, various amendments have been made to the Act to respond to issues as they arose. These ad hoc amendments have resulted in an Act that has become difficult to understand and navigate. A first principles review of the Act is well overdue."[3]

An updated NZ National Drug Policy was released in August 2015. The overall goal of the Policy for the period 2015–2020 is stated to be:

> "To minimise alcohol and other drug-related harm and promote and protect health and wellbeing."[4]

The Policy statement indicates that:

> "The idea of harm minimisation encompasses the prevention and reduction of health, social and economic harms experienced by individuals, their families and friends, communities and society from AOD [alcohol and other drug] use."[4]

Strategies outlined in the Policy to support this overall goal encompass "priority areas" for action in relation to:

- Problem limitation
- Demand reduction
- Supply control

"Priority area 3" linking to these strategies is: "Getting the legal balance right". Specific actions against this priority area include:

- "Work with the Expert Advisory Committee on Drugs (EACD) to ensure harm minimisation is a central feature of drug classification assessments."
- "Review the regulation of controlled drugs for legitimate purposes (such as medicines) alongside reviews of the Medicines Act 1981 and other therapeutics legislation."
- "Develop options for further minimising harm in relation to the offence and penalty regime for personal possession within the Misuse of Drugs Act 1975."[4]

"Getting the legal balance right" could be seen as the elusive underlying objective of the legislative approaches described in the remainder of this chapter.

17.2   Scheduling of Substances under the MoDA

The MoDA contains a number of schedules to provide varying levels of control and penalties under the Act. It must be admitted that for historical reasons there are numerous inconsistencies in the existing scheduling. In 2000 an Amendment to the principal Act established an Expert Advisory Committee on Drugs (EACD) to advise on scheduling matters. At that time the amendment was also intended to provide a "fast-track" mechanism for listing new substances which were giving concern, *via* an Order in Council. In making recommendations to the responsible Minister (the Minister of Health), the matters to be considered by the EACD in providing advice are stated in the Act to be:

- The likelihood or evidence of drug abuse.
- Specific effects of drug, including pharmacological, psychoactive and toxicological effects.
- The risks, if any, to public health.
- The therapeutic value of the drug, if any.
- The potential for the drug to cause death.
- The ability of the drug to create physical or psychological dependence.
- The international classification and experience of the drug in other jurisdictions.
- Any other matters that the Minister considers relevant.

Drugs and other substances are assessed and then recommendations may be made on the basis that substances considered as posing a:

- Very high risk of harm should be scheduled as Class A.
- High risk of harm should be scheduled as Class B.
- Moderate risk of harm should be scheduled as Class C.

The EACD can make recommendations to the Minister for the inclusion of new substances in the schedules of classification available under the Act, or for increasing the level of control on already-listed substances. The Committee may choose to provide advice on removing or reducing the classification of a substance, but the "fast track" provision *via* an Order in Council is not available for these purposes and legislative amendment is required.

The process of assessing drug harms or potential harms in order to provide a rational basis for drug scheduling is a challenging business. It is worth noting that Professor David Nutt lost his position as chair of the UK Advisory Council on the Misuse of Drugs (ACMD), the equivalent of the NZ EACD, for advocating changes in the scheduling of certain substances based on a reassessment of relative harms and risk, as proposed in his article: *"Estimating Drug Harms: A risky business?"*.[5,6] See Chapter 15 for further UK scheduling discussions.

The framework provided to the EACD in the MoDA for scheduling drugs and substances created several challenges, including that:

- The scheduling criteria are intended to be evidence-based. Therefore, the appearance of a novel drug or substance on the New Zealand drug scene, that may be being used "recreationally", could lead to a scheduling hiatus until information or evidence of harm (or of lack of potential for harm) could be gathered; and
- A substance that is not a medicine but does not present at least a moderate risk of harm existed in a legal "no man's land" between the Medicines Act and the Misuse of Drugs Act.

The introduction of the Psychoactive Substances regime described latter in this chapter addressed these issues.

17.3 Controlled Drug Analogue Provisions of the MoDA

The MoDA contain provisions relating to "controlled drug analogues", sometimes referred to in news media as "designer drugs".

This provision was added through an amendment in 1988 as a means to counter a growing trend for the appearance of variants of known, sought-after drugs, with a slightly modified chemical structure. Drugs are generally listed in legislation as a specific chemical entity or name, *e.g.* "STP, DOM [2-amino-1-(2,5-dimethoxy-4-methyl)-phenylpropane]".

This means that in the absence of an analogue provision, if a new drug or substance of concern appears, it cannot be controlled until it is added to the current legislation, a process that can be time-consuming. In NZ the controlled drug analogue definition covers any substance that "...has a structure substantially similar to that of any [already-listed] controlled drug".

The Act makes no reference to physiological or psychoactive effect; it is based purely on the chemical structure. This analogue provision in the MoDA provide a legislative "safety-net"; substances considered to be analogues are treated as Class C controlled drugs, regardless of the classification of the already-listed parent controlled drug.

Such wording in legislation can lead to potential lack of certainty around how broadly the phrase "a structure substantially similar to" should be interpreted. The definition has the potential to capture a range of substances that either have not yet been used for human consumption or currently have not been synthesised. Until recently, no case law guidance was available on the legal interpretation of the "substantially similar" provision, but early in 2015 a judicial ruling on the interpretation of this point was delivered in the course of a major series of trials involving large-scale production and dealing in new drug derivatives. At the time of writing, this ruling was still suppressed from publication by court order, as legal proceedings are continuing.

17.4 The Appearance of "Party Drugs" and the Introduction of "Restricted Substance" Class

The legislative framework described in the preceding sections provides the background to the situation in which BZP (benzyl-piperazine) and TFMPP (trifluoromethylphenylpiperazine) and some other piperazine analogues became very widely available in NZ in the early 2000s and became referred to as "party drugs". These compounds were usually sold in the form of tablets with an impressed logo and generally resembled tablets that had previously been marketed as "ecstasy" (generally taken to refer to methylenedioxymethamphetamine or MDMA) tablets. MDMA was then and still is classified as a Class B controlled drug in NZ. BZP alone is a relatively mild stimulant, estimated to have about a tenth the potency of dexamphetamine. BZP was often encountered combined with TFMPP. Taken together the two substances were said to produce effects resembling MDMA. At overdose levels, reported adverse effects included palpitations, agitation, nausea, vomiting and seizures.

Initially the EACD indicated to the Minister that there was insufficient scientific evidence at that time of at least a moderate risk of harm to schedule BZP as a controlled drug. The legislative response in due course was to create a new Fourth Class under the Misuse of Drugs Act, called "Restricted Substances". Relatively limited restrictions on supply were provided for, including an age limit and limits on advertising similar to those contained in legislation governing the sale of alcohol or tobacco. Later, following further local studies, the EACD recommended rescheduling BZP and related piperazines as Class C controlled drugs, although there was vigorous debate on whether the adverse effects reported reached the threshold of moderate risk of harm.[7] This recommendation was enacted. It is worth noting that although one local medical report included data on toxic effects experienced by 61 patients who presented to the Emergency Department of Christchurch Hospital, many of the patients had taken other substances as well, particularly alcohol. One patient had apparently also consumed ecstasy, LSD, nitrous oxide (NOS) and "magic mushrooms".[8]

The creation in the MoDA of this new "Restricted Substance" regulatory framework represented a significant move from the historical response to drug misuse in the NZ context based on a prohibition model, through the introduction of a framework for regulating supply of recreational substances. Other countries were also grappling with similar issues around the same time. In the UK a major review appeared entitled: "*Drug classification: Making a hash of it?*".[9] An article in the *New Scientist* magazine in 2006, in which the approach taken by NZ was noted, included the (not entirely accurate) comment:

> "New Zealand, however, has taken a different and arguably more enlightened approach…in response the government introduced a new Class of drug called 'non-traditional designer substances', also known as Class D".[10]

Since being made a controlled drug, BZP along with TFMPP has continued to circulate in the "party drug" scene in NZ as an illegal substance, sometimes mixed with other controlled drugs. Nevertheless, the occurrence of BZP and TFMPP is currently at relatively low levels, presumably as a result of their controlled drug status, and also because of the appearance of a number of other, new substances in the market.

The ability to schedule Restricted Substances was little-used in subsequent years and a very small number of substances that were recommended for inclusion in that classification by the EACD (*e.g.* dimethylamylamine or DMAA) were not enacted.

17.5   Consequences

The development of a ready market for legal party drugs in NZ has seen a significant number of novel drugs and substances appear as suppliers and entrepreneurs have attempted to capitalise on the demand. Some but not all of these new substances have been found to infringe the "controlled drug analogue" provisions of the MoDA.

In this respect the situation in NZ has much in common with many other western countries, although, for reasons that are not entirely clear, NZ seems to encounter a disproportionately high number of novel substances compared to other jurisdictions. For the forensic drug chemist it can be difficult to prove the identity of such compounds to court standard (which normally requires a direct comparison against an authenticated reference standard), even if analytical data are available in the scientific literature, which may not be the case. In the absence of controlled clinical studies, assessment of potential harm as a basis for scheduling of novel substances in a legislative framework is similarly highly problematic. The question of potential for harm after long-term use by humans is clearly very difficult to answer if the only scientific data available are from limited, short-term animal studies, if that.

17.6   The Rise of "Synthetic Cannabinoid" Products

The next major development affecting the availability of psychoactive substances in NZ was the introduction of herbal smoking mixes, also known as "herbal highs", which tended initially to be referred to by use of brand names, particularly "Kronic" or "Spice". Interest in these products grew gradually, attracting media interest and wide public awareness in NZ by 2011, with multiple brands becoming available. As alluded to in the previous section, the NZ market was an early and enthusiastic adopter of such products and for a time NZ also became a base for internet sales and international distribution. These herbal blends contained one or more additional active ingredients – synthetic cannabinoids or cannabinomimetic substances, which have similar properties to tetrahydrocannabinol (THC), the active ingredient in cannabis plant. A few of these added active ingredients were similar enough in chemical structure to be regarded as analogues of THC and controlled under the MoDA controlled drug analogue provisions, but the majority were structurally sufficiently different from THC or any other controlled drug that they were not controlled at the time they appeared on the market. There was little literature available about the risks associated with use of these substances, which in some cases had been synthesised as model substances for medicinal chemistry studies investigating the effects of cannabinoids. The topic of synthetic cannabinoids is dealt with more fully in Chapter 8 in this book.

17.7 Introduction of "Temporary Class Drug Orders"

Although the scheduling route provided for under the MoDA through EACD recommendation to the Minister was intended to provide a "fast-track" mechanism for listing new substances that gave rise to concern, in practice the classification of new substances was still slow. An additional issue that arose at this time of proliferation of new drugs, drug analogues and recreational substances, was the lack of robust scientific information on which to assess the risk profile and appropriate scheduling of novel substances. Concern over the appearance on the party drug market of a significant number of products containing cannabinomimetic substances led to the introduction of the provision for "Temporary Class Drug Orders" as a short-term solution. This measure provided for limited term "emergency" banning of substances for 12 months, with one extension of an additional 12 months possible, to allow time for the collation of evidence on which to base formal scheduling decisions. This regime resembled that used in some other jurisdictions, notably in the UK, under similar circumstances. In due course, 35 substances were banned as Temporary Class Drugs over a period beginning in August 2011. The vast majority of these substances were cannabinomimetic agents.

17.8 The Psychoactive Substances Regime

The 2010 Law Commission report stated:[3]

> "We advocate a new approach, which would prohibit any new psycho-active substance from being manufactured, produced or imported without prior approval. Upon application for approval, a variety of regulatory options would be available; prohibition would be the last resort. This would provide much greater protection for the public."

In July 2013 the Psychoactive Substances Act 2013 came into effect in New Zealand. This superseded the Temporary Class Drug scheme as all psychoactive substances are restricted by default, except where specifically licensed. The Act defines "psychoactive substance" broadly as:[11]

> "..a substance, mixture, preparation, article, device or thing that is capable of inducting a psychoactive effect (by any means) in an individual who uses the psychoactive substance."

However, this definition is qualified by clauses excluding substances already scheduled under other legislation, including the MoDA and Medicines Act and also excluding alcohol and tobacco.

The Psychoactive Substances Act was intended to follow the "new approach" advocated by the Law Commission and more cautiously explored in the NZ context in the earlier introduction of the Restricted Substances classification; moving away from the "prohibition model" through the introduction of a framework for regulating supply of recreational substances that were considered to be of "low risk of harm" to the user. Although this generated significant interest amongst those interested in drug law reform, the major shift in social policy that was involved in the introduction of this Act was not widely debated amongst the general public at the time. Arguably, this failure to articulate the underpinning harm reduction objectives and explain the intent to take the manufacture and sale of substances of interest to those seeking products for recreational use away from the black market, led to or at least contributed significantly to the eventual stalemate around the implementation of the Act's provisions.

The objective of the regime introduced in the Act and intended subsequent Regulations was to introduce a strict regulatory regime requiring toxicity testing and quality control standards for products containing recreational substances not unlike those applied to pharmaceutical products. Approved products shown to be of low risk were to be allowed to be sold legally under significant restrictions. These restrictions included:

- Minimum age to purchase (18 years).
- Type of premises able to sell such products (*e.g.* excluding convenience stores, grocery stores or supermarkets, a store selling alcoholic drink . . ..).
- A ban on certain forms of promotional activities and sponsorships.
- Restrictions and requirements limiting advertising.
- Restrictions and requirements relating to labelling and packaging.
- Health warnings.
- Restrictions and requirements on display of products.

Because it was anticipated that products undergoing the required testing, including toxicological testing, and progressing through the licensing and approval process might take 18 months or more to reach the market, an interim, transitional arrangement was allowed, with substances on the market, and apparently not giving rise to undue concerns about adverse effects, allowed to remain on the market. It has been claimed that following the restrictions introduced in the Act coming into force, the number of retail outlets offering "legal highs" fell from as many as 4000 to fewer than 170 licensed premises nationwide. The number of branded products was cut from around 200 to fewer than 50. There was also some evidence that the number of severe presentations to emergency departments and severe issues reported to the National Poisons Centre had reduced since the Act came into force.[12]

Nevertheless, public disquiet increased, focusing not on the principle of the shift from a prohibition model to a regulatory model, but on the desire not to have a retail outlet in the local neighbourhood and more particularly on the use of animals in toxicology testing. In 2014 the government responded to mounting public pressure by introducing an amendment that repealed the parts of the original Act providing for interim product licensing, halted the legal sale of all psychoactive substances and effectively banned animal testing of prospective products. The actual wording in the amended Act states that "...the advisory committee must not have regard to the results of a trial that involves use of an animal".

There is an additional provision that the Psychoactive Substances Regulatory Authority and the advisory committee may take into account an overseas animal trial that finds a product "may pose more than a low risk of harm to individuals using the product". In other words, they may only take into account evidence from animal trials to ban a product, not to approve it.

It is generally considered that it would not be possible to have a product approved under the psychoactive substances regime without animal testing. Somewhat ironically, the products allowed to remain on sale under the interim, transitional arrangement were "herbal high" products that would almost certainly have failed to gain approval because of the health hazards associated with smoke inhalation.

The full regulatory framework required to implement the regime is nearing completion. The Psychoactive Substances Regulatory Authority is established within the Ministry of Health. A code of manufacturing process is in place and regulations defining requirements for manufacturing, importing, research and products approvals have been completed. Regulations for wholesaling and retailing approved products are under development. However, it is a moot point as to whether it will be possible to demonstrate for any proposed product, without animal testing, that it "should pose no more than a low risk of harm to individuals who use it". So for now NZ has returned to the historical "cat and mouse" cycle of the appearance of a new drug or substance of concern, followed with a time lag, by prohibition. Nevertheless, the "default" legal status of many of the new substances that may appear on the illicit market in future is now likely to be that of an "unapproved substance" under the Psychoactive Substances regime, rather than the legal "no man's land" that applied to many such substances prior to the introduction of the regime.

Figure 17.1 illustrates the evolution over time of drug control legislation in NZ.

The emergence of "party pills", the appearance of new designer analogues and development of herbal blends containing pharmacologically active additives have challenged the New Zealand regulatory framework for drug control and raised fundamental issues, including the effectiveness of prohibition compared to restriction as a harm reduction strategy. This raises interesting issues of chemistry, drug policy, risk, harm and individual responsibility.

NEW ZEALAND MISUSE OF DRUGS ACT 1975 – 2014

1975
Misuse of Drugs
Act 1975
(Principal Act)

1987
Misuse of Drugs
Amendment
Act 1987
commenced
1988

*Introduction
of controlled
drug analogue
provisions*

2000
Misuse of Drugs
Amendment Act
2000

*Introduction
of EACD to
advise on drug
classsification*

2005
Misuse of Drugs
Amendment Act
2005

*Introduction
of "Restricted
Substances"
provisions*

2011
Misuse of Drugs
Amendment Act
(No.2) 2011

*Introduction
of "Temporary
Class Drug
Orders"*

2013
Psychoactive
Substances Act
2013

New regime

2014
Psychoactive
Substances
Amendment Act
2014

*Repeal of
interim product
licensing; animal
testing effectively
banned*

2015

Figure 17.1 The evolution over time of drug control legislation in NZ.

Declaration of Interest

Dr Bedford is a member of the NZ Expert Advisory Committee on Drugs (EACD). Opinions expressed in this chapter are his personal opinions and do not necessarily reflect the views of the EACD.

References

1. Available online at: http://www.legislation.govt.nz/act/public/1981/0118/latest/DLM53790.html?src=qs (accessed 2 April 2015).
2. Available online at: http://www.legislation.govt.nz/act/public/1975/0116/latest/DLM436101.html?src=qs (accessed 2 April 2015).
3. Law Commission, 2010, *Controlling and Regulating Drugs* (Issues paper 16), Available online at: http://www.lawcom.govt.nz/project/review-misuse-drugs-act-1975?quicktabs_23=issues_paper (accessed 2 April 2015).
4. Inter-Agency Committee on Drugs, 2015, *National Drug Policy 2015 to 2020*, Wellington, Ministry of Health, Available online at: http://www.health.govt.nz/publication/national-drug-policy-2015-2020 (accessed 30 September 2015).
5. D. Nutt, 2009, *Estimating Drug Harms: A risky business?* Centre for Crime and Justice Studies, Briefing 10, October 2009, Available online at: http://www.crimeandjustice.org.uk/publications/estimating-drug-harms-risky-business (accessed 2 April 2015).
6. Media report appearing at the time available online at: http://www.theguardian.com/politics/2009/oct/30/drugs-adviser-david-nutt-sacked (accessed 26 June 2015).
7. Expert Advisory Committee on Drugs (EACD), 2004, *The Expert Advisory Committee on Drugs (EACD) Advice to the Minister on: Benzylpiperazine (BZP)* (Reference No: 20045663). Available online at: http://www.health.govt.nz/system/files/documents/pages/eacdbzp.pdf (accessed 2 April 2015).
8. P. Gee, S. Richardson, W. Woltersdorf and G. Moore, 2005, Toxic effects of BZP-based herbal party pills in humans: a prospective study in Christchurch, New Zealand, *The New Zealand Medical Journal* 118 (1227):1–10. http://www.nzma.org.nz/__data/assets/pdf_file/0005/17888/Vol-118-No-1227-16-December-2005.pdf (accessed 2 April 2015).
9. House of Commons Science and Technology Committee [UK] 2006, *Drug classification: making a hash of it?* Fifth report of

Session 2005-06 (HC 1031), Available online at: http://www.publications.parliament.uk/pa/cm200506/cmselect/cmsctech/1031/1031.pdf (accessed 2 April 2015).

10. G. Vince, Legally High, *New Sci.*, 2006, **2571**, 40–45(30 September 2006).

11. Available online at: http://www.legislation.govt.nz/act/public/2013/0053/latest/DLM5042921.html?src=qs (accessed 2 April 2015).

12. R. Brown, 2014, Our 'psycho' psychoactive substance legislation, *Matters of Substance* (Information journal of the NZ Drug Foundation) November 2014 issue. Available online at: http://www.drugfoundation.org.nz/matters-of-substance/november-2014 (accessed 2 April 2015).

18   Use of Reference Materials in Toxicology

Jennifer Button

Chiron AS, Trondheim, Norway
Email: jenny.button@chiron.no

18.1   Importance of Standardisation and Harmonisation

The terms standardisation and harmonisation are often used synonymously. There is a subtle, but important, difference between the two. Standardisation is concerned with conformity, whereas harmonisation is more about consistency. Both share equal importance in clinical and forensic analysis. In laboratory medicine they offer many advantages, including:

- Ease of implementing consensus guidelines
- Intra- and inter-methods or laboratory comparisons
- Application of standardised reference intervals or "cut-offs"
- Reduced variables in multicentre studies

18.2   The Language of Metrology

The science of measurement, known as metrology, has its own language. Understanding the terms and definitions used is critical to the correct application of various ISO standards and guides in which

Forensic Toxicology: Drug Use and Misuse
Edited by Susannah Davies, Atholl Johnston and David Holt
© The Royal Society of Chemistry 2016
Published by the Royal Society of Chemistry, www.rsc.org

they are referenced. A common language is also essential to enable scientists around the globe to understand each other. To aid interpretation and the international exchange of metrology concepts, the International Vocabulary of Metrology, or VIM for short (derived from the French; Vocabulaire International de Métrologie), was born. VIM was produced by working group 2 of the Joint Committee for Guides in Metrology (JCGM) and has been published as ISO Guide 99 and JCGM 200:2008.[1] It is important to recognise that metrology has a wide application and even when trying to create a universal language it can sometimes be difficult to create common concepts and terms which relate equally to physical, chemical and biochemical measurements. Whilst terminology has been revised in VIM 3 to better accommodate chemical and biochemical measurement, it was written primarily for physical measurement. In recognition of this, Eurachem produced a guide, *Terminology in Analytical Measurement – Introduction to VIM 3*, to provide additional context and examples for those working in analytical laboratories, and more specifically those dealing with chemical, biological and clinical measurements.[2]

In addition to VIM, there is a separate guide, ISO Guide 30, which deals with the terms and definitions used in connection with reference materials.[3] In particular, this addresses the terms used in certificates of analysis and associated certification reports. More specifically, terms related to materials, measurement and testing, certification and issuance of reference materials and statistical terms used in the characterisation of reference materials. This guide falls under the responsibility of the ISO Committee on Reference Materials (REMCO), the international committee that leads on reference material matters. There are differences between the definitions within ISO Guide 30 and VIM. The definitions within ISO Guide 30 are better tailored to enhance the understanding of reference materials and their uses. However, there is a desire to see harmonisation of different terminology guides.

The terms and definitions referred to within this chapter are derived from both VIM and ISO Guide 30. The vocabulary will be addressed as new terms and concepts are introduced in the relevant sections.

18.3 International System of Units

Metrology involves determining quantity values and expressing them using an international system of units (SI), to enable metrological concepts to be universally exchanged and understood. It is a requirement of ISO/IEC 17025:2005 that reference materials are traceable to the SI:

5.6.3.2 Reference materials – ISO/IEC 17025:2005[4]
Reference materials shall, where possible, be traceable to SI units of measurement, or to certified reference materials. Internal reference materials shall be checked as far as is technically and economically practicable.

The international system of units (SI) recognises seven base units founded on seven base quantities (Table 18.1). For chemical measurement the base quantities are mass and amount of substance and their respective base units, the kilogram (kg) and the mole (mol). Increasingly, laboratories are moving away from molar units in favour of mass units, except where current international uniformity of reporting dictates otherwise [*e.g.* lithium (mmol L^{-1}), methotrexate (μmol L^{-1}), vitamin D (nmol L^{-1}) and thyroxine (pmol L^{-1})]. Metre is the base unit of length (m) and volume is expressed in m^3. Litre is a submultiple of this (1 litre $= 1$ dm^3) and is a widely accepted measurement unit in analytical chemistry.[2]

Despite attempts to harmonise units, there remains much diversity between laboratories. This can be particularly problematic for clinicians receiving results for the same analyte from different laboratories. Switching between mass units such as $ng\,mL^{-1}$ and $mg\,L^{-1}$ should be quite noticeable, but changes introduced by switching to molar units can be more subtle and readily overlooked. Whilst subtle, these differences can still result in misdiagnosis or incorrect treatment decisions. Units such as ppm, ppb and ppt, which stand for parts per million (10^{-6}), billion (10^{-9}) and trillion (10^{-12}), respectively, are widely used in analytical and environmental chemistry, but

Table 18.1 The seven base quantities of the International System of Quantities.[1]

Base quantity	Base unit	Symbol
Length	Metre	m
Mass	Kilogram	kg
Time	Second	s
Electric current	Ampere	A
Thermodynamic temperature	Kelvin	K
Amount of substance	Mole	mol
Luminous intensity	Candela	cd

are ambiguous and IUPAC recommend against their usage.[5] Such units can easily be replaced with SI compatible quantities.

18.4 Metrological Traceability

The term metrological traceability, sometimes abbreviated to just traceability, is defined in VIM as follows:

2.41 (6.10) Metrological Traceability – VIM[1]
Property of a *measurement result* whereby the result can be related to a reference through a documented unbroken chain of *calibrations*, each contributing to the *measurement uncertainty*.

Traceability to a common metrological reference is a pre-requisite to the comparison of measurement results between procedures, and across time and space. There is a universal acceptance that these "anchor points" define what is correct. The reference can be a practical realisation of a defined measurement unit, *i.e.* the kilogram, a reference measurement procedure, or a measurement standard, *i.e.* certified reference material. Metrological traceability requires a calibration hierarchy, the ideal end-point of which is a base unit of the SI.

Each level of the metrological traceability chain is represented by a measurement procedure or measurement standard, which transfers its value to the level below, descending from the highest metrological reference to the measurement result in the laboratory. Values assigned to measurement standards have uncertainty. Each value inherits uncertainty from the measurement standards and procedures in the higher levels of the calibration hierarchy. Therefore, measurement uncertainty increases the further away the measurement is from the reference. A generic metrological traceability chain is shown in Figure 18.1. The Eurachem/CITAC Guide *Traceability in Chemical Measurement* discusses in detail how metrological traceability is established.[6]

18.5 Measurement Uncertainty

ISO/IEC 17025 has placed greater emphasis on the measurement uncertainty of test results. Measurement uncertainty gives a quantitative measure of the quality of a result. It is defined in VIM as:

2.26 (3.9) Measurement Uncertainty – VIM[1]
Non-negative parameter characterizing the dispersion of the *quantity values* being attributed to a *measurand*, based on the information used.

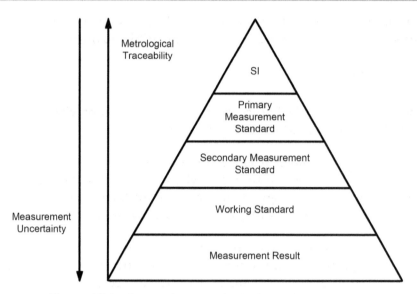

Figure 18.1   A generic metrological traceability chain.

No measurement is exact. Measurement results deviate from the true value as a consequence of systematic (bias) and random error. A measurement is, at best, an estimate of the amount of measurand present and is only complete when accompanied by a statement of measurement uncertainty. Any measurement procedure will comprise several sources of measurement uncertainty, *e.g.* incomplete definition of the measurand, non-representative sampling, instrument limitations (sensitivity and specificity), personal bias, environmental conditions, *etc.* In analytical chemistry, measurement uncertainty is typically expressed as the expanded uncertainty (U), sometimes referred to as overall uncertainty. This gives an interval within which the true answer is believed to lie, *i.e.* value ± uncertainty, and is usually expressed as % confidence of the interval. The expanded uncertainty is calculated by multiplying the standard uncertainty by an appropriate coverage factor k, and depends on the level of confidence required. In most cases a coverage factor of 2 is used, which gives an expanded uncertainty at a confidence level of approximately 95%.[7]

The *Guide to the Expression of Uncertainty in Measurement* (GUM) is considered the definitive document on the subject, and addresses how to evaluate and express uncertainty in a broad range of measurements.[8] A joint Eurachem/CITAC guide *Quantifying Uncertainty in Analytical Measurement* translates these concepts to the field of analytical chemistry.[9] LGC, the UK's designated National Metrology Institute for chemical and bioanalytical measurement, has also produced a user-friendly guide to evaluating measurement uncertainty in clinical chemistry, which includes worked case examples.[7]

18.6 (Certified) Reference Materials

Reference materials (RMs) and certified reference materials (CRMs) are commonly used for the calibration and quality control of measurement systems. The following definitions for RMs are given in ISO Guide 30 and VIM, respectively:

2.1.1 Reference Material (RM) – ISO Guide 30[3]
Material, sufficiently homogeneous and stable with respect to one or more specified properties, which has been established to be fit for its intended use in a measurement process.
5.13 (6.13) Reference Material (RM) – VIM[1]
Material, sufficiently homogeneous and stable with reference to specified properties, which has been established to be fit for its intended use in measurement *or in examination of nominal properties.*

The definitions differ only in reference to nominal properties. In this context, the term "nominal property" is used to refer to a characteristic which has no defined quantity, yet still portrays important information, like gender, colour or amino acid sequence. In other words, the definition for RMs is intended to incorporate both quantitative and qualitative analysis. Since measurement by definition relates to quantitative determinations, the term examination is more appropriate when referencing qualitative or "nominal properties". For CRMs the uncertainty of nominal properties may be expressed in terms of probabilities or confidence level:

2.1.2 Certified Reference Material (CRM) – ISO Guide 30[3]
Reference material (RM) characterised by a metrologically valid procedure for one or more specified properties, accompanied by an RM *certificate* that provides the value of the specified property, its associated *uncertainty*, and a statement of *metrological traceability.*
5.14 (6.14) Certified Reference Material (CRM) – VIM[1]
Reference material, accompanied by *documentation* issued by an authoritative body and providing one or more specified property values with associated *uncertainties* and *traceabilities*, using valid procedures.

The definitions are still ambiguous and open to misinterpretation. Who qualifies as an "authoritative body?" A National Metrology Institute (NMI)? An ISO Guide 34 accredited producer? An ISO/IEC 17025 accredited testing and calibration laboratory? Does any reference material accompanied with documentation stating uncertainty and metrological traceability qualify as a CRM? Currently this term is being applied to materials of varying degrees of characterisation. Some believe only a small number of true CRMs, produced by NMIs, exist, whilst others believe this term to relate to an abundance of commercially available measurement standards. Sometimes it can depend on the accreditation body. This is partly because some of the terminology concepts were inherited/adapted from the world of physics and the standardisation of mass, length and time measurements. The world of chemistry and biology is far more complex and further refinement of some of definitions is required to reflect a situation that is not so black and white. A grading or five-star system such as that used to rate hotels may be easier for end users to comprehend, and offers better differentiation. At the top end of the scale would be a well-defined measurand, characterised by more than one NMI through a CCQM laboratory intercomparison and, if clinically relevant, listed on the JCTLM database (see Sections 18.22–18.23). The opposite end of the spectrum would be a reference material developed to meet an urgent need, such as a new psychoactive substance (see Supplementary Reading).

18.7   Quality Control Materials

ISO Guide 30 describes a quality control material (QCM) as a reference material used for quality control of a measurement.[3] QCMs are primarily used to monitor a pre-validated measurement system to verify it has attained the intended quality and does not go outside statistical control. The ideal QCM should be similar to the test samples such that it reacts with the measuring system in a manner that mirrors the sample, and reflects what is happening with the measurement procedure when analysing the intended sample type. The QCM should contain the analyte(s) of interest at a suitable concentration(s), *i.e.* at or near cut-off concentrations or clinical decision limits. The QCM should be homogenous, preferably stable over time and of sufficient quantity. It is prudent to check lot-to-lot consistency. Plotting of quality control values aids monitoring of performance over time and provides a visual, early indication of any emerging problems. QCMs should be independent of those supplied by the reagent or kit manufacturer, or in the case of in-house prepared QCMs, independent of the reference material used for calibration (see Section 18.9).

Whilst commercial QCMs are available for well-established analytes, there is a dearth of material for research purposes or novel emerging compounds. Often the time frame for commercial production, dictated by strict requirements such as those outlined in IVD Medical Devices Directive 98/79/EC, is not in keeping pace with demand. This, and the increased introduction of open platforms, such as LC-MS, into analytical laboratories also generates need for in-house preparation of QCMs. In-house preparation of QCMs may, by some, be seen as cost saving and offers the advantage of customisation of analyte combinations and concentrations not commercially available.

In recognition of the growing requirements for in-house preparation of quality control, ISO REMCO introduced ISO Guide 80, which provides specific guidance on this topic.[10] The guide is aimed at laboratory staff who are required to prepare and use materials for in-house quality control applications. It outlines essential characteristics of QCMs and the processes by which they are made. The general principles of ISO Guide 34 and ISO Guide 35 (see Section 18.8.3) are also applicable, although the requirements are less stringent than for CRMs. Whilst CRMs are concerned with trueness and are intended for method validation and calibration, the purpose of a QCM is to demonstrate that a measurement system is reproducible and yields consistent results. Only limited characterisation to determine property values is therefore needed. Metrological traceability and uncertainty are not essential requirements of QCMs. Sourcing of bulk material, subdivision, homogeneity, characterisation and stability are the main steps outlined. Guidance on use of QCMs is outside the scope of this guide and this chapter.

18.8 Accreditation for Reference Material Producers

The result of any analysis can only ever be as good as the reference material used. Therefore it is important that laboratories have access to high quality, reliable reference materials. Laboratories should have confidence in the manufacturer(s) from whom they source reference materials, and manufacturers should be able to demonstrate their competence to customers and regulatory authorities. Accreditation is one way in which this competence may be demonstrated. It is not an absolute requirement that reference materials come from an accredited manufacturer, but the benefit of using an accredited manufacturer is that by definition they are believed to be competent and working to an internationally acknowledged standard. It is important to note that just because a reference material producer holds an accreditation, it does not necessarily mean that everything they produce or sell fall under that accreditation. In the same way, not all tests performed by an ISO/IEC 17025 accredited laboratory fall under their scope of accreditation.

The United Kingdom Accreditation Service (UKAS) recommends the use of accredited reference material producers where they exist.[11] The accreditations most relevant to reference material producers are ISO 9001, ISO/IEC 17025 and ISO Guide 34.

18.8.1 ISO 9001[12]

ISO 9001, entitled *Quality Management Systems*, is a standard designed to aid organisations of all types and sizes in implementing and operating effective quality management systems. It ensures that the needs of the customer and the regulatory requirements of the product are met. ISO 9001 addresses documentation requirements, management responsibility, resource management and product realisation. Measurement, analysis and continuous improvement are all essential elements of this standard. Whilst there are several documents in the ISO 9000 series, ISO 9001 is the only one to which an organisation can be accredited. ISO/IEC 17025 and ISO 15189 incorporate essential elements of ISO 9001 but add technical competency requirements specific to their application.

18.8.2 ISO/IEC 17025[4]

ISO/IEC 17025 outlines the general requirements for the competence of testing and calibration laboratories. It sets standards for the accuracy and reliability of analyses performed by a laboratory, including calibration, validation and quality control. There is a need for laboratories to understand metrological traceability, comparability and uncertainty measurement.

18.8.3 ISO Guide 34[13]

ISO Guide 34 outlines the general requirements for the competence of reference material producers. It is intended as a reference for the implementation of quality management systems, administrative and technical operations. It may also be used by customers, regulatory authorities and accreditation bodies in confirming competence. The guide is the responsibility of ISO REMCO. ISO Guide 34 is now considered the "gold standard" for reference material producers, and more and more producers are attaining this accreditation. ISO Guide 34 requires testing in an ISO/IEC 17025 accredited testing laboratory. Therefore ISO/IEC 17025 is a prerequisite of ISO Guide 34 and not an additional goal. There are many closely corresponding subclasses between ISO/IEC 17025 and ISO Guide 34, and a table of cross references is supplied in Annex C of the Guide. ISO Guide 34 is divided into two main sections, one that focuses on the organisation and management system and the other that focuses on the technical production requirements. ISO Guide 34 is just one of six guides within the ISO Guide 30 series, all published by ISO REMCO. It is the only one to which a producer can be accredited, although the different guides are intrinsically linked:

- ISO Guide 30, Terms and definitions used in connection with reference materials.
- ISO Guide 31, Reference materials – Contents of certificates and labels.
- ISO Guide 32, Calibration in analytical chemistry and use of certified reference materials.
- ISO Guide 33, Uses of certified reference materials.
- ISO Guide 34, General requirements for the competence of reference material producers.
- ISO Guide 35, Certification of reference materials – General and statistical principles.

ISO Guide 34 will be published as a full ISO standard. It is currently available for comment as a Draft International Standard (DIS), referred to as ISO/DIS 17034. This is likely to come into force in 2016.

18.9 Requirements for Independent Sources

Several ISO guidelines discuss the requirements for independent sources of reference materials for calibration and control:

ISO 15189, note[14]
Consideration be given to the use of control materials of an independent third party supplier, *i.e.* independent from those supplied by the reagent or instrument manufacturer.

ISO/IEC 17025:2005, Section 5.6.3.1[4]

Such reference standards of measurement held by the laboratory shall be used for calibration only and for no other purpose, unless it can be shown that their performance as reference standards would not be invalidated.

ISO Guide 99:2007 (VIM) Section 5.13, note 6[1]

In a given measurement, a given reference material can only be used for either calibration or quality assurance.

ISO Guide 35 Section 3.1, note 4[15]

A reference material can only be used for a single purpose in a given measurement.

ISO Guide 31, January 2012 draft[16]

A single reference material cannot be used for both calibration and validation in the same measurement procedure.

UKAS Guidance and Policy on the Selection and Use of Reference Materials, note 3[11]

Laboratories should be aware that it is not normally considered appropriate to use the same reference material for the calibration and validation/verification of a method.

ILAC-G9:2005[17]

In the wider sense, the validity of measurements can be assured when. . ... independent evidence of performance is available.

In the latter, the context is with reference to proficiency testing, but can also be extrapolated to independence of reference materials.

UKAS also recognise that in some instances it is difficult (or even impossible) to find two independent sources of reference materials and take a more pragmatic view. They advise:[11]

> "Laboratories will need to ensure that any bias in the calibration standard/quality control material is detectable by the quality control measures within their systems. . . .If the laboratory uses the same source/ batch for calibration and quality control then they will need to ensure that the possibility of contamination is considered and measures are put in place to pick this up."

The ideal situation is to obtain reference materials for calibration and control from two completely independent suppliers. Whilst it is not impossible that both may have obtained candidate reference material from the same source, the resultant reference material will have been independently purified, if required, verified and certified. Obtaining two different lots of solution from the same supplier cannot be considered truly independent if derived from the same neat source. If an impurity is not detected by the methods used, wrong correction factors are applied to hydrates/salts, balance issues are encountered or insolubility is inadvertently overlooked, it is likely these errors will be duplicated in subsequent lots. Likewise, some laboratories address the issue by having one analyst prepare the calibration material and another the control material. If both use the same procedure, then there should be good agreement, but it is questionable whether any errors in the reference material or procedure would be identified.

It is noteworthy that both sources need not be obtained from an accredited supplier. One source is a sufficient anchor point. If the two are not in agreement, then a problem is identified (see Section 18.7).

18.10   Certificates of Analysis

The first, and most important, point to note is that many materials are supplied with certificates, data sheets or other documentation, but this does not necessarily mean that they are certified reference materials. Careful study should be made of the certificate of analysis (CoA) to ensure that the material is fit for purpose. There are two ISO documents that deal with the requirements of CoA: ISO Guide 31, Reference materials – Contents of certificates and labels; and ISO 15194, Requirements for certified reference materials and the content of supporting documentation.[16,18] ISO Guide 31 states "Without the certificate, the material, however costly its production, is valueless".[16] It therefore goes without saying that the production of the certificate is equally as important as production of the reference material itself. What's more, a reference material should never be separated from its certificate. ISO Guide 30, Terms and definitions used in connection with reference materials, defines three document types that accompany reference materials:

2.3.2 Reference Material Certificate – ISO Guide 30[3]
Document containing the essential information for the use of a CRM, confirming that the necessary procedures have been carried out to ensure the validity and metrological traceability of the stated property values.
2.3.3 Reference Material Certification Report – ISO Guide 30[3]
Document giving detailed information, in addition to that contained in a reference material certificate, *e.g.* the preparation of the

material, methods of measurement, factors affecting accuracy, statistical treatment of results, and the way in which metrological traceability was established.

2.3.4 Product Information Sheet – ISO Guide 30[3]
Document containing all the information that is essential for using a RM other than a CRM.

In recent years there has be a decline in the issue of certification reports. These are comprehensive in nature and thus are time consuming to compile and costly to reproduce. For customers who repeat purchase standards, supplying such documentation on each occasion is unnecessary. Instead, more information is now being incorporated in the certificate of analysis/reference material certificate. Certification reports should still be made available upon request. The ready availability of the certification report dictates, to a certain extent, how comprehensive the CoA need be. Inclusion of analytical data in reference material certificates can be helpful to the end user, provides confidence in the analysis and in its interpretation. After all, the end user is in the business of analysis and interpretation! However, there is no absolute requirement nor correlation between number of pages and quality. In fact, frequently those materials at the top of the calibration hierarchy have the most concise CoA.

The information deemed essential on a certificate of analysis includes:[16]

- Name and address of the manufacturer
- Material identification, including batch/lot number
- Description of the material
- Intended purpose
- Instructions for use
- Storage conditions
- Product certification, including full description of methods used and uncertainty
- Date of document issue and material expiry date

18.11 Structural Confirmation

Structural elucidation, the process of determining a compound's chemical structure, involves used of nuclear magnetic resonance (NMR), mass spectrometry or Fourier transform infrared spectroscopy (FTIR).[19] Depending on the type of compound, polarimetry may be required for isomer determination. Microcrystalline tests can be used to distinguish between enantiomers. These are chemical reactions between a compound and a specific reagent which result in the formation of unique microcrystals. They are quick and cheap to perform and easy to interpret, but qualitative only. Results derived from identification techniques can be compared to monographs, peer reviewed literature, commercial libraries and previous batches of reference material where they exist.[19]

18.12 Purity

Purity determination should ideally be made by more than one technique, as not all impurities are detectable using a singular method. The mass balance approach is one of the most common methods of determining purity. This is an orthogonal approach taking into consideration chromatographic purity, residual solvent, residual water and residual inorganics. Chromatographic purity should be determined by two techniques, such as high-performance liquid chromatography (HPLC) and gas chromatography (GC). Differential scanning calorimetry may be used to determine purity and also provides information on melting point. Karl Fischer analysis provides a direct measure of residual water. Residual solvents are generally evaluated by thermogravimetric analysis (TGA). Combustion of the sample, sometimes referred to as ashing, enables determination of the non-volatile, inorganic impurities. This is a destructive process, so is not favoured for low-yield products. It should be noted that not all compounds are amenable to all techniques. In particular, those that thermally degrade cannot be analysed by techniques such as GC and TGA. In the mass balance approach (eqn (18.1)), the impurities determined by the methods described are summed and subtracted from 100% to give the mass fraction purity of the main analyte in the sample. This should be traceable to the SI.

Purity factor

$$= (100 - \text{wt}\,\% \ \text{residual solvent} - \text{wt}\,\% \ \text{water} - \text{wt}\,\% \ \text{inorganic})$$
$$\times \frac{\text{Chromatographic purity}}{100\,\%} \quad (18.1)$$

It is important not to be misled by poorly characterised materials with a high apparent purity. Impurities will only be detected if looked for. Therefore, conversely, the more analysis done, the more impure a compound can appear. It is not uncommon for only TLC or HPLC to have been performed on chemicals. The purity may be given as >98%, but it is possible the impurities may not visualise with UV or indicator used on TLC. For chromatographic techniques, impurities may not resolve from the compound of interest, may be insoluble in the medium used, not volatilize by GC or not be detectable by the wavelength used on UV. A multi-pronged approach is recommended to compensate for the limitations of each technique.

Quantitative NMR spectroscopy (qNMR) is gaining increasing popularity as a method for determining purity. qNMR has good resolving power and, in contrast to the mass balance approach, is independent of the analyte's chemical structure and provides a direct measure of the analyte of interest, negating the need to determine other classes of impurity. The technique is based on a signal comparison of the analyte with a CRM, such as a National Institute of Standards and Technology (NIST) traceable internal standard, which provides metrological traceability to the SI and low measurement uncertainty. Good equivalence has been demonstrated between the mass balance approach and qNMR, supporting the use of qNMR alone.[20] There is also likely benefit in combining these two methods, to minimise hidden bias and gain added confidence in the overall purity determination.

Reference materials should ideally have purity of greater than or equal to 98% for quantitative use and greater than or equal to 90% for qualitative use.[19] For forensic analysis, reference material manufacturers should consider purification when the purity falls below this specification. This may be carried out by prep-HPLC, distillation, extraction, salt formation and/or recrystallisation. In instances where the purity cannot be improved or represents the "best available", the European Network of Forensic Science Institutes (ENSFI) Drugs Working Group recommend the use of correction factors when the purity of the reference material is known and less than 98%.[21]

18.13   Guidance for Selection and Use of Reference Materials

Given below is a non-exhaustive list of considerations when selecting a reference material:

- Manufacturer's accreditation
- Property values (purity, concentration)
- Isotopic purity (for stable isotope labelled materials)
- Metrological traceability

- Uncertainty of property values
- Methodology (identity, purity, verification of concentration – consider range and selection of techniques)
- Intended use
- Period of validity/stability
- Commutability
- CE mark if intended for *in vitro* diagnostic use

18.14 Distinguishing between Chemicals and Reference Materials

In the jungle of products and suppliers it can sometimes be difficult to identify the wheat from the chaff, not least because chemicals and reference materials are often marketed side by side, with only price and/or quantity as a differentiator. Under such circumstances, chemicals can appear an attractive option. Commercially available chemicals, whilst sharing the same description and CAS number, lack the full characterisation that reference materials undergo. The purity of a chemical is usually given as a minimum, for example $\geq 98\%$. This is often derived from TLC or a singular chromatographic technique, and does not necessarily take into consideration impurities such as residual solvent, water or inorganics. Chemicals may be accompanied with documentation, and that may even be described as a certificate of analysis, but that does not mean to say that it complies with ISO requirements for a CRM. Documentation for chemicals usually only lists basic physical properties. Typically, metrological traceability and uncertainty are omitted. Chemicals can be used but in the absence of certification data provided by the supplier it is the responsibility of the end user to determine fitness and undertake further characterisation as appropriate.

Reference materials for toxicology are typically retailed as dilute, calibrated solutions (0.1 or 1 $mg\,mL^{-1}$) or in neat form in milligram quantities, most commonly 10–100 mg. If a product is sold in gram quantities it is a good indication that the material is likely chemical grade. Of course, there are exceptions to this rule.

18.15 Weighing Environment

The environment in which a reference material is weighed can impact the accuracy of the weighing. Ideally, the room should be temperature and humidity controlled. A change in ambient temperature can result

Table 18.2 Balance tolerances.

Balance	7-Place	6-Place	5-Place	4-Place
Resolution	0.0001 mg	0.001 mg	0.01 mg	0.1 mg
Minimum weight	1 mg	3 mg	20 mg	125 mg

in a change in weight: a 1 °C change in temperature can result in a 0.1 mg change in the reading of a 100 g weight.[22] The weighing bench should be vibration and sag resistant. Earthing of the bench will prevent against the build-up of electrostatic charge, although many modern balances are fitted with static charge detectors which generate opposite charge to neutralise static.

The ideal weighing surface is a floor mounted, stand alone, polished granite work bench. This should not be located in a high transit area, nor be in the direct path of air conditioning or other drafts. Care should be taken not to site a balance near extraction cabinets, or other instrumentation, which may generate vibrations and cause air flow and temperature fluctuations from outlets for integrated cooling fans.[22] Direct sunlight should also be avoided due to localised temperature fluctuation. Where these criteria cannot be met, an assessment should be made to determine whether it is fit for the task being undertaken. Balances and weights should be calibrated by a UKAS accredited company and checked daily prior to use. The readability of a balance (display resolution) is an important contributor to measurement uncertainty. If a balance does not have sufficient resolution for the quantity to be measured, it should be recognised that the measurement uncertainty will be unacceptably high (see Table 18.2).

18.16 Neat *versus* Solutions[23]

A wide range of reference materials are available as calibrated solutions, as well as in neat form. In many instances, selection is a matter of personal choice, although for certain applications one form may be preferred over the other. Bulk drug laboratories often prefer neat, whilst toxicology laboratories prefer solutions. Neat is preferred for methods such as FTIR. Solutions are less labour intensive since they remove the arduous weighing step. They also eliminate the risk of in-house errors associated with correction for purity, salt or water content. The solubility, appropriateness of the solvent and stability is also pre-determined by the manufacturer with solutions. The single-use nature of ampoules can be easier from a storage and stock checking perspective. For expensive materials, calibrated, dilute solutions are a means of offering a standard in an affordable and user-friendly presentation, as compared to low-quantity neat material. Neat material may also be difficult to weigh if volatile, static, deliquescent (readily adsorb water vapour from the air) or oils. It should however be remembered that not all compounds are stable in solution.

18.17 Free Base *versus* Salts[23,24]

Reference materials are supplied in a variety of different chemical forms. Most commonly encountered are freebase, salts (hydrochloride, sodium, sulfate, *etc.*) and hydrates (addition of water). Freebase is the pure, basic, deprotonated form of an amine as opposed to its salt form. Freebase may be preferred as no correction factors are required in weighing, and it may be perceived you get more for your money when purchasing 10 mg freebase as compared to 10 mg of salt. However, some freebase forms are unstable and thus are better obtained in salt form. Salts can also improve the solubility in aqueous solvents. The physical state of a compound may change between freebase and salt form. Liquids can be more difficult to handle, so in instances where the freebase is a liquid, the salt may be preferred.

When purchasing reference materials as ready-made solutions, the concentration is normally corrected for the presence of salt, and this should be documented on the CoA. When preparing solutions in-house from neat material, errors can result from incorrect calculation and adjustment for salt or hydrate. This is particularly the case when the ratio of freebase : salt deviates from 1 : 1, *i.e.* amphetamine sulfate (2 amphetamines to 1 sulfate). It is recommended that all calculations be independently verified. The term hydrate is used to describe a one-to-one ratio of water to drug. Dihydrates and trihydrates are used to describe two-to-one and three-to-one ratios, respectively.

18.18 Recovering Reference Materials from their Container[24]

It can sometimes be difficult to recover all the reference material from the container. Material can distribute across the surface area of the bottle and lid, or can become static. Conversely, materials may be deliquescent or otherwise difficult to manage. The narrow opening of ampoules or other containers can also restrict access. For these reasons it is a common practice for reference material suppliers to overfill containers to ensure the end user can recover the retailed quantity. A 10 mg pack size will always contain in excess of 10 mg, but the overfill is variable. Ampoules containing 1 mL are typically overfilled by 0.1–0.2 mL. The overfill should be indicated on the certificate of analysis but should not be assumed to be exact. For quantitative purposes the end user should always transfer the desired volume or quantity of material to another container before diluting to the required concentration. Flushing out the ampoule or container will result in a higher than anticipated concentration.

When weighing powders, gently tap the container on the work surface so that any material distributed on the lid or sides collects at the bottom. Weigh the material using a five-place analytical balance calibrated using weights traceable to the SI and higher order standards (*i.e.* NIST) through an unbroken chain of comparisons. Depending on the nature of the material being weighed, specialist anti-static functions on the balance can assist. Weighing both the container and weighing boat before and after material transfer acts as a double check of the total weight of standard removed from the vial. It should be remembered that some material may be lost on the spatula during weighing and thus the quantity removed from the container may be greater than that transferred to the weighing vessel. Before opening ampoules it is important to ensure that there is no liquid in the top portion. Hold them by the top and flick gently to ensure that all of the liquid moves to the bottom before snapping the vial. The desired volume should be removed with a volumetric pipette or standard GC syringe.

18.19 Dealing with Low Quantity Reference Materials[24]

Reference materials may be synthesised or extracted/isolated from nature. The yield can vary significantly, depending on the natural abundance and complexity of the production process (*i.e.* number of synthesis steps). The cost of the raw materials and efficiency of the reactions are also important determinates of quantity of material produced. If not optimized, a large quantity of the raw material invested in a reaction can go to waste. These factors can impact the retail pack size and price. Frequently this relates to stable isotopically labelled internal standards (SILIS), glucuronides and other rare metabolites.

In toxicology, neat material retailed in 1–5 mg pack sizes offer an affordable way of obtaining otherwise expensive material. However, such small quantities can present challenges when trying to recover the material for production of quantitative standards. The same can be said of liquid or oil standards, which can dissipate across the surface area of the container and lid, giving an appearance of an empty vial. In recognition of the weighing challenges that small pack sizes present, some suppliers offer standards in a dried-down, ready to reconstitute presentation. In such instances, the producer has accurately weighed a larger starting quantity, prepared a solution, aliquotted it and evaporated the solvent to dryness to give a known quantity of residue in each vial.

On request, most reference material suppliers should be able to provide an exact calibrated weight of material per vial. This can be used to flush out the container with a known volume of solvent, to obtain a solution suitable for quantitative analysis, otherwise referred to as total transfer (dilution of the reference material *in situ*). Exact weights may not be available from distributors who re-sell pre-packed material, or may incur a charge associated with administration and new weighing. If this service is not available, then the following procedure may be used: control weigh the container with the material inside; flush it out with the desired solvent; dry the container to remove residual solvent and let it acclimatise; after an hour, recheck the weight to ascertain how much material was inside. The total weight of standard removed from the vial can be calculated by subtracting the dry weight from the original weight.

18.20 Correct Selection of Solvent[24]

The correct choice of solvent for the preparation of a reference material solution is essential. Solubility issues may manifest as products failing to dissolve, partial dissolution or the compound falling out of solution, *i.e.* when refrigerated. Increasing the volume of solvent, gentle warming or sonication can help to dissolve the compound, but are not always ideal resolutions. Degradation or conversion of a compound may be catalysed by impurities in a particular batch of solvent, a particular vessel type or by other compounds present in a multi-component standard mixture.

Some compounds, such as tacrolimus and zopiclone, are known to undergo methanolysis. Where solutions are to be used for spiking, for example as internal standards, calibrators and controls, the solvent used must be miscible with the matrix to be spiked. It is also important to be aware that, if the solvent content is too high, precipitation can occur when spiking into a matrix such as blood. The matrix may also affect compound stability. When you buy reference materials in solution you can be confident that the solvent selected has been investigated and is appropriate for the analyte. Methanol and acetonitrile are the most common solvents used for production of toxicology reference material. Some glucuronides are supplied in water/methanol mixes.

18.21   Stability[24]

The assessment of the stability of a reference material is addressed in ISO Guide 34, and in more detail in ISO Guide 35 – General and statistical principles for certification. It is the responsibility of a reference material producer to determine the degree of stability of the reference materials they supply. The stability assessment should incorporate determination of long-term storage and stability under transport conditions. This is achieved by real time and accelerated stability testing. The information obtained is used to specify validity period and storage conditions, which are documented on the certificate. Stability after opening is more difficult to determine due to laboratory specific variables, such as usage pattern, container, storage conditions, volume, dilution, solvent/matrix, *etc.* For this reason the reference material producer's responsibility ends after the materials first opening. This is outlined in section 5.7.4 of ISO Guide 34:[13]

> "The reference material producer shall ensure that the integrity of each individual reference material unit is maintained until the seal has been broken or up to the point when presented for analysis. The producer cannot be held responsible for the material once the seal has been broken."

Nevertheless, the reference material producer does have post-sale responsibility to re-test the materials sold throughout the course of their lifespan and update customers about changes in shelf-life and repercussions on use.

Ideally, solutions should be used shortly after opening the ampoule to avoid concentration changes due to evaporation. The use of Certan® vials can prolong their life by minimising evaporation and contamination. They offer the advantage of a sealed ampoule with the flexibility of a screw cap bottle. Neat material should not be stored for any length of time after opening, due to the risk of water absorption from atmospheric humidity.

18.22 Databases for Certified Reference Materials

18.22.1 COMAR Database[25]

COMAR (Code d'Indexation des MAtériaux de Référence) is an international database for Certified Reference Materials (CRMs). It was developed to aid users of such materials in navigating the diverse and complex market, and identifying materials that are reliable and fit for purpose. The COMAR database was established in the early 1970s by Laboratoire National de Metrologie et d'Essais (LNE) in France. Since then it has grown and developed as a result of the contributions of a network of 20 international metrology institutes. In 2003, launch of a web-based version of the database, http://www.comar.bam.de, significantly improved accessibility. This is hosted and maintained by Bundesanstalt für Materialforschung und-prüfung (BAM), the Federal Institute for Materials Research and Testing in Germany. This database is searchable by catalogue number, description, certified properties, matrix and field of application. COMAR is a resource that supports the identification of suitable CRMs and producers but is not intended as a retail site.

18.22.2 JCTLM Database[26]

The JCTLM (Joint Committee for Traceability in Laboratory Medicine) was established in 2002 by CIPM (International Committee of Weights and Measures), IFCC (International Federation for Clinical Chemistry and Laboratory Medicine) and ILAC (International Laboratory Accreditation Cooperation). The committee was formed in response to the implementation of the European Community Directive 98/79/EC on *in vitro* diagnostic (IVD) medical devices.[27] The aim was to provide a worldwide platform for guidance on harmonisation and traceability of measurements in laboratory medicine. This led to the birth of the JCTLM database, available *via* http://www.bipm.org/jctlm/.

The database, which is administered by BIPM, is a compilation of "higher order" reference materials, measurement procedures and services. The term "higher order" is derived from the IVD Directive, which states that "The traceability of values assigned to calibrators and/or control materials must be assured through available reference measurement procedures and/or available reference materials of a higher order". These "super" reference materials are produced by institutes or international organisations with responsibility for metrological traceability. The materials have been thoroughly characterised using the best available measurement procedures, sometimes through inter-laboratory comparison between national metrology institutes arranged *via* the Consultative Committee for Amount of Substance: Metrology in Chemistry and Biology (CCQM). The uncertainty should therefore be the smallest attainable.

The database contains more than 200 CRMs covering approximately 130 analytes, but is not a comprehensive list of those available. Only CRMs that have been put forward by the manufacturer, have undergone international review, and been deemed to fulfil the requirements of ISO 15194 and 15193, are considered for the JCTLM database. For ease of use, reference materials are categorised as follows: blood gases, blood grouping, coagulation factors, drugs, electrolytes, enzymes, metabolites and substrates, microbial serology, non-electrolyte metals, non-peptide hormones, nucleic acids, proteins, vitamins & micronutrients or other. Whilst IVD manufacturers and regulators are the prime intended audience of the JCTLM database, as more clinical laboratories move towards mass spectrometry and in-house method development, its target audience widens. When "off the shelf kits" are introduced into the laboratory, end users should ask the manufacturer what steps have been taken to ensure metrological traceability.

18.23 CCQM[28]

The Consultative Committee for Amount of Substance: Metrology in Chemistry and Biology (CCQM) was first established in 1993, and assumed responsibility for developing, improving and documenting equivalence of CRMs and reference measurement procedures for chemical and biological measurements. To this aim, the CCQM established eight working groups: organic, inorganic, gas, electrochemical, bioanalysis, surface analysis, key comparison and calibration and measurement capacities (CMC), and strategic planning. There are two further *ad hoc* working groups on microbial measurements and the mole. The CCQM has 30 members and 10 observers. Between 1999 and 2012, under the direction of BIPM, CCQM organised more than 100 key comparisons and pilot studies.

These are interlaboratory studies, which enable NMIs to assess and demonstrate their measurement capabilities and establish equivalence of their measurement standards.

18.24 Import and Export of Controlled Drug Reference Materials[29]

Production, possession, use, import/export and distribution of narcotic drugs, psychotropic substances and precursor chemicals are controlled *via* International Drug Control Conventions. The substances to which these controls apply are outlined in the so-called Yellow, Green and Red lists, respectively, readily available on the International Narcotics Control Board (INCB) website (www.incb.org). The INCB monitors and supports compliance with these Conventions. In addition, there are European and National controls, giving rise to sometimes diverse control measures being adopted globally. Each country typically maintains a list of substances controlled domestically. For drug control status in the UK, reference should be made to the notes in Parts I, II, III and IV of Schedule 2 to the Misuse of Drugs Act 1971 and in Schedules 1–5 to the Misuse of Drugs Regulations 2001. A simplified, but inexhaustive, list can also be found on the UK Home Office website, and clarification can be sought from the Drugs Licensing and Compliance Unit Communications Officer (DFLU.ie@homeoffice.gsi.gov.uk).

Producers, re-sellers or end users of controlled drug reference materials require a licence from their National Competent Authority – the Home Office, Drugs Licensing and Compliance Unit in the case of the UK – to possess, manufacture and supply. In addition, a licence is required for each import or export, *i.e.* it is valid for one consignment only. If a substance is controlled at site of origin but not in the receiving country, an official letter of no objection (LONO) is required from the recipient in order for the sender to apply for the necessary export licence. In most, but not all, countries there exist agents who are able to assist laboratories with the importation and exportation of controlled drugs. Where this service is not provided, or the compounds of interest are not available *via* such a supplier, there may arise a situation where an end user has to act as importer. Applications for import and export of controlled substances are made *via* the National Drugs Control System (NDS). See Figure 18.2.

At the time of writing, the licence application fee for import or export of a controlled substance in the UK is £24. Licence applications

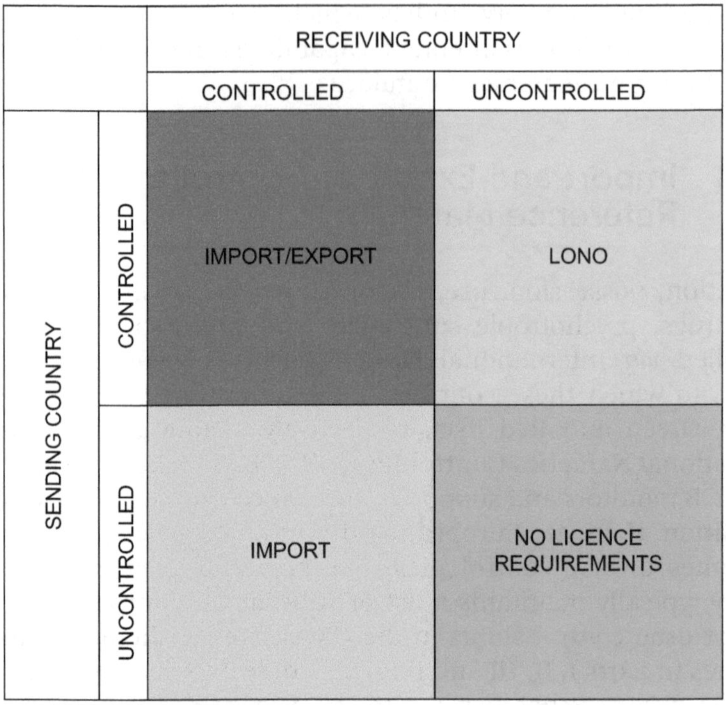

		RECEIVING COUNTRY	
		CONTROLLED	UNCONTROLLED
SENDING COUNTRY	CONTROLLED	IMPORT/EXPORT	LONO
	UNCONTROLLED	IMPORT	NO LICENCE REQUIREMENTS

Figure 18.2 Overview of licence requirements for import/export of controlled substances.

are typically processed within 2 weeks of submission. Licences to import controlled substances into the UK are valid for a maximum period of 3 months. Licences to export controlled substances from the UK are valid for a maximum period of 2 months. This term is reduced if the corresponding import licence expires earlier.

Companies or individuals who are engaged in production and supply of controlled substances are required to complete an annual return, which the Home Office collates and returns to the INCB. Annual statistical returns must be submitted by the 31st January each year. End users do not ordinarily need to complete a full return unless engaged in import/export activity.

18.25   Import and Export Exemptions

The import/export procedures were originally designed around the bulk pharmaceutical trade, where transport of kilogram, metric tonne or even container loads is not an infrequent occurrence. Nonetheless, microgram to milligram quantities of reference material are also caught up in the regulations.

In the UK there are some exemptions for licensing where the drug is to be used for forensic analysis, but these are limited.[30] The Misuse of Drugs Regulations 2001 has provision for "exempted products", which it defines as "a preparation or other product consisting of one or more component parts, any of which contains a controlled drug", where:

(a) the preparation or other product is not designed for administration of the controlled drug to a human being or animal;
(b) the controlled drug in any component part is packaged in such a form, or in combination with other active or inert substances in such a manner, that it cannot be recovered by readily applicable means or in a yield which constitutes a risk to health; and
(c) no one component part of the product or preparation contains more than one milligram of the controlled drug or one microgram in the case of lysergide or any other *N*-alkyl derivative of lysergamide.

All three definitions must be satisfied to qualify for exemption. Most commonly this is applied to drugs contained within a biological matrix, *i.e.* calibrator or control material in serum, plasma, blood or urine. In contrast, reference materials supplied in solvent are considered easily recoverable. It could be argued that, for many drugs, 1 mg or 0.1 mg represents far less than a typical active dose, and thus is unlikely to be considered a risk to human health, but to date this exemption has not been fully explored.

Other countries have adopted different approaches to aiding the movement of forensic reference materials. In Germany, solutions containing 0.1 mg mL^{-1} or less are considered uncontrolled and may cross the border without import or export licence. Italy have an exempt threshold of up to 0.5 mg mL^{-1}. In the USA, producers of reference materials can apply for DEA exemptions for concentrations of up to 1.0 mg mL^{-1}. In Canada a similar process is in operation. A LONO is still required if controlled in the exporting country. In the UK, deuterium-labelled compounds are included within control measures, whilst in other countries a deuterated narcotic could fall outside of control. In Poland, controlled substances in a matrix are captured in their legislation. Exemptions are not the only measures in place to assist the legitimate supply of reference materials. Some countries provide a helping hand with managing applications for controlled drug licences free of charge. Others, such as Australia, allow ongoing authority to import. This enables import of the specified items for a designated period using Import Consigned Records. In 2007 and 2011 the INCB investigated and made recommendations for the removal of obstacles affecting the availability of internationally controlled reference materials.[29,31,32]

Supplementary Reading

R. P. Archer, R. Treble and K. Williams, Reference materials for new psychoactive substances, 2011, 3(7–8), 505–514.

J. Button and R. Treble, A Proactive Approach to Reference Material Production for New Psychoactive Substances. The Column (LCGC online version), 21 June 2013, 9(11), 14–17. Available from: http://images2.advanstar.com/PixelMags/lctc/digitaledition/June21-2013.html#14.

Scientific Working Group for the Analysis of Seized Drugs (SWGDRUG) Recommendations, Version 7.0, August 2014.

References

1. Working Group 2 of the Joint Committee for Guides in Metrology (JCGM/WG 2). International vocabulary of metrology – Basic and general concepts and associated terms (VIM), JCGM 200:2008. 2008. Available from www.bipm.org.
2. *Eurachem Guide: Terminology in analytical measurement – Introduction to VIM 3. 2011*, ed. V. J. Barwick and E. Prichard, ISBN 978-0-948926-29-7, Available from www.eurachem.org.
3. International Organisation of Standardisation (ISO). Reference materials – Selected terms and definitions, ISO Guide 30:2015 3rd edn, 2015.
4. International Organisation of Standardisation (ISO). General requirements for the competence of testing and calibration laboratories, ISO/IEC 17025:2005 2nd edn 2005.
5. International Union of Pure and Applied Chemistry (IUPAC), Physical and Biophysical Chemistry Division, *Quantities, Units and Symbols in Physical Chemistry*, 3rd edn, RSC Publishing, 2007.

6. S. L. R. Ellison, B. King, M. Rösslein, M. Salit and A. Williams. Eurachem/CITAC Guide: Traceability in Chemical Measurement – A guide to achieving comparable results in chemical measurement. 2003. Available from www.eurachem.org.
7. V. Barwick. Evaluating Measurement Uncertainty in Clinical Chemistry – Case Studies. LGC Limited (2012). Available from http://www.lgcgroup.com/LGCGroup/media/PDFs/Our%20science/ NMI%20landing%20page/Publications%20and%20resources/ Reports/Clinical_worked_examples_report_Final.pdf.
8. Working Group 1 of the Joint Committee for Guides in Metrology (JCGM/WG 1). Uncertainty of measurement – Guide to the expression of uncertainty in measurement (GUM:1995), JCGM 100:2008 (2008). Available from www.bipm.org.
9. S. L. R. Ellison and A. Williams. Eurachem/CITAC Guide CG4 Quantifying uncertainty in analytical measurement. 3rd edn, 2012.
10. S. Wood and A. Botha, The new ISO Guide 80: Guidance for the in-house preparation of quality control materials (QCMs), *Accredit. Qual. Assur.*, 2014, **19**, 477–480, DOI: 10.1007/s00769-014-1084-1.
11. United Kingdom Accreditation service (UKAS). Guidance and policy on the selection and use of reference materials, TPS 57. 2nd edn, March 2011.
12. International Organisation of Standardisation (ISO). Quality management systems, ISO 9000:2015 5th edn, 2015.
13. International Organisation of Standardisation (ISO). General requirements for the competence of reference material producers, ISO Guide 34:2009 3rd edn, 2009.
14. International Organisation of Standardisation (ISO). Medical laboratories – Requirements for quality and competence, ISO 15189:2012 3rd edn, 2012.
15. International Organisation of Standardisation (ISO). Certification of reference materials – General and statistical principles, ISO Guide 35:2006. 3rd edn, 2006.
16. International Organisation of Standardisation (ISO). Reference materials – Contents of certificates and labels, ISO Guide 31:2000. 2nd edn, 2000.
17. International Laboratory Accreditation Cooperation (ILAC). Guidelines for selection and use of reference materials, ILAC-G9:2005. 2005 Available from www.ilac.org.
18. International Organisation of Standardisation (ISO). In vitro diagnostic medical devices – Measurement of quantities in samples of biological origin – Requirements for certified reference

materials and the content of supporting documentation, ISO 15194:2009 2nd edn, 2009.

19. R. E. Franckowski, Authentication of Reference Materials. 2013. NIST Emerging trends in synthetic drugs workshop. April 30, 2013 Available from http://www.nist.gov/oles/upload/5-Day-2 Authentication-of-Standards-DEA-NIST-2013-REF-Franckowski.pdf.

20. S. R. Davies, K. Jones, A. Goldys *et al.*, Purity assessment of organic calibration standards using a combination of quantitative NMR and mass balance, *Anal. Bioanal. Chem.*, 2015, **407**(11), 3103–3113.

21. European Network of Forensic Science Institutes (ENFSI) Drugs Working Group. Guidelines of the use of reference materials in forensic analysis (ENSFI/002) 2007.

22. T. Scorer, M. Perkin and M. Buckley. Measurement Good Practice Guide No. 70 – Weighing in the Pharmaceutical Industry. June 2004.

23. J. Button. Reference materials – why quality matters. The Column (LCGC online version) 22 June 2012 Volume 8 Issue 11:11–21. Available from http://digital.findanalytichem.com/nxtbooks/ advanstaruk/thecolumn062212/#/11.

24. J. Button. Using reference materials – commonly encountered errors and how to avoid them. The Column (LCGC online version) 7 November 2012 Volume 8 Issue 20:10–15. Available from http:// digital.findanalytichem.com/nxtbooks/advanstaruk/ thecolumn110712.

25. T. Steiger and R. Pradel, COMAR: The international data base for certified reference materials – An overview, *Accredit. Qual. Assur.*, 2015, **20**, 47–52.

26. D. Armbruster and R. Miller, The Joint Committee for Traceability in Laboratory Medicine (JCTLM): A global approach to promote the standardisation of clinical laboratory test results, *Clin. Biochem. Rev.*, 2007, **28**, 105–113.

27. Directive 98/79/EC of the European Parliament and of the Council of 27 October 1998 on in vitro diagnostic medical devices. Official Journal of the European Communities. 1998.

28. Bureau International des Poids et Mesures (BIPM). Consultative Committee for Amount of Substance: Metrology in Chemistry and Biology (CCQM) Available from http://www.bipm.org/en/ committees/cc/ccqm/.

29. J. Button, Frustrated with delivery times for reference materials? Don't delay, find out why today!, *TIAFT Bull.*, 2013, **XLIII**(3), 38–41.

30. UK Home Office, Forensic Analysis and Reference Standards-Licensing Factsheet https://www.gov.uk/government/uploads/system/uploads/attachment_data/file/397245/Forensic_Analysis__Reference_Standards-_FAQs_-_January_2015.pdf.
31. International Narcotics Board (INCB), Guidelines for the import and export of drug and precursor reference standards for use by national drug testing laboratories and competent national authorities, United Nations publication, ISBN: 978-92-1-048121-2, 2007.
32. International Narcotics Board (INCB), Additional courses of action in support of the implementation of the 2007 INCB Guidelines for the import and export of drug and precursor reference standards – for use by national drug testing laboratories and competent national authorities, 2012.

19 Nail Analysis in Forensic Toxicology

Nikolas P. Lemos

The University of California, San Francisco & Office of the Chief Medical Examiner, City and County of San Francisco, CA, USA
Email: nikolas.lemos2@ucsf.edu

19.1 Introduction

In forensic and analytical toxicology laboratories around the world, forensic toxicologists rely heavily on the analysis of blood for the detection, confirmation and quantification of drugs and/or their metabolites when an acute or recent exposure to a substance (often within hours) is suspected or important in the outcome of the case investigated. Urinalysis is often utilized by the same laboratories when information regarding more distant exposure to drugs and substances (often within days) is suspected or critical to the evaluation of the case at hand.

It may, however, be of interest or may be needed in certain areas of drug monitoring science to determine analytes in biological fluids and matrices other than in these conventional ones. Traditionally, systematic forensic toxicology used a wide selection of body fluids and tissues in order to optimize the detection of drugs and poisons with the analytical techniques available to the forensic toxicologist, but this approach fell out of use due to the advent of more sensitive methods and techniques that were suitable for the detection of low concentrations of analytes in blood. There has also been an increased

Forensic Toxicology: Drug Use and Misuse
Edited by Susannah Davies, Atholl Johnston and David Holt
Published by the Royal Society of Chemistry, www.rsc.org

need for faster turnaround times in obtaining analytical results while utilizing more cost-efficient procedures in laboratories evaluating post-mortem forensic toxicology cases as well as human performance forensic toxicology ones.

Specifically, in post-mortem forensic toxicology laboratories, interpretation of drug concentrations in alternative biological tissues and fluids was problematic because such interpretation usually involved comparison to blood concentrations which not always resulted in scientifically valid conclusions being reached despite the time-consuming and laborious analyses involved. In contrast to the post-mortem forensic toxicology experiences where it was not always possible or desirable to sample alternative biological fluids and matrices, drug monitoring in such specimens has been successfully performed in response to a completely different set of needs than those arising in forensic autopsy cases.[1] Drug analysis in specimens such as amniotic fluid, breast milk and foetal blood may provide evidence of foetal drug exposure.[2] To account for drug elimination routes other than the renal one, biliary and faecal drug concentrations have also been successfully monitored.[3] Keratinous matrices such as hair and nail may provide an insight into an individual's past licit or illicit drug use history or exposure,[4] while in other types of drug monitoring cases, analyses in alternative specimens such as bronchial secretions, nails and tears may confirm the presence of a drug at the site of intended action.[5,6]

Drug monitoring in alternative biological specimens offers significant advantages when compared to conventional specimens such as blood and urine. For example, alternative biological specimens may offer extended windows of drug retention as well as non-invasive specimen collection by minimally trained personnel. Other scientific areas in which alterative biological specimens become useful include bioavailability studies, doping control, drug organ targeting and localized effect monitoring. This chapter focus on the analysis of nail in forensic toxicology.

19.2 The Nail

A part of the human skin, nail and its immediate surroundings have been extensively studied.[7-13] The main purpose on nail in humans is to provide protection of the finger's terminal phalanx.[14] It is also there to assist in touch and for picking up small objects as well as for scratching. The nail plate consists of keratin and is derived from an invagination of the epidermis. Zaias[15] reported as early as 1963 that this invagination is first visible in the 9 week embryo and that the nail is virtually formed by the 20th week.

The nail fold consists of a roof, a floor and lateral walls, whilst the nail bed represents that part of the terminal phalanx which lies below the exposed nail plate. There has been some debate regarding the area of the nail fold and nail bed which takes part in the formation of the nail plate. Zaias *et al.*[16] described in 1968 that the nail plate is derived from the matrix and that there was some probability that a small part of the nail roof of the nail fold also takes part in the formation of the nail plate. These observations of Zaias *et al.* were based on early experiments on the squirrel monkey, a primate with a flat nail very similar to that of a human. Experiments on human subjects, however, showed that there was some activity in the nail bed and so the formation of the nail in the squirrel monkey was somewhat different from that of humans.[17] As early as 1954, the opinion was put forward that the nail is not formed from one source but in three layers, which were termed "dorsal", "intermediate" and "ventral" nails.[18] The intermediate nail is the main portion and is derived in great part from the matrix. The ventral nail is derived from the nail bed distal to the lunula, while the dorsal nail is derived from the roof and a small portion of the floor of the nail fold. These two models of nail formation oppose the original theory of nail formation as described by Pinkus[19] in 1927, quoting Boas (1894).[20] This theory described three parts of the nail bed. Firstly, the proximal part extends as far forward as the distal margin of the lunula, which is considered to be the nail forming or fertile part of the nail bed. Secondly, the part on which the nail plate rests and which takes no place in the formation of the nail (sterile part). This part extends from the distal margin of the lunula to the line where the anterior edge of the nail separates from the bed (also known as the onychodermal band). Lastly, the "sole horn" (sohlenhorn) part which brings horny substance to the nail but does not form any part of it.

More recent research allows for a more complete picture of the structure of the nail unit to be described.[21–23] According to these reports the nail unit consists of six main components:

1. The portion responsible for generating the nail, the nail matrix.
2. The actual product, the nail plate.
3. The portion responsible for ensheathing, composed of the visible cuticle (or eponychium) and the true cuticle.
4. The portion responsible for support, the nail bed and phalanx.

5. The portion responsible for anchoring, the ligaments between the phalanx and the distal and lateral grooves.
6. The portion responsible for framing, the nail walls (or nail folds).

Reports have been published addressing issues regarding the presence of vascular networks surrounding the nail,[24] as well as the suspected regeneration of the nail bed.[25] Research on the structure and behaviour of the nail unit is on-going, with the interest of dermatologists remaining high and that of forensic toxicologists rising.

19.3 The Nail in Forensic and Analytical Toxicology

Nails have been used for the purpose of determining voluntary or involuntary exposure to substances for decades. Traditionally, nail has been used as an analytical specimen in studies examining the efficacy of antifungal agents such as itraconazole,[6,26] fluconazole[27,28] and terbinafine,[29–31] as reviewed by Zaias *et al.* in 1996.[32]

More than 20 years earlier, in 1972, Harrison and Clemena[33] showed that it is possible to estimate the quantity of many trace elements in fingernail clippings by mass spectrometry. Pounds *et al.*[34] published a method for the analysis of arsenic in nail in 1979. Following that publication, nail became the preferred specimen in cases of suspected industrial or criminal poisoning due to arsenic and other heavy metals.

Subsequently, analytical protocols for the qualitative and quantitative determination of several drugs in nails have been offered by researchers from around the world. Specifically, fingernail and/or toenail specimens were successfully used in the detection of various amphetamine-related compounds,[4,35,36] cocaine and its metabolites,[37–39] cannabis,[40] morphine-type opiates and opioids such as methadone,[41,42] ketamine and metabolites,[43,44] and phencyclidine,[45] as well as caffeine, nicotine and cotinine in the nails of new-borns.[46] Finally, nail specimens have been successfully analysed to detect steroids as possible doping agents in sport[47,48] and ethyl glucuronide as a biomarker of alcohol use, abuse or binge drinking behaviour[49] as reviewed by Cappelle *et al.* in 2015.[50]

19.4 Challenges and Further Considerations

When attempting to incorporate nail analysis in toxicology studies, several challenging areas should be considered and addressed in order for the adopted analytical protocol to be successful and any drawn inferences to be scientifically valid.

Scientists must comprehend the limitations of such analysis, which stem from the inadequately investigated and poorly understood mechanism of drug incorporation in nail. Models of drug incorporation in fingernails appear to suggest that a three-dimensional model of drug incorporation is the most likely one, consisting of incorporation from sweat, from the vertical growth of the nail bed as well as from the horizontal growth of the nail matrix.[51,52] Drug incorporation in nail is further complicated by variations due to such factors as individual nail growth rate, gender, age, pregnancy and season of the year. Growth rate differences between fingernail and toenail specimens have been empirically established, but little or no research exists that systematically examines the effects of common factors such as diet or cosmetic treatments.

Before analysing fingernail specimens which are normally associated with touch, it is important for a forensic or analytical toxicologist to employ decontamination methodologies that allow for the detection of what is in the nail instead of what is on the nail. Similarly, toenails often suffer from drug contamination from sweat and require decontamination as well. Typically, the decontamination involves a multistep procedure based on sonication of the nail in a small volume of detergent capable of dissolving and removing surface fats and lipids, followed by three sonication cycles in water designed to remove the detergent and finally multiple sonication cycles in an organic solvent such as methanol. It is critical to screen the wash solutions for the target drugs and to continue sonication cycles in the selected solvent, *e.g.* methanol, until the final wash tests negative for the drug or drugs of interest. Often, nails from regular drug users will require three methanolic washes before all superficial contamination is washed away and the final wash tests negative, but there have been forensic cases where fingernails from subjects regularly handling drugs required as many as seven methanolic washes before the final methanolic wash produced a negative drug test.

After all superficial contamination has been removed, drugs should be extracted from the nail for the purpose of instrumental drug screen and confirmation/quantitation as necessary. Before the nail matrix is broken apart, thus releasing any drugs into solution, the weight of the decontaminated and air-dried nail material should be accurately recorded to enable concentration measurements to be later possible, especially since the next step in the procedure typically results in the complete destruction of the nail specimen.

Once weighed, the nail keratin matrix needs to be broken apart in order to release any drugs that are incorporated inside it, enabling their detection and quantitation. Careful consideration must be exercised when determining how to achieve this goal. Nails may be dissolved in concentrated acid or base followed by a clean-up step involving liquid–liquid or solid phase extractions, but one should be aware that such chemically harsh treatments often alter the identity of the drugs of interest into their hydrolysed derivatives. For example, after employing a harsh chemical treatment, such as alkaline hydrolysis of the nail, one should not expect to detect heroin (diacetylmorphine) or 6-monoacetylmorphine but should instead expect to mostly detect morphine even in the nails of confirmed heroin users. Alternative techniques such as pulverization of the nail keratin matrix at room temperature or cryogenic grinding of the nail in liquid nitrogen have been successfully employed to overcome this drug-identity-changing obstacle. Such techniques, although mechanically harsh, are chemically inert and result in the production of a fine nail powder often smaller than 6 μm in particle size from which drugs can easily be extracted in an organic solvent without their chemical identities being changed. Cryogenic grinding of nail (and hair) in liquid nitrogen is commonly used in some jurisdictions in order to establish if court-ordered abstinence is adhered to in probation or in child-custody cases.

The ability of a laboratory to detect the relatively low drug concentrations often found in nail (compared with those measured in blood or urine specimens) is directly dependent on the availability and sensitivity of its drug screening and confirmatory techniques. Gas chromatography coupled with mass spectrometry operated in positive or negative electron ionization (EI) mode has been extensively used for this purpose and for many analytes of interest (amphetamines, cocaine, cannabis, *etc.*), with opioids being the biggest exception, for which typically liquid chromatography with various detections modes has been the preferred analytical method. Significant work remains to be performed in the analysis of nail for drugs in forensic toxicology utilizing the analytical sensitivity of liquid chromatography coupled to tandem mass spectrometry, although recent reports for ethyl glucuronide, the alcohol-use biomarker, indicate that pg/mg concentrations in nail are easily measurable using such advanced analytical equipment.

19.5 Conclusions

It is without doubt that nail analysis in forensic and analytical laboratories can be a powerful tool that may provide additional information to the scientist regarding the possible long-term exposure to drugs. However, blood and urine remain the preferred analytical specimens in laboratories interested in answering questions regarding recent exposure to drugs. In cases where the timeframe of interest exceeds the ability of blood and urine to capture any drug exposure, alternative biological specimens such as hair and nail become extremely useful. Hair has been extensively studied and can provide drug exposure information ranging from weeks to months based on its length, but is prone to significant challenges by fractions of the scientific and legal communities who raise concerns regarding its melanin bias and growth rate differences – concerns that do not, for the most part, affect the nail. Drug concentrations in nail are often measured in pg/mg or ng/mg and, therefore, require very sensitive and often expensive analytical equipment, as well as carefully designed specimen decontamination and extraction procedures. Nail can be a useful adjunct to specimens in forensic investigations as fingernails often provide information on drug exposure spanning six or more months prior to collection and toenails may extend the window of detection to 9–12 months, given their slower rate of growth.

Finally, additional research in nail in both post-mortem and human performance toxicology case settings should be funded and conducted in order to better understand and characterize this specimen in the hope that we may soon be able to expand the scope of analytes detected in nail and to compare drug concentrations in nail to those measured in other keratinous specimens such as hair. Another potentially interesting research area could evaluate the discriminating power of nail between single *versus* multiple exposures to drugs.

In conclusion, nail remains an exciting and promising alternative biological specimen that is easy to collect and store at room temperature and which has been the source of toxicological information to the scientific community for over 50 years. It is increasingly the subject of forensic and analytical research around the world and may, as the cost of the sensitive equipment drops, soon become the specimen of choice in certain post-mortem and human performance forensic toxicology case investigations in which questions regarding drug exposure at certain timeframes fall beyond those that can typically be answered by blood, urine or hair analyses.

References

1. S. Pichini, I. Altieri, P. Zuccaro and R. Pacifici, Drug monitoring in nonconventional biological fluids and matrices, *Clin. Pharmacokinet.*, 1996, **30**, 211–228.

2. G. M. Pacifici and R. Nottoli, Placental transfer of drugd administered to the mother, *Clin Pharmacokinet.*, 1995, **28**, 235–269.

3. A. Agarwal and N. P. Lemos, Significance of bile analysis in drug-induced deaths, *J. Anal. Toxicol.*, 1996, **20**, 61–63.

4. O. Suzuki, T. Inoue, H. Hori and S. Inayama, Analysis of methamphetamine in hair, nail, sweat, and saliva by mass fragmentography, *J. Anal. Toxicol.*, 1989, **13**, 176–178.

5. G. A. Wong, T. H. Pierce, E. Goldstein *et al.*, Penetration of antimicrobial agents into bronchial secretions, *Am. J. Med.*, 1975, **59**, 219–223.

6. M. Wilemsen, P. De Doncher, J. Willems, R. Woestenborghs, V. Van de Velde, J. Heykants, J. Van Cutsem, G. Cauwenbergh and D. Roseeuw, Posttreatment itraconazole levels in the nail: New implications for treatment in onychomycosis, *J. Am. Acad. Dermatol.*, 1992, **26**, 731–735.

7. B. Forslind and N. Thyresson, On the structure of the normal nail. A scanning electron microscope study, *Arch. Dermatol. Forsch.*, 1975, **251**, 199–204.

8. P. Fleckman, Anatomy and physiology of the nail, *Dermatol. Clin.*, 1985, **3**, 373–381.

9. D. Dykyj, Anatomy of the nail, *Clin. Podiatr. Med. Surg.*, 1989, **6**, 215–228.

10. E. G. Zook, Anatomy and physiology of the perionychium, *Handb. Clin.*, 1990, **6**, 1–7.

11. C. M. Ditre and N. R. Howe, Surgical anatomy of the nail unit, *J. Dermatol. Surg. Oncol.*, 1992, **18**, 665–671.

12. D. J. McCarthy, Anatomic considerations of the human nail, *Clin. Podiatr. Med. Surg.*, 1995, **12**, 163–181.

13. T. Hirai and M. Fumiiri, Ultrasonic observation of the nail matrix, *Dermatol. Surg.*, 1995, **21**, 158–161.

14. P. D. Samman, *The Nails in Disease*, W. Heinemann Medical Books, London, 1978.

15. N. Zaias, Embryology of the human nail, *Arch. Dermatol.*, 1963, **87**, 37.

16. N. Zaias and J. Alvarez, The formation of the primate nail plate. An autoradiographic study in the squirrel monkey, *J. Invest. Dermatol.*, 1968, **57**, 120.

17. L. A. Norton, Incorporation of thymide-methyl-H^3 and glycine-2-H^3 in the nail matrix and bed of humans, *J. Invest. Dermatol.*, 1971, **56**, 61–68.

18. B. L. Lewis, Microscopic studies of foetal and mature nail and surrounding soft tissue, *Arch. Dermatol.*, 1954, **70**, 732.

19. F. Pinkus, in *J. Handbuck der haut und geschlechtskrankeiten*, ed. Jadassohm, Julius Springer, Berlin, 1927, pp. 267–289.

20. I. E. U. Boas, Zur morphologie der wirbeltirkralle morphol. Jb. Bd., 1894, **21**, 281.

21. R. Dawber and R. Baran, *Diseases of the Nails and their Management*, Oxford, Blackwell Scientific Publications, 1984.

22. P. D. Samman, *The Nail in Disease*, Chicago, Year Book, 1986.

23. N. Zaias, *The Nail in Health and Disease*, New York, Spectrum Publications, 1980.

24. R. Wolfram-Gabel and H. Sick, Vascular networks of the periphery of the fingernail, *J. Hand. Surg.*, 1995, **20**, 488–492.

25. K. Ogo, Does the nail bed really regenerate? *Plast. Reconstr. Surg.*, 1987, **80**, 445–447.

26. G. Gauwenbergh, H. Degreef, J. Heykants, R. Woestenborghs, P. Van Rooy and K. Haeverans, Pharmacokinetic profile of orally administered itraconazole in human skin, *J. Am. Acad. Dermatol.*, 1988, **18**, 263–268.

27. R. J. Hay, Pharmacokinetic evaluation of fluconazole in skin and nails. Inter, *J. Dermatol.*, 1992, **31**(suppl. 2), 6–7.

28. J. Faergemann and H. Laufen, Levels of fluconazole in normal and diseased nails during and after treatment of onychomycosis in toe-nails with fluconazole 150 mg once weekly, *Acta Dermatol. Venereol*, 1996, **76**, 219–221.

29. T. Matsumoto, H. Tanuma, S. Kaneko, H. Takasu and S. Nishiyama, Clinical and pharmacokinetic investigations or oral terbinafine in patients with tinea unguium, *Mycoses*, 1995, **38**, 135–144.

30. F. Schatz, M. Brautigam, E. Dobrowolski, I. Leffendy, H. Haberl, H. Mensing *et al.*, Nail incorporation kinetics of terbinafine in onychomycosis patients, *Clin. Exp. Dermatol.*, 1995, **20**, 377–383.

31. R. Nandwani, A. Parnell, M. Youle, C. J. N. Lacey, E. G. V. Evans, J. Midgley *et al.*, Use of terbinafine in HIV-positive subjects: pilot studies in onychomycosis and oral candidiasis, *Br. J. Dermatol.*, 1996, **134**(suppl. 46), 22–24.

32. N. Zaias, B. Glick and G. Rebell, Diagnosing and treating onychomycosis, *J. Fam. Pract.*, 1996, **42**, 513–518.

33. W. W. Harrison and G. G. Clemena, Survey analysis of trace elements in human finger nails by spark source mass spectrometry, *Clin. Chim. Acta.*, 1972, **36**, 485.

34. C. A. Pounds, E. F. Pearson and T. D. Turner, Arsenic in fingernails, *J Forensic Sci. Soc.*, 1979, **19**, 165–173.

35. O. Suzuki, H. Hattori and M. Asano, Nails as useful materials for detection of methamphetamine or amphetamine abuse, *For. Sci. Intern.*, 1984b, **24**, 9–16.

36. V. Cirimele, P. Kintz and P. Mangin, Detection of amphetamines in fingernails: an alternative to hair analysis, *Arch. Toxicol.*, 1995, **70**, 68–69.

37. D. Tiess, R. Wegener, I. Rudolph, U. Steffen, B. Tiefenbach, V. Weirich and F. Zack, 1994, Cocaine and benzoylecgonine concentrations in hair, nails and tissues: a comparative study of ante and post mortem materials in a case of acute lethal cocaine intoxication, In SOFT/TIAFT 94 Proceedings, ed. Spiehler, Oct 31-Nov 4, Tampa, FL, USA.

38. J. D. Ropero-Miller, D. Garside, B. A. Goldberger, W. F. Hamilton and W. R. Maples, 1998, Nails: another useful matrix for the identification of cocaine exposure, In Proceedings of the 50th Anniversary Meeting of the American Academy of Forensic Sciences.

39. D. A. Engelhart and A. J. Jenkins, Identification of cocaine, opiates and their metabolites in nails from postmortem cases, in SOFT/TIAFT 98 Proceedings, ed. Backer, Rao and Spiehler, Oct 5–9, Albuquerque, NM, USA, 1998.

40. N. P. Lemos, R. A. Anderson and J. R. Robertson, Nail analysis for drugs of abuse: extraction and determination of cannabis in fingernails by RIA and GC-MS, *J. Anal. Toxicol.*, 1999, **23**, 147–152.

41. J. Y. Kim, S. H. Shin and M. K. In, Determination of amphetamine-type stimulants, ketamine and metabolites in fingernails by gas chromatography-mass spectrometry, *Forensic Sci. Int.*, 2010, **194**, 108–114.

42. N. P. Lemos, R. A. Anderson and J. R. Robertson, The analysis of methadone in nail clippings from patients in a methadone maintenance program, *J. Anal. Toxicol.*, 2000, **24**, 656–660.

43. N. P. Lemos, R. A. Anderson, R. Valentini, F. Tagliaro and R. T. Scott, Analysis of morphine by RIA and HPLC in fingernail clippings obtained from heroin users, *J. Forensic Sci.*, 2000, **45**, 407–412.

44. J. Y. Kim, J. C. Cheong, J. I. Lee, J. H. Son and M. K. In, Rapid and simple GC-MS method for determination of psychotropic phenylalkylamine derivatives in nails using micro-pulverized extraction, *J. Forensic Sci.*, 2012, **57**, 228–233.

45. A. J. Jenkins and D. A. Engelhart, Phencyclidine detection in nails, *J. Anal. Toxicol.*, 2006, **30**, 643–644.

46. F. Mari, L. Politi and E. Bertol, Nails of newborns in monitoring drug exposure during pregnancy, *Forensic Sci. Int.*, 2008, **179**, 176–180.

47. M. H. Choi, Y. S. Yoo and B. C. Chung, Measurement of testosterone and pregnenolone in nails using gas chromatography-mass spectrometry, *J. Chromatogr. B*, 2001, **754**, 495–501.

48. H. G. Brown and D. Perrett, Detection of doping in sport: detecting anabolic-androgenic steroids in human fingernail clippings, *Med. -Leg. J.*, 2011, **79**, 67–69.

49. L. Morini, M. Colucci, M. G. Ruberto and A. Groppi, Determination of ethyl glucuronide in nails by liquid chromatography tandem mass spectrometry as a potential new biomarker for chronic alcohol abuse and binge drinking behavior, *Anal. Bioanal. Chem*, 2012, **402**, 1865–1870.

50. D. Cappelle, M. Yegles, H. Neels, A. L. N. van Nuijs, M. De Doncker, K. Maudens, A. Cocaci and C. L. Crunelle, Nail analysis for the detection of drugs of abuse and pharmaceuticals: a review, *Forensic Toxicol.*, 2015, **33**, 12–36.

51. C. Hang, X. Ping and S. Min, Long-term follow-up analysis of zolpidem in fingernails after a single oral dose, *Anal. Bioanal. Chem.*, 2013, **405**, 7281–7289.

52. M. M. Madry, A. E. Steuer, T. M. Binz, M. R. Baumgartner and T. Kraemer, Systematic investigation of the incorporation mechanisms of zolpidem in fingernails, *Drug Test. Anal.*, 2014, **6**, 533–541.

20 Hair Testing in Forensic Toxicology

Donna M. Cave* and Robert Kingston

Lextox Drug and Alcohol Testing, Cardiff, UK
*Email: donna.cave@lextox.co.uk

20.1 Introduction

Hair testing has been successfully used worldwide and continues to show great promise and potential to demonstrate usefulness for many years to come. In the field of toxicology, hair testing is used in three main areas: criminal investigations, family proceedings and drug treatment/workplace drug testing.[1,2] Its unique long window of detection, regularly going back months, and sometimes years, allows it to be a very powerful and useful technique.

Over the years the main improvements and advances in hair testing have been the capability of achieving lower cut-off levels by the use of more sensitive instrumentation. However, this in itself brings with it the question of what exactly does a low (trace) amount of drug in hair really mean? The biggest challenge hair testing faces is the distinction between positive results being obtained from passive exposure as opposed to actual use. If cut-off levels applied to hair testing continue to become lower, this challenge is amplified.

Forensic Toxicology: Drug Use and Misuse
Edited by Susannah Davies, Atholl Johnston and David Holt
© The Royal Society of Chemistry 2016
Published by the Royal Society of Chemistry, www.rsc.org

20.2   Cut-off Levels

Low cut-off levels are needed for hair testing and this is especially true for the detection of cannabis metabolites, 11-hydroxy-Δ^9-THC (THC-OH) and 11-nor-9-carboxy-Δ^9-THC (THC-COOH), and the alcohol marker ethyl glucuronide (EtG) as they are detectable in hair at considerably low levels. This is due to the poor incorporation of these metabolites into the hair shaft. Even when active use is declared by a donor, these metabolites are quite often not detected as they have not incorporated into the hair shaft following use.

The use of cut-offs is an interesting area of debate, as again these are not always applied in the same way by different hair test laboratories. Traditionally the cut-off value was the level above which a test becomes detected and a concentration level can be reported. At Lextox, if a level detected is below the cut-off level, it is reported as not detected. What cut-off level is to be used? The only real guidance comes from the Society of Hair Testing (SoHT), where they define a maximum cut-off to show chronic use for the classical drugs of abuse, *i.e.* amphetamine, cannabis, cocaine and opiates.[2]

This, you may think, is straightforward; however, for example with cannabis the SoHT states that the cut-off for Δ^9-THC should be less than or equal to 0.05 ng mg^{-1} and that the metabolite THC-COOH should have a cut-off of 0.2 pg mg^{-1}. When being asked to compare results between laboratories the cut-off used for Δ^9-THC can vary between 0.1 ng mg^{-1} and 0.01 ng mg^{-1}, resulting in many cases, in the eyes of the customers, having greatly different results. In addition, even though it is quite clearly stated by the SoHT that the THC-COOH is needed to confirm use, many commercial hair testing laboratories do not test for this compound as few can achieve the low cut-off level suggested.

It has recently been seen that there has been a growing number of hair testing laboratories that do not strictly adhere to their reporting cut-offs. Many laboratories are reporting concentration levels below their reporting cut-off and defining it as a negative result, whereas others are reporting the parent compound below the reporting cut-off if the corresponding metabolite is seen detected above. All of which can lead to even more confusion for the customer.

20.3   Laboratories and Accreditation

Why do solicitors use different hair testing laboratories? This happens when the donor does not agree with the results and wants to challenge them, so the B sample is analysed or another sample is collected by a different company covering the previously analysed time period. This leads to other complications when comparing results, such as different assumed hair growth rates (leading to the question, do results even overlap?), different cut-off levels used and even different metabolites targeted. This has led to the increasing request of a schedule of agreement and disagreement between the companies and the use of telephone conference calls between experts.

Even with ISO accreditation, laboratories are still using slightly different methods. The main advantages of ISO are traceability of results, documentation and uncertainty of measurement. ISO does not impose uniformity of methodology.

The ISO 17025 standard for testing has become the minimum standard which hair testing laboratories in the UK must obtain before being considered credible by solicitors when placing work in the medico-legal sector. However, this standard does not mean that the laboratories are all getting the same results or even using the same methods, as the only criteria used by UKAS (United Kingdom Accreditation Service) to assess laboratories is the results of proficiency test (PT) schemes or incurred samples; this comes back to the fact that there is no standard incurred hair samples with a known drug concentration. UKAS does not impose any procedures or methods on the laboratory, only that those procedures are followed and that the results are traceable. The only guidance would come from any SoHT consensus or published papers on the methods. The ISO 17025 is usually accredited for individual tests, which leaves the possibility for a laboratory to claim to be ISO 17025 with only one test, *e.g.* cocaine, accredited.

UKAS auditors usually insist that the laboratory publish reporting cut-offs, uncertainty and bias of all ISO17025 accredited tests. However, these parameters are usually far beyond the understanding of the typical customer. For example, if an uncertainty of $\pm 30\%$ is quoted, that means that if the original result was 1.0 ng mg^{-1} then a repeated result could be anywhere between 1.3 and 0.7 ng mg^{-1}, which could result in a detected result becoming not detected or *vice versa*. The idea of a bias being added to the results is a concept that would horrify most customers in the black and white world of the typical solicitor.

Perhaps the best way to test an analytical method is to produce samples with known or spiked levels of the target drug compounds and take them through the whole analytical process to obtain a final concentration which may be compared with the spiked value. As hair is a non-homogeneous matrix and in a non-liquid form it is not possible to easily generate a spiked quality control sample, as would be the case for, say, urine where liquid reference standards can be simply spiked into urine to produce a fairly homogeneous standard after mixing.

To attempt to overcome the problem of measuring the extraction of drugs from the inside of the hair strand, incurred or positive real casework samples may be used to create an incurred sample containing all the target test compounds at a level near cut-off. This may involve careful selection of previously analysed samples. As calibrators and quality controls are not produced by extraction from the hair, when a sample batch is analysed there is no control/indication of the extraction efficiency from the hair strand. So if there are no positive samples on the batch the question may be asked, has the extraction worked? The incurred sample can give a reassurance that the extraction efficiency is within control limits. This sample may also be used to estimate the analytical uncertainty of measurement, by repeating the sample on different days the variation of the measurement, as it encompasses the whole method, may be estimated. The accuracy (trueness) may not be estimated as the "real" concentration of the drug in the hair is not actually known.

All drug testing laboratories gain assurance for the accuracy of their results by analysing external PT samples. Unfortunately, in the world of hair testing, the range of schemes are severely limited with only one scheme (SoHT) giving an incurred sample, with the others supplying soaked or spiked powdered hair samples. Unsurprisingly, the range of results for the SoHT scheme using incurred samples is far wider than that of the other schemes. However, these schemes' results can be useful to investigate result trends. The PT samples are perhaps the nearest hair testing laboratories get to a known incurred sample, assuming that the consensus mean of all the participating laboratories is close to the true value; however, the quantity of sample supplied is limited, making it impractical to analyse these samples as daily quality control samples. There may be sufficient quantities left over after making the normal PT measurement to be used in initial validations to gain accuracy (bias) information of any new method or instrument.

One way the uncertainty of measurement may be calculated in an individual laboratory is by using a combination of incurred/chopped/powdered samples along with the laboratory produced incurred sample using eqn (20.1). The combined root-mean-square coefficient of variation (CV_{rms}) will be calculated from this equation:

$$CV_{rms} = \sqrt{1}/n(CV1_{\text{incurred PT}})^2 + (CV2_{\text{incurred other}})^2$$

$$+ (CV3_{\text{spiked at cut-off}})^2 + (CV4_{\text{spiked PT}})^2 \tag{20.1}$$

Table 20.1 Expected uncertainties at cut-off concentrations calculated using the modified Horwitz equation.

Compound	Proposed cut-off $(ng\,mg^{-1})$	Expected uncertainty based on the modified Horwitz equation (%)
6-Acetylmorphine	0.2	27.3
Benzoylecgonine	0.05	33.6
Δ^9-THC	0.01	42.9
Methadone	0.2	27.3
Codeine	0.2	27.3
Dihydrocodeine	0.2	27.3
Diazepam	0.04	34.8

The calculation has been used to determine the uncertainty of measurement (Uc) to produce a combined CV_{rms} before the coverage factor of 2 is applied; n is the number of uncertainty measurements, $CV_{incurred\ PT}$ is the CV from repeat runs of incurred PT samples; $CV_{incurred\ other}$ refers to the CV obtained from incurred sample prepared from old casework samples and analysed with each batch of casework samples; $CV_{spiked\ at\ cut-off}$ refers to the CV obtained from repeat analysis of blank hair spiked at QC1 (cut-off level). If there is no incurred PT scheme available, then the CV obtained from spiked PT samples can be added to the calculation.

To check if the uncertainty obtained is reasonable, the uncertainty of any measured concentration may be estimated using the Horwitz eqn (20.2). Some calculated examples of expected uncertainties at different cut-off are shown in Table 20.1.

$$VC = 2^{(1 - 0.5\ \log\ C)} \tag{20.2}$$

where VC = variation coefficient (%) and C = analyte concentration $(kg\ L^{-1})$.[3]

20.4 Collection Process

Even if the methods employed within a laboratory are all standardised, there can still be variation introduced at the collection stage. The analysis of hair collected from collectors who have had specific training and have a thorough understand of the procedure is always recommended.[2] There will inevitably be some variation between collections; however, as long as a fully trained collector is used, then this should not have a great impact on the results obtained.

The collection of straight European hair has been well documented; however, with changes to the demographic in the UK, and particularly in the London area, the requirement to collect Afro-Caribbean and dreadlocked hair has increased. When hair is received into the laboratory the alignment and orientation is checked; any doubt can be checked using a manual friction test. This test is used to identify the direction of growth of individual hairs and can be used as a method to provide evidence of the specific orientation of a sample of hair where this is not visibly apparent. The test is based on the knowledge that the external surface of hair is constructed from overlapping scales' cells[4] that form in a consistent manner, with the exposed raised edge of each scale apparent at the distal edge.

The test relies on the fact that when a hair is agitated between the thumb and forefinger along the length of the hair, the hair scales rasp against the finger and the hair moves through the fingers. In practice this means that when you hold a hair vertically between the finger tips and move the finger tips against each other, the proximal end of the hair will always move away from the fingertip. Other methods have been trialled in this laboratory, *e.g.* high-powered microscopes, but the manual friction test has been found to give the most consistent results.

The cut/root end of the sample is then "aligned" by removing any stray hairs and moving the strands at the cut-end so that start point for all the hairs is the same before sectioning, to achieve the most accurate assignment of approximate time periods.

Are we collecting hair from the correct person? This is usually taken care of by way of a printed name and signature by the donor, and more importantly someone or something identifying the donor; this is most easily achieved by the collector taking a photograph of the donor at the time of collection. This of course does not guarantee that the donor is the correct person, but does achieve a permanent record of the donor who supplied the hair sample, which can then be included into the expert report along with the results obtained.

20.5 Queries on Reported Results

The expectation of hair testing is that a sample of hair will be collected non-invasively and the result obtained will be able to provide a history of what a donor has taken, when and exactly how much. In reality, hair testing is best used as a guide to what an individual may have been taking in the past. Even now the majority of queries that arise are exactly how much a donor has used and are the results obtained consistent with the donor's declaration. Hair testing cannot be used to establish exact amounts used;[5] however, by assessing trends and patterns and looking at all the evidence a qualified expert will be able to say with confidence if it is likely the results are consistent with a particular declaration. As previously mentioned, the majority of hair testing cases are in family court proceedings and therefore a likelihood conclusion is acceptable. The most accurate interpretation of a donor's history of drug abuse is obtained using untreated hair. Unfortunately, due to the modern way of living in the 21st century, cosmetic treatments are now widely used.[6] In addition to this, it is now such a well-known fact by the people being tested that hair dye can lower drug levels, that the majority of hair samples received are treated. This just increases to the challenge expert witnesses face in reaching a conclusion of patterns of use. As hair treatments, such as dyes and bleach, can greatly lower results obtained, it is always advisable for the testing laboratory to visually examine both the hair sample prior to analysis and extracts during the analysis process for signs of treatment, as it is not always correctly declared by the donors.

There are two main questions that are asked in relation to alcohol testing results. Firstly, whether the use of hair care products or treatments used could have impacted the results. Namely, the use of products contributing to levels of fatty acid ethyl esters (FAEE). The second most commonly asked question is to provide an explanation as to why not all markers tested for, EtG, FAEE, CDT and LF, correlate and provide the same interpretation, *i.e.* why are they not all either not detected or detected. It is always advised to test for as many parameters as possible when trying to establish a person's alcohol consumption, to strengthen the evidence. However, in the majority of cases this can provide what may seem like contradicting results. It is very plausible to obtain some alcohol markers detected in combination with not detected markers, especially when hair and blood are combined. Each test has its own limitations and benefits, and combining them provides a more accurate overall interpretation. Best practice when analysing for more than one marker is for the expert to provide a summary of all the tests results covering the whole tested time period, which should include possible explanations of why the markers may not match, as each case is unique and the reason why may change from case to case, for example depending on what hair products and treatments have been used.

Other commonly asked alcohol questions are what else can cause a detected result other than consuming alcohol? Can the hair be segmented like it can be for drug testing, and are the results obtained consistent with the declared consumption?

With regards to queries of drug results issued, the most frequently asked question is: can the results be attributed to medications declared? It is standard practice to ask what medications/substances a donor has used when a hair sample is collected. This is because medications can be detected in hair if tested for. For example, when the painkiller co-codamol is declared as being used, codeine has the potential to be detected. However, the misconception seems to be that common medication can give rise to commonly abused drugs such as cannabis and cocaine. If sensitive conformation techniques are used, then this will not be the case; however, one can only assume that this question is asked time and time again due to some laboratories still using non-specific screening techniques. Although this may not be in the field of hair, as many courts now order an array of biological testing (urine and saliva testing), this concept may still be attributed to hair.

The next frequently asked question is: can you tell what amount of drug the donor is using from the level obtained in the hair? Although hair testing is very useful in showing trends and patterns over a long period of time, it cannot be used to back-calculate the actual amount of drug being used, and this is something hair will never be able to do. This is the case with not only hair testing but also urine and saliva testing. Repeat testing, regardless of the matrices being tested, will give the fullest information on a donors drug use/abstinence.

The next two questions asked are: can the results be due to previous use or passive exposure? These two aspects are probably the most challenging aspect of interpreting drug results obtained from analysing hair samples. Each case will need to be assessed individually and the likelihood on what exactly the results show should be reached. Extra background information regarding a donor's declaration is useful when interpreting such results, even if it is just to say that the results do not support a particular declaration.

20.6 Expert Report and Interpretation

Although hair testing has been used successfully for many years in the family and criminal courts there is still a great need for expert interpretation of the results, given the complexity of collection, laboratory analysis and treatments that the hair may have undergone. Seldom are results issued by laboratories without the accompaniment of a full interpretation of what the results mean; less than 10% of all results issued at Lextox do not have an accompanying expert interpretation. An expert interpretation also minimises the possibility of miscarriages of justice where stand-alone results could have the potential to be "misinterpreted". Owing to the complexity of the interpretation of the hair testing results the courts almost always request that an expert report is prepared, rather than simply issuing a detected or not detected result. Another reason for this could be that currently when hair results are reported there is a large number that are still challenged by the donor.

It is very common for a donor to put forward potential scenarios that could have caused a result in their opinion once a report is issued. The most challenging question in hair testing today, as previously mentioned, is the possibility of results being obtained due to passive exposure rather than use. Although there are many scientific papers published detailing this challenge, the fact still remains that in some instances it can be quite difficult, even for an expert, to be sure with a degree of certainty that use is the source.

Within family court proceedings in the UK, a likelihood conclusion is acceptable; however, in a criminal court where the conclusion has to be beyond all reasonable doubt, this could be seen as a problem. Can hair testing results really give a "beyond reasonable doubt" conclusion? This does not mean that hair testing should not be used in criminal cases. It is just advisable to be used along with other forms of evidence.

When hair testing is ordered by a court it is normally in respect of a child at risk; therefore, all adult parties associated (for example, within the home/same environment) have the potential to be tested. Cases have been seen where one parent has declared using cannabis, with their sample providing detected results for both cannabis constituents and metabolites, and the other parent has denied using cannabis, but still yielding positive low-level results for cannabis constituents. Now although the use of cannabis cannot be ruled out due to low metabolite incorporation into hair, in a case like this it is plausible that only one parent has used, with the other parent merely being exposed. Cannabis constituents detected in hair due to exposure has been detailed;[7] however, the detection of a cannabis metabolite in adult hair solely due to cannabis exposure has still yet to be proven.

20.7 Types of Analysis Requested

Nowadays, solicitors and social workers representing their clients have a close working relationship and are often very clued up on as to what they may or may not be using in terms of illegal drugs, quite often self-declared by the donors themselves. As a result, more tests are now being ordered for a handful of specially chosen drugs rather than a full "panel"-type analysis that was the trend a few years ago. The advantage of this is to keep court costs low; however, the drawback is that untargeted drug use will go unnoticed. However, the way in which drugs are used has changed over recent years, with many people now using "legal highs" or NPS (new psychoactive substances) often purchased over the internet. This pattern of use could change the way in which hair testing is conducted in the future. Almost all of these NPS purchased over the internet are a combination of compounds mixed together. As such, the user often has no idea what they are actually taking. Therefore, keeping up-to-date with the drug-using community, hair samples may need to be screened/panel tested for substances such as these, rather than a targeted analysis on specifically chosen compounds.

Sectional analysis is preferred to an overview analysis even with the higher cost associated, as it gives the court a far better overall picture, and is much more useful. This is very beneficial to the customers and is something that should continue to be offered routinely by testing laboratories.

Workplace drug testing is a rapidly increasing industry all over the world, with many employers now implementing such practices. However, what about the future of such testing? Why not start earlier and implement such testing in schools and colleges? This will ensure that the future generation understand the importance of compliance and they will also tackle the issue of drug taking at a much earlier stage.

20.8   Cannabinoids Results Seen in Hair

Cannabis is used worldwide, and is the most commonly abused illicit drug.[8,9] At Lextox, 64% of the samples analysed tested positive for the main active constituent of cannabis, Δ^9-THC. However, there are important factors to be aware of. Not only are cannabinoids poorly incorporated into hair,[7] due to their acidic nature,[10] but the fact that smoking is the most popular route of administration of cannabis[7] also needs to be another consideration when interpreting hair results.

After seeing thousands of cannabis results, a general pattern is observed where sectional analysis almost always shows that the level of Δ^9-THC detected increases the further away from the scalp you go.

This has been previously observed.[11] From this pattern of results it could be assumed that a donor has reduced their use. However, concern over an interpretation of reducing use is increasing due to the number of samples we see that follow this pattern; therefore, is there another explanation that could account for this phenomenon?

A simple explanation could be that the external exposure to cannabis smoke has more of an effect than first realised. If a donor regularly smokes a consistent amount of cannabis the expectation should be a consistent level of cannabinoids being detected along the length of a sectioned hair sample. However, the older hair would have been exposed to cannabis smoke for many more months than the newly gown hair, and as such may have increased levels of Δ^9-THC.

Another consideration could be the degradation of Δ^9-THCA into Δ^9-THC. Δ^9-THCA is a non-psychoactive precursor of Δ^9-THC.[10] When heated, Δ^9-THC is partially released from Δ^9-THCA by decarboxylation (the removal of a CO_2).[12] However, as Δ^9-THCA is found in cannabis smoke it has the potential to contaminate the outside of the hair. It has been shown in a simulated smoking process that only 30% of Δ^9-THC is recovered from Δ^9-THCA. Once Δ^9-THCA has made contact and remains on the outside of the hair it has potential over time to break down into Δ^9-THC. Δ^9-THCA is very unstable, decomposing rapidly in the presence of oxygen, light and acids. In addition, Δ^9-THCA may be decarboxylated during the laboratory process (*e.g.* high temperature) and artificially increase the THC concentration. Again this could play a role in the increasing levels of Δ^9-THC seen within distal sections analysed.

It is well documented that Δ^9-THC has been detected in both sweat and sebum.[13] Therefore as the hair is constantly being bathed in sweat and sebum, this could be another potential source of increasing amounts of Δ^9-THC in distal sections of hair. This phenomenon is also seen with the alcohol marker fatty acid ethyl esters (FAEE).

More studies need to be conducted on this phenomenon and caution should always be applied when trying to establish trends of cannabis use, especially when it appears a user may be decreasing their use, as this may not be the case.

Hair is most routinely analysed for the cannabis constituents, Δ^9-THC, cannabinol and cannabidiol, and the cannabis metabolites 11-nor-carboxy-Δ^9-THC and 11-hydroxy-Δ^9-THC. Cannabis constituents have been found in hair even after multiple wash procedures.[10] A way to easily eliminate any positive results being obtained from exposure alone would be to only analyse hair samples for evidence of cannabis metabolites, as their detection in adults unequivocally proves cannabis use.[14,15] However, owing to the acidic nature of the cannabis metabolites, they have a weak incorporation rate into the hair matrix, resulting in very low concentrations being present.[10] Until methods are able to confidently detect cannabis metabolites at the concentrations found in hair (low picogram levels),[9,11,16] then the analysis for cannabis constituents in combination with cannabis metabolites is beneficial.

20.9   Alcohol Markers and Hair Testing

Hair alcohol testing is much newer than hair drug testing, with hair alcohol abstinence testing still in its infancy. The SoHT published the alcohol consensus in an attempt to standardise and give unity and meaning to hair testing results. The biggest challenge with hair alcohol testing is that alcohol is a legal substance and is the most frequently abused substance.[17] Therefore, the importance is often not to prove alcohol has merely been consumed but to prove the quantity consumed.[17] Now it is very important to remember that a hair test will not be able to "back-calculate" to an exact amount of alcohol consumed. There are too many variables to allow this to ever be the case. However, a positive result above the SoHT cut-off levels give a strong indication that alcohol is being consumed to chronic excessive levels, and in conjunction with other forms of testing, such as blood testing and clinical assessments, it enables courts to make informed decisions. Another important consideration with the interpretation of alcohol markers in hair is the effect both hair treatments and products have on the results obtained. As with drug levels, chemical hair treatments can reduce the alcohol markers in hair; however, conversely, hair products can increase the level of FAEE markers detected, if products used contain ethanol. A case example has been seen at the Lextox laboratory, where a donor has declared consuming approximately 100 units of alcohol per week. No hair products were declared as being used; however, the hair had been dyed 10 days prior to the sample being collected. A proximal 3 cm portion of hair analysed returned a total FAEE result of 1.70 $ng\,mg^{-1}$, which falls in the chronic excessive range and is consistent with the donor's declaration; however, the EtG result was zero, which falls in the abstinence range. The effects of the dye in this case are likely to have removed the EtG from the hair. Conversely, we have seen a case where the donor's declaration was no alcohol consumed, with a declaration of no hair treatments; however, the use of gel had been declared. The results obtained for a 3 cm proximal section of hair was a zero EtG, which falls in the abstinence range and is consistent with the donor's declaration; however, the total FAEE result obtained was 2.02 ng mg^{-1}, which falls in the chronic excessive range. If the product used by the donor contained ethanol, then the likelihood would be they have not consumed chronic excessive amounts of alcohol, even with one positive marker. These examples show the importance of not using a single marker in isolation when trying to establish a donor's alcohol consumption.

Another good indicator of high level of alcohol consumption is cocaethylene. Cocaethylene is produced when cocaine is used in conjunction with alcohol. From samples analysed at the laboratory when a donor tests positive for chronic excessive alcohol consumption in combination with cocaine, in 81% of these cases cocaethylene was also detected. In cases where cocaethylene was not detected, for just over half of the cases the cocaine was only just above the SoHT cut-off of 0.5 ng mg^{-1}. Cocaethylene will not be increased due to the use of hair products containing ethanol. Therefore, a detected result, like EtG, is a indicator of excessive alcohol consumption. However, it is important to remember that this marker will only be detected when alcohol and cocaine are used in conjunction.

20.10 Patterns of Results Obtained

It is quite clear to see that cannabis is the most widely abused drug in the UK ,with the percentage of positive case samples being much higher than any other drug analysed at Lextox. From samples that have been analysed at Lextox, which originate from all over the UK, amphetamine, cocaine, cannabis and methadone use all seem fairly constant between 2013 and 2015 (see Table 20.2). Heroin use as seen from analysed casework samples has halved in the last 18 months compared to 2013.

Positive mephedrone samples analysed at Lextox has also almost halved in the past two and a half years, suggesting that mephedrone use in the UK has halved. This could be due to the increasing rise of other new psychoactive substances (commonly referred to as "legal highs") readily available on the internet. There has been a marked increase in the amount of positive ketamine case samples analysed at Lextox in 2015 compared with the two previous years. There has also been a rise in methamphetamine use in the UK in the last two years compared with a very low percentage of positive samples seen at

Table 20.2 Percentage of positive samples analysed at Lextox for the full years of 2013, 2014 and the first six months of 2015.[a]

Compound	2013	2014	2015
6-Acetylmorphine	38%	21%	20%
Amphetamine	44%	36%	37%
Benzoylecgonine	45%	47%	52%
Cathinone	0.5%	0.5%	0
Δ^9-THC	64%	62%	65%
Ketamine	5%	3%	24%
Mephedrone	17%	11%	9%
Methamphetamine	0.2%	2%	1%
Methadone	19%	19%	17%

[a]For example, in 2013, 64% of all samples analysed tested positive for Δ^9-THC.

Lextox in 2013, although the number of positive methamphetamine samples seen is still low.

Khat became a class C controlled drug in the UK in June 2014.[18] One of the main constituents of khat is cathinone. Only 0.5% of samples analysed were positive for cathinone in 2013 and 2014, with no samples being positive in the first 6 months of 2015, suggesting it is not a drug of choice for people who are in the UK family court system.

References

1. M. A. Huestis *et al.*, Cannabinoid Concentrations in Hair from Documented Cannabis Users, *Forensic Sci. Int.*, 2007, **169**, 129–136.
2. G. A. A. Cooper, R. Kronstrand and P. Kintz, Society of Hair Testing guidelines for drug testing in hair, *Forensic Sci. Int.*, 2012, **218**, 20–24.
3. W. Horwitz, Evaluation of analytical methods used for regulation of food and drugs, *Anal. Chem.*, 1982, **54**, 67A–76A.
4. R. Kronstrand and K. Scott, Drug Incorporation into Hair, in *Analytical and Practical Aspects of Drug Testing in Hair*, ed. P. Kintz, CRC Press, Boca Raton, 2007, p. 1.
5. C. Jurado, Hair Analysis for Cocaine, in *Analytical and Practical Aspects of Drug Testing in Hair*, ed. P. Kintz, CRC Press, Boca Raton, 2007, p. 114.
6. G. Skopp, L. Pötsch and M. R. Moeller, On cosmetically treated hair – aspects and pitfalls of interpretation, *Forensic Sci. Int.*, 1997, **84**, 43–52.

7. J. Thorspecken, G. Skopp and L. Pötsch, In Vitro Contamination of Hair by Marijuana Smoke, *Clin. Chem.*, 2004, **50**, 596–602.
8. S. W. Toennes *et al.*, Comparison of Cannabinoid Pharmacokinetic Properties in Occasional and Heavy Users Smoking a Marijuana or Placebo Joint, *J. Anal. Toxicol.*, 2008, **32**, 470–477.
9. S. Pichini *et al.*, Identification and quantification of 11-nor-Δ9-tetrahydrocannabinol-9-carboxylic acid glucuronide (THC-COOH-glu) in hair by ultra-performance liquid chromatography tandem mass spectrometry as a potential hair biomarker of cannabis use, *Forensic Sci. Int.*, 2015, **249**, 47–51.
10. V. Auwärter *et al.*, Hair Analysis for Δ9-tetrahydrocannabinolic acid A-New insights into the mechanism of drug incorporation of cannabinoids into hair, *Forensic Sci. Int.*, 2010, **196**, 10–13.
11. D. Thieme, H. Sachs and M. Uhl, Proof of cannabis administration by sensitive detection of 11-nor-delta(9)-tetrahydrocannabinol-9-carboxylic acid in hair using selective methylation and application of liquid chromatography – tandem and multistage mass spectrometry, *Drug Test. Anal.*, 2014, **6**, 112–118.
12. J. Jung *et al.*, Studies on the metabolism of the Delta9-tetrahydrocannabinol precursor Delta9-tetrahydrocannabinolic acid A (Delta9-THCA-A) in rat using LC-MS/MS, LC-QTOF MS and GC-MS techniques, *J. Mass Spectrom.*, 2009, **44**, 1423–1433.
13. M. Uhl and H. Sachs, Cannabinoids in hair: strategy to prove marijuana/hashish consumption, *Forensic Sci. Int.*, 2004, **145**, 143–147.
14. S. Dulaurent *et al.*, Simultaneous determination of Δ9-tetrahydrocannabinol, cannabidiol, cannabinol, and 11-norΔ9-tetrahydrocannabinol-9-carboxylic acid in hair using liquid chromatography-tandem mass spectrometry, *Forensic Sci. Int.*, 2014, **236**, 151–156.
15. H. Sachs and P. Kintz, Testing for drugs in hair Critical review of chromatographic procedures since 1992, *J. Chromatogr. B*, 1998, **713**, 147–161.
16. B. Moosmann, N. Roth and V. Auwärter, Hair Analysis for THCA-A, THC and CBN after passive *in vivo* exposure to marijuana smoke, *Drug Test. Anal.*, 2014, **6**, 119–125.
17. F. Pragst and M. A. Balikova, State of the art in hair analysis for detection of drug and alcohol abuse, *J. Clin. Chem.*, 2006, **370**, 17–49.
18. M. Stark, J. Payne-James and M. Scott-Ham, *Symptoms and Signs of Substance Misuse*, CRC Press, Boca Raton, 2015, p. 75.

21 Drugs in Oral Fluid

Peter Akrill

Alere Toxicology plc, Abingdon-on-Thames, UK
Email: peter.akrill@alere.com

21.1 Introduction

The analysis of oral fluid samples for drugs of abuse has been reported for over 30 years. With improvements in method and instrument sensitivity and the development of point of collection screening devices, the popularity of using oral fluid as a sample type for toxicological investigations has increased in recent years. This is particularly the case in workplace testing, drug addiction rehabilitation and roadside testing for drugged drivers.

The main advantage of oral fluid as a matrix in analytical toxicology is the ease of sample collection, where a simple collection can be carried out under observation while respecting the dignity of the donor. This has made the use of oral fluid samples for forensic investigations attractive. However, there are many considerations that need to be made prior to using oral fluid analysis within investigations and studies.

It is important to have an understanding of the physiology of oral fluid and the mechanisms by which drugs enter into the oral fluid. Additional considerations are an understanding of the impacts of routes of administration, particularly in relation to the deposition of drugs taken orally, smoked or snorted in the oral or nasal cavities.

Forensic Toxicology: Drug Use and Misuse
Edited by Susannah Davies, Atholl Johnston and David Holt
© The Royal Society of Chemistry 2016
Published by the Royal Society of Chemistry, www.rsc.org

The methods used for the collection of oral fluid samples can also impact on the detection of different drugs of abuse, through the stimulation of the production of saliva, retention of drugs of abuse on collection devices, instability of certain analytes and variation in the volume of oral fluid collected.

This chapter investigates these issues as well as looking at other important factors when considering the use of oral fluid, such as the detection windows and the detection of inadvertent exposure to drugs.

21.2 What is Oral Fluid?

Oral fluid is the term given for the fluid that is present in the oral cavity, which can be collected for analysis. Oral fluid is predominantly saliva, although blood, other crevicular fluids, cellular and food debris may be present in oral fluid.

Saliva is produced by glandular excretions mainly by the submandibular and parotid glands, with the sublingual glands also producing some saliva. During normal production, about 70% of the saliva is produced by the submandibular glands and 25% by the parotid glands. When saliva production is being stimulated, only 40% is produced by the submandibular glands and 55% by the parotid glands.

The salivary glands are well supplied with blood from the arteries and contain two distinct regions, the acinar region and the ductal region. The secretary cells are found in the acinar region. Water and proteins are secreted in this region and the fluid produced is isotonic with plasma. Following secretion the saliva then passes through the ductal region. This is where sodium and chloride are re-absorbed from the saliva and potassium and bicarbonate are secreted into the saliva. Upon arrival at the mouth the salt content of the saliva is higher than that of plasma.

Following stimulation, saliva passes rapidly through the ductal region, which results in decreased sodium re-absorption. This in turn increases the pH of the saliva. The increase in pH can affect the diffusion of drugs into the saliva.

21.3 Mechanisms of Drug Entry into Saliva

21.3.1 Passive Diffusion

Passive diffusion through cell membranes is the primary route by which drugs of abuse enter the saliva from the blood. Small molecules can pass into the saliva by passive diffusion of the non-protein-bound species through the cells in the salivary glands. This is a concentration gradient driven process based on non-protein-bound drug concentration in the plasma, which favours the non-ionised, lipophilic drugs. For many drugs this means that the concentration of the unbound parent drugs and the lipophilic metabolites are more representative of the blood concentrations. The degree of diffusion is influenced by the pK_a of the drug or metabolite and the pH of the saliva and plasma. As stimulated saliva has a higher pH than unstimulated saliva, salivary stimulation can result in both increased or decreased drug and metabolite concentration in the saliva, depending on the pK_a of the drug or metabolite. Kato *et al.* demonstrated the effect of stimulation on the concentration of cocaine in oral fluid.[1] More than five times as much cocaine and metabolites was present in unstimulated oral fluid than with oral fluid produced by citric acid stimulation. Codeine concentrations were shown to be between 1.3 and 2 times higher in unstimulated oral fluid when compared to oral fluid collected with a collection device, and 3.6 times higher when compared to oral fluid collected with citric acid stimulation.[2] Methadone concentrations in saliva have also been shown to be affected by salivary stimulation, where a decrease in methadone concentration was seen in oral fluid collected following stimulation. It was suggested that this is due to the change in pH of the stimulated saliva in addition to the increase in production of saliva from the parotid gland, from which methadone is not excreted.[3]

21.3.2 Ultrafiltration

Ultrafiltration is the transfer of drugs through pores and cell junctions in cell membranes within the salivary gland or oral mucosa into the saliva. This mechanism is particularly related to very small molecules, with molecular weights below 100 Daltons, and is the primary route by which ethanol passes into the saliva. For drugs and metabolites with molecular weights above 300 Daltons, passive diffusion is the primary mechanism.

21.3.3 Oral Deposition

With many drugs taken orally, smoked or snorted, deposition of drugs within the nasal or oral cavity is a significant source of drugs in oral fluid. Oral fluid samples collected shortly after drug use may contain drugs at far higher concentrations than would be expected based on the blood concentrations and the published saliva to plasma ratios. Several hundredfold increases in saliva to plasma ratios shortly after smoking or snorting of cocaine have been demonstrated when compared to intravenous use.[4,5] Similar increases in heroin saliva to plasma ratios have also been reported when comparing smoked heroin to intravenous heroin.[4] A study of a population of drug users concluded that the concentrations of opiates and cocaine detected strongly supported the proposition that smoking or inhaling drugs results in a deposition of drugs in the oral and nasal cavities.[6]

21.4 Windows of Detection

The length of time drugs of abuse are detectable in oral fluid will vary, depending upon the amount and purity of drug used, the route of administration, inter-individual variation, and the analytical cut-off or detection limit used, amongst other things.

Studies have generally been conducted on single-dose administrations and may underestimate the windows of detection experienced in real world situations. Jufer showed cocaine remained detectable for up to five times longer and benzoylecgonine remained detectable for up to ten times longer in individuals who had received repeated doses of cocaine when compared to individuals who received a single dose. The data show that cocaine use is detectable for up to 72 h in one individual after repeated dosing.[7]

The route of administration has also been shown to affect the window of detection of drugs in oral fluid. Following a single dose of cocaine, Jenkins demonstrated that the mean detection window for cocaine in volunteers who had cocaine administered intravenously was 514 min and for donors who were administered cocaine by smoking was 446 min.[4] The maximum concentration of cocaine and metabolites detected was higher for volunteers who were administered cocaine by smoking than intravenously. In the same study, the window for the detection of 6-acetylmorphine following heroin use was twice as long in the individuals who smoked heroin than the individuals who were administered heroin intravenously. It was also demonstrated that, regardless of the route of administration (smoking or intravenous), cocaine and heroin use were evident in samples collected within 2 min of the dose being given.

Two reviews have determined typical detection times of drugs of abuse in oral fluid.[8,9] The detection times are based mainly on single-dose studies and are summarised in Table 21.1.

Table 21.1 Detection windows for drugs of abuse in oral fluid.

Analyte	Detection window limit (h)	Ref.
Amphetamine	50	8, 9
Methamphetamine	24	8, 9
MDMA	24	9
THC	34	9
Cocaine	12	9
Benzoylecgonine	24	8, 9
6-Acetylmorphine	8	8, 9
Morphine	24	8, 9
Codeine	36	8
Methadone	24	8

21.5 Oral Fluid Sample Collection and Storage

The sample collection process for oral fluid samples is a critical step in ensuring the integrity of the sample. One of the main advantages is that the collection can be carried out under observation of the individual collecting the sample whilst respecting the dignity of the donor. This observed collection when carried out in accordance with a prescribed collection process makes it very difficult for the donor to defeat the collection process.

Oral fluid samples are collected in a variety of different ways. Sample collection by direct spitting has been reported,[10,11] although users expressed a dislike of providing a sample in this way. A more common and acceptable sample collection technique is by the use of a collection device which absorbs the oral fluid from the mouth.

Some collection devices contain chemicals that can stimulate saliva production which can alter salivary pH, resulting in changes to drug concentration levels. However, the European guidelines for workplace drug testing in oral fluid state that a collection device should not contain additives that stimulate saliva production.[12]

It has been demonstrated that drugs and metabolites that contain ester bonds in their structures, such as cocaine and 6-acetylmorphine, can undergo possible hydrolysis and degradation when stored after sample collection and refrigerated at room temperature.[13] No hydrolysis or degradation was observed following the addition of sodium azide and citrate buffer at pH 4.

An additional consideration that should be taken when using a collection device is the recovery of the particular drugs being analysed from the collection device. It has been shown that recovery of cannabis from certain oral fluid collection devices can be quite low.[14]

In an attempt to address the sample stability and drug recovery issues, most oral fluid collection devices are placed into a buffer solution within the collection kit after sample collection. These collection buffers contain components designed to keep sample hydrolysis and degradation to a minimum, whilst also facilitating the recovery of drugs from the collection devices. However, it is unlikely that analyte hydrolysis and drug retention will be completely overcome.[15]

In an attempt to avoid the issues related to the recovery of drugs from the oral fluid collection device, there are oral fluid collection systems that use a coloured mouthwash to collect the oral fluid. After rinsing the mouth, the mouthwash is spat into a collection beaker. The colour intensity of the mouthwash can be analysed by spectrophotometric techniques to determine the level of dilution of the mouthwash by the collected oral fluid. This gives an indication of the volume of oral fluid collected. Whilst a collection system like this removes the problems associated with drug recovery, a study comparing collection devices reported that users did not find it easy to use.[16]

The addition of buffers and dyes to these collection devices and systems has the potential to cause suppression or enhancement in LC-MS based analytical methods.[17] When validating an analytical method, the ion suppression or enhancement should be determined.[18] It is possible to remove most of these significant matrix effects through the use of sample preparation techniques such as solid phase extraction.

21.6 Detection of Inadvertent Exposure to Drugs

As oral fluid samples are collected from the mouth, the detection of inadvertent (or passive) exposure to drugs is sometimes raised as an area of concern for those from whom a positive result can have serious consequences. The scenario most commonly mentioned is passive exposure to cannabis smoke.

The likelihood of positive oral fluid results following passive exposure to cannabis has been investigated. In two studies where four individuals were subject to severe passive smoke exposure conditions, some Δ^9-tetrahydrocannabinol (THC) was detectable in oral fluid samples above the European Workplace Drug Testing Guidelines[12] cut-off of 2 ng mL^{-1} for up to 30 min after the smoking had ceased.[19,20] However, in these studies the sample collection devices were kept within confined spaces where the smoking occurred. Sample collection was also carried out in this area. The authors of the second of these studies demonstrated that some contamination of the collection devices occurred prior to sample collection.

The same authors repeated the study with cigarettes containing twice as much THC, but collection devices were kept outside of the smoking area and samples were collected once donors had been removed from the smoking area.[20] This represents the likely scenario for samples collected in a workplace testing environment. In this study, no sample collected after the end of smoking produced THC concentrations above the recommended workplace cut-off from those passively exposed for THC.

THC and its main metabolite, 11-nor-Δ^9-tetrahydrocannabinol-9-carboxylic acid (THC-COOH), in addition to the cannabinoids cannabidiol (CBD) and cannabinol (CBN), were analysed in a passive exposure study reported by Moore *et al.*[21] The study investigated the detection of these compounds following passive exposure within a cannabis coffee shop in the Netherlands. Ten volunteers were exposed to cannabis smoke in two coffee shops for 3 h. Oral fluid samples were collected at intervals during the exposure time and immediately at the end of the exposure time. A subsequent oral fluid sample was collected between 12 and 22 h after leaving the coffee shop. Seven of the 10 volunteers had levels of THC above the workplace cut-off at the time they left the coffee shop, but THC was not detected above the recommended workplace cut-off in any sample collected after this time. CBN was only detected in six of the 10 samples collected as the end of the exposure. No CBN was detected in any subsequent samples. No CBD or THC-COOH was detected during or after the exposure. THC, CBD and CBN were detected at significant levels on collection devices left exposed in the smoky atmosphere.

Based on the data from these passive exposure studies, a THC-positive oral fluid sample is unlikely to occur after moderate or severe passive exposure to cannabis smoke in a normal workplace testing scenario, when the recommended workplace cut-offs are applied. Analysis of THC-COOH as a cannabis marker is also beneficial in assessing active or passive exposure.

21.7 Analysis of Oral Fluid Samples for Drugs

The low concentration of drugs present in oral fluid has meant that the analysis of oral fluid samples for drugs of abuse using traditional techniques has been challenging. Recent technological improvements increasing the sensitivity of analytical instruments have resulted in the wider adoption of oral fluid analysis.

21.7.1 Screening Tests

The screening immunoassays are aimed at specific drug groups, with the response calibrated to a particular target drug within that drug group at a particular concentration (cut-off concentration). Screening kit manufacturers provide sensitivity and specificity data based on samples spiked with the target drug at concentrations close to the cut-off. The sensitivity is a measure of the assay's ability to detect positives and the specificity is a measure of the assay's ability to detect negatives. As the cut-offs for the oral fluid collection devices are at a concentration at least one order of magnitude lower than the urine-based screening tests, the reproducibility at the cut-off is more variable than with urine. Therefore the spiked samples usually contain the target drug at 50% below the cut-off and 100% above the cut-off, which is further from the cut-off than usually used for urine-based kits.

21.7.1.1 Point of Collection Testing (POCT)

Screening tests that can be run at the time of sample collection have been developed and improved over recent years. They are predominantly immunochromatographic screening tests employing lateral flow technology. The main mechanism of these tests involves the oral fluid rehydrating labelled antibodies within the kits, which then pass over a line of immobilised drug conjugate. The subsequent binding between the antibodies and the conjugate creates a colour response indicating the absence of drug in the sample. When drug is present in the sample the binding between the antibody and the drug results in little or no binding with the immobilised drug conjugate, resulting in no colour response and indicating a positive result.

Independent studies have been conducted to evaluate the performance of several drug testing devices, particularly in relation to their suitability in detecting drugged drivers.[22,23] These studies demonstrated a huge variation in the sensitivity and specificity of the devices tested, and reflected the performance of the devices at the time the studies were conducted. As manufacturers make improvements to their devices and produce new versions, it is important to assess the sensitivity and specificity of the specific devices to be used.

To ensure that the performance evaluation gives a true representation of the device performance in a "real life" situation, any laboratory-based evaluation should be conducted on a matrix that produces responses as close to real oral fluid as possible.

21.7.1.2 Laboratory-based Screening Tests

Commercially produced laboratory-based immunoassay kits are available for the analysis of oral fluid samples. These screening kits are predominantly either Enzyme Linked Immunosorbent Assay (ELISA) based or Homogenous Enzyme Immunoassay (HEIA) based.

21.7.2 Confirmation Analysis

Where unequivocal identification and quantification of the drugs present in oral fluid samples is required, it is usual to carry out analysis by gas chromatography-mass spectrometry (GC-MS) or liquid chromatography-mass spectrometry (LC-MS). Owing to the low volume of sample available for analysis is it preferable for confirmatory analyses to detect as many different analytes as possible within one analysis. Methods for the simultaneous detection of multiple analytes have been published for GC-MS.[24] The added advantage of using an LC-MS based confirmatory method is the increased sensitivity when using low volumes. Wood *et al.* reported an LC-MS based method which identifies and quantitates multiple illicit drugs of abuse.[25]

21.8 Oral Fluid Guidelines and Standards

In an attempt to standardise laboratory processes and general drug testing practice, oral fluid-based testing guidelines have been produced. The only standard that has been produced in relation to drugs of abuse in oral fluid is the Australian Standard AS 4760-2006.[26] This provides requirements and guidance that covers sample collection, handling, storage and dispatch; initial analysis both as POCT and laboratory-based screening tests; and further confirmatory analysis as well as target compounds cut-offs. The European Workplace Drug Testing Society has also produced similar guidance for testing in workplace settings.[12]

The United Nations Office on Drugs and Crime has produced guidelines for the analysis of controlled drugs in oral fluid (in addition to hair and sweat). These guidelines propose sample extraction and analytical techniques for adoption globally to analyse illicit drugs in oral fluid.[27]

21.9 Conclusion

The analysis of oral fluid samples for drugs of abuse is useful for the determination of recent drug use. The ease of sample collection is the main advantage of oral fluid analysis; however, the analysis of oral fluid samples is more challenging than more traditional matrices, due to the low concentrations found in oral fluid samples and the adverse effects that sample and collection buffer constituents can have on analytical instruments.

Care needs to be taken when interpreting oral fluid data, as the concentration of drugs in an oral fluid sample can be influenced by the route of drug administration and the method of sample collection, as well as sample storage conditions.

References

1. K. Kato, M. Hillsgrove, L. Weinhold, D. A. Gorelick, W. D. Darwin and E. J. Cone, Cocaine and metabolite excretion in saliva under stimulated and nonstimulated conditions, *J. Anal. Toxicol.*, 1993, **17**, 338–341.
2. C. L. O'Neal *et al.*, Correlation of saliva codeine concentrations with plasma concentrations after oral codeine administration, *J. Anal. Toxicol.*, 1999, **23**, 452–459.
3. A. M. Bermejo, A. C. S. Lucas and M. J. Tabernero, Saliva/plasma ratio of methadone and EDDP, *J. Anal. Toxicol.*, 2000, **24**, 70–72.
4. A. J. Jenkins, J. M. Oyler and E. J. Cone, Comparison of heroin and cocaine concentrations in saliva with concentrations in blood and plasma, *J. Anal. Toxicol.*, 1995, **19**, 359–374.
5. E. J. Cone, J. Oyler and W. D. Darwin, Cocaine disposition in saliva following intravenous, intranasal and smoked administration, *J. Anal. Toxicol.*, 1997, **21**, 465–474.
6. M. D. Osselton, S. M. Robinson, A. R. Cox and S. R. Reddick, An evaluation of oral fluid drug testing for the detection of opiate and cocaine use in a population of drug users, Poster presented at SOFT meeting September 2001.
7. R. Jufer, S. L. Walsh, E. J. Cone and A. Sampson-Cone, Effect of Repeated Cocaine Administration on Detection Times in Oral Fluid and Urine, *J. Anal. Toxicol.*, 2006, **30**, 458–462.

8. N. Samyn, A. Verstaraete, C. van Haeren and P. Kintz, Analysis of Drugs of Abuse in Saliva, *Forensic Sci. Rev.*, 1999, **11**, 2–18.
9. A. G. Verstraete, Detection Times of Drugs of Abuse in Blood, Urine, and Oral Fluid, *Ther. Drug Monit.*, 2004, **26**, 200–205.
10. H. A. Bird and M. E. Pickup, Salivary or blood specimen for evaluation of drug concentrations in rheumatoid arthritis, *Clin. Rheumatol.*, 1985, **4**, 220.
11. C. L. O'Neal, D. J. Crouch, D. E. Rollins and A. A. Fatah, The effects of collection method on oral fluid codeine concentrations, *J. Anal. Toxicol.*, 2000, **24**, 536–542.
12. G. Cooper, C. Moore, C. George and S. Pichini, Guidelines for European workplace drug testing in oral fluid, *Drug Test Anal.*, 2011, **3**, 269–276.
13. M. Ventura, S. Pichini, R. Ventura, P. Zuccaro, R. Pacifici and R. de laTorre, Stability studies of principal illicit drugs in oral fluid: preparation of reference materials for external quality assessment schemes, *Ther. Drug Monit.*, 2007, **29**, 662–665.
14. D. J. Crouch, Oral fluid collection: the neglected variable in oral fluid testing, *Forensic Sci. Int.*, 2005, **150**, 165–173.
15. M. Ventura, S. Pichini, R. Ventura, S. Leal, P. Zuccaro, R. Pacifici and R. de la Torre, Stability of drugs of abuse in oral fluid collection devices with the purpose of external quality assessment schemes, *Ther. Drug Monit.*, **31**, 277–280.
16. K. Langel, C. Engblom, A. Pehrsson, T. Gunnar, K. Arineimi and P. Lillsunde, Drug testing in oral fluid – Evaluation of sample collection devices, *J. Anal. Toxicol.*, 2008, **32**, 393–401.
17. R. Dams, M. A. Huestis, W. E. Lambert and C. M. Murphy, Matrix effect in bio-analysis of illicit drugs with LC-MS/MS: influence of ionization type, sample preparation and biofluid, *J. Am. Soc. Mass Spectrom*, 2003, **14**, 1290–1294.
18. F. T. Peters, Method validation using LC-MS, ed. A. Polettini, *Applications of LC-MS in Toxicology*, Pharmaceutical Press, London, UK, 2006, ch. 4, pp. 71–95.
19. S. Niedbala, K. Kardos, S. Salamone, D. Fritch, M. Bronsgeest and E. J. Cone, Passive Cannabis Smoke Exposure and Oral Fluid Testing, *J. Anal. Toxicol.*, 2004, **28**, 546–552.
20. R. S. Niedbala, K. W. Kardos, D. F. Fritch, K. P. Kunsman, K. A. Blum, G. A. Newland, J. Waga, L. Kurtz, M. Bronsgeest and E. J. Cone, Passive Cannabis Exposure and Oral Fluid Testing. II. Two Studies of Extreme Cannabis Smoke Exposure in a Motor Vehicle, *J. Anal. Toxicol.*, 2005, **29**, 607–615.

21. C. Moore, C. Coulter, D. Uges, J. Tuyay, S. van der Linde, A. van Leeuwen, M. Garnier and J. Orbita Jr, Cannabinoids in oral fluid following passive exposure to marijuana smoke, *Forensic Sci. Int.*, 2011, **212**, 227–230.

22. S. M. R. Wille, N. Samyn, M. del Mar Ramirez-Fernandez and G. De Boeck, Evaluation of on-site oral fluid screening using Drugwipe-5 +, RapidSTAT and Drug Test 5000 for the detection of drugs of abuse in drivers, *Forensic Sci. Int.*, 2010, **198**, 2–6.

23. (a) T. Blencowe, A. Pehrsson, P. Lillsunde, K. Vimpari, S. Houwing, B. Smink, R. Matthijssen, T. Van der Linden, S.-A. Legrand, K. Pil and A. Verstraete, An evaluation of eight on-site oral fluid drug screening devices using laboratory confirmation results from oral fluid, *Forensic Sci. Int.*, 2011, **208**, 173–179; (b) S. Strano-Rossi, E. Castrignano, L. Anzilloti, G. Serpelloni, R. Mollica, F. Tagliaro, J. P. Pascali, D. di Stefano, R. Sgalla and M. Chiarotti, Evaluation of four oral fluid devices (DDS, Drugtest 5000, Drugwipe 5+ and RapidSTAT) for on-site monitoring drugged driving in comparison with UHPLC-MS/MS analysis, *Forensic Sci. Int.*, 2012, **221**, 70–76.

24. T. Gunnar, K. Arineimi and P. Lillsunde, Validated Toxicological determination of 3 drugs of abuse as optimized derivatives in oral fluid by long column fast gas chromatography/electron impact mass spectrometry, *J. Mass Spectrom.*, 2005, **40**, 739–753.

25. M. Wood, M. Laloup, M. del Mar Ramirez Fernandez, K. M. Jenkins, M. S. Young, J. G. Ramaekers, G. De Boeck and N. Samyn, Quantitative analysis of multiple illicit drugs in preserved oral fluid by solid-phase extraction and liquid chromatography-tandem mass spectrometry, *Forensic Sci. Int.*, 2005, **150**, 227–238.

26. Standards Australia, Procedures for specimen collection and the detection and quantitation of drugs in oral fluid, *AS 4760-2006*, Standards Australia, Sydney, Australia 2006.

27. United Nations Office on Drugs and Crime, The analysis of drugs in oral fluid, *Guidelines for Testing Drugs under International Control in Hair, Sweat and Oral Fluid*, United Nations, New York, USA, 2014, ch. 3, pp. 51–54.

22 Sweat as an Alternative Biological Matrix

Nadia De Giovanni

Catholic University of the Sacred Heart, Rome, Italy
Email: nadia.degiovanni@policlinicogemelli.it

22.1 Introduction

Forensic toxicology covers an important area of forensic investigations concerning death and poisoning related to use/abuse of drugs. Laboratories performing toxicology tests aim to detect drugs of abuse and other psychoactive substances in biological matrices in order to identify possible impairment due to drug use, or to detect or exclude drug addiction. Drug testing in fact is used in many contexts, including crime investigations, safety and performance monitoring, workplace surveillance and treatment assessment, so covering important issues in the medical legal diagnosis.[1]

The choice of biological specimens suitable for the specific purpose, the selection of the analytical strategies and finally the interpretation of the results obtained by toxicological analyses are essential areas of forensic investigations.[2–4] The wide variety of specimens today available, the highly specific and sensitive analytical methods, and the high competence and know-how of toxicologists provide much information and correct interpretation.

The legal consequences of the results compel the management of significant and reliable data, hence the analytical toxicological laboratory should respect some important requirements suggested by

Forensic Toxicology: Drug Use and Misuse
Edited by Susannah Davies, Atholl Johnston and David Holt
© The Royal Society of Chemistry 2016
Published by the Royal Society of Chemistry, www.rsc.org

national guidelines and/or international standards.[5–8] With this premise, it is essential to obtain the assurance of the identity of the biological sample that must be traceable in all operating steps, from acquisition to disposal, establishing a "chain of custody" that documents with appropriate forms all stages: preanalytical, handling, storage, analysis and reporting.[1] In this perspective, the organization of the laboratory must ensure the correct procedure, from regulating access to the laboratory to final report delivery. The sampling must be preceded by a step of acceptance comprising the identification of the subject (who will sign an "informed consent"), the purpose of the investigation and the type of analysis. To the biological sample, collected respecting privacy but ensuring identity, is assigned a unique identification code to which reference is made in all subsequent stages.

The relevance of any finding in forensic toxicology is determined, in the first instance, by the nature and integrity of the specimen submitted for analysis. This means that there are several specific challenges to select and collect specimens. Moreover, the specimen should be divided into aliquots, one of which must be preserved for a predetermined time to be used in case of dispute. The amount of sample taken must then be sufficient to carry out the analysis in relation to the purpose of the survey, and for any eventual retesting. The storage of the sample must be appropriate to maintain the organoleptic characteristics, so ensuring the incorruptibility of the substances possibly present.[9]

22.1.1 Biological Specimens for Forensic Purposes

The purpose of the investigation, the legislation and the availability are the main requirements for the choice of the most suitable biological fluid. Moreover, a thorough knowledge of the type of substance to be identified, the time, manner and type of intake, as well as knowledge of the pharmacokinetic and pharmacodynamic parameters of the substances involved, have to be considered. The choice of biological samples is then a crucial analytical step because it influences the significance of analytical data, since each matrix reflects a different temporal window, being able to provide different information on the time of intake.[10]

Thus, important differences between the various biological samples consist in the time window within which the drug can be identified. The time of elimination of xenobiotics varies: from a few hours for blood and saliva, 2–3 days for urine, a few days for sweat, and even a few months for the keratinic matrices.[11] The certification of a psycho-physical state of the subject at the time of a particular event needs biological samples other than those chosen for the determination of the state dependence on a substance.

Other differences to be considered are related to the ease and invasiveness of the collection, the costs and difficulties of analysis, the validity and reproducibility of the results, and the possibility of interferences. Blood and urine are historically the most used biological matrices to objective detect use/abuse of drugs both in corpses and living persons, but today a number of other specimens can help toxicologists to answer the questions posed by judges.

22.1.2 Alternative Matrices

Scientific research has managed over the years to establish the possibility of using other biological matrices that can complement the traditional ones, providing important input to the interpretation of the data. Hair analysis was first used to document chronic drug exposure in 1979;[12] since then, the use of alternative samples has gained much attention in many forensic contexts. A number of alternative matrices had been studied so far, to obtain different information and allow easy, rapid and non-invasive collection. For this purpose, saliva, sweat and keratinic matrices represent the biological samples most extensively studied in recent decades. Each of these alternative biological matrices show different characteristics in terms of time window, so allowing different data to be obtained.[11,13] Saliva can be used as an alternative to blood, displaying a narrow window of a few hours, thus giving information about any psycho-physical impairment. On the other hand, keratinic matrices exhibit the longest window, related to the length of the strand of hair; a history of drug abuse can be built from the analysis of hair by examining different portions of the shaft, each one reflecting a particular period of time.

Sweat has a time window longer than saliva, but shorter than hair. It is in some ways comparable to urine, but with a wider time window, hence providing different information from other biological matrices.[8,12] It is not surprising, therefore, that a number of papers dealing with the determination of several drugs in sweat have been published, owing to the fact that collection devices are becoming more efficient to trap sweat, and chromatographic equipment is becoming more and more sensitive, with mass spectrometry (MS) and tandem mass spectrometry (MS/MS) being the most widely used analytical tool, combined with gas chromatography (GC) or liquid chromatography (LC).

Sweat is currently used in a clinical setting, such as in the test for cystic fibrosis.[14] Although applications in forensic toxicology are still limited, its use is desirable in order to obtain information on a continuity condition of taking drugs, avoiding in this way the periodic urine samples.

The two main limiting factors for the use of sweat are the amount that can be collected and the concentrations of analytes lower than in urine. The volume of sample obtainable is a critical point, especially with a forensic application that needs to preserve one part sample to repeat the test in the case of contestation. Nevertheless, the international scientific literature has developed numerous approaches for the analytical determination of xenobiotics in sweat.

22.2 Skin and Sweat

The skin is the largest organ of the body and protects from external attacks, regulates body temperature and permits the sensations of touch, heat and cold. Skin is made of three layers: epidermis, dermis and hypodermis. The epidermis is the outer layer of the skin and is thin and tough. Most of the cells in the epidermis are keratinocytes originating from cells in a deeper layer. The outermost portion of the epidermis is called the *stratum corneum*, is relatively waterproof and protects the body from bacteria, viruses and other substances. The dermis, the skin's next layer, is a thick layer of fibrous and elastic tissue that gives the skin flexibility and strength. It contains nerve endings, sweat glands and sebaceous glands, hair follicles and blood vessels. The sweat glands produce sweat in response to heat and stress while the sebaceous glands secrete sebum into hair follicles. The blood vessels of the dermis provide nutrients to the skin and help regulate body temperature. The deeper subcutaneous tissue called hypodermis is made of fat and connective tissue. The skin actively metabolizes drugs with multienzyme protein systems which oxidize and combine exogenous compounds; it transfer various substances through the outside and represents the collection area of the sweat.

Sweat is a biological fluid produced by a major homeostatic mechanism necessary to maintain a constant body temperature; it causes loss of heat during evaporation and hence sweating is one of the fundamental mechanisms of thermoregulation.[8] It is also a defence mechanism of the skin, the excretion fluid of chemosignals and the waste of metabolites. Moisture loss *via* the skin and elimination of insensible (non-visible) sweat takes place during normal breathing at a rate of about 0.3–0.7 L day^{-1}; sensible sweat refers to perspiration at rates of 2–4 L h^{-1}. An adult produces an extremely variable amount of sweat, depending on physical activity, emotional states and environmental temperature. About half the total volume of sweat is eliminated from the trunk of the body; the remaining fluid is lost from the legs or upper extremities and head in approximately equal amounts.[15]

Sweat is mainly composed of water (approximately 99%), in which several minor components are dissolved: electrolytes (Na$^+$, K$^+$, Cl$^-$), organic substances (urea, uric acid, creatinine, pyruvic acid, lactic acid, ammonia, immunoglobulins, acid volatile fat, cholesterol and a small amount of glucose) and inorganic molecules such as sulfates and phosphates.[8] The pH of sweat generally ranges between 4 and 6.8, with an average pH of 6.3 in resting individuals. During exercise the sweat flow rate increases, resulting in an increase in pH up to 6.8. It is therefore a hypotonic fluid, colorless, with acid reaction and variable composition of salts, in relation to different physiological conditions of the organism. The presence of exogenous substances has been demonstrated for several years in this body fluid.

22.2.1 Sweat Secretion

Sweat is secreted by about three million glands in the deep layer of the skin, distinguished as apocrine or eccrine depending on their location. Although they are both coiled, tubular glands, there are several differences between them, including the size of the lumen, which is slightly smaller in the eccrine glands, and the type of the secretion.

The apocrine glands are most commonly located in the armpits, genital area and around the nipples and do not participate in thermal regulation. They reach deep into the layer of the dermis and secrete their fluids into the tiny canals of the hair follicles. They are slightly larger and produce a thicker, stickier and more odorous sweat containing proteins, fats and other aromatic substances responsible for the organoleptic features. There is no scientific evidence regarding the purpose of the apocrine glands. In animals, they seem to act as scent glands that aid in attracting the opposite sex, which some believe is true for humans as well. These glands are primarily activated by stress or excitement.

The eccrine glands are present over almost the entire body surface, with a higher density in the palm of the hand, in the soles of the feet, in the armpit, scalp area and in front and they have a thermoregulatory function. They are also known as "exocrine glands" and they secrete substances directly onto the skin's surface *via* the pores instead of into the bloodstream. The eccrine sweat is usually odorless, composed primarily of water and small amounts of mineral salts containing high amounts of sodium. The main purpose of the eccrine glands is then to control the temperature of the body; in fact as the body temperature increases, the sweat secretion increases as well, allowing the temperature to drop. Exercise, stress and excitement can also activate these glands. Another benefit of the sweat produced by these glands is that it helps to protect the body against harmful bacteria and viruses.

The surface of the human body also contains sebaceous apocryphal glands; they produce sebum, an oily mixture of fats that keeps the skin moist and soft and acts as a barrier against foreign substances. Sweat and sebum both contribute to the formation of the hydrolipid film that protects the skin surface from chemical (detergents, solvents, inks, *etc.*) and bacterial aggression.[8]

22.2.2 Mechanism of Incorporation of Drugs

A small amount of a drug absorbed, distributed and metabolized is excreted in sweat with a mechanism not completely clear, though the pH of sweat and pK_a of the drugs are believed to be of the utmost importance.[8]

In 1844 the presence of quinine in perspiration fluid was verified, and this can be considered the first observation on outward transcutaneous drug delivery.[16] To date, the presence of various drugs on the skin surface has been reported.

Although it is poorly understood how nonvolatile chemicals exit the body through the skin, and the mechanism of transfer of drugs from the bloodstream to the sweat is still not definitely clarified, at least two distinct pathways for drug molecules to reach the skin surface have been proposed. Passive diffusion from blood into sweat glands and transdermal passage of drugs across the skin are best considered.[17–21] The role of the *stratum corneum* as a temporary drug reservoir has been also demonstrated.[16]

The majority of authors suggest that drugs in the blood are transferred into sweat by a passive diffusion process governed by the same factors as the secretion into saliva and are then released onto the surface layer of the skin. The factors affecting excretion of drugs into sweat are based on the chemical and physical properties of the drug, such as molecular weight, pK_a, protein binding and lipophilicity.[22] This diffusion through the skin also depends on physiological and pathological conditions of the skin, such as hydration, temperature, age and body area. A low basal pH favors higher sweat to plasma ratios of basic drugs. Hence, non-ionized basic drugs easily pass from the blood to sweat and may be accumulated owing to the lower pH of sweat compared with blood. Transdermal migration of drugs through the skin is another possible mechanism of transfer of xenobiotics; however, this is limited by multiple barriers (subcutaneous fat, dermis, epidermis and *stratum corneum*).

Specific characteristics of the sweat collection devices, the site of sweat collection and the application of heat to increase the amount of drug secreted can all influence the transdermal migration of drugs.[23] Drug binding to various skin fractions and re-absorption of drugs from the skin have also been observed. Therefore, a continued presence of drugs on the skin surface results in the time period when blood or urine levels are already undetectable. The time interval between drug consumption and detection on the skin surface depends on the nature of the particular drug and on the sensitivity of the analytical method used. In chronic abusers, drug molecules are permanently present on the skin due to the temporary reservoir of the *stratum corneum*.

Despite the lipid bilayer of the skin, fat-soluble drugs can overcome the barrier and enter the circulating system, facilitating a passage of substances in both directions, both from the outside to the blood and *vice versa*.

22.3   Analytical Issues

The analytical detection of drugs of abuse excreted with sweat includes various steps. First of all is the collection of the fluid, which today is relatively easy to perform using devices able to capture and retain molecules of interest. The extraction of xenobiotics from these collection devices, the analytical determination by chromatographic techniques and the identification and quantitation of the drugs are the next steps. All these procedures need development and validation for the assurance of the quality of the results. Analytical studies found in the international literature outline the requirements of technical issues.

22.3.1   Sample Collection and Time of Permanence

One of the most critical aspects in relation to the analysis of sweat is the collection step, which can be performed in different ways. Several collection methods have been proposed for sweat sampling: one of the more traditional is the rubbing of the skin with cotton balls, gauze or filter paper. Dry gauze covered with waterproof plastic, pads impregnated with salts and cotton swabs have also been used. In order to increase the production of sweat, the local application of certain chemicals, such as pilocarpine, has been also exploited.[15,20]

Today, sweat can be collected as liquid perspiration on sweat wipes[24,25] or over time using sweat patches.[26–29] The wipe technology is not very common, however, and is often linked to a non-instrumental immunoassay for the rapid detection of drugs; this will be referred to in the next section of this text. Problems linked to low volume, low concentration of drugs and possible external contamination limit its employ.

Sweat patches, on the contrary, received much attention for some advantages with respect to wipes. In recent years, methods have been developed and marketed to facilitate the collection of the fluid. Special attention needs to be paid to the standardization of sampling methods and the time, in order to avoid misleading drug concentration results; care should be taken to avoid contamination of the fluid during sampling with extraneous materials that would interfere with the performance of the assay.

Once the patches are applied on the skin there is no more chance of contamination, as the outer layer is impermeable to adulterants. Sweat is then usually collected using an adhesive absorbent patch that is placed on the surface of clean skin. Careful preparation of the skin is necessary prior to placement of a sweat patch to minimize external drug contamination or bacterial degradation of the drug once it has been retained.[8] The devices on the market are usually polyurethane membranes retained by adhesive film applied to the skin. They have three components: the absorption pad, a release liner and a polyurethane/adhesive layer. The absorption pad is approximately 3 cm wide, 5 cm long, and consists of inert, medical grade cellulose that retains the nonvolatile components of sweat (including drugs of abuse and their metabolites) collected from the surface of the skin. The membrane is formed by two parts of cellulose on which is visible a number of recognitions essential for proper application of the "chain of custody", required to identify the sample with certainty. It is a barrier, but the passage of oxygen, carbon dioxide and water vapor is allowed for normal perspiration. Salts, solids and larger molecules, such as drugs that pass through the skin, are trapped in the absorbent pad, where they are temporarily stored. The release liner allows the easy removal of the collection pad from the adhesive layer after patch use.

The patch can be applied to different areas of the body, preferably the abdomen, back or on the deltoid arm after thoroughly cleansing the area with disinfectant; it is kept in place for a few days and subsequently removed and stored in special packages at low temperature $(-20\,^{\circ}\text{C})$.

Sweat patches might be used for short periods (hours or days) in a spot fashion very similar to urinalysis, or for longer periods (weeks). The optimal time of permanence of the patch in the sweat collection has not yet been definitely established. Owing to the relatively small volume (mL) of insensible sweat secreted from a small absorbent area (typically 3–5 cm), patches are typically worn for several days. Literature data refer to experiments related to the minimum length of wear necessary to detect recent or concurrent drug use with increasing longer-term wear.[30] Short-term (between 0.5 and 2 h) and long-term patches (1–14 days), identical except for duration of wear, demonstrated that the minimum duration that patches must be worn to detect recent or concurrent drug use is about 2 h. Analyte concentrations increase significantly with increasing lengths of wear up to one week.

In 1993, PharmChem Laboratories proposed a normalized solution for sweat collection, by commercializing a sweat patch technology approved by the American Food and Drug Administration for the analysis of cocaine, opioids, phencyclidine, amphetamines and cannabinoids (PharmChek™ Sweat Patch). Sweat patches were employed to study the presence of various xenobiotics;[27,31–35] the parent drug was always identified to a higher extent than the metabolites.

Over the years, other devices have been developed for collection of sweat specimens. A patch called "Fastpatch" has been developed with a slightly larger cellulose pad than a standard PharmChek, employing heat-induced sweat stimulation; the experiment used shorter wear periods, and possibly longer time-periods of detected use.[36] The "Hand-held Fast Patch" applied to the palm of the hand and the "Torso Fast Patch" applied to the abdomen or the sides of the trunk (flanks) have been checked for their performances.[23] Concentrations of drugs generally higher in sweat specimens collected with the "Hand-held" than for the "Torso Fast Patch" were observed. Drug concentrations were also considerably higher than those reported for the PharmChek sweat patch.[23]

22.3.2 Sample Preparation and Analysis of Sweat

Once the sweat has been collected, screening and confirmation analysis must be performed in order to identify and possibly quantitate the analyte and/or its metabolites trapped in the pad. Some screening kits have been marketed in addition to collecting the sample, to provide extemporaneous "on-site" analysis for drugs of abuse, always keeping in mind that the confirmation with more specific instrumental techniques after treatment and preparation of the samples is mandatory. Immunoassays are commonly used for the purpose of allowing the exclusion of negative specimens from further confirmatory testing. A rapid on-site test for the screening of drugs of abuse in sweat has been developed, the non-instrumental "Drugwipe" marketed by Securetec (Germany). It is a pen-sized test strip based on an immunological technique performed on sweat wiped from the surface of the body;[37,38] it is able to screen the more common drugs of abuse in sweat, achieving results within 2–3 min. The chance to screen sweat samples has been also exploited using a radio-immunoassay (RIA) technique[39] and immunoassay using a "Lab-on-a-Paper" skinpatch system.[40]

The confirmation of preliminary results are usually performed by chromatographic analysis, considered the technique of choice for this purpose; the high sensitivity of analytical instrumentation in fact allows the detection of drugs at very low concentrations, as is necessary with sweat.[41] Various scientific papers report procedures for extraction and detection by GC-MS or LC-MS/MS techniques.

The first step of the confirmation analysis includes the extraction and purification of drugs of abuse from wipes and/or patches. Simple extraction of the drugs from the patch can be done by shaking the patch with methanol–acetonitrile, evaporating the methanol and reconstituting with the mobile phase for LC-MS analysis. This method was followed by several researchers for the determination of fentanyl,[42] sunitinib and its metabolite,[43] methylphenidate and ritalinic acid.[44] Solid phase extraction (SPE) has been often used for the purification of drugs after their preliminary elution from the cellulose patches. Concheiro et al.[32] developed and validated an LC-MS/MS method for the fast and simultaneous quantification of 14 drugs and metabolites in sweat patches, with good selectivity and sensitivity, and applied the method to weekly sweat patches from an opioid-dependent buprenorphine-maintained pregnant woman. Micro-extraction was also successfully applied to sweat,[41] allowing the reduction of organic solvent consumption to a minimum and providing a rapid, simple and cost-effective approach.

GC-MS methods have been tested to examine sweat extracts, to monitor an individual's drug exposure for the purposes of drug treatment programs, for employment initiatives, and for forensic investigations.[23,34,35,39,45] This technique has been used to study the effect of sweat patch location (back and shoulder) on cocaine levels after controlled intravenous cocaine exposure in 12 subjects.[46] The results were affected by the site of sweat collection, and in fact the levels of cocaine and metabolites in patches were eightfold higher on the back than on the shoulders.[23] Re-absorption (back transfer), degradation or hydrolysis, and loss of cocaine to the environment were also demonstrated, so accounting for substantial loss of cocaine from skin sweat collection patches during patch wear.[46]

LC-MS is, however, the most popular technique developed and validated to detect drugs in sweat after patch collection[32,47,48] because of little sample pre-treatment even though ensuring high performance. Other techniques were only sometimes used to identify xenobiotics in sweat, such as a capillary electrophoresis (CE) method developed for the determination of taurine in sweat, with simplicity and rapidity.[49]

Finally, it is important to mention a relatively new analytical method used to detect drugs in sweat excreted through the pores in the skin; it can be transferred to surfaces, leaving an impression referred as latent fingerprint. The sweat in a latent fingerprint can contain orally ingested drugs and their metabolites.[50] Various methods such as immunoassays,[51,52] fluorescence[51] and Raman spectroscopy[53] have been applied in order to detect drugs of abuse in sweat retrieved from such fingerprints. When an individual handles a substance, he can deposit a contaminated fingerprint; hence the detection of a drug in a fingerprint may indicate that an individual has come into contact with a drug, although it is not a definite proof of use of the drug. To test drug intake in criminal investigations, caffeine was detected in blood and fingerprint by LC-MS with the aim to determine whether an ingested drug could be detected in a subject's fingerprints.[54]

22.4 Monitoring Drug Exposure through Sweat: Forensic Applications

Forensic toxicology can be involved in several situations that may have medicolegal review, such as crime under the influence of alcohol/drugs, date rape, driving under the influence, psychiatric disorders and determination of the cause of death.[1,55] Sometimes, one should be very cautious in investigating exposure to psychoactive drugs related to some kind of crime, due for example to late sampling of biological specimens. Sweat may offer a non-invasive alternative to present sampling for continuous monitoring of drug exposure in living people, as it can be collected for a programmed period with minimal disturbance for the sampled individual. Obviously post-mortem toxicology cannot benefit this matrix. The present section refers to some of the possible applications of sweat analysis in forensic areas explored by the scientific literature.

22.4.1 Driving Under the Influence of Alcohol/Drugs

Driving under the influence of alcohol or illicit drugs (DUID; also see Chapter 13) continues to be a concern in developed countries, contributing to many motor vehicle crashes.[56] To control high risk driving habits, studies to easily identify DUID have been developed on various biological matrices. Blood is the biological sample of choice for its ability to detect the impairment of drivers. It is difficult to collect, however, especially at the moment of the event, and requires particular strategies for preservation, analysis and storage. Different alternative biological matrices have been evaluated to detect DUID, and in particular oral fluid can be considered a valid substitute for blood. Sweat has also been studied, and the literature refers to experiences on sweat in the field of "drugs and driving".[57] At the initiative of the European Commission, the Rosita (acronym for "Roadside Testing Assessment") project was developed in order to evaluate alternative biological matrices and the possibility to use on-site tests able to easily and rapidly locate DUID. Analytical devices for field analysis of biological samples, including sweat, showed encouraging results,[58] mainly because the collection of sweat has been accepted better than urine. The drivers were often impressed by the result of the rapid drug tests, and if at the beginning they denied having taken the drug, a positive result often made them confess, and the roadside tests resulted in a deterrent effect.[57] Unfortunately, only a few participants at the project chose to study sweat; hence the evaluation cannot be considered exhaustive.

The need to obtain information on the impairment of drivers while driving excludes the collection of sweat by patches, but forehead wipes could be useful. Some considerations made on sweat collected from injured drivers and checked for the presence of Δ^9-tetrahydrocannabinol[25] disclosed possible environmental contamination. A positive sweat response could easily be from passive exposure to smoke or from a contaminated hand wiping the forehead. These results were confirmed later[57] when a global overview on the issue of drugs and driving highlighted that sweat is only acceptable but not good enough for possible environmental contamination, the delayed appearance of drugs and the absence of the standardization of sampling. The non-invasive collection which is relatively easy to perform and can be achieved under close supervision is, however, one of the most important benefits in a driving-under-the-influence situation.

In conclusion, although skin testing may identify a substance abuser, supporting conventional biological matrices in order to perform assessments as part of their driving ability, it seems less appropriate to detect recent or actual drug exposure for the variable time delay before the drug will be excreted in sweat. Probably for these reasons, no recent studies about the use of sweat in the context of DUID have been published.

22.4.2 Workplace Drug Testing

Workplace drug testing (also see Chapter 11) is a well-established application of forensic toxicology and it aims to reduce workplace accidents caused by affected workers.[59] Since the use/abuse of drugs is widespread in the entire world population, the safety of workers must also include the monitoring of this problem. In many countries it is now envisaged to check the toxicology of workers, especially those that could put at risk their own or another's life (such as drivers of public transport), also from the toxicological point of view. Urine drug testing has historically been the most common technique for detecting drug use in the workplace. Although widely utilized for its many advantages, problems such as the lack of long-term drug detection, the inability to correlate results with drug impairment and blood drug concentrations, possible specimen adulteration by donors, and privacy issues do not make the testing of urine a guarantee for a drug-free workplace. The use of alternative biological specimens, such as hair, oral fluid or sweat, presents several advantages over urinalysis, mainly the fact that sample collection can be easily performed without infringing the examinee's privacy, so the subject is more likely to perform the test. With this background, sweat could represent a suitable organic matrix alternative and/or complementary method to urine sampling to highlight any use/abuse of drugs.[10] Sweat provides a cumulative measure of drug exposure, having the ability to monitor drug intake for a period of days to weeks, detecting both parent drugs and metabolites. It is a non-invasive specimen collection, and collection device is relatively tamper-proof. However, the low concentrations of drugs and the small amount of sample available require the use of highly sensitive techniques.[60]

The U.S. Department of Health and Human Services (HHS) drug testing standards were published in 1988 and successively revised. HHS proposed to establish scientific and technical guidelines for the Federal Workplace Drug Testing Program in order to permit laboratory testing of hair, oral fluid and sweat patch specimens in addition to urine specimens for many substances of abuse. An extensive evaluation by the Division of Workplace Programs of the HHS was performed to determine the utility of sweat in federally regulated programs. In 2008, the Substance Abuse and Mental Health Services Administration (SAMHSA)[61] reviewed the comments to define the requirements for such testing, including: specimen collection procedures, custody and control procedures that ensure donor specimen identity and integrity, testing facility, initial and confirmatory test cut-off concentrations, analytical testing methods, result review and reporting, evaluation of alternative medical explanations for presence of drug or metabolite in the donor's specimen, and laboratory certification issues. The addition of accurate and reliable workplace drug testing using hair, oral fluid and sweat patch specimens will complement urine drug testing, and aid in combating industries devoted to suborning drug testing through adulteration, substitution and dilution.

Some disadvantages complicate the employ of sweat, however, including the large variation in sweat production, different for each individual, the unknown specimen volume, the limited collection devices available, the high inter-subject variability, the risk of accidental removal, and the risk of contamination during application/removal.

There are not so many studies evaluating this matrix for the purpose of the workplace, but the great advantages of highly limiting the risk of adulteration and/or replacement of the sample must be positively considered.

22.4.3 Drugs of Abuse in Pregnancy and in Pediatric Patients

A peculiar application of forensic toxicology is represented by the use of drugs during pregnancy. Drug and alcohol misuse by pregnant women has been extensively associated with adverse mental, physical and psychological outcomes in their exposed fetus and/or children. This is a serious social problem, not only because of effects on the health of the woman and child, but also because drug or alcohol dependency detracts from child care and enhances the prospect of child neglect and family breakdown.[62] Biomonitoring drugs in biological samples collected from the pregnant mother and/or from the fetus or newborn may ensure the health of children and prevent developmental problems. It has been estimated that more than 75% of infants exposed to drugs show medical problems, compared to 27% of unexposed infants.[63]

Detection of *in utero* drug exposure has traditionally been accomplished through urine testing; however, the window of detection is short, reflecting drug use for only a few days before delivery. For this reason, nonconventional samples have been studied to understand their potentiality to quickly identify addiction during pregnancy.

Sweat was considered as an effective method for monitoring abstinence or illicit drug use relapse in the high-risk population of pregnant drug-dependent women. For this purpose, sweat was collected from pregnant drug-dependent volunteers who wore patches and replaced them approximately weekly from study entry until delivery; samples analyzed by GC-MS disclosed a high percentage of positivities.[27] Polydrug use was evident by the presence of both cocaine and opiate metabolites in many patches.

Some scientists, however, highlighted that although sweat had advantages, drug detection in maternal sweat accounted only for acute consumption that occurred in the hours previous to collection, so giving poor information concerning fetal exposure.[64] Sweat can be used not only to disclose drug abuse, but also to check the adherence to a substitution program with buprenorphine; sweat specimens collected in addition to plasma specimens during various periods of gestation allowed the detection of buprenorphine and its metabolite.[65] Despite the potential application, maternal sweat has rarely been used to determine fetal exposure to drugs.

The non-invasiveness of the collection of sweat through patches encouraged clinicians to investigate a possible use with pediatric patients; there is also the benefit of monitoring psychoactive drugs therapeutically administered to assess the short- and long-term history of drug use.[48,66] The first survey proved that atomoxetine and its metabolites can be measured in patches using a LC-MS/MS method. The second, applied for the monitoring of methylphenidate, demonstrated that sweat patches can be used for non-invasive monitoring of its consumption and misuse in situations where detection of recent abuse is of interest.

22.4.4 Drug Facilitated Sexual Assault

Samples obtained from "drug facilitated sexual assault" (DFSA) victims typically consist of blood and urine. Only a few studies referring to the employ of sweat to identify drug use during sexual assault are reported in literature, even if the delay often observed in sampling biological matrices in this field would benefit from sweat characteristics. While more information regarding the concentrations of DFSA agents and their metabolites in sweat is needed, ion mobility spectrometry (IMS) appears to have the sensitivity and resolution needed to make a qualitative, presumptive identification of the agents. The combination of IMS with facile sweat collection was demonstrated to offer rapid results on a screening basis in a non-invasive manner, guiding initial treatment and investigations.[67]

Cannabis is the most frequently detected illicit drug that has been linked with drug-facilitated sexual assault cases.[68] γ-Hydroxybutyrate (GHB) is also a substance typically used for the purpose of DFSA, but its detection is mainly complicated by a high variability in the endogenous levels.[69] Owing to this high variability, a comparison with a baseline level would be required. Such a baseline would be unlikely to be available in the scenario envisioned; thus sweat would have limited utility in screening for GHB.

Drug facilitated sexual assault is also discussed in Chapter 1.

22.4.5 Doping

Another forensic toxicological issue concerns prevention and protection of health in a sporting environment, pertaining not only to doping centres. Screening of athletes for performance-enhancing substances is monitored by the World Anti-Doping Agency (WADA; also see Chapter 12). WADA, established in 1999 as an international independent agency composed and funded equally by the sport movement and governments of the world, coordinates the fight against doping in sport. It works to help the individual sports' federations to improve the test-doping, research and education against this phenomenon. WADA periodically draws up a list of prohibited substances that athletes cannot take, working towards a vision of a world where all athletes compete in a doping-free sporting environment.[70]

The substances and methods on the list are classified by different categories (*e.g.* steroids, stimulants, gene doping) and sweat testing has been exploited as alternative matrix to detect some substances, with some advantages over blood or urine. In particular, the collection of sweat is easier than blood and urine, with less opportunity for sample adulteration; it has the benefit of being non-invasive, shows a longer time-detection window and poses fewer ethical problems for sample collection than does blood or urine testing. Sweat, together with urine, is actually the preferred sample for doping control, with each biofluid having its advantages and limitations.

A review on the detection of doping agents in unconventional samples (including sweat) was published in 2000.[71] More recently, compounds prohibited at all times (*e.g.* clenbuterol, β-2 agonists, estrogen-receptor modulators) have been successfully tested and clearly interpreted in alternative specimens.[72] In 2013 a brief report on sweat characteristics[22] underlined the need for non-invasive biofluids that are ethically appropriate, cost-efficient and toxicologically relevant as alternatives to plasma/serum in many contexts, including doping control in sports medicine. The types of drugs determined in sweat, the shortcomings for normalization of the sampled volume, sample preparation and the analytical equipment used for their identification and/or determination have been again discussed.[73]

22.5 Interpreting Analytical Results

The interpretation of the presence/absence of an exogenous substance in sweat is still under discussion for the lack of comprehensive scientific evidence covering all areas, ranging from the perfect knowledge of the mechanisms of incorporation, the individual variables, the development of sampling, storage, analysis, *etc.*

In the US, SAMHSA[5] issued for several years the recommendations regarding the use of the sweat test and its applicability even in court. The guidelines detailed operating procedures regarding the collection of sweat, analysis and interpretation of results. Recently, guidelines to support operators wishing to use sweat to identify use/abuse of drugs were published by United Nations Office on Drugs and Crime (UNODC).[8] Guidelines encourage the analysis of multiple specimens to better define the time and manner of exposure to one or more xenobiotics and to confirm laboratory results in cases of doubtful clinical and anamnestic histories. The document underlines that qualitative results in sweat are generally well understood, while the interpretation of quantitative data is still under debate, especially with regard to the dose–concentration relationship. It is then necessary to explore sweat peculiarities more deeply in order to routinely employ this matrix; the information actually obtained from the analysis of such a sample may, however, be used as a complement to other body fluids.

The method of sampling is crucial because it can directly influence the time window: a simple withdrawal from the skin, by means of a wipe, gives information on a very recent intake; the permanence of a patch for a few days, on the other hand, can provide information about previous intakes. The optimal residence time of patches on the skin, suitable for a proper diagnosis of use/abuse of xenobiotics, is a critical point in the interpretation of analytical data. In fact, despite the guidelines of SAHMSA[5] that indicate a period of time equal to 3–7 days is necessary to allow the absorption of xenobiotics on the "patch", still there are conflicting opinions on the optimal time of wearing. We must consider in fact the possibility of establishing a balance that allows the drug to be absorbed by the skin, redistributed and deleted by different modes. The observation that cocaine can be lost from patches, making quantitative cumulative collection more difficult, allowed the suggestion that it can be reabsorbed back to the skin from the patch or it can be metabolized or degraded on the surface of the skin.[46] On the other hand, chronic exposure to fat-soluble substances such as cannabinoids can cause a buildup in adipose tissue, with a subsequent release, giving rise to false positive results for a recent exposure. The concentration of drugs found in sweat is then strongly dependent on the method used for collecting it and on the time of permanence; high variability also depends on the site of the body where the collection occurs. Storage stability is also an important issue to be considered for the interpretation of analytical results because of possible degradation of drugs carried by sweat or their loss during storage.[74] Primarily the parent drug has been found by most analysts, while metabolites are trapped in sweat in a low percentage.

The risk of contamination during application and removal of the patch[31] is a further point to consider. The removal of sweat through the patch does not allow the measurement of the volume of fluid secreted, so preventing the quantitative determination of substances, allowing the analysis of sweat for a qualitative assessment only. The concentration is then expressed in nanograms and calculated for each individual patch (nanograms/patch). Appenzeller *et al.*[75] proposed a method to overcome this problem through the determination of sodium in the patch, similar to the standardization of urine by the creatinine content. Similarly to the latter, however, sodium in sweat can be altered by various diseases.

Once the preliminary stage of sampling is completed, the specimen should be subjected to a series of analytical procedures for the selective extraction of drugs and their identification. It would be best to use a preliminary screening test followed by confirmation with different and more specific techniques. To date, the only screening test to make an impromptu analysis of xenobiotics in sweat is constituted by the quick test "Drugwipe"; however, it only allows the detection of a short-consumption term and can be subjected to a high risk of external contamination.

Chromatographic techniques coupled with mass spectrometry are preferentially used for the confirmation, possessing the characteristics of sensitivity and specificity required for such assessments. The more critical step is the sample preparation; in fact the low amount of substances which are excreted in sweat must be extracted with high yield.

To correctly interpret analytical results, appropriate cut-off values should be considered. The value of the cut-off for both screening analysis and confirmation has been discussed over the years. The cut-off levels proposed by the manufacturer PharmChem range from 1.5 to 10 ng per patch for screening and from 0.5 to 10 ng per patch for the confirmation, depending on the drug; SAMHSA, limited to workplace drug testing, proposed the range 4–25 ng per patch for screening and 1–25 ng per patch for confirmation, again depending on the drug.[5,8]

22.6 Conclusions

Sweat can be considered the simplest matrix to monitor drug levels without discomfort to the patients, as peripheral skin secretions are easy to collect. A typical time window, characteristic and somewhat different from other matrices, ensures specific information, providing a detection of up to a week or more after intake. Hence sweat can be considered a non-invasive means to obtain an estimate of drug exposure for at least one week and decreases the likelihood of missing recent use. When applied for long periods, patches have probably a deterrent effect through long-term monitoring.

Among the most beneficial aspects of this sampling are the non-invasive method, the low biological risk to which the operator is exposed, the minor embarrassment of the patient and the possibility of direct supervision. The privacy of the patient is guaranteed, as well as the chain of custody and identification, supported by a unique identification number on the sweat patches. Moreover, sample adulteration is difficult and removal or manipulation of the patch before and after collection can be easily detected by the supervisor of the sampling. These are great benefits for laboratories that must give answer for the purpose of the law.

On the other hand, some negative characteristics restrict its application in the forensic field. One of the main disadvantages with working with sweat is the volume of sweat collected over a period of time. The amount of sweat produced varies in fact between and within individuals, is usually low and is difficult to measure the exact volume. The environment, daily activities, emotional and health status of an individual will also affect sweat production. Special care is required during application and removal of the patch, because of the risk of contamination. The lack of suitable quality control and standardization of procedures, the scarcity of screening techniques and not exhaustive knowledge of the relationship between concentration and time–dose-rate assumption are other technical problems.

The amount of sample collected does not always allow for repetition of the analysis, the inhomogeneity of the sample influences negatively the reproducibility, and the low concentrations of analytes require extremely sensitive techniques for their determination. These are all critical points for forensic application, particularly because the volume of the sample should be partially preserved to repeat the test in case of a positive result or in the event of a dispute.

Analysis of sweat may find applications in the treatment and monitoring of drug abuse as the patches provide a long-term continuous monitoring of drug exposure. Parent drugs are generally excreted at higher concentrations than metabolites, thus allowing detection of molecules with a short half-life in blood and poorly detected in urine.

Only by adopting the accepted strategy in forensic toxicology, based on screening and confirmation on two different aliquots of the same sample, can analytical results obtained in alternative matrices achieve the high degree of reliability which is required in a forensic context. Because of current limited knowledge of the modalities/kinetics of incorporation and elimination of xenobiotics, the interpretation of analytical results may still be difficult.

In conclusion, the information actually obtained from the analysis of drugs of abuse in sweat can be of interest to forensic toxicology in some areas, as workplace drug testing, to monitor pregnant women, in a doping context and may be used as a complement to other body fluids.

References

1. J. F. Wyman, Principles and procedures in forensic toxicology, *Clin. Lab. Med.*, 2012, **32**, 493.

2. O. H. Drummer, Forensic Toxicology, *EXS*, 2010, **100**, 579.
3. M. R. Meyer and H. H. Maurer, Forensic and clinical toxicology, *Bioanalysis*, 2014, **6**, 2187.
4. F. T. Peters, H. H. Maurer and F. Musshoff, Forensic Toxicology, *Anal. Bioanal. Chem.*, 2011, **400**, 7.
5. www.samhsa.gov.
6. www.soht.org.
7. www.tiaft.org.
8. UNODC, Guidelines for "Testing drugs under international control in hair, sweat and oral fluid." United Nations, New York, 2014.
9. www.gtfi.it.
10. Y. H. Caplan and B. A. Goldberger, Alternative specimens for workplace drug testing, *J. Anal. Toxicol.*, 2001, **25**, 396.
11. A. G. Verstraete, Detection times of drugs of abuse in blood, urine, and oral fluid, *Ther. Drug Monit.*, 2004, **26**, 200.
12. E. Gallardo and J. A. Queiroz, The role of alternative specimens in toxicological analysis, *Biomed. Chromatogr.*, 2008, **22**, 795.
13. D. Vearrier, J. A. Curtis and M. I. Greenberg, Biological testing for drugs of abuse, *EXS*, 2010, **100**, 489.
14. B. Aqil, A. West, M. Dowlin, E. Tam, C. Nordstrom, G. Buffone and S. Devaraj, Implementation of a quality improvement program to improve sweat test performance in a pediatric hospital, *Arch. Pathol. Lab. Med.*, 2014, **138**, 920.
15. D. A. Kidwell, J. C. Holland and S. J. Athanaselis, Testing for drugs of abuse in saliva and sweat, *Chromatogr. B Biomed. Sci. Appl.*, 1998, **713**, 111.
16. G. Skopp and L. Pötsch, Perspiration versus saliva - basic aspects concerning their use in roadside drug testing, *Int. J. Leg. Med.*, 1999, **112**, 213.
17. K. Dolan, D. Rouen and J. Kimber, An overview of the use of urine, hair, sweat and saliva to detect drug use, *Drug Alcohol Rev.*, 2004, **23**, 213.
18. M. A. Huestis, K. B. Scheidweiler, T. Saito, N. Fortner, T. Abraham, R. A. Gustafson and M. L. Smith, Excretion of Delta9-tetrahydrocannabinol in sweat, *Forensic Sci. Int.*, 2008, **174**, 173.
19. J. A. Levisky, D. L. Bowerman, W. W. Jenkins and S. B. Karch, Drug deposition in adipose tissue and skin: evidence for an alternative source of positive sweat patch tests, *Forensic Sci. Int.*, 2000, **110**, 35.
20. N. De Giovanni and N. Fucci, The current status of sweat testing for drugs of abuse: a review, *Curr. Med. Chem.*, 2013, **20**, 545.

21. D. J. Crouch, R. F. Cook, J. V. Trudeau, D. C. Dove, J. J. Robinson, H. L. Webster and A. A. Fatah, The detection of drugs of abuse in liquid perspiration, *J. Anal. Toxicol.*, 2001, **25**, 625.

22. K. S. Raju, I. Taneja, S. P. Singh and Wahajuddin, Utility of noninvasive biomatrices in pharmacokinetic studies, *Biomed. Chromatogr.*, 2013, **27**, 1354.

23. M. A. Huestis, J. M. Oyler, E. J. Cone, A. T. Wstadik, D. Schoendorfer and R. E. Joseph Jr, Sweat testing for cocaine, codeine and metabolites by gas chromatography-mass spectrometry, *J. Chromatogr. B Biomed. Sci. Appl.*, 1999, **733**, 247.

24. N. Samyn, G. De Boeck and A. G. Verstraete, The use of oral fluid and sweat wipes for the detection of drugs of abuse in drivers, *J. Forensic Sci.*, 2002, **47**, 1380.

25. P. Kintz, V. Cirimele and B. Ludes, Detection of cannabis in oral fluid (saliva) and forehead wipes (sweat) from impaired drivers, *J. Anal. Toxicol.*, 2000, **24**, 557.

26. C. Gambelunghe, R. Rossi, K. Aroni, M. Bacci, A. Lazzarini, N. De Giovanni, P. Carletti and N. Fucci, Sweat testing to monitor drug exposure, *Ann. Clin. Lab. Sci.*, 2013, **43**, 22.

27. B. R. Brunet, A. J. Barnes, R. E. Choo, P. Mura, H. E. Jones and M. A. Huestis, Monitoring pregnant women's illicit opiate and cocaine use with sweat testing, *Ther. Drug Monit.*, 2010, **32**, 40.

28. A. J. Barnes, M. L. Smith, S. L. Kacinko, E. W. Schwilke, E. J. Cone, E. T. Moolchan and M. A. Huestis, Excretion of methamphetamine and amphetamine in human sweat following controlled oral methamphetamine administration, *Clin. Chem.*, 2008, **54**, 172.

29. M. C. Chawarski, D. A. Fiellin, P. G. O'Connor, M. Bernard and R. S. Schottenfeld, Utility of sweat patch testing for drug use monitoring in outpatient treatment for opiate dependence, *J. Subst. Abuse Treat.*, 2007, **33**, 411.

30. H. J. Liberty, B. D. Johnson and N. Fortner, Detecting cocaine use through sweat testing: multilevel modeling of sweat patch length-of-wear data, *J. Anal. Toxicol.*, 2004, **28**, 667.

31. D. A. Kidwell and F. P. Smith, Susceptibility of PharmChek drugs of abuse patch to environmental contamination, *Forensic Sci. Int.*, 2001, **116**, 89.

32. M. Concheiro, D. M. Shakleya and M. A. Huestis, Simultaneous analysis of buprenorphine, methadone, cocaine, opiates and nicotine metabolites in sweat by liquid chromatography tandem mass spectrometry, *Anal. Bioanal. Chem.*, 2011, **400**, 69.

33. S. L. Kacinko, A. J. Barnes, E. W. Schwilke, E. J. Cone, E. T. Moolchan and M. A. Huestis, Disposition of cocaine and its

metabolites in human sweat after controlled cocaine administration, *Clin. Chem.*, 2005, **51**, 2085.

34. N. Fucci, N. De Giovanni and S. Scarlata, Sweat testing in addicts under methadone treatment: an Italian experience, *Forensic Sci. Int.*, 2008, **174**, 107.
35. N. Fucci and N. De Giovanni, Methadone in hair and sweat from patients in long-term maintenance therapy, *Ther. Drug Monit.*, 2007, **29**, 452.
36. H. J. Liberty, B. D. Johnson, N. Fortner and D. Randolph, Detecting crack and other cocaine use with fastpatches, *Addict. Biol.*, 2003, **8**, 191.
37. N. Samyn and C. van Haeren, On-site testing of saliva and sweat with Drugwipe and determination of concentrations of drugs of abuse in saliva, plasma and urine of suspected users, *Int. J. Legal Med.*, 2000, **113**, 150.
38. R. Pacifici, M. Farré, S. Pichini, J. Ortuño, P. N. Roset, P. Zuccaro, J. Segura and R. de la Torre, Sweat testing of MDMA with the Drugwipe analytical device: a controlled study with two volunteers, *J. Anal. Toxicol.*, 2001, **25**, 144.
39. D. E. Moody, A. C. Spanbauer, J. L. Taccogno and E. K. Smith, Comparative analysis of sweat patches for cocaine (and metabolites) by radioimmunoassay and gas chromatography-positive ion chemical ionization-mass spectrometry, *J. Anal. Toxicol.*, 2004, **28**, 86.
40. J. M. Ruano-López, F. Laouenan, M. Agirregabiria, S. Rodriguez, J. Krüger and D. Flavin, Laboratory skinpatches and smart cards based on foils, *Conf. Proc. IEEE Eng. Med. Biol. Soc.*, 2011, **2011**, 7647.
41. M. Barroso, I. Moreno, B. da Fonseca, J. A. Queiroz and E. Gallardo, Role of microextraction sampling procedures in forensic toxicology, *Bioanalysis*, 2012, **4**, 1805.
42. S. Schneider, Z. Ait-M-Bark, C. Schummer, P. Lemmer, M. Yegles, B. Appenzeller and R. Wennig, Determination of fentanyl in sweat and hair of a patient using transdermal patches, *J. Anal. Toxicol.*, 2008, **32**, 260.
43. N. A. Lankheet, C. U. Blank, H. Mallo, S. Adriaansz, H. Rosing, J. H. Schellens, A. D. Huitema and J. H. Beijnen, Determination of sunitinib and its active metabolite N-desethylsunitinib in sweat of a patient, *J. Anal. Toxicol.*, 2011, **35**, 558.
44. E. Marchei, M. Farré, R. Pardo, O. Garcia-Algar, M. Pellegrini, R. Pacifici and S. Pichini, Usefulness of sweat testing for the detection of methylphenidate after fast- and extended-release drug administration: a pilot study, *Ther. Drug Monit.*, 2010, **32**, 508.

45. B. R. Brunet, A. J. Barnes, K. B. Scheidweiler, P. Mura and M. A. Huestis, Development and validation of a solid-phase extraction gas chromatography-mass spectrometry method for the simultaneous quantification of methadone, heroin, cocaine and metabolites in sweat, *Anal. Bioanal. Chem.*, 2008, **392**, 115.
46. N. Uemura, R. P. Nath, M. R. Harkey, G. L. Henderson, J. Mendelson and R. T. Jones, Cocaine levels in sweat collection patches vary by location of patch placement and decline over time, *J. Anal. Toxicol.*, 2004, **28**, 253.
47. R. A. Koster, J. W. Alffenaar, B. Greijdanus, J. E. VanDerNagel and D. R. Uges, Application of sweat patch screening for 16 drugs and metabolites using a fast and highly selective LC-MS/MS method, *Ther. Drug Monit.*, 2014, **36**, 35.
48. E. Marchei, E. Papaseit, O. Garcia-Algar, A. Bilbao, M. Farré, R. Pacifici and S. Pichini, Sweat testing for the detection of atomoxetine from paediatric patients with attention deficit/hyperactivity disorder: application to clinical practice, *Drug Test. Anal.*, 2013, **5**, 191.
49. D. L. da Silva, H. H. Rüttinger, Y. Mrestani, W. F. Baum and R. H. Neubert, Development of capillary electrophoresis methods for quantitative determination of taurine in vehicle system and biological media, *Electrophoresis*, 2006, **27**, 2330.
50. O. S. Wolfbeis, Nanoparticle-enhanced fluorescence imaging of latent fingerprints reveals drug abuse, *Angew. Chem. Int. Ed. Engl.*, 2009, **48**, 2268.
51. P. Hazarika, S. M. Jickells, K. Wolff and D. A. Russell, Multiplexed detection of metabolites of narcotic drugs from a single latent fingermark, *Anal. Chem.*, 2010, **82**, 9150.
52. P. Hazarika, S. M. Jickells and D. A. Russell, Rapid detection of drug metabolites in latent fingermarks, *Analyst*, 2009, **134**, 93.
53. J. S. Day, H. G. Edwards, S. A. Dobrowski and A. M. Voice, The detection of drugs of abuse in fingerprints using Raman spectroscopy I: latent fingerprints, *Spectrochim. Acta A Mol. Biomol. Spectrosc.*, 2004, **60**, 563.
54. K. Kuwayama, K. Tsujikawa, H. Miyaguchi, T. Kanamori, Y. T. Iwata and H. Inoue, Time-course measurements of caffeine and its metabolites extracted from fingertips after coffee intake: a preliminary study for the detection of drugs from fingerprints, *Anal. Bioanal. Chem.*, 2013, **405**, 3945.
55. H. H. Maurer, Role of gas chromatography-mass spectrometry with negative ion chemical ionization in clinical and forensic

toxicology, doping control, and biomonitoring, *Ther. Drug Monit.*, 2002, **24**, 247.

56. European Commission Road Safety, 2010.

57. J. M. Walsh, J. J. de Gier, A. S. Christopherson and A. G. Verstraete, Drugs and driving, *Traffic Inj. Prev.*, 2004, **5**, 241.

58. S. Steinmeyer, H. Ohr, H. J. Maurer and M. R. Moeller, Practical aspects of roadside tests for administrative traffic offences in Germany, *Forensic Sci. Int.*, 2001, **121**, 33.

59. K. Pidd and A. M. Roche, How effective is drug testing as a workplace safety strategy? A systematic review of the evidence, *Accid. Anal. Prev.*, 2014, **71**, 154.

60. E. Gallardo, M. Barroso and J. A. Queiroz, LC-MS: a powerful tool in workplace drug testing, *Drug Test. Anal.*, 2009, **1**, 109.

61. D. M. Bush, The U.S. Mandatory Guidelines for Federal Workplace Drug Testing Programs: current status and future considerations, *Forensic Sci. Int.*, 2008, **174**, 111.

62. J. Płotka, S. Narkowicz, Z. Polkowska, M. Biziuk and J. Namieśnik, Effects of addictive substances during pregnancy and infancy and their analysis in biological materials, *Rev. Environ. Contam. Toxicol.*, 2014, **227**, 55.

63. M. A. Huestis and R. E. Choo, Drug abuse's smallest victims: in utero drug exposure, *Forensic Sci. Int.*, 2002, **128**, 20.

64. J. Lozano, O. García-Algar, O. Vall, R. de la Torre, G. Scaravelli and S. Pichini, Biological matrices for the evaluation of in utero exposure to drugs of abuse, *Ther. Drug. Monit.*, 2007, **29**, 711.

65. M. Concheiro, H. E. Jones, R. E. Johnson, R. Choo and M. A. Huestis, Preliminary buprenorphine sublingual tablet pharmacokinetic data in plasma, oral fluid, and sweat during treatment of opioid-dependent pregnant women, *Ther. Drug Monit.*, 2011, **33**, 619.

66. E. Marchei, M. Farrè, M. Pellegrini, O. García-Algar, O. Vall, R. Pacifici and S. Pichini, Pharmacokinetics of methylphenidate in oral fluid and sweat of a pediatric subject, *Forensic Sci. Int.*, 2010, **196**, 59.

67. L. T. Demoranville and J. R. Verkouteren, Measurement of drug facilitated sexual assault agents in simulated sweat by ion mobility spectrometry, *Talanta*, 2013, **106**, 375.

68. M. Laloup, N. Samyn and V. Maes, Bio-analysis of forensically relevant drugs in alternative matrices by liquid chromatography-tandem mass spectrometry, *Verh. K. Acad. Geneeskd. Belg.*, 2008, **70**, 347.

69. S. Abanades, M. Farré, M. Segura, S. Pichini, A. Pastor, R. Pacifici, M. Pellegrini and R. de la Torre, Disposition of gamma-hydroxybutyric acid in conventional and nonconventional biologic fluids after single drug administration: issues in methodology and drug monitoring, *Ther. Drug Monit.*, 2007, **29**, 64.
70. www.wada-ama.org.
71. L. Rivier, Techniques for analytical testing of unconventional samples, *Baillieres Best Pract. Res. Clin. Endocrinol. Metab.*, 2000, **14**, 147.
72. D. Thieme, Potential and limitations of alternative specimens in doping control, *Bioanalysis*, 2012, **4**, 1613.
73. A. Mena-Bravo and M. D. Luque de Castro, Sweat: a sample with limited present applications and promising future in metabolomics, *J. Pharm. Biomed. Anal.*, 2014, **90**, 139.
74. R. J. Dinis-Oliveira, F. Carvalho, J. A. Duarte, F. Remião, A. Marques, A. Santos and T. Magalhães, Collection of biological samples in forensic toxicology, *Toxicol. Mech. Methods*, 2010, **20**, 363.
75. B. M. Appenzeller, C. Schummer, S. B. Rodrigues and R. Wennig, Determination of the volume of sweat accumulated in a sweat-patch using sodium and potassium as internal reference, *J. Chromatogr. B Analyt. Technol. Biomed. Life Sci.*, 2007, **852**, 333.

23 Smart Drugs

Angela Wing Gar Kwan

Bart's and The London School of Medicine and Dentistry,
Queen Mary University, London, UK
Email: a.kwan@smd10.qmul.ac.uk

23.1 Introduction

"Smart drugs" in this chapter will refer to a prescription drug used primarily to improve cognitive functioning in a normal person. Other terms used for this concept include cognitive enhancing drugs, neuroenhancing drugs, "brain steroids" and "nootropics". In public health literature it can be described as "non-medical use" and "prescription drug abuse".[1] Drugs used with this intent include methylphenidate, modafinil, piracetam, donepezil, galantamine, selegiline and amphetamine derivatives. This indicates the range of drugs used as smart drugs, being used therapeutically as treatment for attention-deficit/hyperactivity disorder (ADHD), narcolepsy, depression and Alzheimer's disease. To understand their use as smart drugs, it is necessary to explore their pharmacology and their therapeutic use as treatment. This chapter will mainly focus on methylphenidate and modafinil as smart drugs as they are both widely discussed within current literature.

23.1.1 Modafinil

Modafinil is a psychostimulant currently approved only for the treatment of excessive daytime sleepiness (EDS) associated with

Forensic Toxicology: Drug Use and Misuse
Edited by Susannah Davies, Atholl Johnston and David Holt
© The Royal Society of Chemistry 2016
Published by the Royal Society of Chemistry, www.rsc.org

narcolepsy as a prescription-only drug. Previously, it had been approved for use in obstructive sleep apnoea and chronic shift-work sleep disorder.[2] However, the licence was revised in 2010 as it was concluded that the potential adverse effects outweighed the clinical efficacy in these conditions.[3] It is a drug that has been shown to target multiple different brain areas and neurotransmitter systems within the brain.[4] The specific mechanisms of action are still not well understood. Nevertheless, inhibition of dopamine reuptake appears to be the process underlying its pharmacological effects.[5] This enhancement of dopamine levels in specific brain areas also exists with the use of other psychostimulants, including methylphenidate, cocaine and the amphetamines. However, modafinil differs from such typical psychostimulants in the specific brain areas and circuits it activates and its neurochemical effect.[4] Its consequent action of wake-promoting is therefore utilised to treat EDS associated with narcolepsy. Research has indicated that its wake-promoting property may also be due to an elevation of hypothalamic histamine levels, on top of its effect on dopamine.[6]

23.1.2 Methylphenidate

Methylphenidate is a classical psychostimulant, also a schedule II drug, used in the treatment of ADHD, which is the most prevalent neurobehavioral condition affecting school-age children.[7] ADHD is a condition that is thought to be due to an imbalance between dopaminergic, noradrenergic and serotonergic neurotransmitter systems, with dopaminergic dysfunction appearing to play the most significant part.[8] Dopamine synaptic uptake inhibition appears to be the mechanism of action most responsible for the response in treatment with methylphenidate.[9] This facilitation of dopamine neurotransmission is thought to produce the drug's psychostimulant effects. Stimulants such as methylphenidate have consistently been shown to reduce the symptoms of ADHD, including hyperactivity, impulsivity and inattention.[10]

23.2 Non-medical Use

Literature suggests that smart drugs such as modafinil and methylphenidate are in widespread use for non-medical reasons.[11] They are used with the aim of neuroenhancement rather than the traditional treatment of the medical conditions described above. Online polls have revealed that the main reasons for their use as smart drugs is in order to improve concentration, improve their focus on tasks or to reduce the effects of sleep deficit and jet lag.[12] A systematic review on the use of modafinil and methylphenidate for neuroenhancement in healthy individuals reveals significant improvements of cognitive function with both drugs.[11] It was concluded that modafinil improves attention in well-rested individuals and is effective in maintaining wakefulness, memory and executive functions in sleep-deprived individuals. In contrast, it was shown that methylphenidate only appears to significantly improve memory, with no evidence of enhancement in any other areas of cognitive function. A recent systematic review, published in July 2015, aimed to evaluate modafinil as a neuroenhancement agent in healthy non-sleep-deprived subjects.[13] The authors concluded that modafinil proved to be particularly consistent in enhancing performance in tasks which were longer and more complex. It appears from this systematic review that modafinil may well become a well-validated smart drug in the near future. However, even with considerable research into the cognitive benefits of smart pills there is no definitive answer, with varying evidence on demonstrable cognitive enhancement.[14] It is difficult to gauge the efficacy of smart drugs on enhancing performance due to individual differences. Other factors including baseline levels of performance are thought to contribute to the different effects a single drug may have on different populations.[15] For example, several smart drugs appeared to be of most benefit in healthy individuals with a low baseline performance. In contrast, in a group of patients with schizophrenia, those with a higher premorbid cognitive ability had the greatest effect after treatment with the same drugs. There appear to be multiple types of tests to assess cognition, from basic to complex, which are able to gauge the possible cognitive benefits of modafinil. These have provided contrasting evidence on the performance-enhancing capacity of smart drugs and Battleday, Brem *et al.* suggest that further assessments must be improved to produce "valid, robust and consistent" results.[13]

Unsurprisingly, these neuroenhancing properties of smart drugs have attracted the attention of both those wishing to improve their optimal cognitive function and others with suboptimal cognitive function due to, perhaps, sleep deprivation. The use of smart drugs within the student population has become a growing trend.[14] The reason for its popularity within this group of society appears obvious. However, there may be many reasons beyond simply improving concentration, attention, alertness and performance. For instance, multiple surveys revealed other motives such as to "get high", "lose weight" and "be able to drink and party longer without feeling drunk".[16] If found to be effective, smart drugs could feature heavily in professions in which shift work and jet lag currently affect performance.[17] Smart drugs have been used by the Army and the Air Force to treat sleep deprivation and increase alertness in fatigued pilots.[18] The US military have approval to use smart drugs to prevent decline of function and studies have shown that modafinil use has promoted flight accuracy.[19] It is thought that errors caused by soldiers due to sleep deprivation might be reduced with the aid of smart drugs. Night-time road accidents due to sleep deprivation in drivers could be reduced by the use of smart drugs. Other professionals such as jurors may have limitations of working and may subsequently benefit from its use.

A similar setting in which it could be used is sleep-deprived doctors and surgeons in the hospital and operating theatre. Smart drugs have the potential to minimise fatigue-related errors and maximise safety, such as in surgery.[20] Therefore, if it is effective, its use could influence patient care and be of a benefit for others rather than just for the user. In this case, it could be considered a safety tool rather than treatment of sleep deprivation.[21] It could be said that smart drugs used in this way are simply to restore cognitive function rather than to exceed normal function, which could be considered more acceptable.[22] A study of the use of modafinil in emergency physicians showed that their ability to stay awake the morning after a night-shift was improved significantly, but led to a negative impact on their sleep afterwards.[23] Therefore, it was concluded that it was difficult to say for certain whether the benefits outweighed the problem of not being able to sleep between work shifts, which could impact them even further. Another study by Gill *et al.* showed that emergency physicians had improved performance on simulated patients after taking modafinil.[23] Sugden *et al.* found that modafinil improved information processing, flexible thinking and decision making under time pressure in sleep-deprived doctors.[24] It was also noted that laparoscopic psychomotor performance was not improved with modafinil administration, which suggests it may not be as useful in surgeons.

23.2.1 Prevalence

The rate of non-medical use of stimulants in any population is unclear. It seems that its use as "study aids" in college students is not uncommon. Surveys indicate that in US colleges, up to 25% of students use smart drugs to improve their academic perform-ance.[25–27] McCabe *et al.* reported a 6.9% lifetime non-medical use of a prescription drug in the largest nationwide study of students and colleges and universities in the US.[28] It is difficult to ascertain the prevalence in the UK as there is scarce research even within Europe. The UK is far behind other countries, especially the US, in being able to predict the prevalence of smart drug use in either the student population or the general population.[29] There is also little under-standing of the motives or source of smart drug use in the UK. Nevertheless, evidence suggests there is a less widespread use of smart drugs.[30,31]

Smart drugs are also becoming increasingly popular, beyond the student population, in business and in academia in order to maintain or optimise function and productivity.[32] Its use is thought to have increased in academic, professional and military settings, though it has been difficult to assess this quantitatively.[33] A *Nature*-sponsored survey revealed that 20% of the respondents have used smart drugs to improve their cognitive function, specifically their concentration, focus and memory.[12] Predominantly, methylphenidate and modafinil were reported to be taken but it included drugs such as beta-blockers. Within this group of respondents, their rationale for using smart drugs was in order to improve concentration, focus or to counteract sleep deficit or jetlag.[11] The Care Quality Commission revealed that over a six-year period from 2007 to 2013 there was 56% rise in the number of prescriptions for methylphenidate, which may be partly due to its potential for misuse.[34] Large US Government surveys reported 8.5% of adults above 12 years have used smart drugs, com-pared to 12.3% of Americans between 21 and 25 years.[35]

23.2.2 Smart Drugs *vs.* Caffeine

Caffeine is heavily used as a stimulant in the general population, and for some it could be considered the norm to use it in order to get through the day. The alertness associated with caffeine use is due to an increase in cyclic adenosine monophosphate levels.[36] Considering its comparable stimulant properties to smart drugs, it is interesting that it is available without a prescription, in contrast to smart drugs such as modafinil and methylphenidate. With a considerable number of the population heavily reliant on caffeine for everyday functioning, it becomes a thought-provoking idea if you replace the use of caffeine with the use of smart drugs. Could these smart drugs one day be as integrated in our lives as caffeine currently is? Likewise, perhaps, you could imagine caffeine being considered as a smart drug and re-quiring a prescription to be used by the general public.

23.3   Safety of Smart Drugs

23.3.1   Therapeutic Use

As with any medication, the use of modafinil and methylphenidate for therapeutic effect carries the risk of adverse effects. Studies have shown that methylphenidate is associated with psychosis, cardio-vascular problems and risk of dependency.[37] Given the different mechanism of action modafinil has, significantly fewer side effects are observed in its use than with methylphenidate and other con-ventional stimulants.[38] Adverse effects of modafinil in the treatment of narcolepsy include low mood, nausea, dizziness, headache, poor sleep and dry mouth.[39,40] Importantly, there appears to be little evidence of dependence with modafinil use.[41] Nevertheless, adverse cardiovascular events, psychosis and serious skin reactions are risks associated with modafinil, which has led to its approval to be restricted only for the treatment of narcolepsy.[29]

23.3.2   Non-therapeutic Use

It is important to consider the health risks and the potential for dependence when discussing the non-therapeutic use of smart drugs. The recent systematic review of 2015 investigating modafinil in healthy non-sleep-deprived subjects revealed that 78% of studies considering the side effects of modafinil showed very little overall effect.[13] Most reported no side effects, with the remaining reporting that only a minority of participants noticed a mild effect on mood and slight side effects such as headache, insomnia, stomach ache and dry mouth. Of note, one study assessed modafinil's abuse potential and corroborated the consensus that it is low risk.[42] It appears that acute modafinil use at sensible doses is relatively safe in the short term.[19] The side effects of long-term use have not been researched sufficiently and are not fully understood at this point in time.[43] However, there is limited evidence on the adverse effects of these smart drugs on people with no illness, due to the risk of causing any serious harm in a trial without any therapeutic gain.[44] Present regulations also only promote research for the management of disease, which explains the small amount of data available.[45] In itself, the lack of literature should be a safety concern to those who take it for non-therapeutic use. However, it has been shown that users are not as wary about the risks of side effects and medical complications associated with buying online as are doctors.[29] One safety issue of concern is the effects smart drugs may have on the developing brain, considering that students comprise of a large proportion of smart drug users. Without prescription, users may be less aware of the safe doses to take and medical contraindications; this could cause serious harm beyond just the side effects of taking the drug.[45] In cases of healthy individuals taking smart drugs, the risk *versus* the benefit is very different to those taking it for its licensed use.[15]

On top of the safety concerns associated with these smart drugs in the general population, there are additional issues when considering its use in certain professions, such as hospital doctors. The long-term use of smart drugs for long shifts could pose occupational health risks which could be dangerous for both the doctors or surgeons and their patients.[46] Owing to the lack of research into the long-term effects of these smart drugs, it could potentially be devastating if a generation of doctors began taking these smart drugs and are later found to suffer from the long-term effects of it, which are currently unknown.[47] Therefore, if medical professionals are subject to serious risks with the use of smart drugs and are causing harm to themselves, they should not be obliged to provide the best possible care, the care they would be able to provide with the use of cognitive enhancers.[48]

23.3.3 Availability

The safety issue arising from the availability of smart drugs due to their being prescription drugs is concerning. Currently, it appears that within the student population, up to 77% of students using these smart drugs for cognitive enhancement obtain them from their peers rather than from a doctor.[49–51] Those obtaining prescriptions for the means of enhancement may either request the smart drug directly for cognitive enhancement or they may feign symptoms of conditions such as ADHD.[52,53] Surveys have revealed that in the US it is not uncommon for patients to request smart drugs for enhancement, nor for doctors to prescribe them purely for that reason, with 36.7% prescribing them monthly or more often.[54] It is seemingly much easier to purchase online, with multiple online pharmacies selling smart drugs.[29] *Nature*'s online poll reveals that compared to the US, a greater proportion of UK users obtain their drugs over the internet.[12] The increasing popularity of smart drug use has led to up to 54% of UK students expressing an interest in trying a smart drug for cognitive enhancement.[55,56] Having witnessed the advertisement of selling smart drugs by students to others at university campuses on social networking, it is likely a proportion of students are acquiring these drugs through these means. Therefore it is important to recognise the dangers associated with the difficulty in obtaining these drugs in legitimate ways. Without doubt, what is being purchased over the internet or from classmates is unclear and there is a real risk that what has been sold is not the same drug as what has been advertised. The possibility that it could be anything from caffeine to a completely unknown substance is dangerous.

23.4 Ethics of Smart Drugs

There is a huge ethical dilemma associated with smart drugs and views upon this subject range widely. Consistently, it has been shown that students are most concerned about the health risks and the unfair advantage when used in exams where others will not be using them.[57,58] Some compare smart drugs in academia to doping in sports, which is a form of cheating as it gives them an unfair advantage over competitors who do not use performance-enhancing drugs.[59] However, it appears that in general, university students in England consider doping in sports more unacceptable than using smart drugs for neuroenhancement.[43] It is thought that this perception may be due, in part, to the lack of knowledge and experience of smart drugs compared to the more widely known doping in sports. In terms of safety, this has been discussed earlier in this chapter and there is still a need for further research on smart drugs in healthy individuals. The ethics to using a drug without thorough research is debatable, as some claim if they are aware of this, they should be free to use them as they wish.

Another ethical issue is distributive fairness, in which these smart drugs are only financially available to those from a privileged socioeconomic background.[57] This unfair access could disadvantage those unable to afford smart drugs. However, others argue that even without the introduction of smart drugs, there is an uneven playing field as resources which affect academic performance are not distributed equally across social classes.[60] For example, not all students have access to a home computer, private tuition or even good childhood nutrition, which can all affect academic performance. Therefore, with this argument, it follows that to prohibit smart drugs could be considered the same as prohibiting private tuition.[58] It becomes a slippery slope on what is considered ethically acceptable for the means of improving academic performance. The treatment–enhancement distinction is also an area of debate. One group argues that these drugs should only be used for treatment of a medical condition, whereas the other group questions what can be defined as treatment or as enhancement. An example of this which currently occurs is the use of statins for patients with healthy cholesterol levels, in order to reduce it to a level that cannot be generally achieved by diet and exercise.[44] The threshold for its use has changed over the years and it could be said that statins are being used for enhancement in healthy people with no therapeutic need. The level that "normal" cognitive function is set at could therefore be disputed. If smart drugs become acceptable in the future, the further question of whether individuals should disclose the use of them is introduced.[33] It could be argued that there is a moral obligation to disclose the use of smart drugs as it could be sufficiently harmful to others, such as in the situation where one does not have access to their usual dose. In certain professions, this could put colleagues or others at risk of harm.

Peer pressure and indirect coercion arises upon the introduction of smart drugs into a community, especially in a competitive environment such as a university.[61] Students may feel that, in order to succeed academically or even to simply remain competitive, they must use smart drugs if the majority of students use them. This becomes a large problem when substances such as methylphenidate are used, which has risks of serious side effects and could lead to dependence. People may be coerced into taking smart drugs in order to function in jobs requiring night shifts in order to stay alert.[58,62] However, it is argued by others that there is a breach of personal freedom if people are prohibited from acting in certain ways in order to protect others from feelings of coercion. To prohibit smart drugs in the academic setting would realistically be very difficult to control. Beyond the fact that controlling its use beyond exam day would be extremely difficult, the cost to police smart drug use effectively would be high and the measures needed intrusive. Considering smart drugs such as methylphenidate are currently controlled drugs, but still have high rates of non-therapeutic use due to a relatively ease of access, smart drug regulation would undoubtedly be problematic.[63]

23.5 Future of Smart Drugs

This chapter indicates how little research has been done in regards to the prevalence, efficacy, safety and the role in society of smart drugs.

The development, assessment and regulation of these drugs must be considered further in order to bring about a product that we can hand to the general public. It is vital that we investigate this further before we can safely say that it has an important part to play in neuroenhancement. The increasing interest to use smart drugs by individuals simply reiterates the importance of studying their use in the healthy population. It is clear that the use of smart drugs in any situation with the aim of enhancing performance is not straightforward. However, the possibilities that smart drugs could offer are highly tempting and, if proven safe, could be a huge advance to the lives of its users and third parties. Nevertheless, it is impossible to ignore the many ethical issues that arise from this discussion, which are likely to remain regardless of future scientific research and they can no doubt be explored even further. The balance between benefit and risk is currently unclear and this must be investigated more before the general population can be advised upon what is safe. It appears that the use of smart drugs must be researched much further, in order to produce a definitive and well examined answer.

References

1. E. Racine and C. Forlini, Cognitive Enhancement, Lifestyle Choice or Misuse of Prescription Drugs? *Neuroethics*, 2010, **4**, 1–4.
2. S. J. Williams, C. Seale, S. Boden, P. Lowe and D. L. Steinberg, Waking up to sleepiness: Modafinil, the media and the pharmaceuticalisation of everyday/night life, *Sociol. Health Illn.*, 2008, **30**(6), 839–855.
3. Modafinil: European Medicines Agency recommends restricted use: MHRA, 2010 [25/05/2015]. Available from: http://webarchive. nationalarchives.gov.uk/20141205150130/http://www.mhra.gov. uk/Safetyinformation/DrugSafetyUpdate/CON090901.
4. M. Mereu, A. Bonci, A. H. Newman and G. Tanda, The neurobiology of modafinil as an enhancer of cognitive performance and a potential treatment for substance use disorders, *Psychopharmacology*, 2013, **229**, 415–434.
5. S. Nishino, Narcolepsy: Pathophysiology and Pharmacology, *J. Clin. Psychiatry*, 2007, **68**, 9–15.
6. J.-S. Lin, Y. Hou and M. Jouvet, Potential brain neuronal targets for amphetamine-, methylphenidate-, and modafinil-induced wakefulness, evidenced by c-fos immunocytochemistry in the cat, *Proc. Natl. Acad. Sci. U. S. A.*, 1996, **93**, 14128–14133.

7. F. A. Lopez, ADHD: New Pharmacological Treatments on the Horizon, *Dev. Behav. Pediatr.*, 2006, **27**(5), 410–416.

8. K. S. Patrick and J. S. Markowitz, Pharmacology of Methylphenidate, Amphetamine Enantiomers and Pemoline in Attention-Deficit Hyperactivity Disorder, *Hum. Psychopharmacol.*, 1997, **12**, 527–546.

9. M. Schweri, P. Skolnick, M. Rafferty, K. Rice, A. Janowski and S. Paul, [3H]Threo-(+/ −)-methylphenidate binding to 3,4-dihydroxyphenylethylamine uptake sites in corpus striatum: correlation with the stimulant properties of ritalinic acid esters, *J. Neurochem.*, 1985, **45**(4), 1062–1070.

10. D. Nutt, K. Fone, P. Asherson, D. Bramble, P. Hill, K. Matthews *et al.*, Evidence-based guidelines for management of attention-deficit/hyperactivity disorder in adolescents in transition to adult services and in adults: recommendations from the British Association for Psychopharmacology, *J. Psychopharmacol.*, 2007, **21**(1), 10–41.

11. D. Repantis, P. Schlattmann, O. Laisney and I. Heuser, Modafinil and methylphenidate for neuroenhancement in healthy individuals: A systematic review, *Pharmacol. Res.*, 2010, **62**, 187–206.

12. B. Maher, Poll results: look who's doping, *Nature*, 2008, **452**, 674–675.

13. R. M. Battleday and A.-K. Brem, Modafinil for cognitive neuroenhancement in healthy non-sleep-deprived subjects: A systematic review, *Eur. Neuropsychopharmacol.*, 2015.

14. E. M. Smith and M. J. Farah, Are Prescription Stimulants "Smart Pills"? The Epidemiology and Cognitive Neuroscience of Prescription Stimulant Use by Normal Healthy Individuals, *Psychol. Bull.*, 2011, **137**(5), 717–741.

15. B. J. Sahakian and S. Morein-Zamir, Pharmacological cognitive enhancement: treatment of neuropsychiatric disorders and lifestyle use by healthy people, *Lancet*, 2015, **2**, 357–362.

16. K. Holloway and T. Bennett, Prescription drug misuse among university staff and students: A survey of motives, nature and extent, *Drugs: Educ. Prev. Policy*, 2012, **19**(2), 137–144.

17. B. J. Sahakian, A. B. Bruhl, J. Cook, C. Killikelly, G. Savulich, T. Piercy, *et al.*, The impact of neuroscience on society: cognitive enhancement in neuropsychiatric disorders and in healthy people, *Philos. Trans. R. Soc. B*, 2015, **370**(1677).

18. S. S. Hall, The quest for a smart pill, *Sci. Am.*, 2003, **289**(3), 54–65.

19. S. Porsdam-Mann and B. J. Sahakian, The increasing lifestyle use of modafinil by healthy people: safety and ethical issues, *Behav. Sci.*, 2015, **4**, 136–141.

20. F. Santoni de Sio, N. Faulmüller and N. Vincent, How cognitive enhancement can change our duties, *Front. Syst. Neurosci.*, 2014, **8**(131).

21. C. Coveney, Awakening Expectations: exploring social and ethical issues surrounding the medial and non-medical uses of cognition enhancing drugs in the UK, PhD thesis, University of Nottingham, Nottingham, 2010.

22. I. Goold and H. Maslen, Must the Surgeon Take the Pill? Negligence Duty in the Context of Cognitive Enhancement, *Mod. Law Rev.*, 2014, **77**(1), 60–86.

23. M. Gill, P. Haerich, K. Westcott, K. L. Godenick and J. A. Tucker, Cognitive Performance Following Modafinil versus Placebo in Sleep-deprived Emergency Physicians: A Double-blind Randomized Crossover Study, *Acad. Emerg. Med.*, 2006, **13**, 158–165.

24. C. Sugden, C. R. Housden, R. Aggarwal, B. J. Sahakian and A. Darzi, Effect of Pharmacological Enhancement on the Cognitive and Clinical Psychomotor Performance of Sleep-Deprived Doctors, *Ann. Surg.*, 2012, **255**, 222–227.

25. C. Teter, S. McCabe, J. Cranford, C. Boyd and S. K. Guthrie, Prevalence and motives for illicit use of prescription stimulants in an undergraduate student sample, *J. Am. Coll. Health*, 2005, **53**, 253–262.

26. C. Teter, S. McCabe, K. LaGrange, J. Cranford and C. Boyd, Illicit use of specific prescription stimulants among college students: prevalence, motives, and routes of administration, *Pharmacotherapy*, 2006, **26**, 1501–1510.

27. B. Prudhomme White, K. Becker-Blease and K. Grace-Bishop, Stimulant medication use, misuse, and abuse in an undergraduate and graduate student sample, *J. Am. Coll. Health*, 2006, **54**, 261–268.

28. S. McCabe, J. Knight, C. Teter and H. Wechsler, Non-medical use of prescription stimulants among US college students: Prevalence and correlates from a national survey, *Addiction*, 2005, **100**, 96–106.

29. C. I. Ragan, I. Bard and I. Singh, What should we do about student use of cognitive enhancers? An analysis of current evidence, *Neuropharmacology*, 2013, **64**, 588–595.

30. A. G. Franke, C. Bonertz and M. Christmann, Non-medical use of prescription stimulants and illicit use of stimulants for cognitive enhancement in pupils and students in Germany, *Pharmacopsychiatry*, 2011, **44**, 60–66.

31. L. Maier, M. Liechti, F. Herzig and M. Schaub, To dope or not to dope: neuroenhancement with prescription drugs and drugs of abuse among Swiss university students, *PLoS One*, 2013, **8**(11), 1–10.

32. A. Chatterjee, The promise and predicament of cosmetic neurology, *J. Med. Ethics*, 2006, **32**, 110–113.
33. M. D. Garasic and A. Lavazza, Performance enhancement in the workplace: why and when healthy individuals should disclose their reliance on pharmaceutical cognitive enhancers, *Front. Syst. Neurosci.*, 2015, **9**(13), 1–5.
34. Care Quality Commission. The safer management of controlled drugs: annual report 2012, 2013. Available from: http://www.cqc.org.uk/sites/default/files/documents/cdar_2012.pdf.
35. J. Snodgrass and P. LeBaron, 2008 National Survey on Drug Use and Health: CAI specifications for programming. English version 2007. Available from: http://www.oas.samhsa.gov/nsduh/2k8MRB/2k8Q.pdf.
36. Department of the Army, *Combat and Operation Stress Control Manual for Leaders and Soldiers*, U.S. Army, Washington, DC, 2009.
37. J. Godfrey, Safety of therapeutic methylphenidate in adults: a systematic review of the evidence, *J. Psychopharmacol.*, 2009, **23**, 194–205.
38. M. Billiard, A. Besset, J. Montplaisir, F. Laffont, F. Goldenberg, J. Weill, *et al.*, Modafinil: a double-blind multicentric study, *Sleep*, 1994, **17**, 107–112.
39. H. Moldofsky, R. Broughton and J. Hill, A randomized trial of the long-term, continued efficacy and safety of modafinil in narcolepsy, *Sleep Med.*, 2000, **1**, 109–116.
40. A. Besset, M. Chetrit, B. Carlander and M. Billiard, Use of modafinil in the treatment of narcolepsy: a long term follow-up study, *Neurophysiol. Clin.*, 1996, **26**, 60–66.
41. D. Jasinski, An evaluation of the abuse potential of modafinil using methylphenidate as a reference, *J. Psychopharmacol.*, 2000, **14**, 53–60.
42. A. Makris, C. Rush, R. Frederich, A. Taylor and T. Kelly, Behavioral and subjective effects of d-amphetamine and modafinil in healthy adults, *Exp. Clin. Psychopharmacol.*, 2007, **15**(2), 123–133.
43. E. Vargo, R. James, K. Agyeman, T. MacPhee, R. McIntyre, F. Ronca *et al.*, Perceptions of assisted cognitive and sport performance enhancement among university students in England, *Perform. Enhancement Health*, 2015, **3**, 66–77.
44. S. E. Hyman, Cognitive Enhancement: Promises and Perils, *Neuron*, 2011, **69**, 595–598.
45. S. Sussman, M. Pentz, D. Spruijt-Metz and T. Miller, Misuse of "study drugs:" prevalence, consequences, and implications for policy, *Subst. Abuse Treat. Prev. Policy*, 2006, **1**, 15.

46. D. M. Gaba and S. K. Howard, Patient safety: fatigue Among Clinicians and The Safety of Patients, *N. Engl. J. Med.*, 2002, **347**(16), 1249–1255.
47. H. Shi, Should physicians be taking cognitive-enhancing drugs? *Univ. West. Ont. Med. J.*, 2010, **79**(2), 10–12.
48. G. G. Enck, Pharmaceutical enhancement and medical professionals, *Med. Health Care Philos.*, 2014, **17**, 23–28.
49. S. McCabe, C. Teter and C. Boyd, Medical use, illicit use and diversion of prescription stimulant medication, *J. Psychoactive Drugs*, 2006, **38**, 43–56.
50. S. Barrett, C. Darredeau and L. Bordy, Characteristics of methylphenidate misuse in a university student sample, *Can. J. Psychiatry*, 2005, **50**, 457–461.
51. D. Rainer, A. Anastopoulos, E. Costello, R. Hoyle, S. McCabe and H. Swartzwelder, Motives and perceived consequences of nonmedical ADHD medication use by college students: are students treating themselves for attention problems? *J. Atten. Disord.*, 2009, **13**(3), 259–270.
52. C. Forlini, S. Gauthler and E. Racine, Should physicians prescribe cognitive enhancers to healthy individuals? *Can. Med. Assoc. J.*, 2012, **185**(12), 1047–1050.
53. S. Outram, The use of methylphenidate among students: the furture of enhancement? *J. Med. Ethics*, 2010, **36**, 198–202.
54. T. Hotze, K. Shaw and E. Anderson, "Doctor, would you prescribe a pill to help me … ?" A national survey of physicians on using medicine for human enhancement, *Am. J. Bioeth.*, 2011, **11**, 3–13.
55. I. Singh, I. Bard and J. Jackson, Robust Resilience and Substantial Interest: A Survey of Pharmacological Cognitive Enhancement among University Students in the UK and Ireland, *PLoS One*, 2014, **9**(10).
56. M. Ibrahim. Students could face drug tests in UK crackdown on study pills, 2012; 59:[1&4 p.].
57. C. Scheske and S. Schnall, The Ethics of "Smart Drugs": Moral Judgments About Healthy People's Use of Cognitive-Enhancing Drugs, *Basic Appl. Soc. Psych.*, 2012, **34**, 508–515.
58. V. Cakic, Smart drugs for cognitive enhancement: ethical and pragmatic considerations in the era of cosmetic neurology, *J. Med. Ethics*, 2009, **35**, 611–615.
59. B. Foddy and J. Savulescu, *Ethics of Performance Enhancement in Sport: Drugs and Gene Doping*, ed. R. Ashcroft, A. Dawson, H. Draper and J. McMillan, John Wiley & Sons, Ltd, Chichester, UK, 2006.

60. M. J. Farah, Emerging ethical issues in neuroscience, *Nat. Neurosci.*, 2002, **5**, 1123–1129.
61. D. Degrazia, Prozac, enhancement, and self-creation, *Cent. Rep.*, 2000, **30**, 34–40.
62. M. J. Farah, J. Illes and R. Cook-Deegan, Neurocognitive enhancement: what can we do and what should we do? *Nat. Rev. Neurosci.*, 2004, **5**, 421–425.
63. K. Graff Low and A. Gendaszek, Illicit use of psychostimulants among college students: a preliminary study, *Psychol. Health Med.*, 2002, **7**(3), 283–287.

24 Substandard and Counterfeit Medicines

Badr Aljohani[a,b]

[a] Bart's and The London School of Medicine and Dentistry, Queen Mary University, London, UK; [b] King Abdullah International Medical Research Center (KAIMRC), King Saud bin Abdulaziz University for Health Sciences (KSAU-HS), P.O.Box 9515, Jeddah 21423, Saudi Arabia
Email: b.aljohani@qmul.ac.uk

24.1 Introduction: Nature of the Problem

Substandard and counterfeit medicines are considered to be a worldwide problem, affecting both developed and developing countries. The toughest realization is that there are many cases where counterfeit drugs have caused serious harm to consumers, including death. These are serious consequences and add a real urgency in the fight against counterfeit drugs. The consequences are well known: the Department of Homeland Security (DHS) advised that counterfeit and substandard medicines could cause a serious threat to America's economic vitality, the health and safety of American consumers, and critical infrastructure and national security.[1]

Many incidents have been reported as a result of using such drugs. In Nigeria, 2500 deaths were reported in 1995 because of fake vaccines. In 1995 and 1998, 89 deaths in Haiti and 30 deaths in India occurred because of using a paracetamol preparation that contained diethylene glycol. In 1999, 30 people died in Cambodia after using a counterfeit antimalarial preparation prepared with sulfadoxine–pyrimethamine.[2]

Forensic Toxicology: Drug Use and Misuse
Edited by Susannah Davies, Atholl Johnston and David Holt
© The Royal Society of Chemistry 2016
Published by the Royal Society of Chemistry, www.rsc.org

Developing countries are a clear target for counterfeiters, because of the high prices for genuine medications and weak supervision from the authorities, but developed countries could also be a target. In 2006, the UK authorities detained around £500 000 worth of anti-flu medication that was counterfeit.[3] In 2012, the Food and Drug Administration (FDA) announced that a fake anti-cancer drug had been found in US clinics (Figure 24.1). It was confirmed that the fake intravenous 400 mg per 16 mL Altuzan® (bevacizumab) had no active ingredients.[4] In 2002, Amgen pharmaceuticals issued a warning about counterfeit Epogen® (an anti-anaemic drug), which had been intentionally relabelled with a higher dosage to sell it at a higher price.[5,6] Another example of critical medication being counterfeited is a case in which heparin, an anticoagulant drug, had its active ingredient substituted for a cheaper alternative. This unfortunately resulted in the suspected deaths of 81 patients. A nationwide recall of heparin was announced.[7] Further investigation carried out by the FDA confirmed that the active ingredient, which was the main contaminant, originated out of 12 Chinese companies and had made its way into 11 countries.[8]

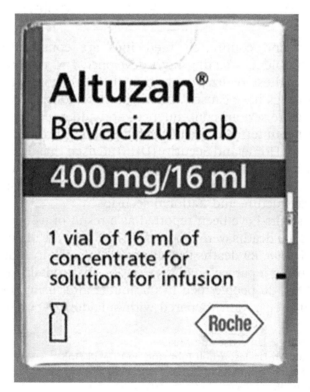

Figure 24.1   Fake anti-cancer drug found in a clinic, USA.

Counterfeit drugs in the USA are growing rapidly, particularly due to ever-increasing suppliers from around the world.[9] In 2006, the World Health Organization estimated that the business in counterfeit drugs could be worth US$75 billion globally by 2010. The size of the problem today is likely to be even greater than these estimates. Dr Lee Jong-wook, General Director of WHO, said "Combating low quality or illegal medicines is now more important than ever".[10]

In 2011, a TV report on the Al-Arabiya News Channel highlighted the problem of the smugglers and counterfeiters of medicines in Yemen. Millions of patients seeking a cure fell victims of expired smuggled, forged and substandard medications. Health authorities discovered that some lifesaving medications, *e.g.* anti-D (Rho) immunoglobulin and vaccines, were only water. Health professionals in Yemen blamed the government, agents and health authorities for lack of supervision. They called it "Trade of Death". Many reasons for smuggling and faking include drug availability, no alternative and high prices. By the year 2000, 80% of medications in the market were smuggled, with the majority containing low or no active ingredients.[11] This could be one of the reasons of a terrorist group called "Houthis, Ansarollah" to appear in Yemen. This terrorist group is linked to and getting support from the Lebanese terrorist group Hezbollah.[12] Another TV report focused on fake medicines in West Africa. Fake medicines were manufactured in Pakistan, India and China and imported to developing countries. This resulted in deaths and blindness in patients. Reasons for the increase of the fake and substandard medicines were lack of government supervision and poor storage conditions. Some medications were sold on the street, with half price for expired medications.[13]

In May 2014, Interpol reported that during a single week, police/customs in 111 countries seized 9.4 million doses (2.4 million in 2011) of fake medicines, including cancer medication, erectile dysfunction pills, anti-malarial and cholesterol medications, worth nearly US$36 million.[14,15]

Recently, the WHO announced a drug alert for counterfeit meningitis vaccines circulating in West Africa (Figures 24.2 and 24.3). The WHO advised increased supervision of the medications, and further verification and checks with drug manufacturers should be made before administration.[16]

Innovators, whose key activities are to create or obtain the rights or exclusivity to their work, make use of intellectual property rights and registration. These rights are appointed in the form of trademarks, copyrights and in the case of the pharmaceutical industry, patents. These ensure that all innovations and efforts are protected to a certain

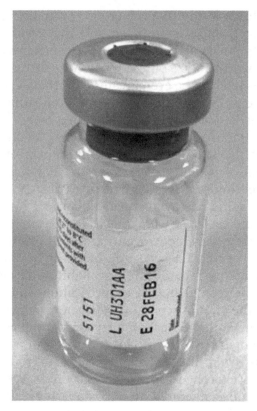

Figure 24.2   Falsified version of meningitis vaccines, West Africa.

degree within the legitimate marketplace. As sufficient as these measures are within the pharmaceutical industry, it is undermined by theft associated with counterfeiting and trafficking of pharmaceutical products.[9]

Counterfeiting may be related to organised crime across the world, assisting the likes of money laundering and terrorism. In 2006, the US attorney's office indicted 18 people in a multimillion dollars conspiracy to smuggle counterfeit Viagra and other goods to fund money for the Hezbollah terrorist group in the Middle East.[17] Profits made could well exceed those compared with narcotics. Drug companies are investing their efforts to fight these criminal acts by applying new technologies, such as short message service (SMS) authentication code systems on packaging. They are working in co-operation with health authorities across the world, like the British Medicines and Healthcare products Regulatory Agency (MHRA), the Australian Therapeutic Goods Administration (TGA) and the FDA to fight against these crimes.

Figure 24.3   Another falsified version of meningitis vaccines, West Africa.

Distribution channels for pharmaceuticals in the European Union is going through a major regulatory update. A new directive was adopted in July 2011 to protect patients and to prevent the distribution of fake medicines.[18] These new directives applied from January 2013. However, it may take up to three years for effective implementation of the new directives.[19]

24.2   Substandard and Counterfeit Medicines

Substandard medicines do not contain the correct amount of active ingredients or meet the standard (innovator's) preparation requirements for quality, safety and efficacy.[18,20] According to the FDA, counterfeit medicine is "fake medicine; it may be contaminated or contain the wrong or no active ingredient; they could have the right active ingredient but at the wrong dose". They may have the wrong manufacturer's name or the country of origin. Counterfeits may include both branded and generic products. Using such drugs may cause harmful effects to patients (Figure 24.4).

Substandard and counterfeit drugs can make their way into the marketplace. This can occur in both branded and generic drugs.

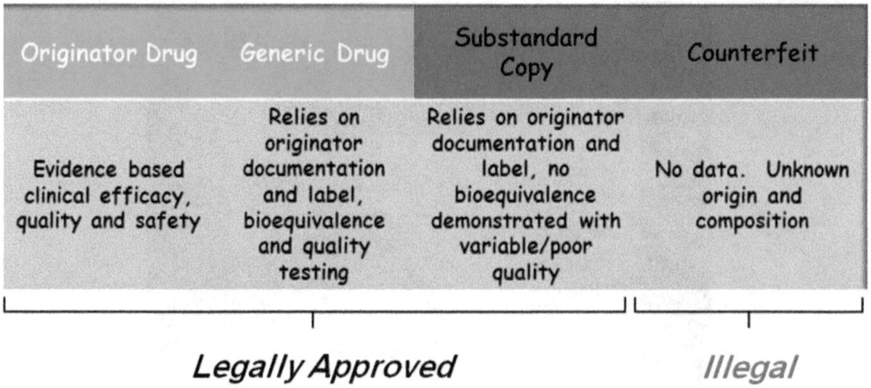

Originator Drug	Generic Drug	Substandard Copy	Counterfeit
Evidence based clinical efficacy, quality and safety	Relies on originator documentation and label, bioequivalence and quality testing	Relies on originator documentation and label, no bioequivalence demonstrated with variable/poor quality	No data. Unknown origin and composition

Legally Approved   *Illegal*

Figure 24.4 Definitions of brand, generic, substandard and counterfeit medicines.

Higher or lower levels of the active ingredient which differ from the required formulation are regarded as a substandard drug.[21]

There are several effects that consumers could experience when using substandard and counterfeit drugs. This all varies on which element of the drug fails to meet the required quality levels. With low quality antibiotics, the patient's treatment plan can be highly unsuccessful and face additional risk of adverse effects. With substandard and counterfeit antibiotics, this can increase the level of antibiotic resistance and there is loss of confidence in the treatment method by both physicians and patients. Some other scenarios include where the non-active ingredients do not meet the required limits and/or contamination of the drug.[22,23]

24.3 Factors Affecting the Spread of Counterfeit and Substandard Medicines

There are many factors encouraging the spread of counterfeit and substandard medicines. These factors include poor regulation in countries that do not establish or have good regulatory mechanisms and weak supervision from authorities. The profit that can be made by selling fake medicines is one of the reasons. According to Blackstone *et al.*, counterfeit medicines are more profitable than selling narcotics.[9] Drug availability is an important reason as well, because if the drug is not available on the market, patients will search for other available resources. Low supply levels are another reason why counterfeit drugs enter the supply chain in the US. Some drugs which are considered lifesaving, such as anti-cancer drugs, are in high demand and offer opportunist counterfeiters the chance to supply a high demand drug at higher than normal prices. However, as these counterfeit lifesaving drugs are sold to consumers, there is greater risk of suffering and even death to patients.

The lack of regulation, especially in free trade zones, is considered an important reason for the prevalence of fake medications. In addition, the packaging and storage conditions may affect the quality of the drug if it is not stored under proper conditions. A study of generic ramipril tablets showed that 24% out of 17 samples failed to meet the drug quality requirements because of improper storage conditions.[21,24] These problems are present in all global regions, but particularly affect most developing countries. According to the FDA, 10% of medications in both industrialised and developing countries are counterfeit. In rich countries, expensive medicines used for body building, such as hormones and steroids, are subject to higher levels of counterfeiting.[2]

Even in a country such as the UK with strict regulations, there is infiltration of counterfeit drugs because of loopholes in transit and supply chains. The common vulnerabilities in the supply chain are due to the many handling levels associated in reaching the end consumer. Because of the many points of entry into the supply chain, detecting substandard or counterfeit products poses a much bigger challenge. Unfortunately, substandard and counterfeit drugs make their way into legitimate supply chains and these also include reputable online pharmacies. There are many points of entry for counterfeiters to enter the supply chain, from key ingredients to manufacturers, to storage vulnerabilities, transportation and several points at the final distribution channels. In the US, 90% of drugs distributed are handled by national wholesalers who deliver direct to pharmacies, hospitals, clinics and physicians. The remaining 10% unfortunately have a more complicated and indirect method of distribution and, due to flaws in their practices, give a greater chance for counterfeits to enter the supply chain. Further regulation and monitoring need to be successfully put in place to avoid such vulnerabilities. The wholesale market comprises three types. Firstly, the primary wholesalers, who typically deal directly with the manufacturers and pose the least risk for substandard and counterfeiting to occur. Secondary, wholesalers comprising large regional wholesalers who can package and repackage. The third comprises thousands of smaller wholesalers.[25]

The internet is the main source of purchasing medications in most of Europe and it was recently revealed that out of a survey of 3100 online pharmacies, only four were certified through the verified Internet Pharmacy Practice.[26] Many online pharmacies are considered a real threat to consumers because they offer potentially harmful medications. High profits can be made selling substandard and counterfeit medications, especially without the need for a prescription. Some patients may prefer an online pharmacy as it can provide an additional level of privacy for some conditions they may find embarrassing, such as impotence. For disabled patients it is also more convenient as they may be home bound.[27] Ordering online medications consisted more of lifestyle drugs such as Viagra, but there is an increase in the use of prescribed medications *via* the internet for more serious diseases, such as cardiovascular, diabetes and cancer. A higher threat is using online pharmacies for purchasing narrow therapeutic index drugs (NTIDs) such as ciclosporin, since patients could be at risk of either toxicity at higher doses or treatment failure with lower doses. Examples of NTIDs which have been found to be available from online pharmacies are:[28] aminoglycosides, amikacin, gentamicin, rifampicin, warfarin, ciclosporin, carbamazepine, lithium, phenytoin, theophylline, phenobarbital, valproic acid, digoxin.

A death was reported due to consumption of a drug which was ordered online. A teenager took one pill to treat his anxiety disorder and this resulted in his death.[29] Many counterfeit online pharmacies are falsely stating that they are based and operate from Canada, as Canada is considered the ninth largest country in pharmaceutical sales, with a global share of 2.5%.[30] In 2005, a study revealed that out of 11 000 online pharmacies registered to be based in Canada, only 214 were legitimate.[9]

24.4 Methods for Detecting Counterfeit Medicines

The war against counterfeit medicines should be united worldwide. There should be a common definition for counterfeit medicines and a universal law against it. The UK government takes many actions against counterfeit drugs. Supplying or sale of counterfeit medicines is considered a criminal offence.

There are many tests carried out to check for counterfeit drugs. Package and general characteristics (size, shape and colour of the drug) analysis is a key step for detecting counterfeits. If there are any changes or doubts raised from these initial examinations, they should be investigated by the pharmacist and reported so that appropriate action can be taken. For further investigation, chemical analysis is considered. Spectroscopic analysis such as Fourier transform infrared, near infrared and Raman spectroscopies can establish the identity of most drugs and discriminate from closely related or structurally similar compounds.[31] Other detection methods use microscopy techniques, such as light microscopy and the scanning electron microscope. Separation techniques are used to detect counterfeits, such as liquid chromatography, capillary electrophoresis and mass spectrometry.[31] HPLC methods with separation are generally an acceptable technique. This is because HPLC methods can generally provide the required specificity. All drug products should have a specific assay to determine the content of the drug.

In 2011, Pfizer started implementing their SMS authentication code system on packaging. Patients and consumers can send an SMS containing the pack code to the company and receive verification of the product's authenticity.[32] On November 28, 2013, President Obama signed the Drug Quality and Security Act into law, which provides for a national track-and-trace system that would allow a specific drug to be followed from the manufacturer to the pharmacy. This should make it more difficult for counterfeiters to enter the supply chain in the US. This law was expected to be implemented in 2015.[33]

Specific tests should be developed on an individual product basis. The dissolution test gives an idea of the release of the drug substance from the drug product when taken orally. Although one-time point measurement is enough for immediate release drugs, suitable test conditions and sampling procedures should be established and reported for non-immediate release drugs. If a drug is not available in the required time after it has been taken orally, it will significantly affect the bioavailability (rate and extent of the drug inside the body). Drug disintegration, tablet hardness, water content and microbial limits should be thoroughly investigated, reported and documented to identify counterfeit and substandard medicines.

24.5 Safety of Drug Use: Branded *versus* Generic

A branded drug is a new medication that is found to improve or treat a certain disease or medical problem. This new medication passes through many phases of developmental and research processes. The phases include the preclinical testing in animals for pharmacokinetics and pharmacodynamics of the tested drug, phase I studies which are conducted on 20–80 healthy volunteers for safety screening,

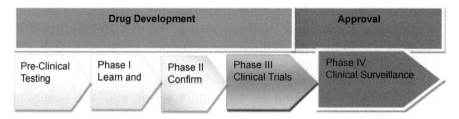

Figure 24.5 General phases for new drug development.

phase II studies to ensure the effectiveness of the drug, phase III studies to obtain more information about its safety and effectiveness (subjects range from 200 to 3000 people) and phase IV which consists of post-approval studies (Figure 24.5). These phases can happen over many years. The product can only be in the market once it has been approved by a regulatory health body such as the FDA. The manufacturer usually obtains a patent for the new drug for a period of up to 20 years.[34] Patenting the product prevents other companies from making and selling the new product. When the patent period has expired, other companies can start making and selling a generic version, but it must be evaluated and approved by the FDA or other regulatory health authorities. Over the last few years, the production of generic medications has increased rapidly due to competition after expiry of the originator drug company's patent.[35]

The FDA approved an average of 101 new generic drugs into the market place each year from 2001 until 2014 (Figure 24.6). These included many drug classes: hormones, antibiotics, analgesics, cardiovascular, respiratory, antimetabolite and many other classes.[36]

Owing to competitive market demands, many generic drug companies place their products on the market as soon as the patent period of the brand medicine has expired, bringing concerns for patients' safety because in some cases they are not safe and effective like their reference counterpart. Some tentative applications may even be approved before the patency has expired. There have been many reported clinical studies highlighting the adverse effects of generic products.[37–39] During recent years, the production of generic medications has increased due to demand for cheaper medicines, especially in developing countries. The generic medications may have lower therapeutic effects and/or toxic effects.[38] Owing to the lack of good manufacturing practice, insufficient bioequivalence and toxicological studies,[37] and even deliberate forgery of the drug, there are an increasing number of patient safety concerns. Small differences in plasma concentrations, less than 4%, may exist in some cases

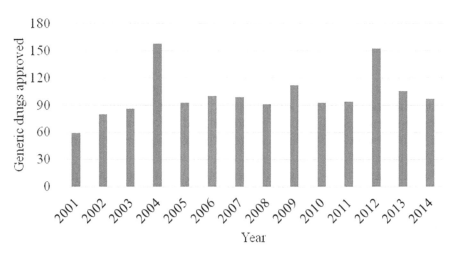

Figure 24.6 FDA generic drug approvals from year 2001 until 2014.

between a brand and its generic equivalent. However, this minor difference is no greater than the difference that may exist between two different manufactured batches of the brand drug manufactured by the same pharmaceutical company. Safety and efficacy trials are only required for new drugs.[40] Therefore the average cost for the production of generic drugs is lower than the brand by approximately 45%.[41]

Bioequivalence is the equivalence of the action of brand and generic medicines. Birkett described bioequivalence as: "Two pharmaceutical products are bioequivalent if they are pharmaceutically equivalent and their bioavailabilities (rate and extent of availability) after administration in the same molar dose are similar to such a degree that their effects, with respect to both efficacy and safety, can be expected to be essentially the same. Pharmaceutical equivalence implies the same amount of the same active substance(s), in the same dosage form, for the same route of administration and meeting the same standards".[42]

A generic drug is bioequivalent when the active ingredients or the active metabolites are absorbed into the body after administration at the same rate and amount as the brand drug. Thus, the need for bioequivalence is evident by the fact that the generic products deliver the same therapeutic effect as the branded counterpart and can safely substitute the brand product. Before a generic drug can be marketed, the manufacturer must prove that it has the same potency and efficacy as the brand medication. If a generic drug passes these tests, it is said to be bioequivalent to the original drug. The generic drug application goes through several stages before approval by the FDA. Key areas being monitored are chemicals, manufacturing processes, bioequivalence tests, dissolution tests, labelling, *etc.*

Bioequivalence gives health professionals and patients the confidence that the generic medicines provide the same therapeutic effect, clinical results and safety profile as their branded counterparts. Bioequivalence studies play an important role during drug development to observe the optimum therapeutic effect and ensure no additional toxicities.[42]

According to the FDA guidelines, bioequivalence is accepted when 90% confidence intervals (CI) for the ratio of target pharmacokinetic parameters of area under the curve (AUC) and maximum concentration (C_{max}) fall in the range 0.8–1.25 (80–125%).[43] The time to maximum plasma concentration (t_{max}) should also be similar. Bioequivalence studies are an important part of drug development for the production of new drug formulations and for generic equivalents. Such studies are important after the approval phase if there are any manufacturing changes.[44] Many countries have established guidelines for the approval of generic drugs. Generic manufacturers are not required to submit data of clinical trials or preclinical tests which are long and expensive procedures, but they must submit proof of bioequivalence tests, in addition to other pharmaceutical information.[45] During bioequivalence studies, some minor differences between brand and generic drugs are allowed. For example, a generic drug may have differences in shape, size or colour compared to the branded product.[46]

Since generic and therapeutic substitution might impact on the clinical outcome, it could create a conflict between the interests of patients, clinicians and those of payers/providers.[47] Patients who are uncertain are warned that substitution that is done only for financial reasons might compromise their quality of care. They may believe that a cheaper medicine must be inferior to the more expensive branded medicines.[48]

Many healthcare providers have been promoting generic and therapeutic substitution in an attempt to contain their costs.[49] In 2013, it was reported that generic prescribing had reached 83.9% of all prescribed items in community pharmacies in England.[50] Furthermore, the Department of Health (DOH) in England considered and then abandoned the idea of automatic generic substitution of medicines by pharmacists.[51,52] Accordingly, pharmacists and other dispensers who receive a prescription containing a branded medicine would be obliged to dispense an equivalent generic version of the medicine instead.

The FDA and the European Medicines Agency (EMA) have set out guidelines to establish the requirement range for bioequivalence for generic drugs, which should be between 80% and 125% of the original innovator drug's bioavailability.[53,54] On the other hand, many authors question the approval of such range limits for NTIDs such as ciclosporin.[55–60] A small dose difference in NTIDs could have serious side effects which could lead to treatment failure and/or toxicity.[61]

24.6 Excipients in Medicines

Excipients are the inactive ingredients in a drug and include binders, fillers, lubricants, sweeteners, preservatives, flavours, colouring, printing inks, *etc*.[62,63] Although excipients are considered to be inactive ingredients that do not have a therapeutic effect, some studies have shown that excipients can cause many side effects.[62] Figure 24.7 shows the differences in sizes, shapes and colours of the same medicine. Excipients do not need to match the innovator's drug formulation. Some evidence shows that different excipients are metabolised differently in the body, such as polyoxyethylated castor oil and polysorbate 80.[21]

Another example of phenytoin (antiepileptic agent) toxicity happened in Australia because of changing the excipient in phenytoin to lactose instead of calcium sulfate. That change affected the solubility

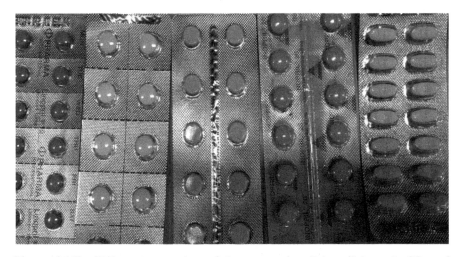

Figure 24.7 Different generics of the same medicine (lisinopril, 20 mg) showing differences in sizes, shapes and colours.

of phenytoin and made it more soluble, which increased its systemic availability, which led to an increased incidence of toxicity.[64]

Another example is of a drug company decreasing the particle size of digoxin powder from 20 μm to 3 μm to formulate digoxin tablets. That caused an increase in the drug absorption up to twofold. Consequently, many patients had signs of toxicity.[65] Therefore, many authors have highlighted the importance of the bioavailability of pharmaceutical products, especially for critical dose and NTIDs.[66,67]

24.6.1 Impurities in Drug Products

Impurities may be defined as any ingredients, substances or contaminations which do not belong to the active or non-active ingredients of the drug. FDA guidelines classify impurities in new drug substances into organic, inorganic or from residual solvent. These impurities may affect the drug product quality and lead to serious adverse effects affecting patients' safety. The origin of these impurities can be from the synthetic procedures for the active ingredient or from the degradation of the inactive ingredients in the drug product.[68,69]

Organic impurities may arise from degradation of the new drug substance or the manufacturing process. The acceptance limits of these impurities should be well specified. If an impurity in a drug product is coming from different sources such as a synthetic product and also a degradation product, it should be monitored and included in the impurity limits. The medication properties can be changed or could result in toxicity by impurities.

24.7 Conclusion

There are many stages where counterfeits or substandard production methods can enter or occur. This could be reduced by refining the stages from manufacturing to distribution and by implementing a regular track-and-trace system where you can follow the whole process of a drug, from manufacturing to distribution and dispensing. Manufacturers may not want to disclose who their suppliers are or their business practices, but transparency in the pharmaceutical industry should be essential to lower the risk to public health, reduce funding terrorism, produce better economic prosperity and reduce the vulnerabilities of counterfeit and substandard drugs entering the supply chain. Pharmaceutical counterfeiting could be the most important criminal activity worldwide, as this funds global terrorism and is a great risk to public safety.

References

1. U.S. Customs and Border Protection Office of International Trade Fiscal year 2012 seizure statistics. http://tinyurl.com/o9tq7e4 (28 May).
2. World Health Organization Counterfeit medicines. http://www.gphf.org/images/downloads/library/who_factsheet275.pdf (21 May).
3. R. Mukhopadhyay, The hunt for counterfeit medicine, *Anal. Chem.*, 2007, **79**, 2622–2627.
4. Food and Drug Administration Another counterfeit cancer medicine found in U.S. - Illegal practice puts patients at risk. http://tinyurl.com/7f7br4j (May 21).
5. Amgen Inc. Important Drug Warning Counterfeiting of Epogen(R). http://www.amgen.com/media/media_pr_detail.jsp?releaseID= 515145&year (28 May).
6. C. C. Tigue and C. L. Bennett, Counterfeit epoetin alfa products, *Commun. Oncol.*, 2006, **3**, 772–773.
7. Food and Drug Administration Protecting consumers from adulterated drugs. http://tinyurl.com/ndqbtcv (29 May).
8. Food and Drug Administration FDA's Ongoing heparin investigation. http://www.fda.gov/newsevents/testimony/ucm115242.htm (29 May).
9. E. A. Blackstone, J. P. Fuhr Jr. and S. Pociask, The health and economic effects of counterfeit drugs, *Am. Health Drug Benefits*, 2014, **7**, 216–224.
10. S. Pincock, WHO tries to tackle problem of counterfeit medicines in Asia, *BMJ*, 2003, **327**, 1126.
11. Al Arabiya News Channel Special mission: Trade of death. https://english.alarabiya.net/webtv/programs/special-mission.html (2 October).
12. D. Schenker, Who's behind the Houthis? http://www.washingtoninstitute.org/policy-analysis/view/whos-behind-the-houthis (29 October).
13. Al Arabiya News Channel Special mission: Trader of death. http://tinyurl.com/plyzncd (9 February).
14. Al Arabiya News Channel Global swoop nets 2.4 million doses of fake drugs: Interpol. http://english.alarabiya.net/articles/2011/09/29/169306.html (26 September).

15. Interpol Thousands of illicit online pharmacies shut down in the largest-ever global operation targeting fake medicines. http://tinyurl.com/pkps4cw (9 February).

16. World Health Organization Falsified meningitis vaccines circulating in West Africa. http://tinyurl.com/nn8t5l4 (15 June).

17. Pfizer Inc. A serious threat to patient safety counterfeit pharmaceuticals. http://www.pfizer.com/files/products/CounterfeitBrochure.pdf (10 February).

18. European Medicine Agency Falsified medicines. http://tinyurl.com/op4klg3 (22 May).

19. A. Williams, Europe prepares to battle the counterfeiters, *Pharm.Technol. Eur.*, 2011, **23**, 28.

20. Food and Drug Administration Counterfeit medicine. http://tinyurl.com/yk2unfc (May 22).

21. A. Johnston and D. W. Holt, Substandard drugs: a potential crisis for public health, *Br. J. Clin. Pharmacol.*, 2014, **78**, 218–243.

22. T. Kelesidis, I. Kelesidis, P. I. Rafailidis and M. E. Falagas, Counterfeit or substandard antimicrobial drugs: a review of the scientific evidence, *J. Antimicrob. Chemother.*, 2007, **60**, 214–236.

23. T. Kelesidis and M. E. Falagas, Substandard/counterfeit antimicrobial drugs, *Clin. Microbiol. Rev.*, 2015, **28**, 443–464.

24. M. H. Khan, K. Hatanaka, T. Sovannarith, N. Nivanna, L. C. Casas, N. Yoshida, H. Tsuboi, T. Tanimoto and K. Kimura, Effects of packaging and storage conditions on the quality of amoxicillin-clavulanic acid – an analysis of Cambodian samples, *BMC Pharmacol. Toxicol.*, 2013, **14**, 33.

25. P. Yadav, Falsified and substandard medicines. http://tinyurl.com/pplpf2w (30 May).

26. American Enterprise Institute For Public Policy Research The Deadly World of Fake Drugs. http://www.aei.org/files/2012/02/27/-appendix-a-master-2_170026856632.pdf (21 May).

27. G. Orizio, S. Rubinelli, P. J. Schulz, S. Domenighini, M. Bressanelli, L. Caimi and U. Gelatti, Save 30% if you buy today". Online pharmacies and the enhancement of peripheral thinking in consumers, *Pharmacoepidemiol. Drug Saf.*, 2010, **19**, 970–976.

28. B. A. Liang, T. K. Mackey and K. M. Lovett, Illegal "no prescription" internet access to narrow therapeutic index drugs, *Clin. Ther.*, 2013, **35**, 694–700.

29. M. Broomhead, 'One pill can kill' warning after death of popular Chesterfield teen. http://tinyurl.com/kkybalj (30 May).

30. M. Jackson, Enforcement in the UK: The international operation – Pangea, in *Combating Counterfeit Medicines: The Solutions*, JPAG, London, UK, 2015.
31. T. Moffat, Methods to analyse counterfeit medicines, *Pharmaceutical. J.*, 2008, **280**, 759.
32. T. Marsh, Counterfeit medicines: the regulatory and industry challenges, in *Counterfeiting of Pharmaceutical Products a Recurring Issue*, JPAG, London, UK, 2011.
33. Food and Drug Administration Compounding quality act. http://tinyurl.com/o7sk4ka (1 June).
34. Pipeline patent intelligence How to Calculate Standard Patent Expiry Dates and Data Exclusivity in Key Territories. http://tinyurl.com/7ke3r2p (5 December).
35. Food and Drug Administration Understanding generic drugs. http://tinyurl.com/bucu8t2 (12 August).
36. Food and Drug Administration ANDA (Generic) Drug Approvals. http://tinyurl.com/6g9swm3 (6 June).
37. C. S. Gautam, A. Utreja and G. L. Singal, Spurious and counterfeit drugs: a growing industry in the developing world, *Postgrad. Med. J.*, 2009, **85**, 251–256.
38. P. N. Newton, M. D. Green and F. M. Fernandez, Impact of poor-quality medicines in the 'developing' world, *Trends Pharmacol. Sci.*, 2010, **31**, 99–101.
39. B. Perks, Faking it, *Chem. World*, 2011, **8**, 56–59.
40. R. L. Nation and L. N. Sansom, Bioequivalence requirements for generic products, *Pharmacol. Ther.*, 1994, **62**, 41–55.
41. Canadian Health Services Research Foundation, Myth: generic drugs are lower quality and less safe than brand-name drugs, *J. Health Serv. Res. Policy*, 2007, **12**, 255–256.
42. D. J. Birkett, Generics – equal or not? *Aust. Prescr.*, 2003, **26**, 85–87.
43. Food and Drug Administration Bioavailability and bioequivalence studies for orally administered drug products, general considerations. http://tinyurl.com/363jwqx (16 October).
44. European Medicines Agency Guideline on the investigation of bioequivalence http://tinyurl.com/p3zfadt (21 February).
45. European Medicines Agency Procedural advice for users of the centralised procedure for generic/hybrid applications. http://tinyurl.com/o3opl79 (15 October).
46. A. S. Kesselheim, A. S. Misono, J. L. Lee, M. R. Stedman, M. A. Brookhart, N. K. Choudhry and W. H. Shrank, Clinical equivalence of generic and brand-name drugs used in

cardiovascular disease: a systematic review and meta-analysis, *JAMA, J. Am. Med. Assoc.*, 2008, **300**, 2514–2526.

47. M. AlAmeri, M. Epstein and A. Johnston, Generic and therapeutic substitutions: are they always ethical in their own terms?, *Pharm. World Sci.*, 2010, **32**, 691–695.
48. P. Meredith, Bioequivalence and other unresolved issues in generic drug substitution, *Clin. Ther.*, 2003, **25**, 2875–2890.
49. M. G. Duerden and D. A. Hughes, Generic and therapeutic substitutions in the UK: are they a good thing? *Br. J. Clin. Pharmacol.*, 2010, **70**, 335–341.
50. Health and Social Care Information Centre Prescriptions dispensed in the community http://www.hscic.gov.uk/catalogue/PUB14414/pres-disp-com-eng-2003-13-rep.pdf (1 June).
51. M. Baker, D. Candy, S. Kownacki, A. McCoig, J. Mossman and T. Solanki, Automatic Generic Substitution – Clinical implications for patients http://tinyurl.com/q6ww89x (1 June).
52. The Pharmaceutical Journal News team, Generic substitution plans abandoned (updated), *Pharmaceutical. J.*, 2010, **285**, 457.
53. European Medicine Agency Note for guidance on the investigation of bioavailabilty and bioequivalence. http://tinyurl.com/36d2q68 (2 June).
54. Food and Drug Administration Approved drug products with therapeutic equivalence evaluations. http://tinyurl.com/y92ahby (2 June).
55. M. R. Bartucci, Issues in cyclosporine drug substitution: implications for patient management, *J. Transpl. Coord.*, 1999, **9**, 137–142, quiz 143-4.
56. D. Cattaneo, N. Perico and G. Remuzzi, Generic cyclosporine formulations: more open questions than answers, *Transplant Int.*, 2005, **18**, 371–378.
57. A. Johnston and D. W. Holt, Bioequivalence criteria for cyclosporine, *Transplant. Proc.*, 1999, **31**, 1649–1653.
58. B. D. Kahan, Considerations concerning generic formulations of immunosuppressive drugs, *Transplant. Proc.*, 1999, **31**, 1635–1641.
59. D. Kamerow, The pros and cons of generic drugs, *Br. Med. J.*, 2011, **343**, d4584.
60. S. Sabatini, R. M. Ferguson, J. H. Helderman, A. R. Hull, B. S. Kirkpatrick and W. H. Barr, Drug substitution in transplantation: a National Kidney Foundation White Paper, *Am. J. Kidney Dis.*, 1999, **33**, 389–397.

61. A. Johnston and D. W. Holt, Immunosuppressant drugs the role of therapeutic drug monitoring, *Br. J. Clin. Pharmacol.*, 2001, **52**(Suppl 1), 61S–73S.

62. C. Wandel, R. B. Kim and C. M. Stein, Inactive" excipients such as Cremophor can affect in vivo drug disposition, *Clin. Pharmacol. Ther.*, 2003, **73**, 394–396.

63. I. Iheanacho and J. Blythe, What are excipients doing in medicinal products?, *Drug Ther. Bull.*, 2009, **47**, 81–84.

64. J. H. Tyrer, M. J. Eadie, J. M. Sutherland and W. D. Hooper, Outbreak of anticonvulsant intoxication in an Australian city, *Br. Med. J.*, 1970, **4**, 271–273.

65. A. Johnston, P. A. Keown and D. W. Holt, Simple bioequivalence criteria: are they relevant to critical dose drugs? Experience gained from cyclosporine, *Ther. Drug Monit.*, 1997, **19**, 375–381.

66. A. Johnston, P. Belitsky, U. Frei, J. Horvath, P. Hoyer, J. H. Helderman, M. Oellerich, S. Pollard, H. Riad, P. Rigotti, P. Keown and B. Nashan, Potential clinical implications of substitution of generic cyclosporine formulations for cyclosporine microemulsion (Neoral) in transplant recipients, *Eur. J. Clin. Pharmacol.*, 2004, **60**, 389–395.

67. D. W. Holt, Digoxin bioavailability, *Lancet*, 1978, **2**, 1103.

68. A. K. Basak, A. S. Raw and L. X. Yu, Pharmaceutical impurities: analytical, toxicological and regulatory perspectives, *Adv. Drug Deliv. Rev.*, 2007, **59**, 1–2.

69. J. Roy, Pharmaceutical impurities–a mini-review, *AAPS PharmSciTech*, 2002, **3**, E6.

25 Detection of Drugs and Drug Metabolites from Fingerprints

Paula García Calavia and David A. Russell*

School of Chemistry, University of East Anglia, Norwich, UK
*Email: d.russell@uea.ac.uk

25.1 Introduction

The surface of human fingers is covered by a unique set of marks, referred to as friction ridge skin.[1] Friction ridge skin, composed of features known as *minutiae*, is formed in the womb through physical tensions and pressures between infant structures in the foetus.[2] As a result, the pattern of friction ridge skin is unique for each individual and the probability of two fingerprints being the same is 64 billion to 1.[2] This uniqueness is also true for identical twins since the two foetuses grow in different parts of the womb, causing their fingerprints to be developed in microenvironments, exposed to distinct intrauterine forces.[3] The unique character of fingerprints, together with their persistence throughout an individual's lifetime, enable them to be used as identification tools for forensic investigations.[2]

Friction ridge skin is composed of several lines, known as ridges, along which sweat pores can be observed.[4] There are three types of sweat in the human body: eccrine, apocrine and sebaceous.[5] Eccrine glands are located all over the body, especially in palmar and plantar surfaces, and they produce mainly water (99%), inorganic salts and organic materials. Apocrine glands are located in axillary, inguinal and genital areas. Sebaceous glands, located all over the body except

Forensic Toxicology: Drug Use and Misuse
Edited by Susannah Davies, Atholl Johnston and David Holt
© The Royal Society of Chemistry 2016
Published by the Royal Society of Chemistry, www.rsc.org

on the friction ridge skin of hands and feet, produce sebum, which is mainly composed of fatty acids, cholesterol and squalene.[5] As a result, when a finger comes into contact with a surface, endogenous compounds from eccrine sweat and also sebum that has been touched by the finger are transferred, leaving an impression of the ridge pattern. This is known as a latent fingermark.[4,5] Exogenous compounds that the finger has interacted with, such as food, cosmetics or drugs, are also transferred to the latent fingermark.

The analysis of latent fingermarks is challenging because of their invisibility to the naked eye.[4] The visualisation of latent fingermarks requires the use of physical and/or chemical methods. The most common physical technique used by police investigators is fingerprint powdering, a simple and cheap procedure in which the powders adhere to the oily components in the fingermarks. The first chemical treatments used to develop latent fingermarks included ninhydrin solution and iodine fuming for porous surfaces. For non-porous surfaces, cyanoacrylate fuming is applied effectively. These physical and chemical methods are extensively and successfully used during forensic investigations.[4] However, the use of latent fingermarks to provide not only identification but also other information about an individual, including lifestyle, has led to the use of other techniques such as mass spectrometry, vibrational spectroscopies and nanotechnology. This chapter gives an overview of the detection of drugs in fingerprints, either endogenously excreted by sweat or as exogenous contaminants. Additionally, the use of fingerprint residue to obtain information about the donor, including gender and age, is also explored.

25.2   Exogenous Drug Residue in Fingerprints

The handling of drugs leads to the deposition of exogenous contact residue on the hands and fingers of the user. Consequently, chemical analysis of fingerprints could provide an indication that a person has manipulated a particular drug.

Drugs of abuse are important analytes in forensic toxicology. Such drugs are usually combined with other non-controlled substances commonly used as adulterants, such as caffeine and sugar. Day *et al.* were one of the first groups to report the detection of drugs of abuse and adulterants in spiked latent fingerprints using Raman spectroscopy.[6] These authors doped the fingerprints of volunteers with codeine phosphate, cocaine hydrochloride, amphetamine sulfate, barbital and nitrazepam together with caffeine, aspirin, paracetamol, starch and talc. The spectra of each compound was successfully identified and differentiated from the others. The presence of additional Raman bands arising from the fingerprint oil did not prevent identification. However, spectral subtraction of the fingerprint oil could be beneficial to avoid falsely attributing these bands to exogenous substances. The main disadvantage with this method for the detection of the exogenous substances is that visual localisation of the dopant on the fingerprints is not possible. However, visualisation of the drugs of abuse on the fingermarks could be achieved using Raman imaging and/or mapping.[6] A more sensitive Raman-based technique, surface-enhanced Raman scattering (SERS), has also been studied. Yang *et al.* used gold nanoparticle-coated magnetic nanocomposites (AMN) modified with inositol hexakisphosphate as both the developing fingerprint powder and the substrate for SERS.[7] The AMN are made of iron oxide nanoparticles covered with 11 nm gold nanoparticles. With this method, the detection of cotinine, a metabolite of nicotine, was possible. SERS allowed for a high sensitivity and a low limit of detection. More importantly, the analysis was performed with a portable Raman spectrometer, an advantage in forensic investigations as this testing could be achieved *in situ*.[7] The detection of the pharmaceutical drug acetaminophen, commonly known as paracetamol, was also possible with this technique.[8]

Another vibrational spectroscopy technique, infrared spectroscopy, has also been widely used for the detection of drugs in fingerprints.[4] Vibrational spectroscopy techniques are highly valued since they are non-invasive and non-destructive, meaning that the fingerprints can still undergo other analyses. Chan and Kazarian used Fourier transform infrared (FTIR) imaging enhanced with a focal plane array (FPA) detector. Using this technology, the model drug caffeine was detected in trace amounts across the surface of a finger. At the same time, a chemical image depicting the fingerprint pattern was obtained.[9] A combination of FTIR with spectral searching algorithms was also used for the detection of caffeine, diazepam and aspirin in a single fingerprint. These drugs were detected as three individual substances, with no ambiguous results.[10]

Mass spectrometry (MS) is an excellent characterisation technique since it measures the molecular weight of the substance being analysed.[11] Different types of MS have been used in the detection of drugs in fingerprints. The first report dates back to 1996, when Buchanan and co-workers inadvertently detected traces of nicotine in fingerprint extracts with gas chromatography-mass spectrometry (GC-MS).[12]

Desorption electrospray ionisation (DESI) MS for analysis of drugs from fingerprints has been reported by the group of Cooks.[13,14] In this technique, charged solvent droplets are sprayed onto the substrate, *i.e.* the fingerprint, creating a thin film of liquid, where the analytes of interest are dissolved. As secondary droplets impact on the surface of the liquid, the dissolved analyte is released. The solvent is then removed by heating under a vacuum, which allows the detection of the ionised analyte by MS.[13] Ifa *et al.* were able to detect and image latent fingerprints based on the distribution of the drugs cocaine and Δ^9-tetrahydrocannabinol (THC), the main psychoactive component in marijuana, using this technique (Figure 25.1). They reported low limits of detection, reaching picogram levels.[14]

Szynkowska *et al.* have effectively applied the technique of time-of-flight secondary ion MS (TOF-SIMS) for the detection and identification of several drugs, namely amphetamine, methamphetamine, methylenedioxymethamphetamine (MDMA; ecstasy) and even arsenic, in contaminated fingerprints on a range of surfaces. Additionally, their distribution over the fingerprint was also possible, which adds forensic value to the evidence.[15,16]

Direct analyte-probed nanoextraction coupled to nanospray-ionisation-MS (DAPNe-NSI-MS) has been described as a direct and sensitive approach for the analysis of drugs in fingerprints. DAPNe-NSI-MS was used to detect cocaine, crystal methamphetamine, ecstasy and caffeine at ultra-trace amounts without the need for sample preparation.[17]

Nanotechnology has also been used for the detection of illicit drugs in fingerprints. The use of aptamers to selectively recognise certain analytes has been investigated. Aptamers, single stranded DNA or RNA, are being used instead of antibodies for the specific detection of target analytes. Aptamers offer design flexibility and biochemical stability, which, together with their tolerance for harsh conditions, make them ideal candidates for selective recognition of analytes of interest.[18] An aptamer specific to cocaine was designed, bound to different types of nanoparticles and tested on latent fingerprints.[18,19] Li *et al.* used gold nanoparticles (AuNPs), together with high-resolution dark field microscopy. The presence of cocaine on the fingerprint was observable by a colour change from green to red upon aggregation of the AuNPs. The quality of the images obtained was sufficient to see second- and third-level detail of the fingerprints, which is highly valuable in forensic casework.[19] Similarly, Wang *et al.* used upconverting nanoparticles, which produce luminescence when excited with near-infrared (NIR) light. The detection of cocaine was

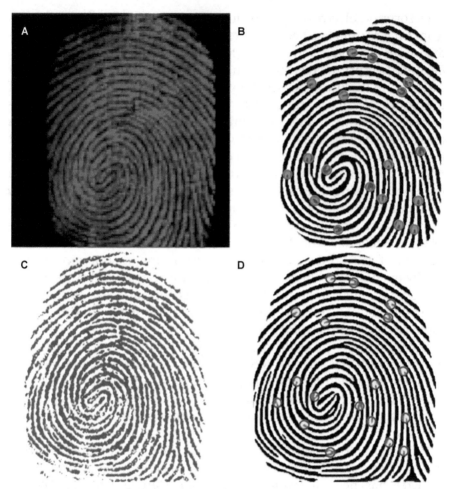

Figure 25.1 (A) DESI image of distribution of cocaine on a latent fingerprint (LFP) blotted on glass. (B) Computer-generated fingerprint from DESI image. (C) Ink fingerprint blotted on paper and optically scanned. (D) Computer-generated fingerprint from optical image. Some of the automatically detected points of interest (minutiae) are represented by dots in (B) and (D). (Reproduced from Ifa *et al.*,[14] with permission from the American Association for the Advancement of Science.)

clear, with no signal being obtained in negative controls. The luminescence images showed great fingerprint detail and no background fluorescence, due to the NIR excitation of the upconverters.[18]

The detection of pharmaceutical drugs in fingerprints has also been studied. Verapamil hydrochloride is an L-type calcium channel blocker, frequently used for the treatment of cardiovascular anomalies.[20,21] Tang *et al.* developed a double imaging technique able to provide an optical image of the fingerprint pattern as well as a molecular image based on the distribution of verapamil on the surface. For this purpose, surface-assisted laser desorption/ionisation time-of-flight mass spectrometry (SALDI-MS) assisted with AuNPs was used. The AuNPs are generated *via* argon ion sputtering and were used to develop latent fingerprints by aggregation in two different ways on the ridges and grooves of the prints. The two types of aggregation led to the development of two contrasting colours, which allowed for the optical visualisation of the fingerprint. The difference in colour between the ridges and grooves was related to the presence of endogenous compounds in the ridges, which supports the formation of AuNPs of smaller size. Furthermore, the detection and spatial localisation of verapamil was possible after MS analysis.[20] In another study, Yagnik *et al.* detected verapamil in fingerprints with a novel multiplex MS imaging procedure, providing chemical confirmation together with an image of the drug's distribution over the fingerprint.[21]

An important consideration when chemically analysing fingerprints in forensic contexts is the use of dusting agents and/or lifters. The visualisation of latent fingerprints in crime scenes is usually performed by forensic investigators with different fingerprint powders. Moreover, fingerprints in hardly-accessible locations are further treated with lifting tapes and transferred to a surface where they can be easily visualised and analysed. It is therefore important to confirm whether the chemical spectrum, either MS or vibrational, of the powders and lifters can hinder identification of the drugs of interest. Vibrational spectroscopy has proven useful for these types of investigations. Day *et al.* explored the detection of drugs of abuse and common adulterants in cyanoacrylate-fumed fingerprints with Raman spectroscopy.[22] Raman spectroscopic analysis was able to identify all of the substances, showing the cyanoacrylate polymer bands do not interfere with identification. However, spectral subtraction could be beneficial and aid in the identification of the drugs.[22] West and Went broadened this study by investigating the interference of both powders and lifting tapes in Raman spectroscopic analysis. These authors doped fingerprints with either over-the-counter analgesics, mainly composed of paracetamol and codeine,[23] or drugs of abuse, including cocaine, ketamine, amphetamine and ecstasy.[24] The application of most powders did not present any drawback for the spectroscopic analysis. In contrast, the powders had an enhancement ability since they reduced the background fluorescence. The only powders that presented problems were those of chemical origin, such as single particle reagents and red fluor-escent magnetic powder, which should be avoided. In addition, lifting adhesive tapes did not interfere in the chemical identification of drugs. Only hinge lifters were found to have a strong Raman spectrum, a problem that could be easily solved by spectral subtraction.[23,24] The combination of Raman spectroscopy with multivariate data analysis has also been studied. Fingerprints doped with ibuprofen, l-arginine and sodium bicarbonate were initially lif-ted with Scotch tape, analysed by Raman spectral mapping and finally by band-target entropy minimization. The combination of these techniques gave good results, as all drugs and adulterants were correctly identified and their spatial distributions over the fingerprint could also be observed.[25]

Similar results have been obtained with FTIR spectroscopy. The use of synchrotron radiation-based FTIR micro-imaging[26] and attenuated total reflectance (ATR)-FTIR[27] was successful for the detection of sugar and aspirin in powdered fingerprints lifted with several tapes, including Mylar foil. Additionally, ATR-FTIR was also a good techni-que in the identification of γ-hydroxybutyrate (GHB) in fingerprints lifted with a gelatine tape.[28]

Mass spectroscopic techniques have been shown to be a useful tool in this area. SALDI-TOF-MS has been widely used for fingerprint analysis. High sensitivity in SALDI can be achieved by addition of matrix assisting particles to the surface-bound sample. The use of hydrophobic silica particles doped with carbon black as a matrix assistant has been explored. The advantage of using these particles is that they can act as both a dusting agent to locate latent fingerprints as well as a SALDI enhancer, allowing the detection of analytes by MS.[29,30] Indeed, this method effectively detected cocaine, heroin, codeine and certain opiates in fingerprints pre-dusted with the silica particles and lifted with commercial lifting tapes.[29] Similar studies were performed with matrix-assisted laser desorption/ionisation time-of-flight mass spectrometry (MALDI-TOF-MS), showing that procaine and pseudoephedrine can be detected on dusted fingerprints.[31]

Both SALDI-TOF-MS and MALDI-TOF-MS have been used for the detection of cocaine, methadone, caffeine, aspirin and paracetamol in cyanoacrylate-fumed fingerprints. Direct analysis by MS proved un-successful. Alternatively, the developed prints need to be dusted with commercial black powders, which in turn improve the visual defin-ition of the fingerprint, and subsequently lifted. The only way to obtain a good transfer of the cyanoacrylate-fumed prints was by ex-posure to acetone vapours before lifting them with commercial lifting tapes. Only after these steps had been performed were the drugs successfully detected (Figure 25.2).[32]

Lim *et al.* compared the SALDI technique to direct analysis in real-time MS (DART-MS). Drugs including 6-monoacetylmorphine (6-MAM), heroin, methadone, nicotine and noscapine were doped in latent fingerprints. Both SALDI and DART produced similar positive results as they were both able to detect the drugs. The main difference

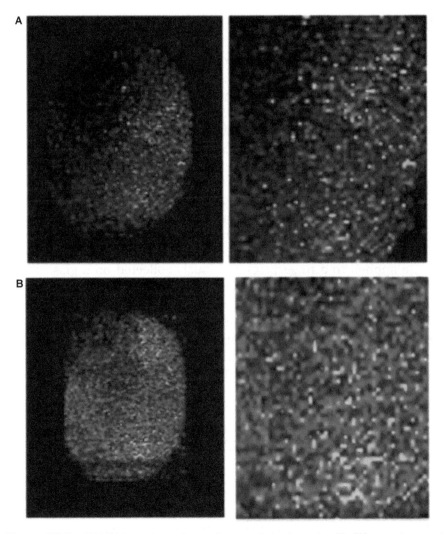

Figure 25.2 SALDI imaging of two fingerprints doped with (A) cocaine and (B) methadone. The images show the intensity mapping of the [M + H⁺] peak of the drugs.
(Reproduced from Sundar and Rowell.[32])

arose in doped fingerprints lifted with commercial tapes. In this case, DART was unsuccessful while SALDI was still able to show the presence of the drugs in the lifted prints. Additionally, SALDI had a further advantage because it could redevelop fingerprints already predusted with general powders by using the silica particles. This holds great potential in providing chemical information of aged preanalysed fingerprints.[33]

25.3 Endogenously Excreted Drugs in Fingerprints

Drugs are metabolised in the body and excreted in various ways, including urine, blood and sweat. Fingerprints are primarily composed of sweat. Therefore, the detection of both the parent drug and its metabolites in the fingerprints of an individual provides evidence of drug consumption, rather than just environmental contamination from having handled the substance.

The first time a drug metabolite was found in the sweat of a fingerprint was reported by Leggett *et al.*[34] The drug investigated was nicotine, since it is a non-controlled and socially accepted drug; hence it was easy to find volunteers. More importantly, cotinine, the main metabolite of nicotine, was known to be excreted in sweat.[34] Gold nanoparticles functionalised with anti-cotinine antibodies *via* protein A were applied to a fingerprint previously collected on a glass slide. The anti-cotinine antibodies only bound to the fingerprints of smokers, following excretion of the cotinine. The bound nanoparticles were visualised by addition of a fluorescently labelled secondary antibody. The authors obtained high-resolution images with second- and third-level detail of fingerprints from smokers (Figure 25.3). Third-level detail refers to the distribution of the sweat pores along the ridges.[34]

The detection of cotinine, as well as nicotine, in the sweat of fingerprints was further studied by different analytical techniques. Hazarika *et al.* used a similar approach to that described above. In this case, gold nanoparticles were substituted by magnetic particles. Magnetic particles are comparable to magnetic fingerprint powder, a popular method of choice for the visualisation of latent fingerprints among police investigators. Commercially available magnetic particles were coated with anti-cotinine antibodies *via* protein A/G. Visualisation of the prints was again possible with a fluorescently labelled secondary antibody. These conjugates were able to simultaneously develop the latent prints and prove consumption of

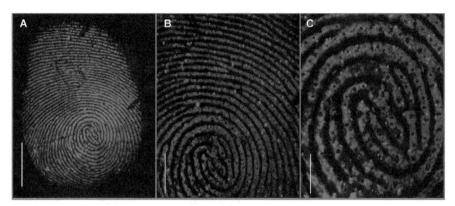

Figure 25.3 Detection of cotinine in a fingerprint from a male smoker using anti-cotinine/gold nanoparticle conjugates and illuminated using a secondary antibody fragment tagged with Alexa Fluor 546. The fluorescence images shown in (A), (B) and (C) are of the same fingerprint, but at varying magnifications. The black dots visible along the ridge pattern (clearly seen in C) are the sweat pores. The scale bars in (A), (B) and (C) correspond to 5 mm, 2 mm and 1 mm, respectively.
(Reproduced from Leggett *et al.*,[34] with permission from Wiley-VCH.)

nicotine by the volunteer in only 15 min.[35] The fingerprints from smokers presented a colour change from grey to brownish-yellow after incubation with the particles, even before the addition of the fluorescent labels. This presents a further advantage in forensic investigations as it eliminates the need for specialised equipment when fingerprints are analysed in the field. This technique was successful for both glass surfaces as well as highly reflective white porcelain surfaces.[35,36] A scheme depicting the detection of cotinine from fingerprints with nanoparticles is shown in Figure 25.4.

Mass spectrometry, specifically SALDI-TOF-MS, has also been used for the detection of cotinine and nicotine in fingerprints.[30] Benton *et al.* showed that SALDI-TOF-MS enhanced with hydrophobic silica particles could be used to detect both cotinine and its parent drug nicotine in the fingerprints from smoker volunteers. These authors reported that detection of nicotine in fingerprints could potentially be used as a marker for smoking, without further testing for cotinine. This conclusion was reached after a study for environmental contamination of nicotine showed that nicotine levels in non-smokers' fingerprints are significantly lower than those of smokers. However, a consideration is the half-life of nicotine in fingerprints, which was found to be degraded over time to levels similar to those found in non-smokers' prints.[30,37]

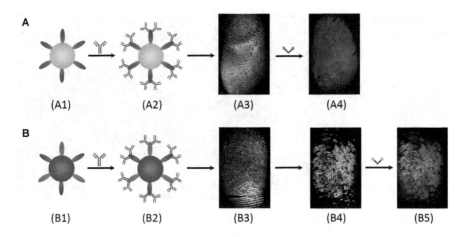

Figure 25.4 Scheme representing the detection of cotinine in latent fingerprints using (A) gold nanoparticles and (B) magnetic particles. The particles, coated with either protein A (A1) or protein A/G (B1), were functionalised with anti-cotinine antibody (A2, B2). These conjugates were then incubated on a latent fingerprint collected on a glass microscope slide (A3, B3). In the case of gold nanoparticles, the fingerprints are still invisible to the naked eye after treatment. However, when magnetic nanoparticles are used, a colour change from colourless to yellowish-brown is developed in the latent fingerprint, which allows its visualisation (B4) under light illumination. Further treatment with a fluorescently tagged secondary antibody produces a fluorescent image for both types of particles (A4, B5). The different colour of the fluorescent images is obtained with the use of different fluorescent dyes: Alexa Fluor 546 for gold nanoparticles and Alexa Fluor 488 for magnetic particles.
(Reproduced from Hazarika and Russell,[4] with permission from Wiley-VCH.)

Other non-controlled drugs have also been detected from fingerprint samples. As an example, caffeine and its metabolites, mainly theobromine, theophylline and paraxanthine, were detected by liquid chromatography-tandem mass spectrometry (LC-MS/MS) over a period of 7 h after coffee ingestion.[38]

A large part of forensic toxicology involves the detection and analysis of drugs of abuse in different biological specimens. Drugs of abuse are known to be metabolised in the human body, producing metabolites that can be excreted *via* eccrine and sebaceous sweat. Also see Chapter 22 for detection of drugs in sweat.

The approach by Hazarika *et al.* based on magnetic particles functionalised with antibodies and tagged with a fluorescent label, as previously shown in Figure 25.4, was developed for the detection of drugs of abuse in latent fingerprints. The particles were coated with antibodies selective to THC, benzoylecgonine (the main metabolite of cocaine), methadone and its main metabolite, 2-ethylidene-1,5-dimethyl-3,3-diphenylpyrrolidine (EDDP). As previously described for cotinine, positive results can be easily seen due to the colour change induced by the magnetic nanoparticles after incubation with the fingerprints. Both brightfield and fluorescence images could be valuable for forensic investigations. These images are not only positive results for a specific drug of abuse, but they are also clear enough to be used to identify an individual (Figure 25.5).[39] The detection of multiple drugs and/or metabolites from one single fingerprint could be beneficial to identify drug addicts using combinations of more than one drug. An example is the combination of cocaine and heroin, known as "speed-balling".[40] To enable multiplexed detection from a single fingerprint, Hazarika *et al.* functionalised magnetic particles with anti-morphine antibody, to target morphine, the major metabolite of heroin, and with anti-benzoylecgonine, the major metabolite of cocaine. The fingerprint was divided into two parts with a delimiting pen. One fragment of the print was treated for morphine and the other fragment for benzoylecgonine. Positive results and clear

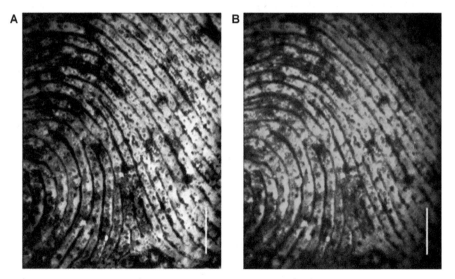

Figure 25.5 Detection of THC in a latent fingerprint. (A) Brightfield image after incubation with anti-THC–magnetic particle conjugates. (B) Fluorescence image after incubation with the conjugates followed by the addition of a fluorescently tagged secondary antibody. The scale bars represent 2 mm.
(Reproduced from Hazarika *et al.,*[39] with permission from Wiley-VCH.)

images were again obtained, proving that it is possible to detect multiple drugs from a single fingerprint.[40]

Drugs of abuse in fingerprints have also been identified by mass spectrometry. SALDI-TOF-MS/MS assisted with a hydrophobic silica dusting agent has been used for the detection of methadone and its main metabolite EDDP in the fingerprints from patients taking methadone as an oral substitute for heroin. Maps showing the distribution of these drugs in the fingerprints were also obtained. These maps, although with poor resolution, can give an insight on the secretion patterns of drugs and metabolites from the sweat pores.[29] Ultra-high-pressure liquid chromatography followed by tandem mass spectrometry (UPLC-MS/MS) has also been used to detect methadone in fingerprints from patients undertaking methadone treatment. Both the parent drug and its main metabolite were found in trace levels, in contrast to fingerprints from volunteers not undertaking the treatment, where no presence of methadone or EDDP was found.[41]

The detection of benzodiazepines in the body is important in forensic toxicology, given that this class of drugs are extensively prescribed and have been involved in a large number of drug facilitated sexual assault (DFSA) crimes.[42] Traditionally, diazepam has been the most widely used benzodiazepine. However, lorazepam is increasingly being used as a substitute for diazepam because of the longer lasting sedative effects. Lorazepam is metabolised in the body to the 3-*O*-glucuronide phase II metabolite.[43] Goucher *et al.* investigated the detection of lorazepam and its glucuronide in fingerprints following oral administration. By using LC-MS/MS, these authors were able to detect and quantify lorazepam and its glucuronide over a period of 12 h. However, detection was only successful when 10 fingerprints from the same volunteer were dissolved and then combined.[43]

Quantitative detection of a drug of abuse, cocaine, from a single fingerprint has been achieved with a competitive enzyme immunoassay (cEIA) by van der Heide *et al.*[44] Immunoassays are commonly used in toxicology for the detection of substances in biological samples, including blood and urine. Such immunoassays are not only easy to perform but also yield great sensitivity and specificity towards the analyte of interest. The study by van der Heide *et al.* showed that cocaine was detected and quantified in a single fingerprint. Further, the concentrations of the cocaine in the fingerprints were related to those concentrations of cocaine found in oral fluid analysis.[44]

Pharmaceutical drugs have also been successfully detected in the excreted sweat of latent fingerprints.[45,46] Lim and Seviour further improved their SALDI method by using magnetised carbon black-doped silica nanoparticles as the matrix enhancer. Latent fingerprints from a patient taking terbinafine, an anti-fungal medication, were located using these particles and subsequently analysed by SALDI-TOF-MS/MS over a period of 14 days. Terbinafine was detected and its spatial distribution was observed *via* SALDI imaging.[46] In another study, a cold medicine containing ibuprofen, chlorpheniramine maleate, methylephedrine hydrochloride and dihydrocodeine phosphate was administered to healthy volunteers. Their fingerprints were collected and analysed by LC-MS/MS over a period of 7 days. The results were then compared to analyses of blood samples. All components of the medicine were detected in fingerprints, except for ibuprofen. Additionally, ephedrine, a metabolite of methylephedrine hydrochloride, was also detected in the analysis. These results were comparable to the ones from blood samples, the main difference being that traces of these drugs could be found in fingerprints during a longer period of time than in blood. This suggests that drugs and metabolites in the sweat glands are excreted more slowly than from the bloodstream.[45]

25.4 Other Chemical Information in Fingerprints

Fingerprints are mainly composed of sweat. As a result, naturally excreted substances in sweat, mainly amino acids and lipids, can be detected in fingerprints by chemical analysis. Their detection and distribution over the fingerprints can be used to obtain an optical image that provides enough detail for identification.

The most abundant proteins found in fingerprint residue come from eccrine sweat and those that take part in the elimination of dead cells from the epidermis, known as desquamation.[47] Immunodetection has been widely studied for the detection of proteins in fingerprint residue. Keratins 1 and 10, cathepsin-D, the antimicrobial peptide dermcidin and the carrier protein human serum albumin (HSA) have all been successfully targeted. Brightfield and fluorescence images can be obtained, depending on the nature of the method employed. The detection of at least two of these proteins simultaneously from the same fingerprint has also been reported.[47,48] More importantly, the use of conventional fingerprint developing methods, such as fingerprint powders and ninhydrin, do not hinder the detection of proteins *via* immuno-techniques. Images with third-level detail have been reported.[49]

Immuno-reagents have also been combined with nanotechnology. Aged and dried fingerprints were found to be especially well developed with gold nanoparticles functionalised with anti-L-amino acid antibodies.[50] In addition, the universally expressed enzyme lysozyme has been targeted by means of upconverting nanoparticles functionalised with a specific lysozyme aptamer.[18] Finally, the combination of antibody targeting with highly sensitive SERS imaging has been shown to provide images of fingerprints with good resolution.[51]

Lipids have also been targeted to develop latent fingerprints. Lipids are excreted by the sebaceous glands so their presence in fingerprint residue often comes from normal actions such as touching the face. For investigation purposes, fingerprints have been "groomed" by touching the forehead and/or side of the nose so that the lipid content would be higher. FTIR imaging has successfully been used for lipid detection. The fingerprints analysed by FTIR had already been treated with cyanoacrylate fuming or lifted with gels.[52,53]

The detection of both amino acids and lipids has been investigated *via* mass spectrometry. Ferguson *et al.* used MALDI-MSI to obtain both an optical image of the fingerprint by the application of the MALDI matrix followed by a chemical image obtained with MSI analysis.[54,55] Other techniques including GC-MS and TOF-SIMS have also been proven useful to detect endogenous compounds.[56,57]

Quantitative inter-individual variation in fingerprint composition has been reported.[58] This characteristic could be important in the specific identification of individuals. Profiling based on either amino acid[59] or lipid[60] composition has been studied, but further research is needed to prove statistical significance. To date, only identification of gender and age based on fingerprint residue has provided positive and promising results. In 1996, Buchanan *et al.* reported that children's fingerprints disappear from surfaces much faster than those from adults. The reason for this was found to be the variation in lipid composition. An adult's fingerprints contain a higher amount of lipids, which are originated from the sebaceous sweat glands. On the other hand, children's lipids arise from the epidermis, leading to the expression of lower molecular weight compounds with high volatility.[12,61] As a result, the best way to develop children's fingerprints is not by targeting lipids but rather acid salts, the major and most prevalent component in a child's fingerprint residue.[62] These differences in composition open the possibility of estimating age from fingerprints. Changes in lipids have been found to be linear with time, with the highest variations occurring during puberty.[63]

Identification of gender from fingerprints can be obtained, depending on the analytes targeted. Asano *et al.* focused on the detection of lipids and reported that certain lipids are found in slightly higher amounts in males. However, this difference was not statistically significant and thus sex determination could not be performed.[64] On the other hand, Ferguson and co-workers showed that direct detection of peptides and proteins could lead to the differentiation of genders with an accuracy varying from 67.5 to 85%.[65]

The composition of fingerprint residue changes over time after the print has been deposited. Studies on how the lipid composition changes have been performed by different researchers.[55,66,67] The amount of squalene, present in all fingerprints, decreases on ageing to the point that is undetectable nine days after deposition in the light and 33 days in the dark.[66] This decrease is supported by the rapid oxidation of squalene, whose oxidation products also have a short half-life.[67] For fatty acids, initially there is a marked increase due to the breakdown of wax esters and glycerides. However, volatilisation and degradation lead to a later decrease in the total amount of fatty acids. The temperature to which fingerprint residue is subjected also plays an important role in the ageing. As an example, the unsaturated oleic acid is increasingly degraded with higher temperatures.[55] This suggests that dating fingerprints based on their deposition time could be possible.

Chemical information from fingerprints, however, is not only limited to naturally excreted substances. The presence of exogenous material can also be detected and directly linked to the fingerprint owner. This provides stronger evidence in a court of law that the person has been in contact with a material that might have been involved in a criminal offence. In this manner, fibres and ink found on fingerprints can be spectroscopically analysed and identified, possibly linking individuals to, for example, counterfeit money.[68,69] Similarly, gunshot residue[16] and cosmetics[28] have also been identified. Other types of biological evidence such as blood, usually present in fingerprints in cases of murder, can be successfully detected by analysis of the haemoglobin. There is a possibility to obtain the blood type, but currently the analysis can only differentiate between human, bovine and equine bloods.[70] Finally, the analysis and identification of lubricants on fingerprint residue can be useful in cases of rape and sexual assault.[71,72]

The chemical analysis of fingerprint residue has many potential applications. An especially important one is the separation of overlapping fingerprints, commonly encountered in crime scenes. This can be done by detection of specific chemical entities in each fingerprint. A selective image of each overlapping fingerprint can be obtained. The chemical compounds targeted can be either endogenous or exogenous in nature. Several studies have shown this application with both mass spectrometry[14,20,73] as well as vibrational infrared spectroscopy imaging.[69,74]

25.5   Conclusions

Fingerprints are mainly composed of eccrine sweat. As a result, naturally excreted substances in sweat, mainly amino acids and lipids, can be detected in fingerprints by chemical analysis. Their detection and distribution over the fingerprints can be used to obtain an optical image that provides enough detail for identification.

Of particular interest in analytical toxicology is the detection of other analytes in fingerprints that can provide information about the individuals as well as their lifestyle. The detection of drug residue left on fingerprints has been studied using a range of different techniques. It has been shown that powdering and lifting of the fingerprints in forensic scenarios prior to chemical testing does not prevent the identification of the exogenous drug residues present.

The detection of drug metabolites endogenously excreted in the sweat deposited as a fingerprint is a particularly exciting development. It has been shown that it is possible to identify drug metabolites in a non-invasive manner to detect and quantify drugs that have been ingested by the fingerprint donor. This type of analysis could be useful in many contexts, including but not limited to forensic crime scenes, workplace drug testing and sports doping.

References

1. A. R. W. Jackson and J. M. Jackson, *Forensic Science*, Prentice Hall, Harlow, 2nd edn, 2008.
2. F. Galton, *Fingerprints*, Macmillan and Co, London, 1st edn, 1892.
3. A. K. Jain, S. Prabhakar and S. Pankanti, *Pattern Recognit.*, 2002, **35**, 2653–2663.
4. P. Hazarika and D. A. Russell, *Angew. Chem., Int. Ed.*, 2012, **51**, 3524–3531.
5. A. Girod, R. Ramotowski and C. Weyermann, *Forensic Sci. Int.*, 2012, **223**, 10–24.
6. J. S. Day, H. G. M. Edwards, S. A. Dobrowski and A. M. Voice, *Spectrochim. Acta, Part A*, 2004, **60**, 563–568.

7. T. Yang, X. Guo, H. Wang, S. Fu, Y. Wen and H. Yang, *Biosens. Bioelectron.*, 2015, **68**, 350–357.
8. R. M. Connatser, S. M. Prokes, O. J. Glembocki, R. L. Schuler, C. W. Gardner, S. A. Lewis and L. A. Lewis, *J. Forensic Sci.*, 2010, **55**, 1462–1470.
9. K. L. A. Chan and S. G. Kazarian, *Analyst*, 2006, **131**, 126–131.
10. P. Ng, S. Walker, M. Tahtouh and B. Reedy, *Anal. Bioanal. Chem.*, 2009, **394**, 2039–2048.
11. S. Francese, R. Bradshaw, L. S. Ferguson, R. Wolstenholme, M. R. Clench and S. Bleay, *Analyst*, 2013, **138**, 4215–4228.
12. M. V. Buchanan, K. Asano and A. Bohanon, in *Forensic Evidence Analysis and Crime Scene Investigation*, ed. J. Hicks, P. R. De Forest and V. M. Baylor, Proceedings of the international society for optical engineers (SPIE), Boston, 1997, vol. 2941, 89–95.
13. D. Ifa, A. Jackson, G. Paglia and R. G. Cooks, *Anal. Bioanal. Chem.*, 2009, **394**, 1995–2008.
14. D. R. Ifa, N. E. Manicke, A. L. Dill and R. G. Cooks, *Science*, 2008, **321**, 805.
15. M. I. Szynkowska, K. Czerski, J. Rogowski, T. Paryjczak and A. Parczewski, *Forensic Sci. Int.*, 2009, **187**, e24–e26.
16. M. I. Szynkowska, K. Czerski, J. Rogowski, T. Paryjczak and A. Parczewski, *Surf. Interface Anal.*, 2010, **42**, 393–397.
17. K. Clemons, R. Wiley, K. Waverka, J. Fox, E. Dziekonski and G. F. Verbeck, *J. Forensic Sci.*, 2013, **58**, 875–880.
18. J. Wang, T. Wei, X. Li, B. Zhang, J. Wang, C. Huang and Q. Yuan, *Angew. Chem., Int. Ed.*, 2014, **126**, 1642–1646.
19. K. Li, W. Qin, F. Li, X. Zhao, B. Jiang, K. Wang, S. Deng, C. Fan and D. Li, *Angew. Chem., Int. Ed.*, 2013, **125**, 11756–11759.
20. H. W. Tang, W. Lu, C. M. Che and K. M. Ng, *Anal. Chem.*, 2010, **82**, 1589–1593.
21. G. B. Yagnik, A. R. Korte and Y. J. Lee, *J. Mass Spectrom.*, 2013, **48**, 100–104.
22. J. S. Day, H. G. M. Edwards, S. A. Dobrowski and A. M. Voice, *Spectrochim. Acta, Part A*, 2004, **60**, 1725–1730.
23. M. J. West and M. J. Went, *Forensic Sci. Int.*, 2008, **174**, 1–5.
24. M. J. West and M. J. Went, *Spectrochim. Acta, Part A*, 2009, **71**, 1984–1988.
25. E. Widjaja, *Analyst*, 2009, **134**, 769–775.
26. A. Banas, K. Banas, M. B. H. Breese, J. Loke, B. Heng Teo and S. K. Lim, *Analyst*, 2012, **137**, 3459–3465.
27. A. Banas, K. Banas, M. B. H. Breese, J. Loke and S. K. Lim, *Anal. Bioanal. Chem.*, 2014, **406**, 4173–4181.

28. C. Ricci and S. G. Kazarian, *Surf. Interface Anal.*, 2010, **42**, 386–392.
29. F. Rowell, S. Hudson and J. Seviour, *Analyst*, 2009, **134**, 701–707.
30. M. Benton, F. Rowell, L. Sundar and M. Jan, *Surf. Interface Anal.*, 2009, **42**, 378–385.
31. K. Kaplan-Sandquist, M. A. LeBeau and M. L. Miller, *Forensic Sci. Int.*, 2014, **235**, 68–77.
32. L. Sundar and F. Rowell, *Analyst*, 2014, **139**, 633–642.
33. A. Y. Lim, F. Rowell, C. G. Elumbaring-Salazar, J. Loke and J. Ma, *Anal. Methods*, 2013, **5**, 4378–4385.
34. R. Leggett, E. E. Lee-Smith, S. M. Jickells and D. A. Russell, *Angew. Chem., Int. Ed.*, 2007, **119**, 4178–4181.
35. P. Hazarika, S. M. Jickells and D. A. Russell, *Analyst*, 2009, **134**, 93–96.
36. A. M. Boddis and D. A. Russell, *Anal. Methods*, 2011, **3**, 519–523.
37. M. Benton, M. J. Chua, F. Gu, F. Rowell and J. Ma, *Forensic Sci. Int.*, 2010, **200**, 28–34.
38. K. Kuwayama, K. Tsujikawa, H. Miyaguchi, T. Kanamori, Y. Iwata and H. Inoue, *Anal. Bioanal. Chem.*, 2013, **405**, 3945–3952.
39. P. Hazarika, S. M. Jickells, K. Wolff and D. A. Russell, *Angew. Chem., Int. Ed.*, 2008, **47**, 10167–10170.
40. P. Hazarika, S. M. Jickells, K. Wolff and D. A. Russell, *Anal. Chem.*, 2010, **82**, 9150–9154.
41. S. Jacob, S. M. Jickells, K. Wolff and N. Smith, *Drug Metab. Lett.*, 2008, **2**, 245–247.
42. J. C. Garriott and A. Mozayani, in *Drug Facilitated Sexual Assault, A Forensic Handbook*, ed. M. A. LeBeau and A. Mozayani, Academic Press, London, 1st edn, 2001, pp. 73–88.
43. E. Goucher, A. Kicman, N. Smith and S. M. Jickells, *J. Sep. Sci.*, 2009, **32**, 2266–2272.
44. S. van der Heide, P. García Calavia, S. Hardwick, S. Hudson, K. Wolff and D. A. Russell, *Forensic Sci. Int.*, 2015, **250**, 1–7.
45. K. Kuwayama, T. Yamamuro, K. Tsujikawa, H. Miyaguchi, T. Kanamori, Y. Iwata and H. Inoue, *Forensic Toxicol.*, 2014, **32**, 235–242.
46. A. Y. Lim and J. Seviour, *Anal. Methods*, 2012, **4**, 1983–1988.
47. V. Drapel, A. Becue, C. Champod and P. Margot, *Forensic Sci. Int.*, 2009, **184**, 47–53.
48. A. van Dam, M. C. G. Aalders, K. van de Braak, H. J. J. Hardy, T. G. van Leeuwen and S. A. G. Lambrechts, *Forensic Sci. Int.*, 2013, **232**, 173–179.
49. A. van Dam, M. C. G. Aalders, T. G. van Leeuwen and S. A. G. Lambrechts, *J. Forensic Sci.*, 2013, **58**, 999–1002.

50. X. Spindler, O. Hofstetter, A. M. McDonagh, C. Roux and C. Lennard, *Chem. Commun.*, 2011, **47**, 5602–5604.
51. W. Song, Z. Mao, X. Liu, Y. Lu, Z. Li, B. Zhao and L. Lu, *Nanoscale*, 2012, **4**, 2333–2338.
52. C. Ricci, S. Bleay and S. G. Kazarian, *Anal. Chem.*, 2007, **79**, 5771–5776.
53. M. Tahtouh, P. Despland, R. Shimmon, J. R. Kalman and B. J. Reedy, *J. Forensic Sci.*, 2007, **52**, 1089–1096.
54. L. Ferguson, R. Bradshaw, R. Wolstenholme, M. Clench and S. Francese, *Anal. Chem.*, 2011, **83**, 5585–5591.
55. R. Wolstenholme, R. Bradshaw, M. R. Clench and S. Francese, *Rapid Commun. Mass Spectrom.*, 2009, **23**, 3031–3039.
56. R. S. Croxton, M. G. Baron, D. Butler, T. Kent and V. G. Sears, *J. Forensic Sci.*, 2006, **51**, 1329–1333.
57. M. J. Bailey, N. J. Bright, R. S. Croxton, S. Francese, L. S. Ferguson, S. Hinder, S. Jickells, B. J. Jones, B. N. Jones, S. G. Kazarian, J. J. Ojeda, R. P. Webb, R. Wolstenholme and S. Bleay, *Anal. Chem.*, 2012, **84**, 8514–8523.
58. R. S. Croxton, M. G. Baron, D. Butler, T. Kent and V. G. Sears, *Forensic Sci. Int.*, 2010, **199**, 93–102.
59. M. de Puit, M. Ismail and X. Xu, *J. Forensic Sci.*, 2014, **59**, 364–370.
60. S. Michalski, R. Shaler and F. L. Dorman, *J. Forensic Sci.*, 2013, **58**, S215–S220.
61. K. M. Antoine, S. Mortazavi, A. D. Miller and L. M. Miller, *J. Forensic Sci.*, 2010, **55**, 513–518.
62. D. K. Williams, C. J. Brown and J. Bruker, *Forensic Sci. Int.*, 2011, **206**, 161–165.
63. A. Hemmila, J. McGill and D. Ritter, *J. Forensic Sci.*, 2008, **53**, 369–376.
64. K. G. Asano, C. K. Bayne, K. M. Horsman and M. V. Buchanan, *J. Forensic Sci.*, 2002, **47**, 1–3.
65. L. S. Ferguson, F. Wulfert, R. Wolstenholme, J. M. Fonville, M. R. Clench, V. A. Carolan and S. Francese, *Analyst*, 2012, **137**, 4686–4692.
66. N. E. Archer, Y. Charles, J. A. Elliot and S. M. Jickells, *Forensic Sci. Int.*, 2005, **154**, 224–239.
67. K. A. Mountfort, H. Bronstein, N. E. Archer and S. M. Jickells, *Anal. Chem.*, 2007, **79**, 2650–2657.
68. N. J. Crane, E. G. Bartick, R. S. Perlman and S. Huffman, *J. Forensic Sci.*, 2007, **52**, 48–53.
69. R. Bhargava, R. S. Perlman, D. C. Fernandez, I. W. Levin and E. G. Bartick, *Anal. Bioanal. Chem.*, 2009, **394**, 2069–2075.

70. R. Bradshaw, S. Bleay, M. R. Clench and S. Francese, *Sci. Justice*, 2014, **54**, 110–117.
71. R. Bradshaw, R. Wolstenholme, R. D. Blackledge, M. R. Clench, L. S. Ferguson and S. Francese, *Rapid Commun. Mass Spectrom.*, 2011, **25**, 415–422.
72. R. Bradshaw, R. Wolstenholme, L. S. Ferguson, C. Sammon, K. Mader, E. Claude, R. D. Blackledge, M. R. Clench and S. Francese, *Analyst*, 2013, **138**, 2546.
73. R. Bradshaw, W. Rao, R. Wolstenholme, M. R. Clench, S. Bleay and S. Francese, *Forensic Sci. Int.*, 2012, **222**, 318–326.
74. T. Chen, Z. D. Schultz and I. W. Levin, *Analyst*, 2009, **134**, 1902–1904.

26 Investigating Drug Metabolism of New Psychoactive Substances Using Human Liver Preparations and Animal Studies

Markus R. Meyer*[a] and Hans H. Maurer[b]

[a] Karolinska Institutet, Stockholm, Sweden; [b] Saarland University, Homburg, Germany
*Email: markus.meyer@ki.se; markus.meyer@uks.eu

26.1 Introduction

As many therapeutic drugs, classic drugs of abuse, new psychoactive substances, and doping agents are more or less lipophilic compounds, they have to be metabolized prior to elimination from the body. Biotransformation or metabolism should lead in general to more hydrophilic compounds, which can then be easily excreted *via* the kidneys. However, metabolites as new chemical entities can sometimes still be pharmacodynamically active and can even lead to toxic effects. Studies on metabolism are therefore important during the development of new therapeutic drugs.

Drug metabolism can be divided into phase I and phase II reactions. Phase I reactions usually lead to the introduction of functional moieties such as hydroxy groups. In addition, oxidation and reduction of existing moieties, such as from a ketone to an alcohol,

Forensic Toxicology: Drug Use and Misuse
Edited by Susannah Davies, Atholl Johnston and David Holt
© The Royal Society of Chemistry 2016
Published by the Royal Society of Chemistry, www.rsc.org

can be observed. In the phase II step, these groups are then usually added to very hydrophilic compounds such as glucuronides or sulfates. These reactions are catalyzed by a wide range of enzymes, mainly cytochrome P450 isoforms (CYP), uridine diphosphate glucuronyltransferase (UGT), or sulfotransferase (SULT). Knowing the chemical identity and properties of the main metabolites of a compound is of great importance for assessing pharmacokinetics, toxic risks, and for developing toxicological and doping control screening procedures.

However, such studies are usually not done with drugs of abuse or new psychoactive substances, in particular before they are put on the market. As it is not possible to perform controlled studies on the *in vivo* metabolism of new psychoactive substance in humans for ethical reasons, *in vitro* approaches or *in vivo* animal experiments must be done. The following chapter will therefore present *in vitro* and *in vivo* strategies describing how metabolism studies on such compounds can be performed. This is particularly of importance for developing toxicological and doping control screening procedures, as some new psychoactive substances were reported to be completely metabolized prior to excretion.[1-3] In consequence, screening procedures focused on the parent compound will always lead to negative results, even after recent consumption of the compound.

The authors would also like to refer to previously published review articles, which discuss the topic in much more detail.[4-9]

26.2   *In vitro* Approaches

In contrast to animal studies requiring specific facilities, expertise and licences, *in vitro* metabolism studies can be performed with laboratory equipment usually available in most analytical toxicology laboratories.

The most important organ for metabolism of foreign compounds, such as new psychoactive substances in humans, is the liver. Again, metabolism can be divided into phase I and phase II steps. Phase I steps include functionalization such as hydroxylation and phase II steps conjugation to very hydrophilic moieties such as glucuronic acid. Metabolically active enzymes can be found membrane-bound at the smooth endoplasmic reticulum (ER) or at the mitochondria. Such enzymes are the CYPs, flavin monooxygenases (FMO), UGT, or monoamine oxidases (MAO). Further metabolic enzymes such as SULT can be found in the cytosol. *In vitro* experiments on the metabolism of new psychoactive substances are therefore usually done using particular subcellular fractions such as liver microsomes, functional hepatocytes, or heterologously expressed single enzymes.[8,10–14]

26.3 Subcellular Fractions

The most frequently used hepatic preparation for general *in vitro* metabolism and structural elucidation of metabolites are liver microsomes. They contain the complete spectrum of enzymes located in the ER, such as CYP, FMO, and UGT, which are amongst the most important drug metabolizing enzymes. Nevertheless, microsomes prepared from insect cells over-expressing individual human enzymes can also be a valuable source of metabolizing enzymes, as they allow separate study of the involvement of a single enzyme or isoenzyme rather than a mixture of enzymes as in liver microsomes. They will be further discussed in a separate paragraph. Human liver cytosol contains the soluble enzymes such as the already mentioned SULT but also the soluble catechol-*O*-methyltransferase (COMT).

It is of importance to note that the different enzymes contained in either the microsomal or cytosolic part may need different co-substrates. The CYP enzymes and the FMO enzymes can only work if they have NADPH + H$^+$ at their side, as they are then able to reduce and cleave molecular oxygen, with one oxygen atom leaving the reaction as water and the second oxygen being transferred to the substrate. The electrons-providing system is transported by the accessory protein CYP reductase. UGT, catalyzing the binding of glucuronic acid to the substrate, needs UDP-glucuronic acid as co-substrate. As SULT catalyzes the transfer of a sulfo group to the substrates, it needs 5′-phosphoadenosine-3′-phosphosulfate as co-substrate. The COMT is dependent on the presence of the co-substrate *S*-adenosylmethionine to allow corresponding methylations.

The mixture of microsomes and cytosol is called the S9 fraction and contains a wide range of metabolically active enzymes. Therefore, particular co-substrates must be added to S9 incubations to allow formation of phase I and II metabolites.[15,16]

However, a brief summary of selected metabolism studies on new psychoactive substances or classic drugs of abuse using human liver microsomes and human liver cytosol will be provided in the following paragraphs. Interested readers are again referred to the publication by Peters and Meyer, where a much more extensive summary can be found.[8]

26.3.1   Human Liver Microsomes

All presented applications here used mass spectrometry either coupled to gas chromatography and/or liquid chromatography to analyze the incubations.

The metabolism of 4-methyl-*N*-ethcathinone (4-MEC), the *N*-ethyl homologue of mephedrone and a member of the β-keto amphetamine (cathinone) group, was studied by Helfer *et al.* using pooled human liver microsome incubations.[17] They could confirm in human liver microsomes all urinary excreted phase I metabolites, with the exception of the *N*-deethyl-dihydro isomers and the 4-carboxy-dihydro metabolite. The metabolism of 4-MEC is summarized in Figure 26.1. Glucuronides and sulfates could not be studied under the applied conditions as corresponding co-substrates were not added. Jin *et al.* incubated CP 47,497, a potent cannabinoid receptor type 1 agonist, in human liver microsomes to characterize its metabolic fate.[18] A total of eight metabolites were detected and structurally characterized, based on mass spectral data. Also in that study, glucuronides and sulfates were not studied due to lack of co-substrates in the incubations. No detailed studies in the literature can be found on the application of pooled human liver microsomes to study the phase I and phase II

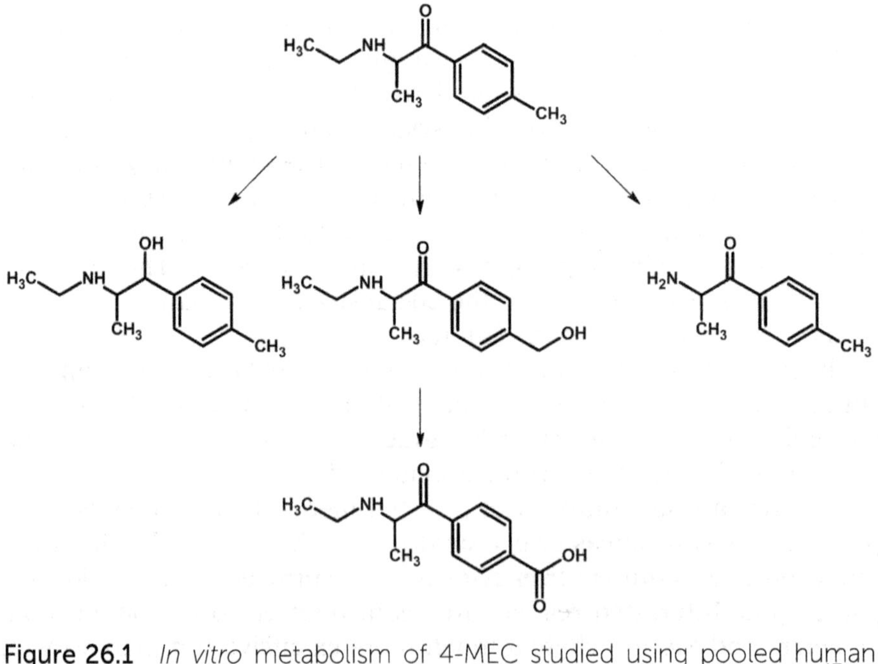

Figure 26.1   *In vitro* metabolism of 4-MEC studied using pooled human liver microsome incubations as published by Helfer *et al.*[17]

metabolism of new psychoactive substances, in particular the formation of glucuronides and sulfates.

26.3.2 Human Liver Cytosol

Schneider *et al.* studied the efficacy of four common *in vitro* assay systems in producing known cocaine *in vivo* metabolites.[19] Besides human liver microsomes, human liver cytosol, human liver S9 fraction, and horseradish peroxidase (an enzyme capable of oxidizing substrates *via* hydrogen peroxide as the oxidizing agent) was applied to investigate the cocaine metabolism. Analytics were again done by mass spectrometry coupled with liquid chromatography. They added to the subcellular systems the co-substrates NADPH, uridine 5'-diphosphoglucuronic acid, or both. They concluded that "For cocaine, whose metabolism is dominated by phase I processes, the metabolic system containing human liver microsomes produced the metabolic profile that best represented the *in vivo* metabolism of cocaine".

26.3.3 Human Liver S9 Fraction

Besides the aforementioned example, other studies can be found using the human liver S9 fraction for both phase I and phase II metabolism studies.[10,20] Thevis *et al.* used the human S9 fraction, besides human liver microsomes, to study the phase I and II biotransformations of the selective androgen receptor modulators S-22 and S-23. They are abused as doping agents because they have similar effects as androgenic drugs such as anabolic steroids. The phase II metabolism reactions were initiated by the addition of uridine 5'-diphosphoglucuronic acid or adenosine 3'-phosphate 5'-phosphosulfate for glucuronidation or sulfation, respectively. They also added D-saccharic acid-1,4-lactone as glucuronidase inhibitor. They concluded that only marginal differences in phase I metabolism between human liver microsomes and S9 fractions were observed and that *in vitro* simulation of metabolic reactions has shown a considerable correlation to *in vivo* obtained metabolites.

26.4 Heterologously Expressed Single Enzymes

26.4.1 Cytochrome P450 Enzymes

The CYPs are for sure the most important metabolizing enzymes for new psychoactive substances or for xenobiotics in general. CYPs can be found in almost every tissue, such as the liver, intestine, and lung.[21,22] As stated above, CYPs are monooxygenases, which act under consumption of molecular oxygen. Heterologously expressed single CYPs are often used to elucidate the involvement of specific human isoenzymes such as CYP2D6 or CYP3A4 in the metabolism of new psychoactive substances. As alternative to single CYP enzymes, human liver microsome incubations in the presence and absence of CYP-specific chemical inhibitors can be used. Comparison of metabolite formation in incubations with and without inhibitor can indicate the involvement of the particular CYP enzymes in the formation of particular metabolites. Such specific inhibitors can be quinidine for CYP2D6,[23–30] fluconazole for CYP2C19,[27] α-naphthoflavone for CYP1A2,[27] and ketoconazole for CYP3A4.[30,31] Recently, a cocktail inhibition approach was developed to simultaneously monitor the involvement of nine different CYPs in specific metabolic reactions.[32,33] Detailed studies on the involvement of particular enzymes in metabolic steps are necessary to predict possible drug–drug or drug–food interactions caused by, for example, inhibition of these enzymes.

Several studies using individual CYP enzymes for studying the metabolism of new psychoactive substances were published from the authors' group.[13,34–39] Some of them will be highlighted in the following. They studied the human CYP enzymes involved in the major metabolic steps of dimethocaine (DMC), a synthetic derivative of cocaine, to measure their kinetics and to estimate their contribution on *in vivo* hepatic clearance by applying the relative activity factor approach.[13] Studies were done using cDNA-expressed CYP enzymes and the formation of metabolites after incubation was analyzed by liquid chromatography-mass spectrometry. The net clearances for hydroxylation and deethylation of DMC were calculated to be 32% for CYP1A2, 5% for 2C19, 51% for 2D6, and 12% for 3A4 and 3% for CYP1A2, 1% for 2C19, <1% for 2D6, and 96% for 3A4, respectively. These data were also confirmed by chemical inhibition tests in human liver microsomes. However, due to multiple contributions of different CYP enzymes, a clinically relevant interaction with single CYP inhibitors and dimethocaine should not be expected.

A further study dealt with the *in vivo* contribution of human CYP enzymes to the hepatic metabolism of glaucine, the main isoquinoline alkaloid of *Glaucium flavum* (Papaveraceae) and a herbal drug of abuse.[37] Kinetics was again studied using recombinant human CYP and the metabolite formation approach with synthesized reference standards of investigated metabolites. Interestingly, they also took care about the influence of simple peak area ratios on the kinetic constants and resulting contribution of P450 isoforms on estimated hepatic clearance. All data were again confirmed by chemical inhibitors. The hepatic clearance after applying the relative activity factor approach was calculated to be 27% and 73% for 2-*O*-demethylation by CYP1A2 and CYP3A4, 82%, 3%, and 15% for 9-*O*-demethylation by CYP1A2, CYP2C19, and CYP2D6, and finally <1% and 99% for *N*-demethylation by CYP2D6 and CYP3A4. They finally concluded that the use of simple peak area ratios for determination of the metabolite concentrations had no relevant impact on the estimation of *in vivo* hepatic clearance of glaucine.

The third study which will be highlighted is a study on the CYP isoenzyme kinetics of methoxetamine, a ketamine-derived drug.[36] Besides identification of human CYP enzymes involved in the initial metabolic steps of methoxetamine *in vitro* and respective enzyme kinetic studies, the authors also compared the value of the metabolite formation and substrate depletion approach for calculating the role of the individual enzymes in the *in vivo* metabolism as net clearance. The enzyme kinetic studies showed that the *N*-deethylation was catalyzed by CYP2B6 and CYP3A4. Concerning the evaluation of the metabolite formation and substrate depletion, they concluded that: "the involvement of enzymes for a specific metabolite formation, the product formation approach shows more sensitive results. A substrate depletion approach can be used to determine the clearance of a substance or for kinetic data if not all metabolites are known". However, the net clearance data using both approaches indicated that CYP3A4 should be responsible for the main part of the CYP-mediated clearance.

26.5 Hepatocytes

Hepatocytes are currently the most upcoming *in vitro* technique for drug metabolism studies. As they contain phase I and phase II metabolic enzymes and the co-substrates needed for the metabolic reactions at physiological concentrations, they have been proposed to most adequately reflect the *in vivo* situation.[40] However, maintaining cultures of mammalian hepatocytes requires equipment and expertise not available in many laboratories studying the metabolism of new designer drugs, so cryopreserved hepatocytes have been successfully used as alternatives to fresh hepatocytes.[14,41–46]

One of the first reports on the use of cryopreserved human hepatocytes for studying the *in vitro* metabolism of drugs of abuse came from Carmo *et al.*, who used them for elucidating the metabolism of 4-MTA[47] and 2C-B.[48] Recent reports on the use of cryopreserved human hepatocytes for studying the *in vitro* metabolism of synthetic cannabinoids were published by the National Institute on Drug Abuse (NIDA).[14,41-46] They typically use *in silico* prediction of metabolites after incubation of drugs with pooled human hepatocytes. Additionally, they partly elucidated the metabolic stability using human liver microsome incubations. Sample analysis was usually done by liquid chromatography-high-resolution mass spectrometry after full-scan mass spectrometry and information-dependent acquisition. They also compared their results with authentic human urine samples and concluded that for AB-PINACA, for example, similar metabolic profiles were found in human urine, confirming the usefulness of human hepatocyte experiments.[41]

Further strategies for studying the metabolism of drugs of abuse and new psychoactive substances in particular were introduced by Kanamori *et al.*[49,50] They investigated the *in vitro* metabolism of 4-bromo-2,5-dimethoxyphenethylamine (2C-B), 5-methoxy-*N*,*N*-diisopropyltryptamine (5-MeO-DIPT), and 2,5-dimethoxy-4-propylthiophenethylamine (2C-T-7) using three-dimensional rat and human hepatocyte culture systems.[50] Their study showed that most *in vivo* metabolites could be found also in the three-dimensional liver cell incubations. However, quantitative metabolite distribution patterns were different from each other. The second study was devoted to the *in vitro* metabolism of [1-(5-fluoropentyl)-1*H*-indol-3-yl](2,2,3,3-tetramethylcyclopropyl)-methanone (XLR-11) and its thermal degradation product.[49] They used two-dimensional cryopreserved human hepatoma HepaRG cells for that purpose. Finally, they compared their results to human urine samples after smoking the drug. They found the primary *in vitro* metabolites of XLR-11 degradant also in the human urine samples and concluded that the *in vivo* metabolism of XLR-11 degradant was well predicted by HepaRG cells and that they produced phase I as well as phase II metabolites. Hence, this *in vitro* system should be an alternative to other *in vitro* and *in vivo* studies.

26.6 *In vivo* Approaches (Animal Studies)

Numerous *in vivo* studies on the metabolism of new psychoactive substances were published by the authors' group in recent years.[17,34-36,51-58] Studies were done using the 24 h pooled urine of male Wistar rats after application of a rather high dose. These high doses were required to also identify the side metabolites in rats, which could be the main metabolites in humans. The studies are aimed to identify targets for toxicological analysis procedures in suspected intoxication or poisoning cases, as well as in forensic settings.

De Brabanter *et al.* used chimeric mice to investigate the metabolism of the synthetic cannabinoid JWH-200 and compared the results to human liver microsomes incubations.[59] They observed that the chimeric mouse model with humanized liver allowed confirmation of most *in vitro* metabolites and that all metabolites are excreted in urine as conjugates, mostly as glucuronides.

Besides animal studies, several reports were published recently on the metabolism in humans either after overdose or controlled administration.[17,60,61]

26.7 Concluding Remarks

The metabolism of new psychoactive substances can be studied using easy to handle subcellular systems but also more complex systems such as hepatocytes, depending on the laboratory equipment, the experience of the staff, and the aim of the study. The best way to study the *in vivo* metabolism is of course controlled human studies, but these are usually not possible due to ethical reasons. *In vitro* studies can be a valuable alternative if conducted thoroughly. Hepatocyte preparations can produce patterns of phase I and phase II metabolites, although they can also be achieved using cheaper subcellular systems if respective co-substrates are added. In summary, *in vitro* metabolism studies are a valuable tool for developing toxicological and doping control screening procedures in urine, especially for compounds completely metabolized prior to excretion.

References

1. A. Grigoryev, S. Savchuk, A. Melnik, N. Moskaleva, J. Dzhurko, M. Ershov, A. Nosyrev, A. Vedenin, B. Izotov, I. Zabirova and V. Rozhanets, Chromatography-mass spectrometry studies on the metabolism of synthetic cannabinoids JWH-018 and JWH-073, psychoactive components of smoking mixtures, *J. Chromatogr. B: Analyt. Technol. Biomed. Life Sci.*, 2011, **879**, 1126–1136.
2. A. Grigoryev, A. Melnik, S. Savchuk, A. Simonov and V. Rozhanets, Gas and liquid chromatography-mass spectrometry studies on the

metabolism of the synthetic phenylacetylindole cannabimimetic JWH-250, the psychoactive component of smoking mixtures, *J. Chromatogr. B: Analyt. Technol. Biomed. Life Sci.*, 2011, **879**, 2519–2526.

3. I. Moller, A. Wintermeyer, K. Bender, M. Jubner, A. Thomas, O. Krug, W. Schanzer and M. Thevis, Screening for the synthetic cannabinoid JWH-018 and its major metabolites in human doping controls, *Drug Test. Anal.*, 2011, **3**, 609–620.

4. M. R. Meyer, Trends in analyzing emerging drugs of abuse–from seized samples to body samples, *Anal. Bioanal. Chem.*, 2014, **406**, 6105–6110.

5. M. R. Meyer and F. T. Peters, Analytical toxicology of emerging drugs of abuse–an update, *Ther. Drug Monit.*, 2012, **34**, 615–621.

6. M. R. Meyer and H. H. Maurer, Current status of hyphenated mass spectrometry in studies of the metabolism of drugs of abuse, including doping agents, *Anal. Bioanal. Chem.*, 2012, **402**, 195–208.

7. M. R. Meyer and H. H. Maurer, Current applications of high-resolution mass spectrometry in drug metabolism studies, *Anal. Bioanal. Chem.*, 2012, **403**, 1221–1231.

8. F. T. Peters and M. R. Meyer, In vitro approaches to studying the metabolism of new psychoactive compounds, *Drug Test. Anal.*, 2011, **3**, 483–495.

9. M. R. Meyer and H. H. Maurer, Absorption, distribution, metabolism and excretion pharmacogenomics of drugs of abuse, *Pharmacogenomics*, 2011, **12**, 215–233.

10. M. R. Meyer, A. Schutz and H. H. Maurer, Contribution of human esterases to the metabolism of selected drugs of abuse, *Toxicol. Lett.*, 2014, **232**, 159–166.

11. M. R. Meyer, A. Robert and H. H. Maurer, Toxicokinetics of novel psychoactive substances: characterization of N-acetyltransferase (NAT) isoenzymes involved in the phase II metabolism of 2C designer drugs, *Toxicol. Lett.*, 2014, **227**, 124–128.

12. M. R. Meyer, L. H. Richter and H. H. Maurer, Methylenedioxy designer drugs: mass spectrometric characterization of their glutathione conjugates by means of liquid chromatography-high-resolution mass spectrometry/mass spectrometry and studies on their glutathionyl transferase inhibition potency, *Anal. Chim. Acta*, 2014, **822**, 37–50.

13. M. R. Meyer, C. Lindauer and H. H. Maurer, Dimethocaine, a synthetic cocaine derivative: studies on its in vitro metabolism catalyzed by P450s and NAT2, *Toxicol. Lett.*, 2014, **225**, 139–146.

14. A. S. Gandhi, M. Zhu, S. Pang, A. Wohlfarth, K. B. Scheidweiler, H. F. Liu and M. A. Huestis, First characterization of AKB-48 metabolism, a novel synthetic cannabinoid, using human hepatocytes and high-resolution mass spectrometry, *AAPS J.*, 2013, **15**, 1091–1098.
15. C. A. Aiub, L. F. Pinto and I. Felzenszwalb, Standardization of conditions for the metabolic activation of N-nitrosodiethylamine in mutagenicity tests, *Genet. Mol. Res.*, 2004, **3**, 264–272.
16. J. W. Allen, R. Langenbach, S. Nesnow, K. Sasseville, S. Leavitt, J. Campbell, K. Brock and Y. Sharief, Comparative genotoxicity studies of ethyl carbamate and related chemicals: further support for vinyl carbamate as a proximate carcinogenic metabolite, *Carcinogenesis*, 1982, **3**, 1437–1441.
17. A. G. Helfer, A. Turcant, D. Boels, S. Ferec, B. Lelievre, J. Welter, M. R. Meyer and H. H. Maurer, Elucidation of the metabolites of the novel psychoactive substance 4-methyl-N-ethyl-cathinone (4-MEC) in human urine and pooled liver microsomes by GC-MS and LC-HR-MS/MS techniques and of its detectability by GC-MS or LC-MS standard screening approaches, *Drug Test. Anal.*, 2015, **7**, 368–375.
18. M. J. Jin, J. Lee, M. K. In and H. H. Yoo, Characterization of in vitro metabolites of CP 47,497, a synthetic cannabinoid, in human liver microsomes by LC-MS/MS, *J. Forensic Sci.*, 2013, **58**, 195–199.
19. K. J. Schneider and A. P. DeCaprio, Evaluation of in vitro metabolic systems for common drugs of abuse. 1. Cocaine, *Xenobiotica*, 2013, **43**, 1043–1054.
20. M. Thevis, E. Gerace, A. Thomas, S. Beuck, H. Geyer, N. Schlorer, J. D. Kearbey, J. T. Dalton and W. Schanzer, Characterization of in vitro generated metabolites of the selective androgen receptor modulators S-22 and S-23 and in vivo comparison to post-administration canine urine specimens, *Drug Test. Anal.*, 2010, **2**, 589–598.
21. M. F. Paine, H. L. Hart, S. S. Ludington, R. L. Haining, A. E. Rettie and D. C. Zeldin, The human intestinal cytochrome P450 "pie", *Drug Metab. Dispos.*, 2006, **34**, 880–886.
22. H. Raunio, J. Hakkola, J. Hukkanen, O. Pelkonen, R. Edwards, A. Boobis and S. Anttila, Expression of xenobiotic-metabolizing cytochrome P450s in human pulmonary tissues, *Arch. Toxicol. Suppl.*, 1998, **20**, 465–469.
23. R. F. Staack, D. S. Theobald, L. D. Paul, D. Springer, T. Kraemer and H. H. Maurer, Identification of human cytochrome P450 2D6 as major enzyme involved in the O-demethylation of the designer

drug p-methoxymethamphetamine, *Drug Metab. Dispos.*, 2004, **32**, 379–381.

24. R. F. Staack, L. D. Paul, D. Springer, T. Kraemer and H. H. Maurer, Cytochrome P450 dependent metabolism of the new designer drug 1-(3-trifluoromethylphenyl)piperazine (TFMPP). In vivo studies in Wistar and Dark Agouti rats as well as in vitro studies in human liver microsomes, *Biochem. Pharmacol.*, 2004, **67**, 235–244.

25. R. F. Staack, D. S. Theobald, L. D. Paul, D. Springer, T. Kraemer and H. H. Maurer, In vivo metabolism of the new designer drug 1-(4-methoxyphenyl)piperazine (MeOPP) in rat and identification of the human cytochrome P450 enzymes responsible for the major metabolic step, *Xenobiotica*, 2004, **34**, 179–192.

26. D. Springer, L. D. Paul, R. F. Staack, T. Kraemer and H. H. Maurer, Identification of cytochrome p450 enzymes involved in the metabolism of 4′-methyl-alpha-pyrrolidinopropiophenone, a novel scheduled designer drug, in human liver microsomes, *Drug Metab. Dispos.*, 2003, **31**, 979–982.

27. F. T. Peters, M. R. Meyer, D. S. Theobald and H. H. Maurer, Identification of cytochrome P450 enzymes involved in the metabolism of the new designer drug 4′-methyl-alpha-pyrrolidino-butyrophenone, *Drug Metab. Dispos.*, 2008, **36**, 163–168.

28. D. Springer, R. F. Staack, L. D. Paul, T. Kraemer and H. H. Maurer, Identification of cytochrome P450 enzymes involved in the metabolism of 4′-methoxy-alpha-pyrrolidinopropiophenone (MOPPP), a designer drug, in human liver microsomes, *Xenobiotica*, 2003, **33**, 989–998.

29. D. Springer, R. F. Staack, L. D. Paul, T. Kraemer and H. H. Maurer, Identification of cytochrome P450 enzymes involved in the metabolism of 3′,4′-methylenedioxy-alpha-pyrrolidinopropiophenone (MDPPP), a designer drug, in human liver microsomes, *Xenobiotica*, 2005, **35**, 227–237.

30. C. Sauer, F. T. Peters, A. E. Schwaninger, M. R. Meyer and H. H. Maurer, Identification of cytochrome P450 enzymes involved in the metabolism of the designer drugs N-(1-phenylcyclohexyl)-3-ethoxypropanamine and N-(1-phenylcyclohexyl)-3-methoxypropanamine, *Chem. Res. Toxicol.*, 2008, **21**, 1949–1955.

31. C. Sauer, F. T. Peters, A. E. Schwaninger, M. R. Meyer and H. H. Maurer, Investigations on the cytochrome P450 (CYP) isoenzymes involved in the metabolism of the designer drugs N-(1-phenyl cyclohexyl)-2-ethoxyethanamine and N-(1-phenylcyclohexyl)-2-methoxyethanamine, *Biochem. Pharmacol.*, 2009, **77**, 444–450.

32. J. Dinger, M. R. Meyer and H. H. Maurer, Development of an in vitro cytochrome P450 cocktail inhibition assay for assessing the inhibition risk of drugs of abuse, *Toxicol. Lett.*, 2014, **230**, 28–35.

33. J. Dinger, M. R. Meyer and H. H. Maurer, Development and validation of a liquid-chromatography high-resolution tandem mass spectrometry approach for quantification of nine cytochrome P450 (CYP) model substrate metabolites in an in vitro CYP inhibition cocktail, *Anal. Bioanal. Chem.*, 2014, **406**, 4453–4464.

34. M. R. Meyer, S. Mauer, G. M. Meyer, J. Dinger, B. Klein, F. Westphal and H. H. Maurer, The in vivo and in vitro metabolism and the detectability in urine of 3′,4′-methylenedioxy-alpha-pyrrolidinobutyrophenone (MDPBP), a new pyrrolidinophenone-type designer drug, studied by GC-MS and LC-MS(n.), *Drug Test. Anal.*, 2014, **6**, 746–756.

35. J. Welter, M. R. Meyer, E. U. Wolf, W. Weinmann, P. Kavanagh and H. H. Maurer, 2-methiopropamine, a thiophene analogue of methamphetamine: studies on its metabolism and detectability in the rat and human using GC-MS and LC-(HR)-MS techniques, *Anal. Bioanal. Chem.*, 2013, **405**, 3125–3135.

36. M. R. Meyer, M. Bach, J. Welter, M. Bovens, A. Turcant and H. H. Maurer, Ketamine-derived designer drug methoxetamine: metabolism including isoenzyme kinetics and toxicological detectability using GC-MS and LC-(HR-)MSn, *Anal. Bioanal. Chem.*, 2013, **405**, 6307–6321.

37. G. M. Meyer, M. R. Meyer, C. S. Wink, J. Zapp and H. H. Maurer, Studies on the in vivo contribution of human cytochrome P450s to the hepatic metabolism of glaucine, a new drug of abuse, *Biochem. Pharmacol.*, 2013, **86**, 1497–1506.

38. M. R. Meyer, C. Vollmar, A. E. Schwaninger, E. Wolf and H. H. Maurer, New cathinone-derived designer drugs 3-bromomethcathinone and 3-fluoromethcathinone: studies on their metabolism in rat urine and human liver microsomes using GC-MS and LC-high-resolution MS and their detectability in urine, *J. Mass Spectrom.*, 2012, **47**, 253–262.

39. M. R. Meyer, J. Dinger, A. E. Schwaninger, D. K. Wissenbach, J. Zapp, G. Fritschi and H. H. Maurer, Qualitative studies on the metabolism and the toxicological detection of the fentanyl-derived designer drugs 3-methylfentanyl and isofentanyl in rats using liquid chromatography-linear ion trap-mass spectrometry (LC-MS(n)), *Anal. Bioanal. Chem.*, 2012, **402**, 1249–1255.

40. R. Laine, Metabolic stability: main enzymes involved and best tools to assess it, *Curr. Drug Metab.*, 2008, **9**, 921–927.

41. A. Wohlfarth, M. S. Castaneto, M. Zhu, S. Pang, K. B. Scheidweiler, R. Kronstrand and M. A. Huestis, Pentylindole/ Pentylindazole Synthetic Cannabinoids and Their 5-Fluoro Analogs Produce Different Primary Metabolites: Metabolite Profiling for AB-PINACA and 5F-AB-PINACA, *AAPS J.*, 2015, **17**, 660–677.

42. A. Wohlfarth, S. Pang, M. Zhu, A. S. Gandhi, K. B. Scheidweiler and M. A. Huestis, Metabolism of RCS-8, a synthetic cannabinoid with cyclohexyl structure, in human hepatocytes by high-resolution MS, *Bioanalysis*, 2014, **6**, 1187–1200.

43. A. Wohlfarth, A. S. Gandhi, S. Pang, M. Zhu, K. B. Scheidweiler and M. A. Huestis, Metabolism of synthetic cannabinoids PB-22 and its 5-fluoro analog, 5F-PB-22, by human hepatocyte incubation and high-resolution mass spectrometry, *Anal. Bioanal. Chem.*, 2014, **406**, 1763–1780.

44. A. S. Gandhi, M. Zhu, S. Pang, A. Wohlfarth, K. B. Scheidweiler and M. A. Huestis, Metabolite profiling of RCS-4, a novel synthetic cannabinoid designer drug, using human hepatocyte metabolism and TOF-MS, *Bioanalysis*, 2014, **6**, 1471–1485.

45. A. S. Gandhi, A. Wohlfarth, M. Zhu, S. Pang, M. Castaneto, K. B. Scheidweiler and M. A. Huestis, High-resolution mass spectrometric metabolite profiling of a novel synthetic designer drug, N-(adamantan-1-yl)-1-(5-fluoropentyl)-1H-indole-3-carboxamide (STS-135), using cryopreserved human hepatocytes and assessment of metabolic stability with human liver microsomes, *Drug Test. Anal.*, 2015, **7**, 187–198.

46. A. Wohlfarth, S. Pang, M. Zhu, A. S. Gandhi, K. B. Scheidweiler, H. F. Liu and M. A. Huestis, First metabolic profile of XLR-11, a novel synthetic cannabinoid, obtained by using human hepatocytes and high-resolution mass spectrometry, *Clin. Chem.*, 2013, **59**, 1638–1648.

47. H. Carmo, J. G. Hengstler, D. de Boer, M. Ringel, F. Carvalho, E. Fernandes, F. Remiao, L. A. dos Reys, F. Oesch and M. de Lourdes Bastos, Comparative metabolism of the designer drug 4-methylthioamphetamine by hepatocytes from man, monkey, dog, rabbit, rat and mouse, *Naunyn Schmiedebergs Arch. Pharmacol.*, 2004, **369**, 198–205.

48. H. Carmo, J. G. Hengstler, D. de Boer, M. Ringel, F. Remiao, F. Carvalho, E. Fernandes, L. A. dos Reys, F. Oesch and M. de Lourdes Bastos, Metabolic pathways of 4-bromo-2,5-dimethoxy-phenethylamine (2C-B): analysis of phase I metabolism with hepatocytes of six species including human, *Toxicology*, 2005, **206**, 75–89.

49. T. Kanamori, K. Kanda, T. Yamamuro, K. Kuwayama, K. Tsujikawa, Y. T. Iwata and H. Inoue, Detection of main metabolites of XLR-11 and its thermal degradation product in human hepatoma HepaRG cells and human urine, *Drug Test. Anal.*, 2015, **7**, 341–345.

50. T. Kanamori, K. Kuwayama, K. Tsujikawa, H. Miyaguchi, Y. Togawa-Iwata and H. Inoue, A model system for prediction of the in vivo metabolism of designer drugs using three-dimensional culture of rat and human hepatocytes, *Forensic Toxicol.*, 2011, **29**, 142–151.

51. C. S. Wink, G. M. Meyer, D. K. Wissenbach, A. Jacobsen-Bauer, M. R. Meyer and H. H. Maurer, Lefetamine-derived designer drugs N-ethyl-1,2-diphenylethylamine (NEDPA) and N-iso-propyl-1,2-diphenylethylamine (NPDPA): metabolism and detectability in rat urine using GC-MS, LC-MSn and LC-HR-MS/MS, *Drug Test. Anal.*, 2014, **6**, 1038–1048.

52. J. Welter, M. R. Meyer, P. Kavanagh and H. H. Maurer, Studies on the metabolism and the detectability of 4-methyl-amphetamine and its isomers 2-methyl-amphetamine and 3-methyl-amphetamine in rat urine using GC-MS and LC-(high-resolution)-MSn, *Anal. Bioanal. Chem.*, 2014, **406**, 1957–1974.

53. J. Welter, P. Kavanagh and H. H. Maurer, GC-MS and LC-(high-resolution)-MS(n) studies on the metabolic fate and detectability of camfetamine in rat urine, *Anal. Bioanal. Chem.*, 2014, **406**, 3815–3829.

54. M. R. Meyer, C. Lindauer, J. Welter and H. H. Maurer, Dimethocaine, a synthetic cocaine analogue: studies on its in-vivo metabolism and its detectability in urine by means of a rat model and liquid chromatography-linear ion-trap (high-resolution) mass spectrometry, *Anal. Bioanal. Chem.*, 2014, **406**, 1845–1854.

55. G. M. Meyer, C. S. Wink, J. Zapp and H. H. Maurer, GC-MS, LC-MS(n), LC-high resolution-MS(n), and NMR studies on the metabolism and toxicological detection of mesembrine and mesembrenone, the main alkaloids of the legal high "Kanna" isolated from Sceletium tortuosum, *Anal. Bioanal. Chem.*, 2015, **407**, 761–778.

56. M. R. Meyer, S. Schmitt and H. H. Maurer, Studies on the metabolism and detectability of the emerging drug of abuse diphenyl-2-pyrrolidinemethanol (D2PM) in rat urine using GC-MS and LC-HR-MS/MS, *J. Mass Spectrom.*, 2013, **48**, 243–249.

57. M. R. Meyer, D. Prosser and H. H. Maurer, Studies on the metabolism and detectability of the designer drug beta-naphyrone in

rat urine using GC-MS and LC-HR-MS/MS, *Drug Test. Anal.*, 2013, **5**, 259–265.

58. G. M. Meyer, M. R. Meyer, D. K. Wissenbach and H. H. Maurer, Studies on the metabolism and toxicological detection of glaucine, an isoquinoline alkaloid from Glaucium flavum (Papaveraceae), in rat urine using GC-MS, LC-MS(n) and LC-high-resolution MS(n), *J. Mass Spectrom.*, 2013, **48**, 24–41.

59. N. De Brabanter, S. Esposito, E. Tudela, L. Lootens, P. Meuleman, G. Leroux-Roels, K. Deventer and P. Van Eenoo, In vivo and in vitro metabolism of the synthetic cannabinoid JWH-200, *Rapid Commun. Mass Spectrom.*, 2013, **27**, 2115–2126.

60. E. L. Menzies, S. C. Hudson, P. I. Dargan, M. C. Parkin, D. M. Wood and A. T. Kicman, Characterizing metabolites and potential metabolic pathways for the novel psychoactive substance methoxetamine, *Drug Test. Anal.*, 2014, **6**, 506–515.

61. B. Moosmann, P. Bisel and V. Auwarter, Characterization of the designer benzodiazepine diclazepam and preliminary data on its metabolism and pharmacokinetics, *Drug Test. Anal.*, 2014, **6**, 757–763.

27 Case Examples and Discussion

John Slaughter

Analytical Services International, St Georges, University of London, London, UK
Email: john.slaughter@bioanalytics.co.uk

27.1 Introduction

Had I been allowed to call this chapter something less prosaic than "Case Examples and Discussion" I think I would have called it "Toxicology Results Will Take Some Time", this being a mantra used by police to explain to the public why they have no cause of death for an unexplained fatality which, were it available, could aid the sale of Sunday newspapers. Far be it from me to suggest that an SIO (senior investigating officer) would use the humble toxicologist to allow him or her more time to progress an enquiry. For my part, I adopted a compromise: urgent, critical results within a few days, but a few weeks to complete the report.

In this last chapter of the book I have been given the opportunity to relate a few cases, aided only by imperfect memory and a pile of scrawled-in casebooks. Inevitably, the cases are, at least in part, unusual; the mundane is neither memorable nor usually of interest. To contrast with my own cases I have interspersed a few that are better known, reported either in the media or in professional literature. These show that, although analytical techniques change, the processes of investigation and human foibles vary less. As to my own cases, the results are true, the interpretation is (hopefully) correct, but

Forensic Toxicology: Drug Use and Misuse
Edited by Susannah Davies, Atholl Johnston and David Holt
© The Royal Society of Chemistry 2016
Published by the Royal Society of Chemistry, www.rsc.org

some of the detail is blurred, either accidentally or by design. If these cases appear sometimes to tail off, or lack a conclusion, this is because, despite asking, the outcome was never revealed; and there was always the next case awaiting completion.

No toxicology result can be properly interpreted without context. Information as to what is believed to have occurred, who the subject is (was), with respect to age, lifestyle and medical history, and who any other protagonists are, is essential. This is both to ensure that relevant tests have been performed and the analytical results are assessed with respect to the circumstances.

It is often easier to find out what is thought to have occurred than to ascertain what tests the investigator (or customer) wants performed. Multiple choice submission forms which say "please select from the following": alcohol (ethanol), drugs of misuse, medicinal drugs, carboxyhaemoglobin, biochemical tests (if available), will probably cover a large number of cases where toxicology is requested. A few examples of medicinal drugs and a note of the latest recreational drugs your laboratory can detect may assist, not forgetting "other". Information on drugs (and poisons) found, prescribed or suspected must be captured.

27.2 Alcohol

Consider how erroneous conclusions might be drawn without the relevant information. Alcohol, as the commonest used and misused drug, provides several examples. Hopefully, having read this book, if provided with a post-mortem blood alcohol concentration you will immediately ask several questions. Is there also a urine or vitreous humour result? Was the specimen preserved; if so, with what? How long had the deceased been dead and in what circumstance? In particular, what was the ambient temperature; had the body been in water? Similarly, with samples from live subjects, timings, preservative and storage are all important.

Some may question the merit of testing specimens from decomposed bodies for ethanol on the basis that any result will not be interpretable. Those with an eye on a possible future court appearance are likely to take the view that a result, albeit qualified with a comment that it appears to offer little or no useful information, is better than an unperformed test qualified by a presumptive assertion that it was not performed because the *a priori* assumption was the result would provide little or no useful information.

A blood alcohol concentration of, say, 40 mg dL^{-1} may not imply severe intoxication, whether it is present in a post-mortem specimen or back-calculated to the time of an incident from a lower concentration measured in blood or urine. Suppose, however, this concentration is present in a young child, the victim of an alleged sexual assault or present in a Road Traffic Act sample with a therapeutic concentration of, say, diphenhydramine or mirtazapine or a few nanograms per millilitre of THC.

Eighty milligrams per decilitre was the highest alcohol concentration detected in four blood specimens taken after Leslie Newson's body was recovered from the driver's cab four days after the Moorgate train crash in 1975.[1] Comparison with blood specimens taken from other bodies retrieved from the train at approximately the same time led to the conclusion that, although 80 mg dL^{-1} represented the highest blood alcohol that could have been present in the driver's body at the time of death, he had, nevertheless, drunk some alcohol. However, even by the time the Railway Inspectorate report[2] was published in March 1976, Keith Simpson had expressed the opinion that all the alcohol detected in the driver's blood could have arisen post-mortem.

Moorgate was a key case in establishing microbiology in forensic toxicology and emphasised the need for specimens to be properly preserved and for multiple specimens (blood, urine and vitreous humour) to be collected and analysed whenever possible. Although the presence of a large number of micro-organisms in a specimen meant it could be confidently stated that "it is quite possible that the measured alcohol concentration is not that present at the time of sampling", or at the time of death, more complex assessments as to whether micro-organisms present were likely to have increased or decreased the alcohol concentration were confounded by possible changes in the viable organisms, bacteria or yeasts, over time.

27.2.1 Case Study

A man had spent the night drinking. It was generally agreed that he had imbibed well, if not wisely. He had last been seen alive at around 2 a.m. when a friend had left him on the bank of a river, he being too drunk and too heavy to have been carried any further. A month later the man's body was recovered from the river. At a coroner's post-mortem examination, unpreserved blood and urine were collected. It was never established whether any vitreous humour was available, but none was collected.

The blood and urine alcohol concentrations were measured on behalf of the coroner and found to be 261 mg dL^{-1} and 405 mg dL^{-1}, respectively. Products of putrefaction, namely tryptamine, indole and β-phenylethylamine, were detected in the blood specimen. In addition, trace amounts of other volatiles were visible on the alcohol GC charts. I entered the case when it was suggested that the deceased had been forced into the river and had not died accidentally.

Simple comparison of the blood and urine alcohol concentrations suggested that the deceased was in the alcohol elimination phase when he died. A drinking pattern was suggested, and a colleague of mine embarked on extensive forward calculations including comparison of the blood and urine alcohol concentrations. I didn't see this as a helpful way forward.

Clearly there was a problem: the body had been in water for a month. Although the water temperature had been close to freezing, some putrefaction had occurred, as was evidenced by the amines and other volatiles in the blood.

Although there was no doubt of heavy alcohol consumption on the night of the disappearance, had the victim died that night? Cullen and Mayes quote a maximum urine alcohol concentration of 59 mg dL^{-1} for persons drowned in the North Sea who, being aircrew and oil rig workers, were highly unlikely to have drunk alcohol recently.[3] The deceased was not known to be diabetic. The urine alcohol concentration supported substantial alcohol consumption in the hours before death even if, had it been the only result, the blood alcohol concentration might have cast doubt on this. Despite reasonable agreement between blood and urine alcohol concentrations, it was possible that either value had altered between death and analysis.

The pathologist gave the cause of death as immersion in water. There appeared to be doubt as to whether this had resulted in classic wet (fresh water) drowning or sudden dry drowning. I considered whether sufficient water could have entered the blood stream, by postmortem diffusion *via* the lungs, to decrease the blood alcohol concentration. I found a paper[4] which described the drowning of 15 intoxicated rabbits, a macabre image, which I thought would have had Beatrix Potter spinning in her grave. Despite consideration of the results of these experiments I could say no more than that the blood alcohol concentration might have decreased by absorption of water *via* the lungs.

Although vitreous humour is cited as the fluid of choice, for alcohol analysis, where decomposition has occurred, this may not apply in cases of drowning. Singer and Jones[5] describe how, after immersion in water for a few weeks, alcohol may diffuse across the cornea into the water and lower the vitreous alcohol concentration. I was thus able to be less critical about the absence of a vitreous specimen.

Witnesses said the deceased had drunk several measures of diluted spirit around 30 minutes before he had last been seen alive. I postulated that, on an empty stomach, the Mellanby effect[6] of a rapidly rising blood alcohol concentration could have led to collapse near the water's edge. If the deceased had died shortly afterwards, that theory was not supported by the blood to urine alcohol ratio, in the absence of post-mortem change, but in reality that was clearly a possibility.

The case went to trial: manslaughter due to recklessness. Interestingly, the expert for the defence presented multifarious calculations which attempted to assess the time of death from the drinking pattern and the blood to urine alcohol ratio. Putting aside the other evidence, there were clearly too many variables to say more than, regrettably, a man had drowned after drinking too much alcohol. An acquittal resulted, and the moral of the case may be that even cases involving only alcohol can be overly complex.

27.3 Carbon Monoxide

A carboxyhaemoglobin (COHb) saturation of, say, 8% could be written off as indicative of a heavy smoker and opinion given that the subject was already dead when the fire started. If, however, a person had doused himself in petrol and struck a match in a motor vehicle, the cause of death is likely to be shock and burns with death ensuing before much carbon monoxide could have been inhaled.

Contrast this with the case where a woman died in a caravan.[7] Her blood COHb saturation was 85%, whereas her husband, who survived, had a blood COHb saturation of 14%, shortly after admission to hospital. The toxicologist felt the COHb concentration in the deceased was too high to have arisen from a gas lamp present in the caravan and was at odds with the survival of the husband, who had presumably slept in the same small space. In tests, the gas lamp failed to produce a significant amount of carbon monoxide. When confronted with these facts, the husband claimed that his wife had committed suicide by inhaling car exhaust fumes and he had staged an accidental death; however, movement of a body in the confined space of the caravan appeared unlikely for a man acting alone. The husband was a university lecturer and later admitted to obtaining two small cylinders (lecture bottles) of carbon monoxide. Having vented one of the cylinders above his sleeping spouse until she succumbed, he cycled to work to return the cylinders, but must have inhaled some carbon monoxide (possibly from a car exhaust) on his return in order to give rise to the COHb saturation in his blood.

27.3.1 Case Study

A woman and her 18-month-old child were found dead in the upstairs front room of a house. Her three-year-old child, also deceased, was found in the upstairs back room. In a cupboard in the kitchen, below the back room, was a gas boiler, recently fitted by a friend of the family. Investigation by gas board engineers soon found that the boiler was missing its flue and furthermore produced abundant carbon monoxide.

Blood taken from the mother a day after her body was found showed a COHb saturation of 64%. Post-mortem examinations on the children were delayed for a week until a paediatric forensic pathologist was available. Blood from both children showed methaemoglobin saturations around 10% after chemical reduction with dithionite. Such a concentration of methaemoglobin is flagged by the CO-Oximeter® as it may cause the displayed COHb saturation to be lowered below its true value by a few percent. In this case the saturation for the three-year old was 58% and for the 18-month old it was 44%, leaving no doubt that all three had died from CO poisoning.

Gresham[8] notes that persons who succumb to CO may be found dead in different rooms, which can raise suspicion of foul play. Reconstruction may assist in assessing CO concentrations in different parts of the premises; however, doors and windows must be in the original position and account taken of any draughts.

Gresham (*op. cit.*) also cites a gas boiler flue blocked by a sparrow's nest. Ornithologically speaking, I can trump this. Two elderly spinsters were found dead at home, one collapsed in an armchair, the other lying on the kitchen floor. The table was laid for tea, Marie Celeste-like, with glasses of sherry and Madeira cake, all of which was submitted for analysis. COHb saturations in excess of 60% were found in the blood of the two deceased. When the Raeburn back boiler was taken apart, the bodies of two crows were found in the flue. Unlike Professor Gresham, I was unable to test the avian victims for COHb.

27.4   Cyanide

Were you to stop members of the public and ask them to name a few poisons, cyanide would doubtless feature many times. Toxicologists dread the request "do a general screen for poisons" as, unlike drugs, this has to be put together *ad hoc*. I have long lost count of the number of foodstuffs I have tested for cyanide, often with weakly positive results. These positives frequently mirrored the blank controls, thus emphasising the need for such comparators.

In 1968, during the trial of the Krays at the Central Criminal Court, a plot was revealed to allegedly assassinate a witness using a spring-loaded syringe filled with prussic acid (hydrogen cyanide solution) concealed in an attaché case.[9] The case is or was in the Black Museum.[10] Bryan Ballantyne was tasked with assessing the impact of injecting cyanide solution intramuscularly, both in terms of any post-mortem appearances and by measuring blood concentrations attainable, which was achieved by experimenting on rabbits. No clear post-mortem findings were observed, a view supported by Francis Camps.[9] Although cyanide poisoning is rare, if it occurs it is likely to be oral (*per os.*). As toxicology laboratories are unlikely to test for cyanide without cause, I have always had faith that pathologists who cannot smell cyanide (in stomach content) rely on a mortuary technician who can.

Ballantyne investigated variations in dose of hydrogen and potassium cyanides. He reports that those animals who survived appeared healthy within an hour of injection. A colleague[11] used to recount the tale of a University of London chemistry lecturer who, anxious to prove Paracelsus's maxim "The doses makes the poison", would conclude one of his lectures by placing a crystal of potassium cyanide on his tongue. One day it appeared that the crystal was larger than usual and gave rise to a blinding headache. His riposte as to how he dealt with this apparently life-threatening emergency was "I had a large Scotch and went for a brisk walk in the park".

Hydrogen cyanide can also be a component of fire fumes and apart from CO is the only other toxic fire product that may be (readily) tested for in blood specimens. Many toxicology laboratories either cannot test for cyanide or will only do so reluctantly in fire deaths, if the COHb saturation is low and there is thus insufficient evidence to show that the subject died in the fire, although that is the primary hypothesis. It has been argued that if a fire produces a significant amount of cyanide early on, this can prove fatal before the subject can inhale sufficient CO to give rise to a potentially fatal saturation.[12,13] As Trombley[14] notes, death from inhalation of HCN may not be as rapid as some may claim, even when its production is apparently carefully controlled and lethality is intended.

The Kings Cross fire in 1987[13] lead to a dispute between laboratories as to the role of cyanide (HCN gas) in the deaths of the 52 victims. Part of the dispute involved the method of determining cyanide concentrations in blood. The Conway diffusion method, which dates from 1954,[15] is perfectly reliable provided three-compartment plastic cells (Obrink diffusion units) are used, with dilute acid in the outer cell to act as a seal. In contrast, it is very difficult to obtain a gas-tight seal between ground glass surfaces when using two-compartment glass cells,[16] no matter which grease you use, and these are not recommended.

27.4.1 Case Study

I investigated the death of a dementia patient on respite care, which occurred in a hospital overseas. Owing to his habit of wandering, the patient was restrained in a chair using a fixed tray, a sort of adult highchair. The chair had polyurethane padding and he was in possession of a lighter and cigarettes. One day the patient set the chair alight, but was unable to escape the heat and fumes, the fire being confined to the chair. Tests on a control chair showed it could be ignited with a lighter, but not with a dropped cigarette. The cause of death was myocardial ischaemia; however, soot was present in the airways. Medication included zopiclone, nitrazepam, lorazepam and promazine. Only promazine and nitrazepam were detected and at low therapeutic blood concentrations.

The COHb saturation was found to be 4%, which could have arisen from smoking; however, burns to the patient's right hand showed he was alive when the fire started and presumably had made an attempt to beat out the flames. Cyanide was detected in the blood at a concentration of 0.6 mg L^{-1}. Although this is less than that typically seen in deaths due to fire fumes with a low COHb saturation, it exceeds that seen in smokers. Additionally, it exceeds that expected from post-mortem production, given that the blood specimen was stored frozen for four weeks prior to analysis.[17] In conclusion, it appeared that the patient had inhaled cyanide from the burning chair by virtue of his head being kept in close proximity to it.

27.5 Potassium and Tablets in the Stomach

It is a toxicological myth that when a person has taken an overdose of medication, tablets will always be visible in the stomach. Most tablets are designed to break up rapidly so that the active ingredient can be absorbed. In fact, manufacturers of cold and flu remedies spend large amounts of money convincing the public that their brand is faster acting than their rival's. Nevertheless, there are instances where tablet matter in the stomach can provide a useful clue that medication may have been taken in overdose or in cases where the patient (deceased) should not have taken tablets at all.

Controlled, modified or slow release tablets, as they are variously called, may have an inert core which binds the active ingredient such that it not released until the tablet reaches the ileum or even the colon. If a person takes a very large overdose, particularly of drugs such as aspirin or paracetamol, a gritty sediment or sludge may be visible in the stomach. Coloured tablets or capsules can dye the stomach content. A young boy died suddenly and foul play was suspected. The stomach content had the appearance of tomato soup, but it was found that his mother was prescribed red sugar-coated dosulepin tablets. On analysis a potentially fatal concentration of dosulepin was found in the boy's blood specimen.

In 1955, an infant of six months died. At autopsy what appeared to be red fruit skins were found in the stomach. These were placed in a jar but disintegrated overnight, showing they were not of plant origin. A public analyst detected uncooked maize starch and eosin dye, both indicative of tablets or capsules.[18] I routinely used to examine stomach content under a polarising microscope for uncooked starch grains ("Maltese crosses") using a crude tally of a few, moderate or large numbers, in conjunction with case circumstances, to decide whether further analysis was necessary.

It was suggested that Seconal, supplied as red gelatine capsules, might be the source of the dye and maize starch. On analysis, approximately 20 mg of quinalbarbital (secobarbital) was found in the stomach content. The boy's father was a naval sick-berth attendant and it was discovered that fifty 100 mg Seconal capsules had been stolen from a drug cabinet to which he had access. For a time the investigation stalled, however, after the father sought a separation order; his wife then provided evidence. This resulted in both parents being tried for murder, with the father alone being convicted.

27.5.1 Case Study

A woman was suspected of assisting the suicide of her terminally ill mother using slow release potassium tablets. Although this medication may be bought over the counter, the daughter had left an audit trail showing her acquisition of such tablets. In addition, between 20 and 30 "tablets" were visible in the stomach at autopsy.

In life, sodium is largely extracellular whereas potassium is largely intracellular. It is pointless to measure the concentration of potassium in post-mortem blood because this will always be elevated due to haemolysis. Nevertheless, to avoid the accusation that I had pre-judged the result, I measured the blood potassium and obtained a result of around half the normal intracellular concentration of 150 mmol L^{-1}. I have always held the belief that, where practically possible, it is prudent to perform a test even where it is anticipated that the result will be unhelpful or even meaningless. This can avoid lengthy discussion and even discomfiture in the witness box when one tries to justify what may be presented to the court as incompetence or laziness.

The concentration of potassium in vitreous humour increases more or less linearly after death and in the past this phenomenon has been used to estimate or calculate the time of death. More recent studies have shown that potential inaccuracies in the measurements and calculations are liable to give either an erroneous estimate of the time of death or imply too great an accuracy in determining it. Nowadays, measurement of vitreous humour potassium concentrations is not recommended for estimating the time of death.

The victim had died in hospital. The time of death, the time of sampling and thus the post-mortem interval (PMI), approximately 48 hours, were known. I therefore used the formulas in reverse, having measured the vitreous potassium concentration at 15 mmol L^{-1}. Although vitreous humour is normally collected to measure the alcohol concentration, and is thus supplied in vials containing sodium fluoride and potassium oxalate, here a plain specimen was available. From the literature[19] I found six formulas designed to estimate the post-mortem interval from the vitreous potassium concentration. These all showed that the measured vitreous potassium concentration fell within the range expected for a PMI of 48 h. The spread of results used to derive the six formulas was wide and therefore it was possible that the deceased had ingested a high dose of potassium recently before death and it could not be asserted that she had not.

I knew the brand of potassium tablet used and the number of "tablets" seen in the stomach. Experiments were performed to see exactly what happened when a control tablet was immersed in water. From the summary of product characteristics (SPC) I knew the nature of the tablet excipients and that potassium was present as the chloride salt. Particles recovered from the stomach and duodenum were examined under a scanning electron microscope with energy dispersive X-ray analysis (SEM-EDX). Bright particles on the screen were found to be almost entirely composed of potassium and chlorine. GC-MS and IR spectroscopic analysis of the cores found them to be composed of cetostearyl alcohol. This is used in the UK in cosmetics and as the inert core of certain slow release tablets. There was no information to show that the deceased had access to any other slow release tablets and consumption of cosmetics was deemed unlikely.

No prosecution resulted, which I assumed was in accordance with the DPP's guidelines that it would not have been in the public interest.[20]

27.6 Cocaine

27.6.1 Case Study

A man was taken to hospital suffering from chest pains and a headache. He said he had drunk a tot of rum from a bottle given to him as a present. After a few hours he was discharged and returned home. Shortly afterwards, another ambulance was called; however, he was certified dead on its arrival. His son and another, who drank smaller amounts of rum, were also taken ill, but survived.

Investigation centred round the bottle of rum, which, when I received it, appeared almost full. I recalled a previous case where liquor bottles had been three-quarters filled with cocaine. Molten wax had then been poured in to form a plug and finally some of the original liquor had been added to deceive casual inspection. As I wrote a description of the current bottle I noticed a white deposit on the inside of the neck.

Cocaine and benzoylecgonine were detected in the post-mortem blood at concentrations of 11 mg L^{-1} and 13 mg L^{-1}, respectively. These concentrations are high, even compared to other fatal oral overdoses. That of cocaine is particularly high, considering there was an interval of a few hours or so between ingestion and death; however, this may have been due to a short PMI of 24 h and only a further two days delay to analysis.

The one-litre rum bottle held 930 mL of clear brown liquid with a smell of rum. Apart from the white deposit on the neck it looked like any other bottle of rum; however, on analysis it was found to contain 246 g of cocaine. The impossible question of how much cocaine the deceased had ingested could be partially addressed by considering the amount of rum apparently missing, assuming the bottle was originally full. A single (25 mL) measure held 6.6 g cocaine, well above the oft-quoted MLD (minimum lethal dose) of one gram.

Another fatality due to internally smuggled cocaine led to some interesting data because of a sampling error. A man's body without identifying documents was found dumped by a roadside. At post-mortem, 60 packets were found in his gut. On analysis these were found to contain, in total, around a kilogram of cocaine; however, one was split and it was estimated that 5–6 g of drug had leaked out.

The pathologist took both left and right preserved femoral blood; however, inadvertently the left femoral sample contained no sodium fluoride. The left femoral blood contained benzoylecgonine at a concentration of 3.6 mg L^{-1}, but no unchanged cocaine. The right femoral blood (fluoride preserved) contained benzoylecgonine at a concentration of 9 mg L^{-1} and cocaine at a concentration of 3.4 mg L^{-1}.

Under the Human Tissues Act 2004, it is not permitted to undertake research on post-mortem samples without permission from the next of kin; however, analysis may be performed which addresses the cause of death. Preserved vitreous humour was found to contain cocaine and benzoylecgonine at concentrations of 5.1 mg L^{-1} and 4.2 mg L^{-1}, respectively. Mackey-Bojack et al.[21] showed that vitreous cocaine concentrations are similar to (or slightly above) blood cocaine concentrations, but vitreous benzoylecgonine concentrations tend to be around half those in the blood. Analysis of the vitreous humour supported the results from the right femoral blood and this case clearly shows the need for samples for cocaine analysis to be preserved with sodium fluoride.

27.7   Post-mortem Redistribution

Nowadays, toxicologists are well aware of the phenomenon of post-mortem redistribution. A bland statement that a certain drug may exhibit post-mortem redistribution may be viewed by the investigator or the pathologist as of little or no help. Consideration and discussion of factors such as the propensity of a drug to exhibit post-mortem distribution (volume of distribution and central to peripheral ratio), site of sampling and interval to death as well as any reported studies should be mentioned, with a view to giving an opinion as to the likelihood that the drug concentration could have increased from a non-toxic (or non-fatal) concentration to that found. This is all very well, you may say, if one is looking for a cause of death and no other cause is obviously available. One's confidence in dismissing the possibility of a substantial increase in a post-mortem blood drug concentration may be enhanced by circumstantial evidence such as a heap of empty tablet blister strips by the body. In contrast, one might wish to be more circumspect if the case is adversarial and the cause of death is critical.

27.8 Decomposed Bodies

Post-mortem redistribution normally concerns blood concentrations, but what if you have been forced to analyse liver as no body fluids were available. Liver is one of the organs cited as the source of drug which can diffuse into the blood after death. Liver should always be sampled from the right lobe, away from the stomach and small intestine; however, it is often difficult to ascertain from which part of the liver the specimen was obtained. So, if fluid drains from the liver it may wash drug away, but if the liver desiccates a drug concentration may increase.

27.8.1 Case Study

I examined liver tissue from a body that had been interred in a garden for 16 years. It was thought that the suspect had stupefied the deceased with his medication, including amitriptyline and triazolam. Amitriptyline and nortriptyline were readily detected at concentrations of around 200 mg kg^{-1} and 25 mg kg^{-1}, respectively. I had considered whether to estimate the concentrations or just report drug and metabolite as detected. I didn't want to be in the position of saying I could have estimated the concentrations, but chose not to and, as there was no cause of death, I felt it was reasonable to say that I could not exclude amitriptyline toxicity as a factor in the death. The amitriptyline to nortriptyline ratio indicated acute ingestion unless either post-mortem change had affected drug and metabolite differently or amitriptyline had diffused from the stomach to the liver.

Following the method of Siegers,[22] a good indication of triazolam at a concentration of approximately 0.02 mg kg^{-1} was obtained. Triazolam (Halcion) is a benzodiazepine hypnotic that was available in the UK from 1978, but withdrawn in 1991. As therapeutic blood concentrations tend to be somewhat less than 0.02 mg L^{-1} it was possible to say that the deceased was possibly under the influence of triazolam when she died. At trial, the drug concentrations assumed much less importance than the presence of the drugs and a conviction for murder resulted.

A hundred years ago, in fatal criminal poisonings, it was often deemed necessary to calculate the total body burden of a poison. Frederick Seddon was tried at the Central Criminal Court in 1912 for the murder of his lodger, Miss Eliza Barrow, by arsenic poisoning. Toxicology was performed by William Willcox on Miss Barrow's exhumed remains and Seddon was defended by Edward Marshall Hall. Willcox weighed Miss Barrow's exhumed body at 67 pounds 2 ounces (30.5 kg). He detected one thirtieth of a milligram of arsenic in 6 grams of muscle tissue. Using a formula, he estimated that the body was 40% muscle and thus multiplied his result by 2000. Marshall Hall pointed out that a small error in the initial determination would have been multiplied many fold. He further suggested that such an error was likely if, in life, Miss Barrow had weighed around 10 stone (63 kg) and muscle, being 77% water, had lost more weight compared to, for example, bone. Despite Marshall Hall's fine grasp of toxicology and Seddon appealing to the judge, Mr Justice Bucknill, as a fellow mason, he was hanged at Pentonville Prison on 18th April 1912.[23,24]

27.8.2 Case Study

Skeletal remains were found in the loft of a barn. They were tentatively identified as those of a local woman who had gone missing a generation earlier. Near the body was a brown glass tablet bottle, empty and with no lid, but with the remains of a paper label adhering. I took the bottle to the documents department and placed it in the VSC (visual spectral comparator). With the aid of the various light filters, I obtained a partial address of a local pharmacy, a date close to that of the putative victim's disappearance and enough letters to convince me that the bottle had once held Tuinal.

A three inch section of femur yielded about a gram of desiccated marrow. Unlike other samples of bone marrow I had examined from more recently deceased bodies, this had an apparent low fat content, presumably due to the post-mortem interval and the dry atmosphere of the hay loft, and gave rise to clean (fat-free) extracts. A methanol extract was derivatised, analysed by GC-MS and showed two clear peaks for amylobarbital and quinalbarbital (secobarbital). There was no value in any quantitative work. Interestingly, in the Seconal poisoning discussed above, a sibling had died suddenly. His body was exhumed with a view to testing for Seconal, but only a skeleton was found, which makes me wonder what analysis of the bone marrow would have shown.[18]

As an aside, a glass milk bottle found in the loft, containing a small amount of debris and plant material, was also submitted. Analysis of an alcohol washing of this bottle detected DDE (dichlorodiphenyldi-chloroethylene), a breakdown product of DDT (dichlorodiphenyltri-chloroethane), an insecticide banned in the UK in 1984. DDT and DDE were not included in a routine toxicology screen, but control samples of the pure compounds were found in the "pesticides collection" and with good quality (clean) GC-MS traces from the bone marrow extract it was possible to say there was no analytical evidence that the deceased had ingested DDT. So, never throw anything away.

27.9 Excited Delirium

27.9.1 Case Study

A naked man was seen dancing in a field. Police, unable to communicate with him, restrained him and, thinking his behaviour too bizarre to be addressed in hospital, took him to the police station, where he was seen to thrash around in a cell. A forensic physician, unwilling to enter the cell and examine him, called paramedics who took the man to hospital. Blood was taken shortly after arrival at hospital, but despite treatment, the man died. A bag of powder, found near to the original incident, was submitted as well as ante- and post-mortem samples.

An ante-mortem blood alcohol concentration of 61 mg dL^{-1} and a post-mortem blood alcohol concentration of 42 mg dL^{-1} corroborated an interval of around an hour between hospital admission and death.

Analysis of the post-mortem blood showed the presence of cathinone drugs, but their precise identity was uncertain except mephedrone was detected at a concentration of 0.05 mg L^{-1}. In addition, benzoylecgonine was detected at 0.11 mg L^{-1}, but no cocaine.

Analysis of the powder suggested it contained PVP (pyrrolidinovalerophenone) and MEC (methylethcathinone). At that time these were very much NPSs (novel psychoactive substances) and no certified reference standards were available. Tentative identification was made by comparison with casework samples; however, as I was dealing with 1.5 g of powder it was possible to make the comparison using NMR spectroscopy as well as GC-MS.

Using the data obtained from analysis of the powder it was possible to say there was every indication that PVP and MEC were present in the blood specimen. Without certified standards, no quantitative analysis on the blood sample could be performed. I could only say that, by inspection, the blood concentrations of both compounds appeared significant, if not substantial, but not trivial. In contrast, by comparison with several DUI (driving under the influence) cases, the blood mephedrone concentration was not high.

By the time the case came to inquest, several cases of PVP toxicity had been reported in the literature. I felt confident to say I had no doubt that the blood specimen contained more than trace amounts of PVP and MEC and these drugs were a possible cause of the deceased's behaviour and therefore a possible factor in the cause of death. The medical cause of death was excited delirium. The jury returned a narrative verdict that the deceased should have been taken to hospital, not to a police station, but were unsure that such action would have saved his life.

Excited delirium was first described in 1985 by Wetli and Fishbain[25] as a possible consequence of binge cocaine use. Subsequently it has been associated with heavy use of methamphetamine, amphetamine or phencyclidine. Signs and symptoms include bizarre and violent behaviour, immense strength, immunity to pain and command, keening (making guttural sounds) and pyrexia, possibly leading to undressing. Police are often called to such incidents and, as the subject is not likely to be compliant, he (rarely she) is likely to be restrained using a Taser®, oleoresin (pepper) spray, physical restraint or a combination of these. As stimulant drugs can sensitise the heart and cause dysrhythmia, the combination of excited delirium and restraint may prove fatal.

Certain practitioners do not recognise the condition "excited delirium" and may describe the phenomenon as serotonin syndrome. Interested parties, although ignorant of Bradford-Hill's criteria for causation, are liable to see police intervention as a causal factor in the death. Inquest juries must ponder the questions: Did the police have an option not to intervene? What would have happened had they not acted? Excited delirium is not by any means a fatal condition,[26] so the pertinent question is how to procure medical attention, with sedation, for the subject. Hopefully, police and other agencies are nowadays more aware of excited delirium and, in the US at least, pocket cards are available listing the signs to look out for and key actions (identify, control, sedate, transport).[27]

27.10 Doping and DFSA (Drug Facilitated Sexual Assault)

Over the past 20 years or so it has been the habit of many police forces to test specimens obtained from a complainant in a sexual assault case for potentially stupefying drugs, which he or she may have in-gested unwittingly. Although some of this testing could be viewed as the result of media pressure, it addresses the question – was there a substance present in the complainant's system, apart from alcohol or any drug taken voluntarily, which could account for any memory loss or out of character behaviour? Such testing also provides a check for any recreational drug not admitted to, which might otherwise feature in a (last minute) defence – "The complainant shared a couple of lines of cocaine with me before she agreed to sexual intercourse".

Doping, or the covert administration of drugs, features in many other cases apart from sexual assault, *e.g.* murder, robbery, domestic poisoning and sedation of the troublesome elderly and children. Allegations of doping are raised as a defence to explain various criminal behaviours, for example, "I only drank two pints. I cannot explain how the police found me in Currys at 2 a.m. with four iPhones in my pocket. My drink must have been spiked".

Doping is not a new phenomenon. Glaister[28] reports the use of chloral hydrate added to beer in order facilitate robbery. This includes a fatality where the assailant stole a bottle of chloral from the pharmacist as the drug was being weighed out. He was seen to empty the contents of a small bottle into the victim's drink. Chloral was detected in the deceased's stomach and bowel but not in his blood (R v Parton, Liverpool Assizes, March 1889).

27.10.1 Case Study

A body was recovered from the sea and the cause of death was drowning; however, multiple injuries were present. The period of immersion was estimated at between 2 and 12 hours and therefore post-mortem change was slight.

I detected minute traces of diazepam and metabolites in the blood, with corresponding compounds in the urine. Of much greater sig-nificance, a blood concentration of lorazepam of 0.07 mg L^{-1} was measured in the blood and lorazepam was present in the urine. Although the blood concentration of diazepam was trivial, that of lorazepam was high therapeutic, or more, and of the order seen in persons apprehended for driving under the influence of the drug.

Two questions immediately arose: could the deceased have taken the drug voluntarily and, if not, how easily could it had been ad-ministered covertly (or against his will)? He was not prescribed lor-azepam when he died, but he was a hoarder. Medical (GP) records for 33 years were available. Most of these were on Lloyd George cards.

Younger readers, whose GPs record each consultation on a computer, will be unfamiliar with these buff cards covered with doctors' spidery handwriting and obscure abbreviations. My attention was drawn to a psychiatrist's letter written nine years before the death where prescription of lorazepam was recommended. Having, however, deciphered the Lloyd George cards, which were both deficient in dates and had been well shuffled, there was no evidence that this recommendation had been followed.

Lorazepam is available in injectable form, albeit not supplied to the public as such. Owing to the injuries and effects of immersion, this route could not be excluded. Lorazepam tablets are small (8 mm × 3 mm × 3 mm and around 100 mg), but like all tablets contain insoluble material. The higher strength (2.5 mg) tablet is yellow and may be easier to hide in food than the 1 mg blue tablet. As with other benzodiazepines, lorazepam is tasteless.

In conclusion, if was perfectly feasible that lorazepam could have been administered covertly and the blood concentration was likely to have caused significant sedation in a person not (highly) tolerant to the drug. A trial for murder ensued, which resulted in conviction at the crown court.

27.10.2   Case Study

In another murder case the possibility of doping or sedation prior to death was raised. A badly burnt body was found on a golf course. The cause of death was severe head injury and the pathologist estimated that death would have ensued within a few minutes of the blows being inflicted.

The deceased was a current or ex-heroin user prescribed a typical maintenance dose of 50 mg methadone per day. Blood samples were sparse and in poor condition. Carboxyhaemoglobin was tested for. Not surprisingly, none was found as the fire was thought to have been set at the deposition site and death would have ensued somewhat earlier. This analysis, however, corroborated a visual inspection of the blood by revealing a haemoglobin concentration of only 5 g dL^{-1}, the normal concentration being around three times higher.

Free morphine concentrations in left and right femoral blood were found to be 0.6 mg L^{-1} and 0.4 mg L^{-1}, respectively, and in the left femoral sample, where there was sufficient blood to measure total morphine, the percentage of free morphine was around 80%. Owing to the apparent poor condition of the blood specimens, I analysed vitreous humour for opiates and detected free morphine at

0.35 mg L^{-1} (60% of total) and 6-monoacetylmorphine (6-MAM), the latter showing heroin to be the source of the morphine. Based on Scott and Oliver's work,[29] the vitreous humour results supported those from the blood specimens and did not indicate significant post-mortem increase in the blood concentrations. Unless the deceased had been highly tolerant to opiates, in the absence of the head injuries it was possible to say that death could have arisen from heroin toxicity.

The blood methadone concentration was low (0.06 mg L^{-1}), considering the dose, which raised the possibility of abstraction, despite daily dosing under supervision, and therefore suggested limited opiate tolerance. The morphine concentrations, plus the presence of 6-MAM, suggested that heroin had been used shortly prior to death. Two scenarios emerged: had the victim been encouraged to take heroin only, to be attacked after he passed out, or had he been forcibly injected? Although I gave evidence at the trial, which resulted in a conviction, I did not find out how the deceased came to take a substantial dose of heroin shortly before he died from severe head injuries.

27.11 Plant Poisons

Over the years, I have repeatedly explained to police officers and others that modern forensic toxicology is configured to test for drugs and alcohol. Drugs may be described as medicinal or recreational and are often tested for in groups, *e.g.* chemically acidic or chemically basic, benzodiazepines or drugs of abuse (configured into an immunoassay panel for preliminary screening). When the request is "test for poisons", it is normally necessary to work through an agreed list, substance by substance; however, this is usually qualified by a separate list of what cannot be tested for without the assistance of specialists from elsewhere. One item usually on the "cannot do" list is plant poisons, but the enquirer can be reliably pointed in the direction of the Jodrell Laboratories, Royal Botanic Gardens, Kew.

27.11.1 Case Study

A man returned from a fishing trip and decided to celebrate his return to dry land by spending the day drinking and taking various drugs. Another man arrived with a liquid, thought to be an extract of *Datura*, which was shared between three men. By the end of the day the fisherman had collapsed and was certified dead. His two companions were hallucinating and were taken to hospital.

I analysed the liquid and found it to contain atropine at a concentration of 4 mg L^{-1} and hyoscine (scopolamine) at a concentration of 12 mg L^{-1}. The bottle had apparently held 2 L initially and 150 mL remained. Post-mortem blood showed an alcohol concentration of 185 mg dL^{-1} and atropine and hyoscine concentrations of 0.09 and 0.04 mg L^{-1}, respectively. In addition there was a therapeutic concentration of dihydrocodeine (0.15 mg L^{-1}) and a comparatively low concentration of mephedrone (0.04 mg L^{-1}). Analysis of the urine reflected that of the blood, with an alcohol concentration of 280 mg L^{-1}. Stomach content (100 mL of fluid) was found to contain 0.82 mg hyoscine and 0.13 mg atropine. Traces of atropine were detected in blood specimens from the two men who survived and atropine and hyoscine were detected in the urine of the person who also provided urine.

Although no plant material was identified as *Datura*, analysis of the liquid supported the hypothesis that wild *Datura stramonium*, also known as Jimson weed, angel's trumpet and thorn apple, had been extracted to produce a liquor or infusion. *Datura*, a plant with a centuries' long history of use in witchcraft and as an aphrodisiac, contains the active ingredients hyoscine, atropine and hyoscyamine. The last is an isomer of atropine not distinguishable in the analysis performed. Clearly, dose control would be absent where plant material was extracted into water which was then drunk *ad libitum*.

Datura can cause hallucinations, both visual and tactile, what is called formication, namely the sensation of insects crawling on the skin. This explains both its occasional recreational use and its association with witchcraft. As an anticholinergic poison, symptoms may be described by the mnemonic "blind as a bat, dry as a bone, hot as a hare, mad as a hatter and red as a beet". Where the poison is absorbed rapidly, for example by drinking a decoction as opposed to chewing seeds, coma may arise rapidly and can lead to respiratory or circulatory failure.

Despite a large number of reported *Datura* poisonings, I found no reported blood concentration of atropine and hyoscine in such cases. The blood concentrations found were well in excess of those seen where either drug is used therapeutically. It was my opinion that *Datura* poisoning was likely to be a significant factor in the cause of death with alcohol, dihydrocodeine and mephedrone potentially adding to the overall toxicity. In addition, a photograph taken around five hours before death was certified showed the deceased collapsed face down over one of his companions. The pathologist found petechiae to the eyelids and said positional asphyxia had occurred.

27.12 Morphine

It is occasionally alleged that a patient receiving morphine, diamorphine or another opiate, as palliative care, has died as a result of an overdose of the prescribed analgesia, deliberately administered by a family member or carer, "Easing the Passing" as Patrick Devlin entitled his book on John Bodkin Adams.[30] For example, it may be alleged that the syringe of the Graseby pump has been unclipped and the whole 24 hour dose injected as a bolus. As these patients are usually on high daily doses of painkillers and the potential range of attainable blood concentrations is wide, toxicology rarely resolves the issue. Pharmacokinetics should not be applied to post-mortem blood concentrations; however, the use of various formulas, followed by a strong caveat, is likely to show that either the normal (intended) dose or the higher (suspected) dose could have given rise to the measured blood concentration. In other words, the toxicologist has done his or her best, but cannot support or refute the allegation.

27.12.1 Case Study

I investigated a variant of this type of case where a moment's inattention by a nurse, compounded by some unclear instructions, may have been a factor in the death of an elderly patient. A woman aged around 90 years, and resident in a nursing home, was in pain which could not be controlled by regular doses of paracetamol. As an aside, paracetamol taken four times a day at a dose of one gram can be a remarkably efficient (and cheap) analgesic. The addition of small amounts of codeine or dihydrocodeine may do little more than line the suppliers' pockets. The patient had difficulty in swallowing and the prescriber opted for modified-release morphine granules in suspension. The prescription was a quarter of a 20 mg sachet twice a day. The recipe was to dissolve the contents of one sachet in 10 mL water, but nowhere did the prescription explain how less than a whole sachet should be administered. Although the nurse prepared the drug correctly, in error she gave the patient the whole dose.

Realising almost immediately her mistake, the nurse sought advice from the prescriber. He said to monitor the patient's breathing and an antidote (naloxone) would not be required. Although the patient's breathing appeared unaffected for a few hours, she died 6.5 hours after morphine was administered. At autopsy, severe pneumonia and a severe urinary tract infection were seen, such that a natural cause of death was evidenced.

The blood morphine concentration was 0.03 mg L^{-1} and this represented approximately 20% of the total. This concentration is therapeutic, albeit higher than expected several hours after a 20 mg dose of oral, controlled morphine, as is the percentage present as free morphine. Both of these facts could be explained by the patient's age and state of health.

I found that a control sachet of morphine granules held 570 mg, which in volume was barely a quarter of a teaspoonful. Division of the powder into four was not a reasonable option. The only sensible method was to make up the whole sachet, administer 2.5 mL, using either a half-teaspoon or preferably an oral syringe, then to discard the rest, an answer which was quite obvious in hindsight, with the benefit of a laboratory bench to perform the experiment.

So was morphine toxicity a factor in the cause of death? In my opinion, possibly, but that could not be proved, even on a balance of probabilities. Specifically, no-one could say what the outcome would have been had the patient only been given the intended 5 mg dose. What was clear, however, was that the prescriber had not clearly communicated with the dispenser of the prepared dose.

27.13 Conclusions

Having compiled my collection of cases, I have inadvertently written an anecdotal history of toxicology, at least covering the past few decades. If I have any advice to pass on it includes: look at the whole picture, ask questions and otherwise seek out information. Remember there are tests apart from GC-MS and LC-MS/MS and as Cyril Polson entitled his lecture to the British Academy of Forensic Sciences in 1975,[31] "Let us remember there are also simple tests". Spot tests, pH paper, smell and appearance (visual comparison) are invaluable for examining allegedly contaminated foodstuffs, household products and residues in cups and glasses. Health and safety forbids me to recommend the "Jackson test",[32] namely if you are sure no toxin is present you must taste the material to show faith in your analytical ability. If you can test a theory by experiment, do so. I recall two unrelated cases where simple reconstruction showed the allegations to be unlikely. In one case a woman claimed she had poisoned her children with a mixture of bleach and aluminium sulfate slug bait. Aluminium sulfate is strongly acidic and when I added household bleach (sodium hypochlorite) to it, clouds of chlorine gas were produced. In the second case, another woman blamed her aberrant behaviour on being administered cocktails comprising brandy and amyl nitrite. Despite the addition of much ice, I found the smell to be acrid and overpowering to an extent that such consumption appeared impossible. (Luckily both experiments were performed in the fume cupboard).

Finally, what advice to the fledgling expert witness about to enter the witness box? Don't make a speech, don't interrupt the judge, don't cross-examine counsel, don't make any jokes and don't say anything in Latin.

References

1. S. Holloway, *Moorgate Anatomy of a Railway Disaster*, David & Charles, Newton Abbot, pp. 142–144, 149–150, 154–155.
2. I. K. A. McNaughton, *Report on the Accident that Occurred on 28th February 1975 at Moorgate Station*, HMSO, London, paragraphs 75, 78, 79, 82, 90, 91, 99, 100.
3. T. Cullen and R. Mayes, *Med. Sci. Law*, 2005, **45**, 196.
4. H. Schweitzer, *Dtsch. Z. Gesampte Gerichtl. Med.*, 1960, **49**, 699.
5. P. P. Singer and G. R. Jones, *J. Anal. Toxicol.*, 2007, **31**, 522.
6. E. Mellanby, *Br. J. Inebriety*, 1920, **17**, 157.
7. L. A. King and D. Jobson, *J. Forensic Sci. Soc.*, 1982, **22**, 137.
8. G. Austin Gresham, *A Colour Atlas of Forensic Pathology*, Wolfe Medical books, London, 1975, pp. 28, 29, 264.
9. B. Ballantyne, J. Bright, D. W. Swanston and P. Williams, *Med. Sci. Law*, 1972, **12**, 209.
10. G. R Williams, *The Black Treasures of Scotland Yard*, Hamish Hamilton, London, 1973, plate 21.
11. A. J. Bailey – *personal communication – passim*.
12. T. T. Noguchi, J. J. Eng and E. C. Klatt, *Am. J. Forensic Med. Path.*, 1988, **9**, 304.
13. P. A. Toseland, *Evidence to Fennell Enquiry* (transcript): D Fennell, *Investigation into the King's Cross Underground Fire*, HMSO, London, 1988, (Cm499).
14. S. Trombley, *The Execution Protocol*, Century (Random House), London, 1993, p. 13.
15. M. Feldstein and N. C. Klendshoj, *J. Lab. Clin. Med.*, 1954, **44**, 166.

16. Bryan Ballantyne and T. C. Marrs, *Clinical and Experimental Toxicology of Cyanides*, Wright, Bristol, 1987, p. 28.
17. Bryan Ballantyne, *Clin. Tox.*, 1977, **11**, 173.
18. G. Bailey (pseud) (C. K. Simpson), *The Fatal Chance*, Peter Davies, London, 1969, pp. 111–122.
19. J. I. Coe, *For. Sci. Int.*, 1989, **42**, 201.
20. K. Starmer, *Med. Leg. J.*, 2014, **82**, 48.
21. S. Mackey-Bojack, J. Kloss and F. Apple, *J. Anal. Toxicol.*, 2000, **24**, 59.
22. J. Siegers, Poster Presentation, 12th meeting IAFS, Adelaide, October 1990.
23. P. H. A. Willcox, *The Detective-Physician*, William Heinemann, London, 1970, pp. 47–55.
24. *Trial of the Seddons*, ed. F. Young, Edinburgh, London, 1914, pp. 108–133, 411.
25. C. V. Wetli and D. P. Fishbain, *J. Forensic Sci.*, 1985, **30**, 873.
26. G. M. Vilke, *J. Forensic Leg. Med.*, 2011, **18**, 291.
27. http://www.crtlesslethal.com/training/excited-delirium-cards-for-patrol (accessed 17 August 2015).
28. J. Glaister, *A Text-Book of Medical Jurisprudence Toxicology and Public Health*, E & S Livingstone, 1902, p. 467.
29. K. Scott and J. Oliver, *J. Forensic Sci.*, 2001, **46**, 4.
30. P. Devlin, *Easing the Passing*, Bodley Head, London, 1985.
31. C. J. Polson, *Med. Sci. Law*, 1976, **16**, 2.
32. J. V. Jackson – *personal communication – passim.*

Subject Index